LASERS IN PHYSICAL CHEMISTRY AND BIOPHYSICS

LASERS IN PHYSICAL CHEMISTRY AND BIOPHYSICS

PROCEEDINGS OF THE 27th INTERNATIONAL MEETING
OF THE SOCIETE DE CHIMIE PHYSIQUE

Edited by

JACQUES JOUSSOT-DUBIEN

THIAIS, 17—20 June 1975

ORGANISING COMMITTEE

J. Joussot-Dubien, Chairman
B. Alpert, D.J. Bradley, M. Clerc, M. Dodin, R. Dupeyrat, J. Faure, L. Lindqvist, G. Mayer, C. Reiss, E. Roux, H. Staerk, O. Svelto, C. Troyanowsky, General Secretary

ELSEVIER SCIENTIFIC PUBLISHING COMPANY
Amsterdam — Oxford — New York 1975

ELSEVIER SCIENTIFIC PUBLISHING COMPANY
335 Jan van Galenstraat
P.O. Box 211, Amsterdam, The Netherlands

AMERICAN ELSEVIER PUBLISHING COMPANY, INC.
52 Vanderbilt Avenue
New York, New York 10017

Library of Congress Cataloging in Publication Data

Société de chimie physique.
Lasers in physical-chemistry and biophysics.

English or French.
Includes index.
1. Lasers in chemistry--Congresses. 2. Lasers in biochemistry--Congresses. 3. Laser spectroscopy--Congresses. I. Joussot-Dubien, J. II. Title.
[DNLM: 1. Lasers--Congresses. TK7871.3 L343 1975]
QD715.S62 1975 541'.3'028 75-29093
ISBN 0-444-41383-X

Printed in The Netherlands

CONTENTS

FOREWORD

International meetings are an inevitable blending of organisational nightmares and of scientific problems intermingled with difficult choices. The 27th Discussion of the Société de Chimie physique has been no exception in this respect.

Although our meetings are always fully published, including discussions, this was perhaps not made sufficiently clear to some authors, who were suddenly faced with my request for their manuscripts, a prospect they did not fully appreciate at short notice. But scientific people are definitely cooperative and everybody managed to comply with my preposterous request. Not only were the discussions lively but all participants accepted to set at once to the task of writing down extemporaneously their questions, answers and comments. So that, for the first time in the history of the Société de Chimie physique, the full manuscript of the book of Proceedings is ready for publication five weeks after the actual conference. It includes the contributions of the three colleagues whom illness or unexpected academic duties prevented from attending, as they handsomely offered to send in their texts. And so they did.

It was indeed a good meeting. We had a fairly complete review of the state of the art in lasers, as related to physicochemical or biophysical research, from solid-state to tunable dye lasers through organic lasing materials (including single crystals) and very promising new rare earths glasses. Nearly all the "picosecond community" was present and some papers even dealt with single subpicosecond pulses. After what there was a contribution in which everybody concerned was told why and how they had probably been wrong, this being done with the help of slides very fairly supplied by the very people under attack. This did not prevent some of the picosecond specialists of commenting aloud on the prospect of femto(10^{-15}) second pulses, a piece of futuristic equipement which raises but two problems:

-1-How to do it?

-2-What to do with it?

But it has to be admitted that picosecond work, now in full bloom, would not have sounded very realistic five years ago. When the technical breakthrough is achieved applications follow fast enough.

For reasons one has to accept all research on isotope separation is not presented at meetings. But there is still some unclassified work being done in this field and part of its results are in this book.

As for the biophysical applications it is one of the fields where the Organising Committee was faced with very difficult choices. We had to decide quite arbitrarily that some good papers were outside the scope of the meeting. All we can say to justify our choice is that what we kept was of a fascinating variety, ranging from genetic surgery to a pretty piece of pure physics, where a check on the validity of the theory of convection found an application, through the knowledge of the velocity spectrum of spermatozoids, in the study of human fertility. Some more "conventional" work was devoted to chromosomes, retinal, photosynthesis, molecular motions in mitochondrial membranes, etc..

Although there was a contribution on the fundamental problems of ultrafast kinetics it is certain that the Organising Committee had to leave out all things related to the study of reaction mechanisms. But it did its best to keep most of the research that is at the same time significant and of general value.

It is our hope that this book will prove useful to many, as a review of what has been achieved and of the present trends, also as an illustration of the wonderful versatility of the laser, an irreplaceable instrument for physicists, chemists and biologists alike. The discussions will certainly shed some light on the main questions research teams ask themselves -or others-.

We received generous financial support from a number of organisations: their help is gratefully acknowledged. The Organising Committee performed a difficult task in a way which earned it the gratitude of the Société de Chimie physique and resulted in a timely, efficient and useful meeting. But our thanks are mainly due to the authors and participants, for providing us with their latest results for publication and for vigourous and good-natured discussions which enrich these Proceedings. This book is their work.

July 1975

C.Troyanowsky
General Secretary

ACKNOWLEDGEMENTS

The 27th International meeting of the Société de Chimie physique was organized with the financial help of the following organizations :

- Commissariat à l'Energie Atomique
- Délégation générale à la recherche scientifique et technique
- European Research Office
- Direction des recherches et moyens d'essais
- Secrétariat aux Universités

and the following firms and industrial groups :

- Compagnie Française de Raffinage
- Compagnie Générale d'Electricité
- Société Kodak Pathé
- Société Thomson-C.S.F.
- Union des Industries chimiques

We should like to renew our thanks to all the above.

LIST OF PARTICIPANTS

ADES C., Laboratoire des ultra-réfractaires, 66120 ODEILLO FONT-ROMEU,F
ALPERT B., Institut de biologie physico-chimique, Paris, F
AMAND B., Institut du Radium, Orsay, F
AMOUYAL E., Laboratoire de chimie physique, Faculté des sciences d'Orsay,F
ANDREONI A. Istituto di Fisica del Politecnico, Milano, Italy
ASTIER R., Ecole Polytechnique, Physique moléculaire, Paris, F
AUWETER H., Physikalische Institut, Universität Stuttgart,Germany

BASTIAENS J., Université catholique de Louvain, Belgium
BATEMAN J.B., European Research Office, London, U.K.
BENSASSON R.V., Chimie physique, Faculté des sciences d'Orsay, F
BERGE P., CEN Saclay, Orme du Merisier, Gif s/Yvette, F
BERNS M., University of California, Irvine, California, U.S.A.
BORDONALI C., A.I.M., Milano, Italia
BONNEAU R., Chimie physique, Université de Bordeaux I, Talence, F
BRADLEY D.J., Imperial College, London, U.K.
BRET G., Société Quantel, Zone industrielle, 91400 Orsay, F
BRIAT B., Laboratoire d'optique physique, E.P.C.I., Paris, F
BURDETT J., Dept Inorganic Chemistry, University of Newcastle-on-Tyne,U.K.
BURLAMACCHI L., University of Florence, Florence, Italia
BURLAMACCHI P., Laboratorio di Elettronica Quantistica, Firenze, Italia

CAPITINI R., SRIRMA, CEN Saclay, Gif s/Yvette, F
CHINSKY L., Institut du Radium, Laboratoire Curie, Paris, F
CLARKE R., Dept of Chemistry, University of Boston, Mass., U.S.A.
CLERC M., DGI/SEPCP CEN Saclay, Gif s/Yvette, F
COSTE A., D.G.I.-CEN Saclay, Gif s/Yvette, F

DELAIRE J., Physicochimie des rayonnements, Faculté des sciences d'Orsay,F
DELHAYE M., Service de spectrochimie infrarouge et Raman du CNRS,Thiais,F
DEMAS J.N.Chemistry Dpt, University of Virginia, Charlottesville,U.S.A.
DE WITTE O., Laboratoires de la CGE, Route de Nozay, 91 Marcoussis
DODIN G., Chimie organique physique, Université de Paris VII, Paris, F
DOERR F., Physikalische und Theoretische Chemie, Technische Univ.,München,G
DUBOIS J.E., Chimie organique physique, Université de Paris VII,Paris,F
DUPEYRAT R., Spectroscopie Raman(DRP), Université de Paris VI, Paris, F
DUKE R.W., Physique des solides, Faculté des sciences d'Orsay, Orsay, F

EWEG J.K., Dpt of Biochemistry, Agricultural University,Wageningen,Neth.

FAURE J., Ecole supérieure de chimie, Mulhouse, F
FAUVARQUE F., Ugine Kuhlmann, 41 rue Pergolèse, 75016 Paris, F
FLAMANT P., Physique moléculaire, Ecole Polytechnique, Paris, F
FLORIAN S., Laboratoire de cinétique du C.N.R.S., Villeurbanne, F
FOLCHER G., C.E.N. Saclay, Gif s/Yvette, F

GACOIN P., Physique moléculaire, Ecole Polytechnique, Paris, F
GAILLARD-CUSIN F., Chimie de la combustion et des hautes températures, Orléans, F
GALAUP J.P., Université scientifique et médicale de Grenoble, Grenoble,F
GAUTHERIN J.C., Société Jobin Yvon, rue du Canal, Longjumeau, F
GIANNINI I., Snam Progetti, Monterotondo(Rome), Italia
GOLDSCHMIDT C.R., Hebrew University of Jerusalem, Israel
GRIFFITHS J.E., Bell Laboratories, Murray Hill, New Jersey, USA
GRIPOIS D., D.R.M.E., 16 boulevard Victor, 75996 Paris Armées, F

HARRIS C.B., Dpt of Chemistry,University of California, Berkeley, U.S.A.
HEISEL F., Physique des rayonnements et électronique nucléaire,Strasbourg,F
HUONG P.V., Spectroscopie infrarouge, Université de Bordeaux I,Talence,F
HUSK R.G., European Research Office, London, U.K.

IPPEN E.P., Bell Laboratories, Holmdel, New Jersey, USA

JARAUDIAS J., CEN-DRA/SRIRMa, Gif sur Yvette, F
JOSIEN M.L., Spectrochimie moléculaire, Université de Paris VI,Paris,F
JOUSSOT-DUBIEN J., Chimie physique, Université de Bordeaux I, Talence,F

KARL N., Physikalische Institut, Universität Stuttgart, Germany
KAUFMANN K., MPI Strömungsforschung, Göttingen, Germany
KELLER R.A., National Bureau of Standards, Washington, USA
KIEFFER F., Physicochimie des rayonnements, Faculté des sciences d'Orsay,F
KIMEL S., Technion Israel Institute of Technology, Haifa, Israel
KINDT T., Iwan N. Stranski-Institut, Technischen Universität,Berlin, W.G.
KLEIN J., Physique des rayonnements, C.R.N.-PREN, Strasbourg, F
KNEBA M., MPI für Strömungsforschung, Göttingen, W. Germany
KUHLE T., Iwan N. Stranski-Institut, Technischen Universität, Berlin, W.G.
KUPECEK P., C.N.E.T., 196 rue de Paris, 92220 Bagneux, F

LALANNE J.R., Centre de recherches Paul Pascal, Talence, F
LAM R.M., Chimie physique, I.N.S.A., Villeurbanne, F
LAMOTTE M., Chimie physique, Université de Bordeaux I, Talence, F
LANGELAAR J. Lab. voor Fysische Chemie,Universiteit van Amsterdam
LASCOMBE J., Spectroscopie infrarouge,Université de Bordeaux I,Talence. F
LAUBEREAU A., Physik Department der Technischen Universität, München,W.G.
LAUDE J.P., Société Jobin Yvon, 12 rue du Canal, 91160 Longjumeau,F
LAUSTRIAT G., Faculté de Pharmacie, Université Louis Pasteur,Strasbourg,F
LAVALETTE D., Institut du Radium, Faculté des sciences d'Orsay,Orsay,F
LEACH S., Photophysique moléculaire, Faculté des sciences d'Orsay,Orsay,F
LEICKNAM, SRIRMa/DRA - CEN Saclay, Gif sur Yvette, F
LESCLAUX R., Chimie physique, Université de Bordeaux I, Talence; F
LINDQVIST L., Photophysique moléculaire, Faculté des sciences d'Orsay,F
LOTH C., Physique moléculaire, Ecole Polytechnique, Paris, F
LOUGNOT D.J., Photochimie générale, Mulhouse, F
LUTZ M., Département de biologie, C.E.N. Saclay, Gif sur Yvette, F

McCRAY J.A., Dept of Physics, Drexel University, Philadelphia, Penn.,USA
MACHETEAU Y., DGI/SEPCP-C.E.N. Saclay, Gif sur Yvette, F
MAGAT M., Physico-chimie des rayonnements, Faculté des sciences,Orsay,F
MARTIN M., Photophysique moléculaire, Faculté des sciences, Orsay,F

MATHIS P., Département de Biologie, CEN Saclay, Gif sur Yvette, F
MAYER G., Optique quantique, Université de Paris VI, Paris, F
MEYER Y., Physique moléculaire, Ecole Polytechnique, Paris, F
MEYER E.G., MPI für Biophysikalische Chemie, Göttingen W. Germany
MIALOCQ J.C., DRA/SRIRMa - CEN Saclay, Gif sur Yvette, F
MIEHE J., Physique des rayonnements et électronique nucléaire,Strasbourg F
MIHASHI K., Centre de Biophysique moélculaire, Orléans, F
MOHAN R., Klinische Physidogie , Universität Düsseldorf, W. Germany
MORAND J.P., Ecole Nationale supérieure de Chimie, Talence, F
MORGAN F., Physics Dpt, York University, Downsview, Ontario, Canada
MOUROU G., Optique appliquée, Ecole Polytechnique, Paris, France

NEVE de MEVERGNIES M., Centre d'étude de l'Energie nucléaire,Mol,Belgium
NICKEL B., MPI für Biophysikalische Chemie, Göttingen, W. Germany

ODOM R., Collisions ioniques, Université d'Orsay, Orsay, F
ORLANDI G., Lab. di Fotochimica e Radiazioni di Alta Energia,Bologna,I

PARIS J., C.E.N. Saclay, Gif sur Yvette, F
PAUGH R., Spectra Physics, Mountain View, California, USA
PETIT A., C.E.N. Saclay, Gif sur Yvette, F
PEYRON M., Chimie physique, INSA, Villeurbanne, F
PILLING M.J., Physical Chemistry Laboratory, Oxford, G.B.
PLANNER A., Dpt of Physics, A.Mickiewicz University, Poznan, P
PLURIEN P., C.E.N. Saclay, Gif sur Yvette, F
POCHON E., DRA/SRIRMa - C.E.N. Saclay, Gif sur Yvette, F
POLIAKOFF M., Inorganic Chemistry, University of Newcastle-on-Tyne, G.B.

REISFELD R., Inorganic & Analytical Chemistry,Hebrew University,Jerusalem,I
REISS C., Institut du Radium, Faculté des sciences d'Orsay, Orsay, F
RENTZEPIS P.M., Bell Laboratories, Murray Hill, New Jersey, USA
RETTSCHNICK R.P.H., Laboratorium voor Fysische Chemie, Amsterdam,
RIGNY P., Division de Chimie, C.E.N. Saclay, Gif sur Yvette, F
ROSENBERG A., Soreq Nuclear Research Centre, Yavne, Israel
ROTH E., Dpt de recherche et Analyse, C.E.N. Saclay, Gif s/Yvette, F
ROTHE U., Iwan N.Stranski Institut, Universität Berlin W. Germany
ROUSSET Y., Lab. d'optique moléculaire, Faculté des sciences, Reims, F
ROUX E., Service de Biophysique, C.E.N. Saclay, Gif sur Yvette, F
RULLIERE C., Chimie physique, Université de Bordeaux I, Talence, F

SAKELLARIDIS P., Université Technique Nationale d'Athènes, Grèce
SCHAFER F.P. MPI für Biophysikalische Chemie, Göttingen, W. Germany
SCHNABEL W., Hahn-Meitner-Institut, Berlin-Wannsee,W. Germany
SCHNEIDER S., Inst. f. Physikalische Chemie, Universität Munchen, W.G.
SHANK C., Bell Laboratories, Holmdel, New Jersey, U.S.A.
SMITH P.W., Bell Laboratories, Holmdel, New Jerseu, U.S.A.
STAERK H., M.P.I. für Biophysikalische Chemie, Göttingen, W. Germany
STEHLE, Société SOPRA, 5 rue J.B. Lafolie, 92 La Garenne Colombes, F
STEIN G., Dept of Physical Chemistry, Hebrew University of Jerusalem, I
STOCKBURGER M., M.P.I. für Biophysikalische Chemie, Göttingen, W. Germany
SUNDHEIM B.R., U.S. Office of Naval Research, London, G.B.
SUTTON J., DRA/SRIRMa - C.E.N. Saclay, Gif sur Yvette, F
SVELTO O., Fisica del Plasma ed Elettronica Quantistica, Milano, Italia
SYCZEWSKI M., Instytut Elektroniki Kwantowej, Warszawa, P

TETREAU C., Institut du Radium, Faculté des sciences d'Orsay, Orsay, F
THORAVAL Y., Institut de physique biologique, Strasbourg, F
TROMMSDORFF H., Spectrométrie physique, Université de Grenoble, F
TROYANOWSKY C., Chimie physique, E.P.C.I., Paris, F
TURPIN P.Y., Institut du Radium, Faculté des sciences d'Orsay, F

Van VOORST J.D.W., Laboratorium voor Fysische Chemie, Amsterdam
VARMA C., Gorlaeus Laboratories, Leiden, The Netherlands
VELGHE M., Photophysique moléculaire, Université d'Orsay, Orsay, F
VERMEGLIO M., DB/SBPh - CEN Saclay, Gif sur Yvette, F
VOGELMANN E., Institut für Physikalische Chemie, Universität Stuttgart,WG

WEBER W., Iwan N.Stranski Institut, Universität Berlin, W. Germany
WILLIAMS D., National Research Council of Canada, Ottawa, Canada
WINDSOR M.W., Laser Photochemistry Lab.,Washington State Univ.,Pullman,USA
WOLFRUM J., MPI für Strömungsforsching, Göttingen, W. Germany

YVON J., Commissariat à l'Energie Atomique, Paris, F

ZEGELIN J., Iwan N.Stranski Institut, Universität Berlin, W. Germany
ZWEEGERS F., Gorlaeus Laboratorium, Leide,, The Netherlands

RECENT DEVELOPMENTS IN DYE LASERS

F.P. SCHÄFER

Max Planck-Institut für Biophysikalische Chemie
(Karl-Friedrich Bonhoeffer Institut)
D 34 Göttingen-Nikolausberg, Germany

Progress in the dye laser field has been manifold recently :

1) New dyes have been found and new experimental findings allow a better understanding of the photodegradation process in dye lasers.
2) Vapor phase dye lasers have been operated.
3) New resonator configurations and tuning methods have been employed.
4) Higher average power and higher reliability have been achieved.
5) UV- and IR-wavelength regions that can be generated with dye lasers by nonlinear methods have been extended and the efficiencies of the conversion processes have increased.

While it is easy to find new dyes that are useful with nitrogen laser pumping (1), good dyes for flashlamp pumping are harder to find and only relatively few dyes will show satisfactory action in cw-dye lasers. Recently 16 new dyes of the cyanine class of dyes have been found by the Kodak group to operate satisfactory in flashlamp-pumped dye lasers in the near infrared (2). We have found a new class of dyes, namely the phenoxazones, which exhibit large solvatochromic shifts. One particular dye nileblue - A - phenoxazone, shows a shift of the center of the laser emission from 605 to 700 nm depending on the solvent (3). Molecular orbital calculations of dyes that have not yet been synthesized show, that it should be possible to obtain dye laser action with this class of dyes up to at least 1000 nm.

Another dye that shows a very large solvatochromic shift is the coumarin dye 3-phenyl-7(3) [1-phenyl-2-pyrazolinyl]) coumarin. The laser wavelength of this dye shifts from 530 nm in cyclohexane to 710 nm in DMSO because of a change of the dipolemoment with excitation which we determined to be 23 Debye. This dye also shows an interesting effect

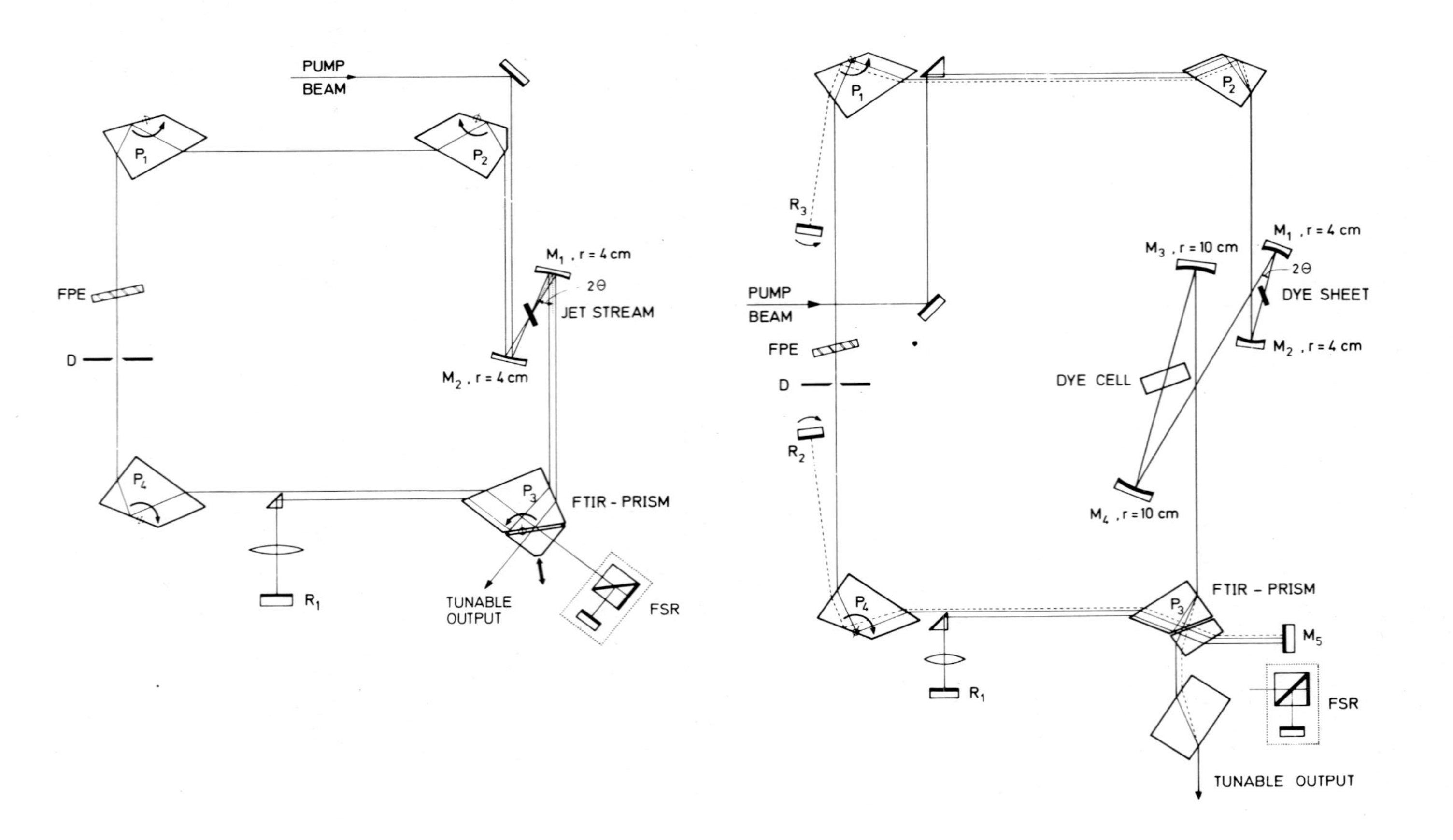
PUMP
BEAM
P$_1$
P$_2$
FPE
D
M$_1$, r = 4 cm
2Θ
JET STREAM
M$_2$, r = 4 cm
P$_4$
P$_3$
FTIR - PRISM
R$_1$
TUNABLE
OUTPUT
FSR
P$_1$
P$_2$
R$_3$
M$_3$, r = 10 cm
M$_1$, r = 4 cm
2Θ
DYE SHEET
PUMP
BEAM
FPE
M$_2$, r = 4 cm
D
DYE CELL
R$_2$
M$_4$, r = 10 cm
P$_4$
P$_3$
FTIR - PRISM
M$_5$
R$_1$
FSR
TUNABLE OUTPUT

of an excited singlet state absorption between 550 and 680 nm (3).

A still disconcerting problem in dye lasers is the photodegradation of the dyes after about a million pump-fluorescence cycles even in the best dyes. There is no hope of improving this situation before the mechanisms of these photochemical processes have been elucidated. Recently a first step in this direction was taken by an American group who investigated the photochemical products formed in 7-diethylamino-4methylcoumarin (4) during dye laser operation. They found five different products, four of which were highly fluorescent and non-absorbing and thus without importance for the dye laser operation, while the fifth, a carboxylic acid that was produced by oxigenation of the methylgroup in 4-position, was strongly absorbing at the laser wavelength and thus quenched the laser action. Filtering the dye solution through an alumina filter that preferentically absorbed the carboxylic acid prolonged the useful life of the dye solution in the laser recirculation system by more than a factor of 3. This observation is consistent with an earlier observation, that the useful life time of a dye in which the methyl-group is replaced by a CF_3-group is much longer, since the CF_3-group cannot be oxidized so easily (5).

Another important recent development is the vapor phase dye laser which holds promise for the future of direct electrical excitation. We obtained dye laser emission in the vapor of the scintillator dye POPOP with nitrogen laser pumping (6) a result that was independently obtained by a group of the Bell Laboratories (7). Similarly, a Russian group reported dye laser operation of POPOP at 250° C with pentane "under high pressure" being present. Since the critical temperature of pentane is 190°C this means that here we have the dye in a supercritical solution and not a true vapor phase laser (8). Meanwhile we have used 3 additional dyes in a vapor phase laser and found a considerable number of dyes that will probably work in the near future. A vapor beam device that can be pumped by flashlamps or an electron beam is under construction in our laboratory.

Since a report on cw-dye lasers will be given in the afternoon session by Dr. Paugh, I will only report on some new developments in our lab. Two prism dye ring laser arrangements are shown in Fig. 1 The first one (Fig. 1 a) uses a Z-shaped arrangement for the leg containing the jet stream. The four Abbe-Prisms are the predispersing elements while fine-tuning is possible with an uncoated etalon of 2 mm thickness and single-mode operation can be achieved with a Fox-Smith-reflector that couples back one of the beams exiting from the frustrated - total - internal - reflection prism. The second one (Fig. 1 b) uses two focussing regions in X-shaped configuration that not only reduces astigmatism but also coma (10). One focal region is for the active dye (e.g. rhodamine 6 G), while the second focal region can either contain a mode-locking dye (e.g. DODTC) or a second active dye (e.g. cresyl violet) that is pumped by the emission of the dye in the first focal region, so that simultaneous emission on two wevelengths can be obtained, if a resonator for the second dye is provided, as is done here using the two mirrors R_2 and R_3. The two laser beams can be made collinear by a compensating plate and then used for frequency-difference generation or similar applications. One important advantage of ring lasers as compared to linear lasers is the low threshold that is immediately apparent when closing the ring. In the case of the laser shown in Fig. 1 b threshold was as low as 10 m W after careful adjustment.

References

(1) D. Basting, F.P. Schäfer and B. Steyer, Appl.Phys.,3(1974)81-88

(2) J.P.Webb, F.G. Webster and B.E. Plourde, IEEE-J,Quant. Electr. QE-11 (1975) 114

(3) D. Basting, Thesis, Marburg, 1974

(4) B.H. Winters, H.I.Mandelberg and W.B.Mohr, Appl.Phys.Lett.25 (1974) 723

(5) E. Schminitschek,J.Trias, M. Taylor and J. Celtro, IEEE-J.Quant. Electron. QE-9 (1973) 781

(6) B. Steyer and F.P. Schäfer : Opt. Commun. 10(1974)210

(7) P.W. Smith, P.F. Liao, C.V.Shank,T.K.Gustafson,C.Lin, and P.J.Maloney, Appl.Phys.Lett.25 (1974) 144

(8) N.A. Borisevič, I.I. Kaloša, and V.A. Tolkaveč, Ph. Priklad. Spectrosk.19,No.6(December),(1973) 1108

(9) B. Steyer and F.P. Schäfer, Appl.Phys. in print.

(10) W.D. Johnston, Jr. and P.K. Runge, IEEE-J Quant. Electron, QE-8 (1972) 724

DISCUSSION

F.P. SCHAFER

J. Wolfrum - What are the conditions for obtaining an 1% conversion for the 3. Stokes line in the stimulated Raman emission of hydrogen gas pumped by a dye laser?

F.P. Schäfer - This result was obtained by W. Schmidt and coworkers Zeiss, Oberkochen and will be published soon.

S. Leach - It is extremely difficult to get a true emission continuum in a flashlamp pumped laser, even with Brewster angle optics and optical wedges. At a spectral resolution of 300 000, very narrow lines can still be seen, making the continuum unsuitable for high resolution absorption spectroscopy. In your ring laser with fibre optics, what spectral resolution was used in determing that your continuum was pure?

F.P. Schäfer - From the completely incoherent feedback used here one would expect a true emission continuum even when you look at it at the very high resolution of 300 000. We have looked at it with a 3/4 Spex-spectrograph with 10 μ slitwidth and any lines would have been noticeable by the over-exposure in the line.

M. Clerc - Could you, please, tell us the duration of the laser pulse used in the stimulated Raman scattering experiment? Knowing this duration, could you calculate or measure the relative importance of the stimulated Brillouin scattering which presumably competes with the stimulated Raman scattering in your experiment.

F.P. Schäfer - The duration of the dye laser pulse was of the order of a μ sec. and to my knowledge no Brillouin scattering was observed by W.Schmidt and coworkers.

T.A. Keller - We have observed what appears to be a continuum with a flash lamp pumped dye laser at a resolution of $\sim$ 300 000. Care must be taken to wedge all components and a curved total reflector seems to help.

F.P. Schäfer - It is true that careful wedging of all optical components and avoidance of all other possibilities of optical feedback is an alternative method of obtaining a true emission continuum albeit a more difficult one.

S. Kimel - 1) Would you please comment on the broad tuning range achieved with stimulated Raman scattering from gaseous hydrogen. The gases have narrow wavelength halfwidth and it would seem difficult to cover the whole of the infrared with second or third Stokes radiation. 2) What special provisions did you make to obtain stimulated emission with such low exciting powers?

F.P. Schäfer - 1) The normal tuning range of e.g. cresyl violet in the visible gives already a tuning range of the 3rd Stokes line in the IR from 2 to 12 μm. 2) W. Schmidt and coworkers obtained very low thresholds of stimulated Raman emission using a confocal resonator for the S.R.E. with broad-band mirrors of 96 to 98%reflectivity for the 1. to 3. Stokes line, 25 atm. of H_2 in the cell and the pumping radiation was focussed so that the pumped volume and the mode-volume of the resonator were matched.

Y. Meyer - There are two main causes for the decrease of dye laser intensity due to irradiation : apparition of permanent absorption or apparition of transient absorption due to photoproducts. For 7 diethylamino 4 methyl coumarine we measured that no transient absorption increases at 460 nm. Thus the reduction of the laser energy is entirely due to the increase of the yellow permanent absorption as suggested by Winthretal, Appl. phy. Letters 1975.

N. Karl - What are the smallest molecules which have been lasing successfully in the vapour phase, or what are the shortest wavelengths obtained from vapour dye lasers? With nitrogen-laser pumping the emission will always be restricted to larger wavelengths than 337 nm. Have there been experiments with a hydrogen laser?

F.P. Schäfer - The smallest molecule was POPOP for N_2-laser pumping, but we expect that with H_2-laser or flashlamp pumping smaller molecules with shorter laser wavelengths will work. Benzofurane seems to be a good candidate according to our measurements, p-terphenyl certainly is another possibility.

D.J. Bradley - We have found that even with careful elimination of subcavities by wedging, there remains intensity variation with wavelength in the output of a broadband dye laser. It seems unlikely that all the laser modes would go with uniform gain so as to produce an uniform spectrum. We use a 3 metre spectrograph and the intensity variation is different from shot to shot.

J. Langelaar - Do you have any information about the phtodegradation of POPOP-vapour as compared to the condensed phase?

F.P. Schäfer - Although we have not measured it exactly the photodegradation as well as the thermal degradation rate were definitively higher than in solution, owing to the elevated temperature.

M.W. Windsor - p-terphenyl has been lased in solution, has it not? What is the wavelengths shift on going to the gas phase?

F.P. Schäfer - p-terphenyl is a very good laser dye in solution, but it cannot be pumped by the N_2-laser in the vapour phase because of the blue shift of 2 000 to 3 000 wavenumbers (not yet measured).

HIGH-PRESSURE VUV GAS LASERS

D. J. BRADLEY
Optics Section, Physics Department, Imperial College, London SW7 2BZ

SUMMARY

Recent developments in the electron-beam pumped Xe_2 laser are reviewed with emphasis on frequency narrowing and tuning. The generation of VUV picosecond pulses by four-wave mixing is described and future developments and applications are briefly considered.

INTRODUCTION

When discussing the possibilities for laser action in the vacuum ultra-violet spectral region, it has been customary to point out that the pumping power requirement scales as the fourth power of the frequency, for a Doppler broadened transition. However with the development of intense high-energy electron-beam sources (1) there is no longer difficulty in obtaining in high pressure gases the necessary very fast rate of excitation for VUV laser action. Indeed it is now at least as easy to obtain narrow-band, tunable frequency, megawatt laser power around 170nm (2), as with dye-lasers operating in the visible spectral region. These new VUV lasers employ high-pressure noble gases as the active media. As long ago as 1960 Houtermans (3) pointed out the potential advantages for high-efficiency laser action of excited diatomic molecules with unstable groundstates. The broad emission continua produced by quasi-molecules of mercury and noble gases (Figure 1) have been extensively studied for several decades. The noble gas continua arise from transitions from the Σ_u^+ excited states to the repulsive ground state and the most intensive fluorescence,centred at $\sim$ 170nm, is produced by Xenon. Stimulated emission of this transition was first demonstrated in liquid Xenon pumped by an electron beam (4). Since then quasi-molecular laser action has been produced in gaseous Xenon (5-9), Krypton (10), Argon (11) or mixtures of these gases (10,12) by transverse excitation with relativistic electron-beams (0.5-2MeV), with current densities $\sim$ 1kA/cm^2. Detailed studies have also been carried out (13-20) of the atomic and molecular processes involved in excitation, including ground state absorption (21). While the laser kinetics are now better understood some of the details are still obscure. More recently the

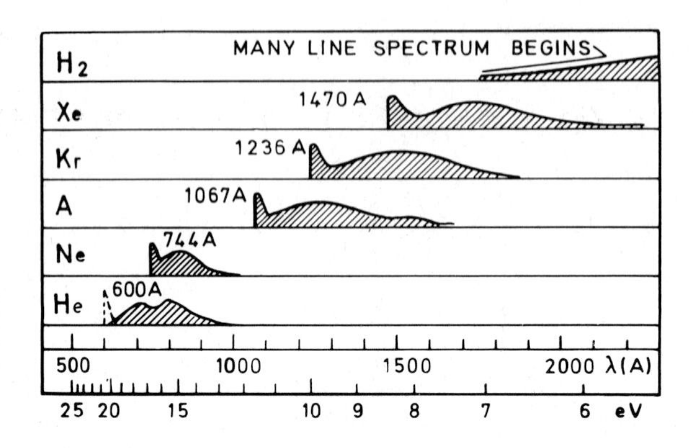

Fig. 1. VUV noble gas continua.

introduction (22) of the coaxial diode electron-beam arrangement has permitted the uniform excitation of the laser gaseous media with high efficiency. As a result the generation of megawatt powers at a high repetition rate is now possible. The purpose of this paper is to review the present state of the art in these tunable-frequency lasers. The Xe_2 laser will be considered as the prototype system since most work has been carried out to date on this laser, for which the materials problem is less severe than for the shorter wavelength Kr_2 and Ar_2 lasers. However it is probable that with sufficient optical engineering development all three noble gas lasers could be brought to the same level of performance as is now achieved with the Xe_2 laser.

Basic Principles

The development of high-current, high-energy electron-beam sources over the last decade has had a major effect on the development of new gas lasers. Largely due to the pioneering work of Martin and his colleagues at the AWRE Aldermaston Laboratory, electron-beams with Kiloampere currents and energies in the megawatt range are now widely available. With these intense beams it is possible to rapidly pump high-pressure gases in such a manner as to

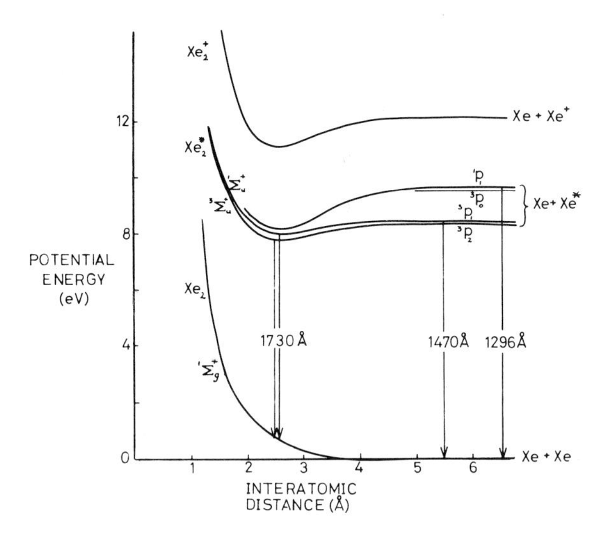

Fig. 2. Potential energy curves of Xe_2.

create population inversion as a result of ionization, recombination, association and energy exchange reaction chains. In particular, with nanosecond pumping the generation of VUV laser action in noble gas excimers is readily achieved. The intermolecular potentials for the ground state rare gas atoms are repulsive but at large separations there is a weak Van der Waals binding force. The electronic excitation energy is rapidly converted to produce molecular excimers which decay with the emission of a strong VUV continuum (23). The relevant energy levels, shown in Figure 2, arise from the $5p^5 6s$ and $5p^5 6p$ atomic configurations (24). The detailed processes occurring under electron-beam excitation of high pressure Xenon are now better understood and can be listed as follows:

<u>$Xe_2^*(^1\Sigma_u^+, ^3\Sigma_u^+)$ Production</u>

$Xe + e_p = Xe^{**} + e_p$ or $Xe^* + e_p$ (1)

$Xe + e_p = Xe^+ + e_p + e$ (2)

$e + Xe^* = e + Xe^{**}$ (3)

$Xe^+ + 2Xe = Xe_2^+ + Xe$ (4)

$Xe_2^+ + e = Xe^{**} + Xe$ (5)

$Xe^{**} = Xe^* + h\nu$ (6)

$Xe^* + Xe + Xe = Xe_2^* + Xe$ (7)

Energy Loss Mechanisms

$$Xe^{*} + e = Xe^{+} + e + e \qquad (8)$$

$$Xe_2^{**} + Xe_2^{**} = Xe_2^{+} + Xe + Xe + e \qquad (9)$$

$$Xe_2^{*} + Xe_2^{*} = Xe_2^{+} + Xe + Xe + e \qquad (10)$$

$$Xe_2^{*} + Xe = Xe + Xe + Xe \qquad (11)$$

Laser Gain and Loss

$$Xe_2^{*} + h\nu = Xe + Xe + 2h\nu \qquad (12)$$

$$Xe_2^{*} + h\nu = Xe_2^{+} + e \qquad (13)$$

$$Xe_2 + h\nu = Xe_2^{*} \qquad (14)$$

Reactions (1) and (2) represent ionization and direct excitation by the high energy primary electrons (e_p). Mechanisms (4), (5), (6) and (7) describe the formation of the molecular ion, followed by dissociative recombination and eventually the filling of the $^1\Sigma_u^+, ^3\Sigma_u^+$ fluorescence levels. The energy loss, or recycling, processes include Penning ionization, which becomes significant only at high densities and under strong pumping conditions, fluorescence quenching, and electron ionization. The important laser loss mechanisms are photoionization, (25) and ground state absorption, with an absorption coefficient of up to 0.1 cm^{-1} at high pressures and a temperature of 423K (21).

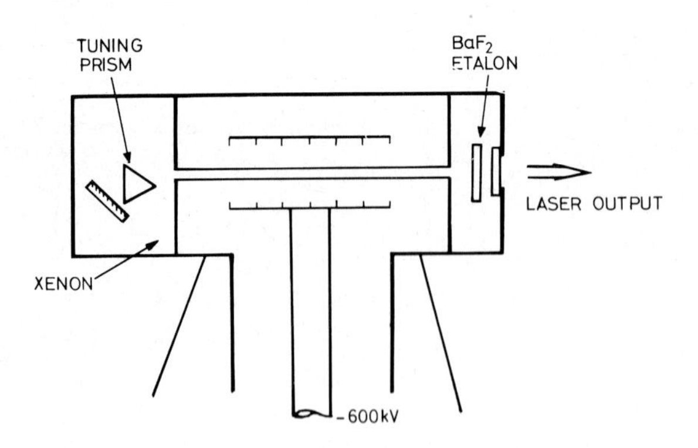

Fig. 3. Coaxial diode VUV Xenon laser arrangement.

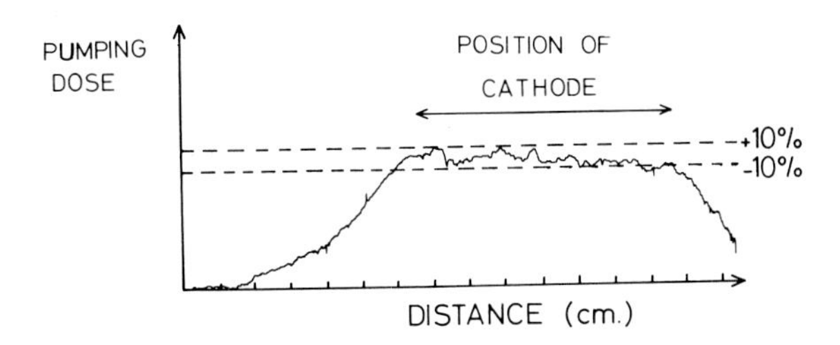

Fig. 4. Microdensitometer trace of cellophane recording of electron energy deposition.

Experimental

The first quasi-molecular ultra-violet lasers were produced by pumping with relativistic electron-beams (500keV - 2 MeV) delivering hundreds of joules in pulses of duration ∼ 50 nanoseconds or longer. Peak powers of 500 MW (10J, 20ns) have been obtained so far (26). An efficient and convenient pumping system was obtained with the coaxial diode, electron-beam arrrangement (2,22) shown in Figure 3. Also with uniform pumping of a well-defined cylindrical geometry investigations of the laser parameters are more easily carried out than in the case of the transverse pumping arrangements. The coaxial field-emission diode consists of a thin-walled (∼ 70 μm) stainless steel tubular anode of ∼ 4mm internal diameter, maintained at earth potential. Both ends of the anode, which is the container for the laser gas, open into high-pressure chambers which contain the optical elements. The concentric cathode, which has an internal diameter of 3.5cm and a length of 10cm, is constructed from titanium sheet, perforated to produce an array of spikes. The diode was originally designed for use with a Febetron 706 pulse generator (Field Emission Corporation) which has a load impedence of 60 Ω (10kA, 600kV). The anode tube radius is matched to the range of the electrons, which depends upon the gas pressure employed. For 600keV electrons a tube of ∼ 4mm diameter gives uniform pumping at a pressure of 10k Torr. To obtain efficient laser action it was necessary to drastically modify the Febetron 706 generator to produce a 20J, 5ns pulse of 500 keV electrons. Pumping is obtained over a 14cm length and calorimeter measurements showed that a total energy of 10J was transmitted into the gas, corresponding to a deposition of 5.5J cm^{-3}. The distribution of pumping energy over the inner surface of the anode surface had been determined for an earlier coaxial-diode of cathode length 7cm (22) using cellophane dosimetry (27). The

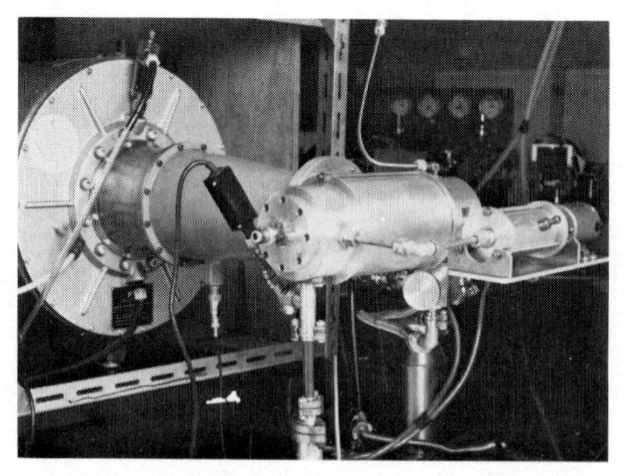

Fig. 5. Photograph of tunable Xenon laser showing modified commercial 500kV power supply and gas circulation system.

microdensitometer trace of Figure 4 shows a variation of less than 10% in dose along the tube. Because of end effects the pumped length exceeds the physical length of the cathode.

The anode tube is evacuated to 10^{-4} Torr. The Xenon gas is frozen in a high-pressure bomb by liquid nitrogen and pumped to $< 10^{-4}$ Torr to ensure that impurity concentrations are $<$ 1ppm. To reduce the water vapour partial pressure the bomb temperature is maintained $< 0^{o}C$ while

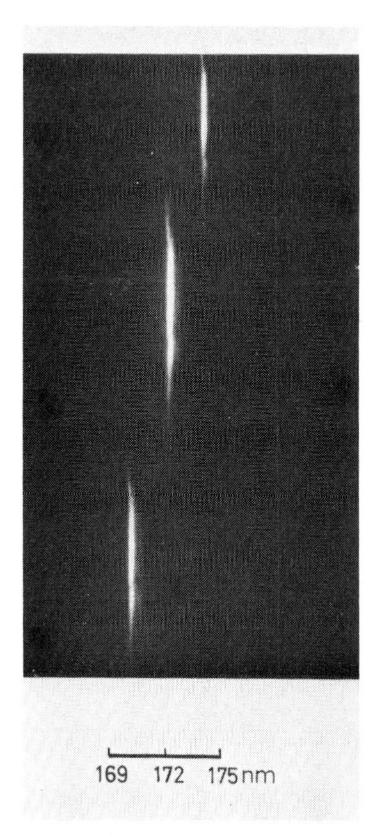

Fig. 6. Spectra of Xe_2 laser showing frequency narrowing and tuning. (The spectrograph plate was moved vertically between recordings and the tuning prism rotated).

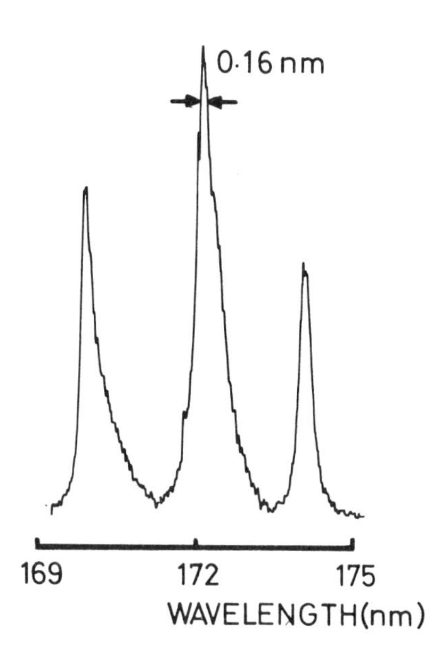

Fig. 7. Microdensitometer trace of spectra of Figure 8.

the tube is filled to the working pressure of 10k Torr. With these procedures stimulated emission intensity measurements are reproducibe to $\pm$ 5% and laser output is maximum. The laser resonator reflectors consist of an Al:MgF_2 coated plane mirror of ~ 85% reflectivity, and a BaF_2 single-plate resonant-reflector, with an effective reflectivity of ~ 20%. When these two reflectors only are employed in an optical cavity of length 25cm, an output laser energy of 10mJ in a 3ns pulse is obtained, corresponding to a peak power of ~ 3MW. The laser intensity is reproducible to 10%. At these high output powers the surface of the BaF_2 etalon suffers damage. From beam divergence measurements a half-angle divergence of ~1 mrad was calculated. By measuring the amplified spontaneous emission when the high-reflectivity mirror was aligned and mis-aligned respectively, and with the resonant reflector removed a nett gain of

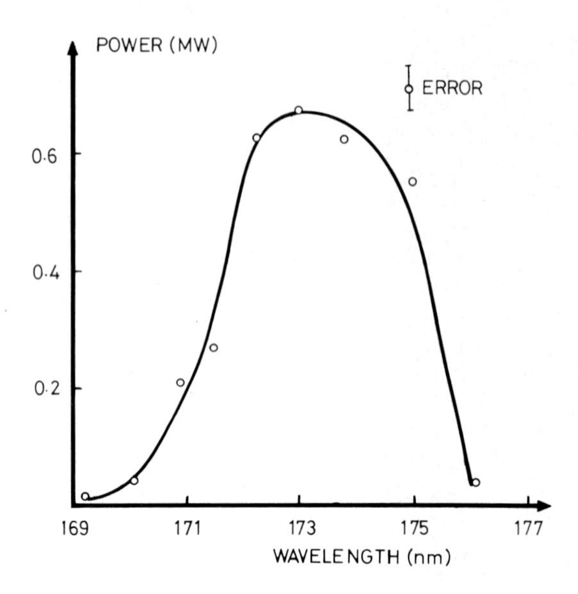

Fig. 8.   Tuning efficiency curve of Xe_2 laser.

0.25 cm^2 was determined. To maintain high laser powers it was necessary to allow a minimum internal of 15 minutes between shots to allow the gas to cool to room temperature. The instantaneous rise in temperature of the Xenon is ~ 700°C and the anode tube temperature increases by ~ 30°C by thermalization with the gas, after a few seconds. Absorption by ground-state Xenon molecules increases rapidly with increasing temperature and gas heating strongly affects the fluorescence and laser intensities. If the delay between firings was increased to 5 minutes the intensity was reduced by ~ 10% only (22). This difficulty has now been overcome by the simple expedient of circulating the Xenon gas through a water-cooled heat-exchanger, which can be seen in the photograph of Figure 5. High-pressure gases have the advantages of high-thermal capacity and low viscosity, and a simple magnetically driven propeller provides adequate cooling. The repetition rate is now limited by the power supply.

Frequency narrowing and tuning

A fused quartz prism is employed as an intra-cavity dispersing element. At the operating wavelength of $\sim$ 172nm the angular dispersion of a prism is comparable to that of a grating of the same area. In addition a prism is less susceptible to damage and has smaller losses. With the prism edge adjusted parallel to the mirror surface, rotation produced continuous tuning of the narrowed bandwidth over a range of 2500 cm^{-1}. Figure 6 shows three typical spectra recorded on SC7 film in a 1 metre, normal incidence, vacuum spectrograph. From microdensitometer measurements (Figure 7) a laser bandwidth of 0.13 nm was determined. There is thus a spectral narrowing by a factor of x 100 from the $\sim$ 15nm fluorescence (14), and by a factor of x 10 from the untuned laser bandwidth of 1.3nm (22).

Fig. 9. Photograph of 50cm length coaxial diode.

The complete tuning range is shown in Figure 8. The laser energy was measured with a thermopile. Each point of the curve is an average of 3 shots and the variation between shots was 10%. The peak power of 0.7 MW corresponds to that obtained from high-power flashlamp pumped dye-lasers, while the tuning range is considerably greater. A narrower -bandwidth should be achievable with longer pumping pulses and a greater number of resonator transits. Multiple-prism arrangements could also be employed. A 50cm long coaxial diode has been constructed (Figure 9) to operate with a pumping voltage of 600kV and up to 100J delivered in 50 nsec. This system is undergoing tests with argon and will be employed to generate a narrow bandwidth, high-power, tunable laser for nonlinear frequency mixing experiments.

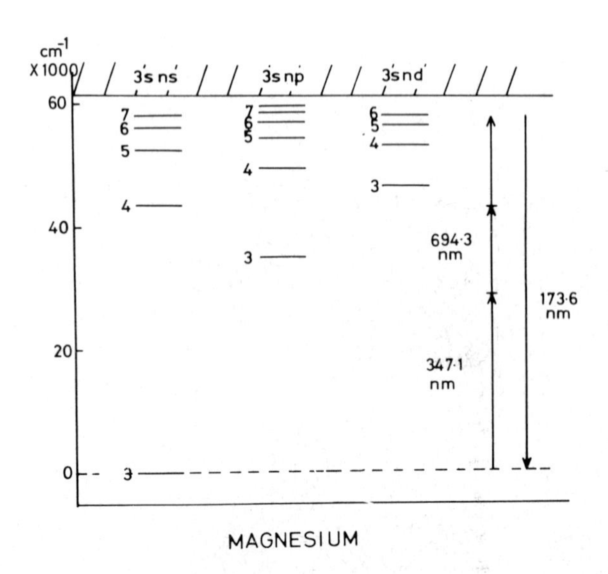

Fig. 10. Energy level diagram of MgI and four-wave parametric mixing

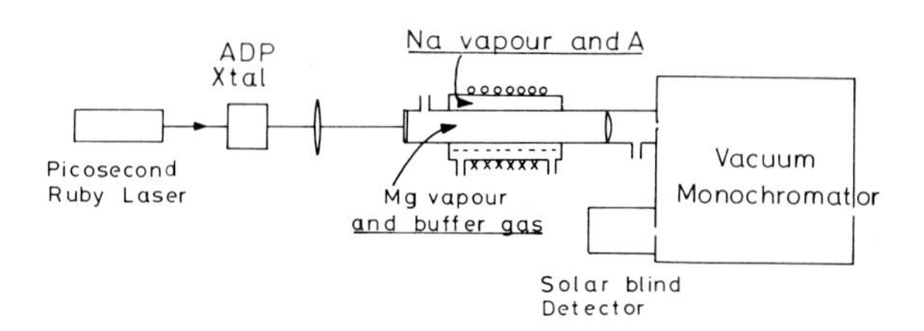

Fig. 11. Experimental arrangement for the generation of the fourth harmonic frequency of a mode-locked ruby laser.

VUV Picosecond Pulse Generation and Amplification

As with dye lasers, the broad bandwidth of the Xe_2, and other noble gas, lasers should allow the amplification of picosecond pulses. While the Xe_2 laser has yet to be mode-locked, picosecond pulses at 173.6nm have been produced (30) by four-wave (31) nonlinear mixing of second-harmonic photons with pairs of fundamental photons of ruby laser picosecond pulses, in magnesium vapour phase-matched with Xenon buffer gas. As can be seen from the magnesium energy level diagram (Figure 10) there is a close two-photon resonance for the $3s^2\ {}^1S_o$ - $3s3d\,{}^1D_2$ transition for one ruby second-harmonic photon and one fundamental frequency photon. The experimental arrangement is shown in Figure 11. The mode-locked ruby laser operating in a low-order single-transverse mode (32) produces a train of pulses of duration 15-20 psec with peak pulse energy of 1mJ (∼ 50MW). The pulses are frequency doubled in ADP with a conversion efficiency of ∼ 10% and both fundamental and second harmonic frequencies are focussed into a magnesium vapour cell, isothermally heated with a sodium/argon heat pipe. The output beam is focussed with a B_aF_2 lens into a vacuum monochromator with either a CsI photomultiplier or a CsTe photodiode. When helium was employed as the buffer gas in the magnesium cell, the efficiency of generation of 173.6nm pulses was insensitive to helium pressure, as is to be expected with the low dispersion of the gas at this wavelength. Maximum efficiency was obtained at a Xenon buffer gas pressure of 8 Torr for a magnesium vapour pressure of 1.6 Torr (cell temperature of 615°C) when a peak power of ~ 200 Watt at 173.6nm was obtained. This corresponds to a power conversion efficiency of 4×10^{-6}. The oscillograms of Figure 12

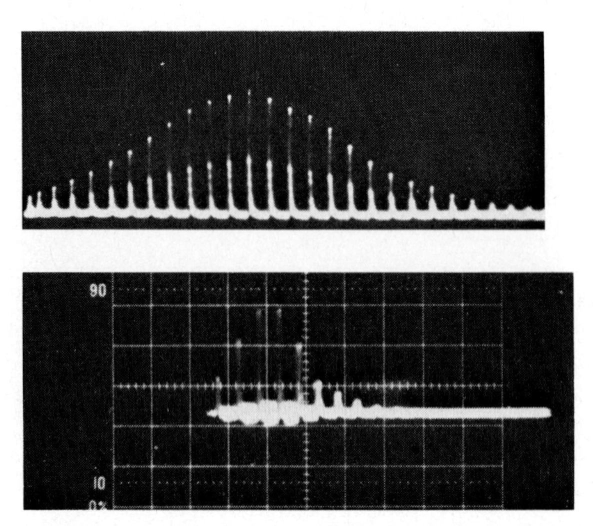

Fig. 12. Top: Oscillogram of ruby laser pulse train.
Bottom: Fourth harmonic pulses.

show the effect of this strongly nonlinear process upon the pulse train profile. Since the overall efficiency of the four-wave mixing process is proportional to the third power of the ruby laser, amplification of the oscillator pulses by x 20 should produce VUV megawatt picosecond pulses, provided saturation or other loss mechanisms do not operate. Further amplification could then be achieved in electron-beam pumped Xe_2 amplifiers. Provided that break down in the laser gas,or two-photon absorption in windows, can be avoided, high-power picosecond pulses could thus be produced for laser plasma pumping. The short wavelength will permit

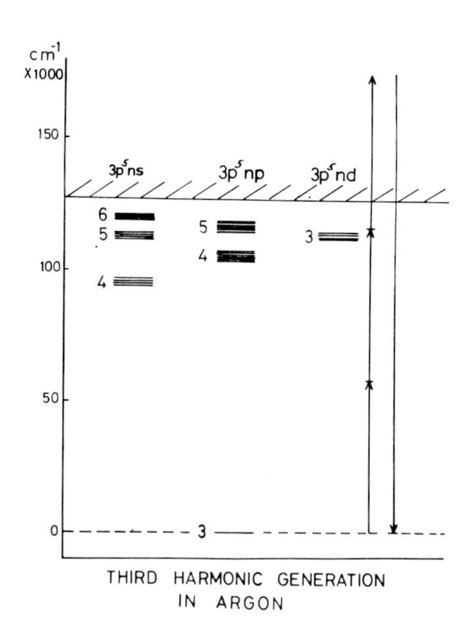

Fig. 13. Energy levels involved in third harmonic generation in Argon of of the Xe_2 laser.

penetration of denser plasma for diagnostic studies of high density, high temperature matter. These VUV pulses could have advantages for pumping of possible X-ray laser media and even for laser compression (33). The ruby fourth-harmonic picosecond pulses can also be employed for studying the detailed time-evolution of the Xe_2 laser pumping mechanisms, so as to clarify the roles played by the $^1\Sigma_u^+$ and $^3\Sigma_u^+$ upper energy levels (15,34) and to determine if the laser bandwidth is homogeneously broadened on a picosecond time-scale.

Discussion

It is clear that the Xe_2 laser is likely to play as important a role in VUV spectroscopy as dye lasers currently are at longer wavelengths and that with further development the Ar_2 and Kr_2 lasers should extend the tunable frequency coverage. With these new lasers it will be possible to employ selective excitation (35) over a wider range of atomic and molecular transitions and to achieve the breaking of bonds by optical means.

Extension to even shorter wavelengths should be possible by frequency tripling or nonlinear mixing in gases and atomic vapours (31). As an example Figure 13 shows the $3p^6 - 3p^5 3d$ two-photon resonance in Argon for the Xe_2 laser wavelength, which should allow the efficient generation of a tunable-frequency coherent source at $\sim$ 58 nm. For all of these applications a repetition rate laser is very desirable and in collaboration with AWRE Aldermaston Laboratory a $\sim$ 10pps, 600keV, 40ns, 40J power supply has been developed and is undergoing tests,(36).

Acknowledgement

The author wishes to thank Dr. M. H. R. Hutchinson, Dr. E. G. Arthurs, D. R. Hull and other members of the Imperial College Optics Section Laser Group whose work is described in this paper. Financial support from the Science Research Council and the UKAEA Culham Laboratory.

References

(1) H. H. Fleischmann, Physics Today, 28, 35 (1975)

(2) D. J. Bradley, D. R. Hull, M. H. R. Hutchinson and M. W. McGeoch. Opt. Commun. 7, 187, (1975)

(3) F. G. Houtermans, Helv. Phys. Acta, 33,933 (1960)

(4) N. G. Basov, V. A. Danilychev, Yu. M. Popov and D. D. Khodkevich, Soviet Physics JETP Letters, 12, 329 (1970)

(5) H. A. Koehler, L. J. Ferderber, R. L. Redhead and P. J. Ebert. Appl. Phys. Letters, 21, 198 (1972)

(6) W. M. Hughes, J. Shannon, A. Kolb, E. Ault and M. Bhaumik, Appl. Phys. Letters, 23, 385 (1973)

(7) J. B. Gerardo and A. Wayne Johnson, IEEE J. Quantum Electron. QE-9, 748 (1973)

(8) P. W. Hoff, J. C. Swingle and C. K. Rhodes, Opt. Commun. 8, 128 (1973)

(9) S. C. Wallace, R. T. Hodgson and R. W. Dreyfus, Appl. Phys. Letters, 23, 128 (1973)

(10) P. W. Hoff, J. C. Swingle and C. K. Rhodes, Appl. Phys. Letters, 23, 245 L973)

(11) W. M. Hughes, J. Shannon and R. Hunter, Appl. Phys. Letters, 24, 488, (1974)

(12) A. Wayne Johnson and J. B. Gerardo, J. Appl. Phys., 45, 867, (1974)

(13) A. Wayne Johnson and J. B. Gerardo, J. Chem. Phys., 59, 1738 (1973)

(14) D. J. Bradley, M. H. R. Hutchinson and H. Koetser, Opt. Commun., 7, 187, (1973)

(15) C. W. Werner, E. V. George, P. W. Hoff and C. K. Rhodes, App. Phys. Letters, 25, 235 (1974)

(16) W. Wieme, J. Phys. B: Atom. Molec. Phys., 7, 850 (1974)

(17) D. C. Lorents, D. J. Eckstrom and D. Huestis. Stanford Research Institute Report. SRIMP 73-12, (1973)

(18) E. H. Fink and F. J. Comes, Chem. Phys. Letters, 30, 267 (1975)

(19) G. Fournier, Opt. Commun., 13, 385 (1975)

(20) A. Wayne Johnson and J. B. Gerardo, Appl. Phys. Letters, 26, 582, (1975)

(21) D. A. Emmons, Opt. Commun. 11, 257 (1974)

(22) D. J. Bradley, D. R. Hull, M. H. R. Hutchinson and M. W. McGeoch Opt. Commun. 11, 335, (1974)

(23) R. E. Huffmann, Y. Tanaka and J. C. Larrabee, Japan J. Appl. Phys., 4, (Suppl. 1), 494 (1965)

(24) R. S. Mulliken, J. Chem. Phys. 52, 5170 (1970)

(25) D. C. Lorents, D. J. Eckstrom and D. Huestis, Stanford Research Institute Report. SRIMP73-2 (1973)

(26) R. O. Hunter, J. Shannon and W. Hughes, Maxwell Laboratories Internal Report - MLR-378 (1974)

(27) A. Malherke, Appl. Optics, 13, 1276 (1974)

(28) P. Jacquinot, J. Opt. Soc. Amer. 44, 761 (1954)

(29) D. J. Bradley, W. G. I. Caughey and J. I. Vukusic, Opt. Commun., 4, 150 (1971)

(30) E. G. Arthurs and M. H. R. Hutchinson. Unpublished.

(31) R. B. Miles and S. E. Harris, IEEE J. Quant. Elect. QE-9, 470, (1973); P. P. Sorokin, J. J. Wynne and R. T. Hodgson, Phys. Rev. Lett., 32, 343, (1974)

(32) D. J. Bradley, M. H. R. Hutchinson, H. Koetser, T. Morrow, G. H. C. New and M. S. Petty. Proc. Roy. Soc.A, 328, 97 (1972)

(33) J. L. Emmett, J. Nuckolls and L. Wood, Scient. Amer. 230, 24, (1974).

(34) J. B. Gerardo and A. Wayne Johnson, Appl. Phys. Lett., 26, 582, (1975)

(35) D. J. Bradley, P. Ewart, J. V. Nicholas and J. R. D. Shaw, J. Phys. B: Atom. Molec. Phys. 6, 1594 (1973); Phys. Rev. Lett. 31, 263 (1973)

(36) C. Edwards, M. D. Hutchinson, J. C. Martin, T. H. Storr. AWRE Report SSWA/JCM/755/99 "Lark - a modest repetive pulse generator", May 1975.

DISCUSSION

D.J. BRADLEY

G. Mourou - I would like to know the dynamic range of the photocron II.

D.J. Bradley - The dynamic range has been tested up to the linear ratio 30 : 1 and it is probably better. With the higher gain we would expect a correspondingly greater dynamic range before saturation effects come into play.

C.R. Goldschmidt - Considering your fast time measurements in the sub-picosecond region, how do you overcome the problem of group dispersion in the optical components as mentioned recently by Topp and Orner in Opt. Comm. 13 (1975), 276?

D.J. Bradley - As physicists, we were of course aware of group dispersion effects. These do not affect our time measurements even in the less than one picosecond range. With broadband sources then differential delays will occur but the streak-camera will of course automatically show these effects up directly unlike the nonlinear measuring techniques such as Duguay shutter.

F.P. Schäfer - With the high power densities of the Xe_2-laser did you experience mirror damage, and if so, what about using rooftop or corner-cube prisms made of BaF_2 or LiF?

D.J. Bradley - With a power density of 30 MW cm^{-2} we did not suffer from mirror damage. We have employed prisms as reflectors but find the $Al:MgF_2$ mirrors more convenient for alignment, etc...

C.B. Harris - In the Xe_2^* laser at 10 ktorr you have approximatly 10 collisions/ns. Do you think that vibrational relaxation will be fast enough to equilibrate the vibrational manifold of the bound state so that most of the excited states (vibrational) will contribute to the output during the pulse?

D.J. Bradley - We are carrying out experiments on the amplification of picosecond pulses and the results of these will directly indicate the homogeneous linewidth on a picosecond timescale.

S. Schneider - With the streak-camera, can you give a figure how much energy you need within one pulse to get a trace with a signal to noise ratio as good as shown on your slides?

D.J. Bradley - The signal to noise ratio is determined by the number of photo-electrons per resolution element. At an overall quantum efficiency of say 5% then the number of photons required in the pulse will be 20 000 per time element for a 30:1 signal to noise ratio. 20 000 photons is equivalent to $\sim 10^{-11}$ millijoules ($\sim$ 10 femto joules).

Mme Vermeil - Is anything known about mixed rare gas systems, mainly potential curves of mixed systems?

D.J. Bradley - Yes. Rare gas systems employing mixtures have been used in lasers but e.g. in argon and xenon mixtures, the xenon lasing wavelength dominates.

M.W. Windsor - Could you speculate a little on how femtosecond pulses would be applied in photophysics or photochemistry?

D.J. Bradley - Mainly at X-UV and X-ray wavelengths where typical lifetimes are picoseconds or shorter. More generally it is difficult to predict in the same way that eight years ago it would have been difficult to foresee the uses of picosecond pulses in laser compression and photobiology I am sure however that they will be applied as any other advances in measurement techniques have been in the past. After all this is the long term justification for basic scientific research.

S. Leach - How important, in the Xe_2 laser, is the loss due to absorption from the weakly bound (van der Waals) ground state? This type of loss should be less important in argon and krypton where the van der Waals well depth is smaller than in xenon.

D.J. Bradley- It seems to be significant if the laser gas is not cooled. You are right that argon and krypton should be better and it is for this reason that we are now developing a tunable argon laser. The main difficulty here is to remove xenon as an impurity.

M. Stockburger - Have there ever been made attempts to pump a Xe_2 VUV laser by an electric discharge? It is well known that the relevant continua are also produced by discharge of that type.

D.J. Bradley - This seems possible and it may be that more efficient excitation could be thus achieved. In part in our first experiments we attempted to do this.

THE FEASIBILITY AND ADVANTAGES OF OFF-RESONANCE LASERS IN CHEMICALLY REACTING SYSTEMS

C. B. HARRIS

Department of Chemistry; and Inorganic Materials Research Division, Lawrence Berkeley Laboratory; University of California, Berkeley, California, USA 94720

SUMMARY

The problem of understanding the semiclassical description of the time evolution of an ensemble of two state systems under the influence of a coherent radiation field is of considerable importance. Previous attempts to deal with these problems have dealt with either broad pulses or ultrashort pulses which allow the use of the rate equations or finite phase memory to be incorporated into the description. In neither case, however, has the effect of incoherent feeding and off-resonance effects in a coherently driven two-level system been analyzed. A closed form solution that includes the effects of relaxation and spontaneous emission between the two levels has been obtained for the general case when the ensemble is being incoherently fed from a population reservoir, as would be the case for example in a chemical laser. In addition to providing a basis for understanding the modifications which occur for such a system, the mathematical formulation predicts that an important effect may be observed. This effect, which is termed "kinetic coherence," is the production of a long-term coherent component that results directly from the kinetic feeding. The magnitude of the component is related to the rate of creating excited states, relaxation pathways and the off-resonance frequency. It will be shown how in principle it is possible to utilize these off-resonance effects in any inhomogeneously broadened system to significantly overcome the losses from T_2 relaxation processes and to provide an experimental system capable of controlling the relative ratio of spontaneous and stimulated emission. Finally, the relationships between chemical kinetics, the off-resonance feature and sustained self-regulation in a system exhibiting gain will be discussed.

INTRODUCTION

The problem of understanding the time evolution of an ensemble of two-level systems under the influence of a coherent radiation field is basic to a variety of fields in optics and in magnetic resonance. In almost all cases, a semiclassical description of the interaction between the particles and radiation has been used [1-3] and the set of coupled differential equations has been solved by the constraints of the strong-field short pulse approximation [4] or, as is done in magnetic resonance, by assuming the total population in the ensemble is constant in time [5]. Little attention has been given to examples where the total population of the two-level system is time-dependent. In some respects the work of Icsevgi and Lamb [6] on pulse propagation in laser amplifiers comes closest to addressing itself to this problem; however they did not explicitly formulate the qualitative and quantitative effects that "kinetic" processes in the ensemble of excited states play in the creation of the coherent superposition of states. In this paper, the role that incoherent feeding and decay play in the production of coherent states will be outlined. Particular attention will be given to the implications of off-resonance fields as a means of generating steady state coherence in the ensemble that results directly from a combination of the incoherent feeding process and the driving field. This effect has been termed "kinetic coherence" [7], and under certain conditions can be several orders of magnitude larger than the usual on-resonance component. Because of the time and space limitations in these proceedings, the development will be outlined only. A complete description of the problem, however, will be published elsewhere [7].

FORMULATION OF THE PROBLEM

(1) Generality of the Magnetic and Optical Case

In Figure 1 the salient rate processes to be considered are diagrammatically illustrated. k_x^+ and k_y^+ represent the incoherent feeding rates into the two-level system $|x\rangle$ and $|y\rangle$, respectively, while k_x^- and k_y^- represent the decay rates in the absence of a driving field. The Hamiltonian responsible for the coherent radiation field, V(t), is applied at a frequency ω which may or may not be at the center frequency, ω_0. The model for feeding and decay contains the following assumptions [7]:

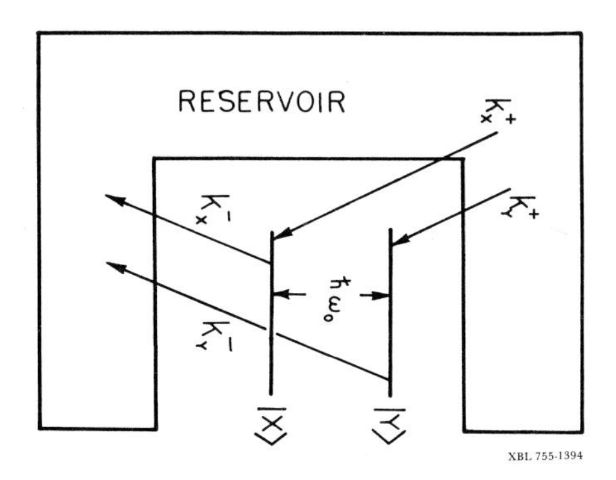

Fig. 1: Diagrammatic representation of a two-level system coupled by a coherent radiation field in the presence of feeding rates into the individual levels k^{+}, and decay rates from the two levels k^{-}.

(a) Feeding only occurs to eigenstates of the ensemble of two-level systems and not to coherent superposition states; however, decay occurs from both the eigenstates and from superposition states depending upon the explicit effects of the driving field. (b) The feeding rates into the eigenstates are constant and independent of the state of the ensemble. (c) The decay rate of the ensemble depends on the state of the ensemble and consequently the total population need not be time or field independent. This model is simple, and approximates many physically realizable situations.

Two general cases can be distinguished at this point. The first is representative of magnetic oscillating dipole fields and is formulated by allowing the coherent driving field to be of the form,

$$V(t) = \gamma\mathcal{H}_1 \hat{S}_i \cos(\omega t + \phi) \qquad [1]$$

$\gamma\mathcal{H}_1$ is the strength of the oscillating field applied at a phase ϕ. Solutions of the coupled differential equations including feeding and decay are applicable in this case to a variety of problems in magnetic resonance, including triplet state spin dynamics [8,9], optically pumped electron and nuclear spin polarization [10], chemically induced nuclear and electron spin polarization [11,12], spin diffusion, and others where the wavelength of the driving field is greater than the sample size, i.e., $\lambda^3 >>$ volume. Without loss of generality, one can consider cases where the wavelength of the radiation field is small relative to the size of the sample (the optical region) and the coherent Hamiltonian is an oscillating electric dipole field:

$$V(t) = -\mu \cdot E_1 \cos(\omega t + \phi - \vec{k} \cdot \vec{r}) \quad [2]$$

$\vec{k}$ is the wavevector of the radiation field and μ is the transition dipole moment associated with $|y\rangle \rightarrow |x\rangle$. If one considers "thin samples," the coupled Maxwell and Schrödinger equations may be avoided, thus allowing an exact solution to be obtained for both steady state and transient behavior of the ensemble in the presence of feeding and decay for all strengths of the driving field and in the presence of phenomenological relaxation terms [13]. This is true for either form of the time-dependent Hamiltonian in Eqs. 1 and 2. Indeed, one can show that the equation of motion for the time-dependent density matrix associated with Eqs. 1 and 2 is of the same form when the explicit time dependence is removed from Eq. 1 and the explicit time and spatial dependence is removed from Eq. 2. The appropriate transformations are

$$U^{-1} = \exp(i\mathcal{H}_0\omega t/\hbar\omega_0) \quad [3]$$

and

$$U^{-1} = U_{k,j}^{-1} = \exp[i\mathcal{H}_0(\omega t - \vec{k} \cdot \vec{r}_j)/\hbar\omega_0] \quad [4]$$

respectively. The resulting time-independent density matrix in the resulting rotating frame is given by

$$\rho^*(t) = U\rho(t)U^{-1} \quad [5]$$

Both Eqs. 3 and 4 satisfy the equation of motion, $i\mathcal{H}\rho^* = [\mathcal{H}^*,\rho]$, where

$$\mathcal{H}^* = U[\mathcal{H}_0 + V(t)]U^{-1} \quad [6]$$

and $\mathcal{H}_0$ is the zeroth order Hamiltonian having eigenvalues of $\pm\hbar\omega_0/2$.

In the optical regime, solutions to the coupled differential equations serve as a potential model for understanding a variety of kinetic problems including chemical laser dynamics [14,15], stark shift optical coherence [16], pulse propagation [6,17] in an amplifying media, to mention only a few.

(2) Outline of the Solution

There are at least two distinct ways of solving the coupled differential equations. The first [7] is to formulate the density matrix in such a way that it operationally includes both processes. Specifically, the density matrix describing a two-level system restrained by assumptions (a)-(c) above is

$$i\hbar\frac{d\rho}{dt} = [\mathcal{H},\rho] - [K,\rho]_{+} + F \qquad [7]$$

where the feeding terms are given by

$$F = i\hbar\begin{bmatrix} k_y^{+} & 0 \\ 0 & k_x^{+} \end{bmatrix} \qquad [8]$$

and the decay terms are incorporated by constructing an imaginary operator:

$$K = \frac{i\hbar}{2}\begin{bmatrix} k_y^{-} & 0 \\ 0 & k_x^{-} \end{bmatrix} \qquad [9]$$

The final solution to Eq. 7 can be shown [7] to be

$$\rho(t) = Q^{\dagger}[\rho(0) - \rho_s]Q + \rho_s \qquad [10]$$

where ρ_s is the density matrix describing the steady state to which the ensemble evolves under a given set of experimental conditions. Q is given by

$$Q = \exp\{i[(\mathcal{H} + K)/\hbar]t\} \qquad [11]$$

The second method [18] results in an expression identical to the above, but is more cumbersome to develop mathematicaly. It retains, however, considerably more physical insight into the problem. In brief, the problem is solved by incorporating the kinetic processes directly into the time-dependent Schrödinger Equation by subpartitioning the ensembles of two-level systems properly in time as not to disregard information concerning the relative phases of the states which are continually being created. One first solves the coupled differential equations for the initial t = 0 population, including the effect of decay and then separately solves the equation of motion for those additional states that are created in a small increment of time, δt. At any time t the solution is simply the initial solution $\rho_0(t)$, plus $\rho_{\delta t}(t) + \rho_{2\delta t}(t) + \ldots \rho_{n\delta t}(t)$, where $t = n\delta t$ and ρ_0, $\rho_{\delta t}$, $\rho_{2\delta t}$, etc., are the t = 0, t = 0 to δt, t = δt to $2\delta t$, ... t = $(n-1)\delta t$ to $n\delta t$ subensembles, respectively. In the limit that $\delta t \to 0$ one obtains a set of solvable coupled integral equations. The analytic solutions [7] are formidable and will not be presented here; however, the qualitative features of the results are simple and revealing, particularly in the rotating frame.

EFFECTS OF FEEDING INTO A COUPLED TWO-LEVEL SYSTEM

(1) <u>On-Resonance Feeding</u>

Consider first on-resonance driving fields in the case where the ensemble is initially inverted, i.e., all the population is in the upper level $|y\rangle$ at t = 0, as illustrated in Figure 2. When the driving field is externally turned on as in magnetic resonance or internally builds up from the reactance field according to Maxwell's Equation as in an optical oscillator or amplifier, the ensemble is driven around the field. In the case of an oscillating magnetic dipole, it executes a transient nutation and dephases because of relaxation. In the absence of feeding at some later

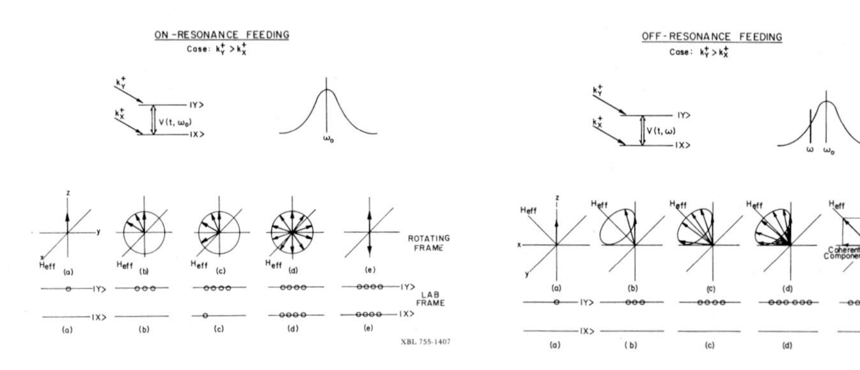

Fig. 2: The relationship between the rotating and laboratory frames on-resonance in the presence of feeding processes.

Fig. 3: The relationship between the rotating and laboratory frames off-resonance in the presence of feeding processes.

time, $\rho_0(t)$, the density matrix describing the initial population contains a random and equal distribution of vectors in the yz plane of the rotating frame, x being associated with $\phi = 0$ of the driving field. In the optical case for a monochromatic wave [6], the reactance field will drive the ensemble to saturation, causing the gain to reach a limiting value. Naturally transient oscillations in the reactance field are possible as is the case with optically induced self-transparency [19]. In the case of an amplifier the limiting steady state intensity approaches a value that is determined by the amount by which threshold is exceeded and by the Lorentz profile for homogeneous relaxation. As we will see, such is not the case for the off-resonance driving field.

When one considers the feeding processes as illustrated in Fig. 2, the ensemble also evolves to a saturated value for on-resonance fields. This can be seen by considering the case where $k_y^+ >> k_x^+$. In such instances, excited states (subensembles) are incoherently created in time, each along the + z axis in the rotating frame because there is no phase relation between the excited states that are created and those that already exist in the ensemble. Each subensemble is in turn driven by the field in the yz plane. In both the optical and magnetic cases, a steady state saturated value is reached, and is represented as a disk of vectors in the yz plane, each vector representing one of the subensembles. Thus, for on-resonance feeding, the driving field equally and incoherently distributes the population into the two levels, $|x>$ and $|y>$. The only exception to this would be if the decay rates were much shorter than the transient nutation frequency in the yz plane. In such a case for a magnetic dipole, a steady state polarization would exist that generates a coherent component 90° out of phase with the driving field. In the optical case, however, the reactance field would presumably accelerate the ensembles to such an extent that it might be difficult to have the nutation rate slower than the decay rate at steady state.

(2) Off-Resonance Feeding

When the driving fields are applied, or develop off-resonance at a frequency ω by an amount $\Delta\omega$, given by

$$\Delta\omega = \omega_0 - \omega, \qquad [12]$$

several new phenomena result from the kinetic feeding and decay processes. Consider the case where feeding into one of the two levels is preferred

$(k_x^+ >> k_y^+)$. In such instances the t = 0 ensemble will precess about an effective field which is at some angle in the xz plane determined by $\Delta\omega$. Because the states are radiatively or nonradiatively decaying at k_x^- and k_y^-, the initial polarization will be lost after a time comparable to $(k_x^- + k_y^-)^{-1}$. The states that are being continually created by k_y^+, however, enter randomly along + z at a rate proportional to k_y^+, and like the t = 0 subensemble, decay as k_x^-, k_y^- or some average value. If the feeding rate is comparable to the decay rates and the driving field is sufficiently large, then a cone of vectors will develop around the effective field. This cone will evolve to some steady state value depending upon the magnitude of various relaxation and kinetic processes. This is illustrated in Figure 3.

The important qualitative points to note are: (i) a steady state nonsaturated value for the polarization is achieved by kinetic feeding; (ii) this polarization appears "locked" on the field; (iii) relaxation is overcome when H_{eff} is large enough because the off-resonance condition is satisfied for all members of a Lorentz absorption regardless of whether the transition is homogeneously or inhomogeneously broadened; (iv) there is a coherent component of the polarization which is simply the projection of the steady state value of the lock polarization onto the xy plane. This has been called kinetic coherence because its magnitude results directly from the kinetic parameters; and finally, (v) the coherent component can be stabilized and maximized when the off-resonance frequency is chosen properly.

The off-resonance value, $\Delta\omega_{max}$, that corresponds to a maximum coherent component $\overline{r_1^s}(max)$ is given by [7]:

$$\Delta\omega_{max} = 1/T_2^* + (1/\sqrt{TT_e})(1 + \omega_1^2 T\tau)^{1/2} \qquad [13]$$

and yields a value for the Feynman, Vernon and Hellwarth [20] r_1 component of

$$\overline{r_1^s}(max) = [r_3^0\omega_1/2][1/T_2^* + (1/\sqrt{TT_e})(1 + \omega_1^2 T\tau)^{1/2}]^{-1} \qquad [14]$$

where T_2^* is the inhomogeneous relaxation time associated with a Lorentz lineshape function, $g(\omega_0)$;

$$g(\omega_0) = [T_2^*/\pi][1 + (\omega - \omega_0)^2 T_2^{*2}]^{-1} \qquad [15]$$

where

$$\int_{-\infty}^{+\infty} g(\omega_0)\,d\omega_0 = 1 \qquad [16]$$

and T_e contains relaxation along the field [21] T_{2e} as

$$1/T_e = [(k_x^- + k_y^-)/2] + T_{2e}^{-1} \qquad [17a]$$

T and τ contain the kinetic parameters and relaxation terms. T is given by

$$1/T = [(k_x^- + k_y^-)/2] + 1/T_2 \qquad [17b]$$

where T_2 is the normal transverse relaxation time. τ is given in terms of the kinetic parameters and T_1 processes as:

$$\tau = [(k_x^- + k_y^-)/2][k_x^- k_y^- + k_y^- T_x^{-1} + k_x^- T_y^{-1}]^{-1} \qquad [18]$$

T_x and T_y allow for both spontaneous emission from |y> to |x>, T_{is}, and normal T_1 type terms, T_{1i}:

$$T_i^{-1} = T_{is}^{-1} + T_{1i}^{-1} \qquad (i = x \text{ or } y) \qquad [19]$$

The relationship between the feeding process and the coherent components in Eq. 14 is given by r_3^0, which is [7]:

$$r_3^0 = [\tau/(k_y^- + k_x^-)][(k_y^+ k_x^- - k_x^+ k_y^-) + (k_y^+ + k_x^+)/T_x - (k_y^+ + k_x^+)/T_y] \qquad [20]$$

Finally, ω_1 is the strength of the driving field and is given by the strength of the transition dipole. For the magnetic case it is

$$\omega_1 = \gamma \mathcal{H}_1 F \qquad [21]$$

while for the optical case it is

$$\omega_1 = \mu \cdot E_1 \qquad [22]$$

When the driving field strength is large, i.e.,

$$\omega_1 T_2^* > 1 \quad [23]$$

$$\omega_1 T > 1 \quad [24]$$

and

$$\omega_1^2 T\tau > 1, \quad [25]$$

the maximum kinetic coherent component is given by a simple expression:

$$r_1^s(\max) = (r_3^0/2)(\sqrt{T_e/\tau}) \quad [26]$$

while for on-resonance. the coherence is given by [7]:

$$r_2^s = [-r_3^0\omega_1][\omega_1^2\tau + \{1/T + \sqrt{T_e/T}\, T_2^{*-1}(1 + \omega_1^2 T\tau)^{1/2}\}]^{-1} \quad [27]$$

The important point to note is that the maximum value on-resonance for an inhomogeneous transition is significantly less than possible values off-resonance. This is illustrated in Table 1, where the ratio of on-resonance to off-resonance coherence is given versus a measure of the homogeneous to inhomogeneous linewidth ratio, T/T_2^*. For example, (cf. Table 1) if a transition is inhomogeneously broadened by hundredfold the homogeneous width, then the maximum coherence on-resonance is less than one percent of the value that could be obtained off-resonance. Therein lies the central importance of off-resonance driving fields.

Table 1: Ratio of off-/on-resonance coherence versus T/T_2^* in units of r_3^0 (see Eq. 20).

T/T_2^*	0	1	10	100
Off-/on-r.c.	0.5/0.5	0.5/0.3	0.5/0.075	0.5/0.009
$\Delta\omega(\max)$ in units of T_e^{-1} (Eq. 17a).	~2.5	~2.5	~2.5	~2.5

APPLICATION TO CHEMICAL LASERS

In some respects chemical lasers [14] are ideally suited to test the above theory and predictions. Many of the inherent limitations that are associated with chemical lasers and chemically reacting systems might be circumvented by the development of "off-resonance" type lasers.

An oversimplified picture of chemical lasers is that a nonequilibrium distribution of final product states is created by selective channeling of energy from a chemical reaction in such a way that a partial population inversion is created between the molecular rotation-vibration states of the products. In this sense the kinetic parameters k_y^+ and k_x^+ in the above description can analytically incorporate the chemical kinetics associated with the formation of products into particular vibrational-rotation states. In the absence of vibrational relaxation, the population of the various states, N_{VJ}, is proportional to microscopic rate constant k_V for product formation [22] and hence determines the extent of population inversion in the various P-branch ($\Delta J = +1$) and R-branch ($\Delta J = -1$) transitions.

One can go quite far in the analogy between chemical lasers and the above model. It is well-known that relaxation processes usually limit the storage of energy in excited states, insofar as stimulated emission competes with collisional deactivation processes, vibration deactivation being the principal shortcoming of most hydrogen halide lasers [23,24]. The details of vibrational, rotational [25] and other relaxation processes are contained in the relaxation terms, k_y^-, k_x^-, T_{1x} and T_{1y}. Consider, for example, a single P-branch transition in the HF laser, $P_2(3)$. In such cases, the levels $|y\rangle$ and $|x\rangle$ are associated with the product states ($V = 2$, $J = 2$) and ($V = 1$, $J = 3$) in HF respectively. k_y^+ and k_x^+ would be related to the bimolecular rate constants k_2 and k_1 for the formation of HF ($V = 2$) and HF ($V = 1$) as:

$$k_y^+ = k_2 \qquad [28]$$

$$k_x^+ = k_1 \qquad [29]$$

Vibration relaxation to all levels except $V = 1$, $J = 3$ from $V = 2$, $J = 2$ is given by

$$k_y^- = \sum_{V \neq 1;} \sum_{J \neq 3} P(2,2|VJ) \qquad [30]$$

where $P(i,j|k,\ell)$ are proportional to the probabilities for the individual vibration-rotation decay channels. Likewise,

$$k_x^- = \sum_{V \neq 2;} \sum_{J \neq 3} P(1,3|VJ) \qquad [31]$$

Finally, P(2,2|1,3) and P(1,3|2,2) are proportional to T_{1x}^{-1} and T_{1y}^{-1}, respectively, in Eqs. 18 and 19. In this manner, most of the salient rate processes are incorporated into a single unified description.

At this point two limiting cases should be distinguished to illustrate the potential advantages and usefulness of off-resonance effects. These are the low and high pressure regimes which are characterized by an inhomogeneous or homogeneous transition, respectively.

At low pressure (< 1 torr) collisions are infrequent enough on the time scale of the stimulated emission output (μsec) that there is no significant equilibration of the translation kinetic energy. To a first approximation this limits laser action to regions of the Doppler profile that overlaps the cavity modes. This severely limits the number of molecules that can contribute to the stimulated emission gain and hence the output of the laser on-resonance is reduced by approximately (nT_2^*/T_2), where n is the number of cavity modes that reach threshold and $(T_2^*/T_2)^{-1}$ is the ratio of the inhomogeneous to homogeneous linewidth. For low enough pressures this reduction can be ten to hundredfold or more. Low pressures, however, have important advantages because the vibrational (and/or rotational) deactivation is significantly reduced. This in principle can result in much larger inversions, particularly for the higher vibrational states. Unfortunately this has not been realized operationally in the normal use of most chemical lasers because of the necessity of having a sufficient number of collisions to homogeneously (pressure) broaden the line in order to obtain a full inversion under the entire line profile. An off-resonance laser overcomes this limitation insofar as the full inhomogeneous distribution contributes to stimulated emission independent of collisional broadening. This would mean that in principle <u>the entire inhomogeneous profile in the state of highest inversion could contribute to stimulate emission</u>. If the effects of spontaneous emission on-resonance were minimized in the time necessary to reach threshold off-resonance, the maximum gain from the system could be realized. In addition, because the gain/molecule is low off-resonance, off-resonance lasers might be ideal low noise amplifiers. On-resonance spontaneous emission in the mode structure of the cavity, however, would have to be absorbed or limited.

Finally, I wish to consider qualitatively at least the advantages "off-resonance" chemical lasers potentially have at higher pressures (50-100 torr or greater). Chemical lasers operating at higher pressures suffer from the competition between stimulated emission, and vibrational and rotational deactivation. Although this is currently thought to limit their applicability as high power oscillators or amplifiers, TEA lasers have demonstrated that with a rapid enough build-up in the population inversions, reasonable powers can be achieved in some systems operating at pressures as high as atmospheric [26]. In such cases the line can be considered homogeneously broadened on a nanosecond time scale because of the very rapid equilibration between the Doppler components of transition. Translational equilibration occurs on a time scale given by the collision frequency at room temperature and atmospheric pressure (~ 4 collisions/nsec) [27].

Although the maximum off-resonance and on-resonance coherence are the same for a given set of rate processes (cf. Table 1), the two classes differ with respect to the strength of the driving field necessary to achieve the optimal coherence. For the on-resonance case the coherent component is always higher than that of off-resonance until saturation is reached. For conventional low powered lasers "off-resonance" operation in the homogeneous regime offers little advantage. For higher powers, however, on-resonance systems begin to "power broaden" so that the effective field strength can be significantly reduced. This can be seen by the factor of $\omega_1^2\tau$ in the denominator of Eq. 27. In the off-resonance mode, however, no power broadening is predicted to occur, so that the maximum coherence can always be obtained in the higher power limit without the complication of saturation and subsequent field dependent broadening.

ACKNOWLEDGEMENTS

This research was supported in part by the National Science Foundation and in part by the U.S. Energy Research and Development Administration.

REFERENCES

1 W.E. Lamb Jr., Phys. Rev., 132(1964)A1429.

2 L.M. Frantz and J.S. Nodvik, J. Appl. Phys., 34(1963)2346.

3 N.G. Basov, R.V. Ambartsumyan, V.S. Zuev, P.G. Kryukov and V.S. Letokhov, Zh. Eksperim. i Teor. Fiz., 50(1966)23 [Soviet Phys. - JETP, 23(1966)16].

4 G.L. Lamb, Phys. Lett., 25A(1967)181.

5 A. Abragam, Principles of Nuclear Magnetism, Oxford, London, 1961.

6 A. Icsevgi and W.E. Lamb Jr., Phys. Rev., 185(1969)517.

7 W.G. Breiland, M.D. Fayer and C.B. Harris, Phys. Rev. A, (1975) in press

8 J. Schmidt, W.S. Veeman and J.H. van der Waals, Chem. Phys. Lett., 4(1969)341.

9 C.B. Harris, J. Chem. Phys., 54(1971)972.

10 M.S. de Groot, I.A.M. Hesselmann and J.H. van der Waals, Molec. Phys., 12(1967)259.

11 C.L. Closs, J. Am. Chem. Soc., 91(1969)4552; R. Kaptein and L.J. Oosterhoff, Chem. Phys. Lett., 4(1969)214.

12 F. Adrian, J. Chem. Phys., 54(1971)3912,3918; J.B. Pederson and J.H. Freed, J. Chem. Phys., 57(1972)1004.

13 F. Bloch, Phys. Rev., 70(1946)460.

14 J.V.V. Kasper and G.C. Pimentel, Phys. Rev. Lett., 14(1965)352.

15 J.C. Polanyi, J. C. Appl. Opt., 10(1971)1717; Appl. Opt. Suppl., 2(1965)109.

16 R.G. Brewer and R.L. Shoemaker, Phys. Rev. Lett., 27(1971)631.

17 J.C. Diels and E.L. Hahn, Phys. Rev. A, 8(1973)1084.

18 M.D. Fayer and C.B. Harris, unpublished results.

19 S.L. McCall and E.L. Hahn, Phys. Rev. Lett., 18(1967)908.

20 R.P. Feynman, F.L. Vernon and R.W. Hellwarth, J. Appl. Phys., 28(1957) 49.

21 A.G. Redfield, Phys. Rev., 98(1955)1787.

22 K.L. Kompa, Fortschrittle der Chemischen Forschung: Topics in Current Chemistry, Vol. 37, Chemical Lasers, Springer-Verlag, 1973.

23 K.L. Kompa and G.C. Pimentel, J. Chem. Phys., 47(1967)857.

24 T.F. Deutsch, Appl. Phys. Lett., 10(1967)234.

25 J.C. Polanyi and K.B. Woodall, J. Chem. Phys., 56(1972)1563.

26 J. Wilson and J.S. Stephenson, Appl. Phys. Lett., 20(1972)64.

27 W.H. Flygare, Accounts Chem. Res., 1(1968)121.

DISCUSSION

C.B. HARRIS

D.J. Bradley - I am obviously missing the point. Homogeneously broadened lasers have been driven off- resonance for many years. For an inhomogeneously broadened line how do you couple the energy from all the excited atoms or molecules into the off-resonance lasing frequency? Are you saying simply that with a strong enough laser field even a transition at a considerable off-(laser)-resonance can be stimulated?

C.B. Harris - First, let me say that for an inhomogeneously broadened transition there is no advantage in producing stimulated emission off-resonance under most circumstances. The exception to this is when the field gets large enough to power broaden a transition on-resonance. Under these conditions the maximum components of stimulated emission on-resonance could be less than that available off-resonance. Secondly, the energy from all the excited atoms or molecules in an inhomogeneous distribution is coupled in the tail of each transition. Since the oscillating field induced in each can become large when the driving field is sufficiently strong it should be possible to get stimulated emission off-resonance in an inhomogeneous distribution. Furthermore, if the off-resonance frequency is chosen properly, the theory suggests that the system would self-regulate and be stable at high powers for an inhomogeneous distribution. The power available from the system under these circumstances would be many times that available on-resonance.

F.P. Schäfer - Referring to your proposed experimental realization, I really cannot see what is new in it. Many experiments of that sort have been done in the dye laser field : a He-Ne or Ar^+-laser or dye laser beam being injected into the resonator of a dye laser far away from the maximum of the gain curve resulting in a high amplification at the injected wavelength and reducing significantly the amplified spontaneous emission, and this laser emission continues even after the starting laser is shut off. What then is new in your proposal?

C.B. Harris - The experiments you have referred to are not of the sort I am proposing here. Dye systems are homogeneously broadened and injection of an initiating field at some frequency off the maximum in the gain curve simply allows the resonant dye system to reach threshold at that point first after which it can self-sustain itself. As I understand it, the amplified spontaneous emission is reduced because the small signal gain is low at these frequencies. The experiments I am proposing are entirely different. The principle of off-resonance lasers in an homogeneous distribution is that the off-resonance driving field in combination with the rate of creating a partial inversion creates a steady state polarization "locked" on the off-resonance field which self-regulates at well defined off-resonance frequencies. The theory suggests that the frequency will be "pushed" further off-resonance as the field increases. Certainly, this is not the case in the experiments you have mentioned.

O. Svelto - I am afraid that I have somewhat missed the physical principle underlying your calculation. To try to understant, I would like to ask two questions : 1) It is known that it is only the 90° out of phase component of the polarization which contributes to the energy exchange between the e.m. field and the matter. Did you take this into account? 2) What is the difference between your theory and the old Lamb theory?

C.B. Harris - The component of the polarization which contributes to energy exchange is the same off-resonance as it is on-resonance. It does of course have a different magnitude. In the rotating frame diagram I have shown the reactance field is in both cases in the xz plane. On-resonance, it is along the x-axis. The development is very similar to Lamb's theory. In particular, the equations which predict the off-resonance effect can be obtained by a further development of the "monochromatic wave" case in the work of Icsevgi and Lamb. The details however of both homogeneous and inhomogeneous relaxation in relation to the strength and frequency of the resonance effect at steady state were developed here.

P. Rigny - I have two questions to make the analogy with NMR clearer to me 1) The z axis in magnetic resonance is determined by an external magnetic field and is fixed in space. How is it chosen in your description? 2) In NMR it is possible to do spin-locking in the rotating frame even on-resonance. Why did you develop the analogy only off-resonance?

C.B. Harris - The answer to your first question is that any two-level system can be transformed to the rotating frame. It is perfectly general. In our case both the time and spacial dependence is removed by a standard transformation. The + z axis in the rotating frame represents the population in one of the two levels and the -z axis represents population in the other level.
In magnetic resonance, spin-locking has been done on and off-resonance. The importance of locking off-resonance in a system where states are continually being created and are continually decaying is that one can achieve a steady state-polarization along the field. On-resonance, the feeding processes do not create a polarization locked along the field unless of course you phase shift the field 90°. But then any new population created after the phase shift is not locked and the component you managed to lock decays rapidly because of the short lifetime of the excited state or via faster relaxation processes (T_1 or T_2). The off-resonance component along the field does not decay when the off-resonance frequency is chosen properly.

J. Wolfrum - What are the proposed practical parameters (pressure, volume, etc...) for an off-resonance chemical laser?

C.B. Harris - I cannot tell you at this time specifically the pressure, the pathlength, the off-resonance field strength, etc... necessary to build an off-resonant chemical laser. I can, I think, give you some general guidelines. First, one would like to keep the pressure low enough to significantly reduce collisional deactivation. A pressure of ~ 1 Torr would probably be sufficient. Secondly, one must prevent spontaneous emission from initiating stimulated emission on-resonance. Third, the initiating off-resonance field must be large enough, such that its ω_1 spans most of the Doppler broadened line width and finally the length of the active media should be much longer because the gain per pass is low in the wing of the transition. If this idea works experimentally the initial field must induce enough of an oscillating dipole in the wing of the transition so that the resulting reactance field off-resonance is larger than any developing reactance field on-resonance.

DYE LASER APPLICATIONS IN PHYSICOCHEMICAL RESEARCH

Robert L. Paugh
Spectra-Physics, Inc., Mountain View, California

Any contribution of new dye lasers to the improvement of physicochemical research will be made primarily through the enhancement of those unique properties of the dye laser which have already made this instrument a research tool in the area of physical chemistry.

There are basically three general areas where the performance of the dye laser can be improved. These are 1) tuning range, 2) linewidth, and 3) power.

Tuning range is fundamentally a dye development problem. To date, very few dye laser manufacturers have done dye development work. Most have waited for companies like Eastman Kodak and others to produce the dye *and then* have moved quickly to quantify the performance parameters of the dye and to pass that information on to actual and potential dye laser users.

On the other hand, the actual performance of a dye in a dye laser is a function of the dye laser design as well. In the design area, extensive work has been done to decrease the losses in the dye laser cavity. This has been accomplished primarily through improvement of the optical design (specifically in the dye laser tuning elements) and through the development of better mirror coatings for dye laser optics. The result is lower lasing thresholds for most dyes and higher output power for a given pump power.

New dyes will continue to be developed. The areas of dye development focus will be in the blue, far red, and infrared portions of the spectrum. Success in these areas will be made easier by the recent introduction of high power UV lasers by Spectra-Physics and others and by the introduction of a high power krypton laser by Spectra-Physics. The availability of high UV, red, and far red pump powers means that new dyes will be of practical use to the physicist or chemist even though those dyes may have low gain and high threshold.

The second area, that of linewidth, has seen significant

improvement in the last year. The electronically-tuned single-frequency dye laser has substantially increased the resolution of dye lasers. Linewidths have been reduced from fractions of an angstrom to less than 10 megahertz (or 10^{-4} cm^{-1}). One of the primary objectives in utilizing tunable lasers is to excite an atomic or molecular transition. Until recently, the criteria for spectral resolution has been determined by the absorption linewidth. However, the advent of tunable single-frequency dye lasers has improved resolution to such an extent that the lasers are being used to select between isotopically shifted absorption lines.

The most recent development in commercial dye laser technology has occurred in the area of output power. The introduction of a second generation tunable single-frequency dye laser makes available to the researcher output powers at least three times that previously obtainable from single-frequency systems. This factor of 3 in power means an increase in signal-to-noise ratio of a factor of 3. In applications where signal averaging techniques are used it means a decrease by a factor of 9 in signal averaging time required to obtain a given signal-to-noise ratio.

Second generation dye lasers, using the jet dye stream, have raised dye laser output power to the point where the devices are now limited by fundamental thermal lensing problems which are caused by the absorption of high power in the jet stream itself.

The significant point is, I feel, that dye lasers exist today which are capable of providing the chemist with photons in quantities of Avogadro's number in reasonable amounts of time. This capability is what is required if light-induced or light-assisted chemical reactions are to be used to produce usable quantities of end products.

The true extent of the contribution of these new dye lasers to the improvement of physicochemical research will be more quantitatively illustrated in other discussions of this conference. However, I would like to review a few applications where I feel contributions have been made.

Single-photon fluorescence and absorption spectroscopy work has been extended beyond what could be done using non-tunable laser sources.

Laser fluorescence techniques have been used to study energy transfer processes. Particular emphasis has been given to the channels of energy flow between the different modes of particular molecules. [1]

Ultra-high-resolution two-photon spectroscopy was really made possible on a broad scale by single-frequency dye lasers. Highly excited atomic and molecular states can be studied with very high resolution using the new method of Doppler-free, two-photon spectroscopy. Excitation is obtained using two counter-propagating laser beams whose Doppler shifts cancel. CW and pulsed dye lasers have been used to investigate Stark splittings, to study shifts in highly excited alkali states, and to study the 1S-2S transition in atomic hydrogen. This latter study permits a high-precision measurement of the ground state Lamb shift. [2]

Two-photon absorption spectroscopy has also been developed to use the non-linear process of simultaneous absorption of two photons to provide information on selection rules and to use the polarization of the light as a tool for identifying the symmetry of the induced transitions. Again the now standard techniques of eliminating Doppler broadening are used here as well. This technique has been used to identify "hidden" electronic bands in biphenyl and to increase the understanding of the lower electronic states of benzene. [3]

Endothermic reactions can also be enhanced using dye laser radiation. The focus here is on the <u>temperature dependence</u> of a reaction rate and the use of laser radiation to increase the temperature and hence increase the reaction rate.

FIGURE 1

$$A + BC \xrightarrow{k} AB + C$$
$$k = D(T)e^{-\varepsilon/KT}$$

Figure 1 illustrates the reaction of "A" to the form "AB". It is clear that as temperature is increased the reaction rate <u>kappa</u> is increased as well.

Coherent anti-Stokes Raman scattering (often called "CARS"), a non-linear optical phenomenon, can be done using two closely tuned dye lasers or a CW laser with acousto-optically induced frequency shifts. CARS is of interest to chemists and others interested in materials in general because it permits the measurement of Raman spectra with a 10^8 to 10^{10} signal-to-noise advantage over spontaneous Raman scattering. It has the further advantage of using sample volumes as small as $10^{-6} cm^3$ and of using low average power levels (as low as 1 mW) with complete rejection of fluorescence. [4]

Negative ion spectroscopy has recently been accomplished using dye lasers. The simple static properties of the electron affinities of atoms and small molecules have been determined using a tunable dye laser to intersect a beam of negative ions. The threshold wavelength for production of neutral particles by the photodetachment of an electron is determined. This wavelength corresponds essentially to the electron affinity of the neutral atom. [5]

Finally, an area of intense interest and activity this past year has been that of isotope enrichment. The narrow linewidths of dye lasers have been extremely useful in taking advantage of isotope shifts. Dr. Benjamin Snavely of the Lawrence Livermore Laboratories in California is presenting a paper at Laser '75 in Munich on "Isotope Separation Using a Dye Laser".

Isotope enrichment can be obtained by two different processes: photophysical isotope enrichment and photochemical isotope enrichment. I would like to discuss the role of the dye laser in each of these processes.

With the photophysical process, one starts with an isotopically mixed sample. A method of selective excitation capable of discriminating between isotopic forms of the same atom or molecule is determined. A subsequent photophysical separation process is applied and the isotopically enriched end product is obtained. See Figure 2.

There are a number of possible processes. Starting with an atomic or molecular vapour, the method of selective excitation could be an electronic state change. The absorption of the light energy to effect this change can result in radiation pressure which causes the absorbing atoms or molecules to deviate from their original path. The result is a physical separation of the desired isotope from the mix and one obtains an enriched product. A dye laser can be used in this technique.

This process is illustrated in Figure 3.

FIGURE 2

PHOTOPHYSICAL ISOTOPE ENRICHMENT

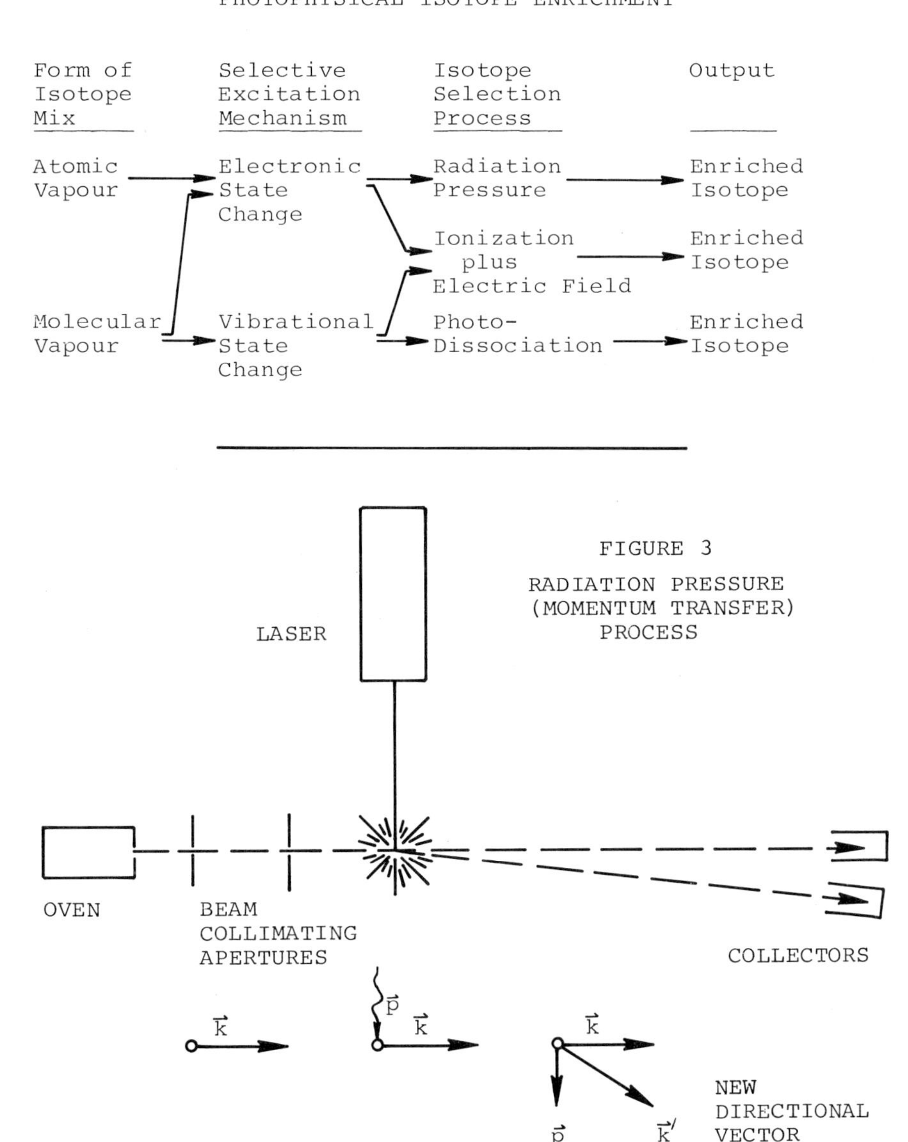

FIGURE 3

RADIATION PRESSURE (MOMENTUM TRANSFER) PROCESS

The absoprtion of light energy may be sufficient to cause ionization of the absorbing atom or molecule. In this illustration an electric field is combined with the ionization process to selectively separate a particular isotope. See Figure 4.

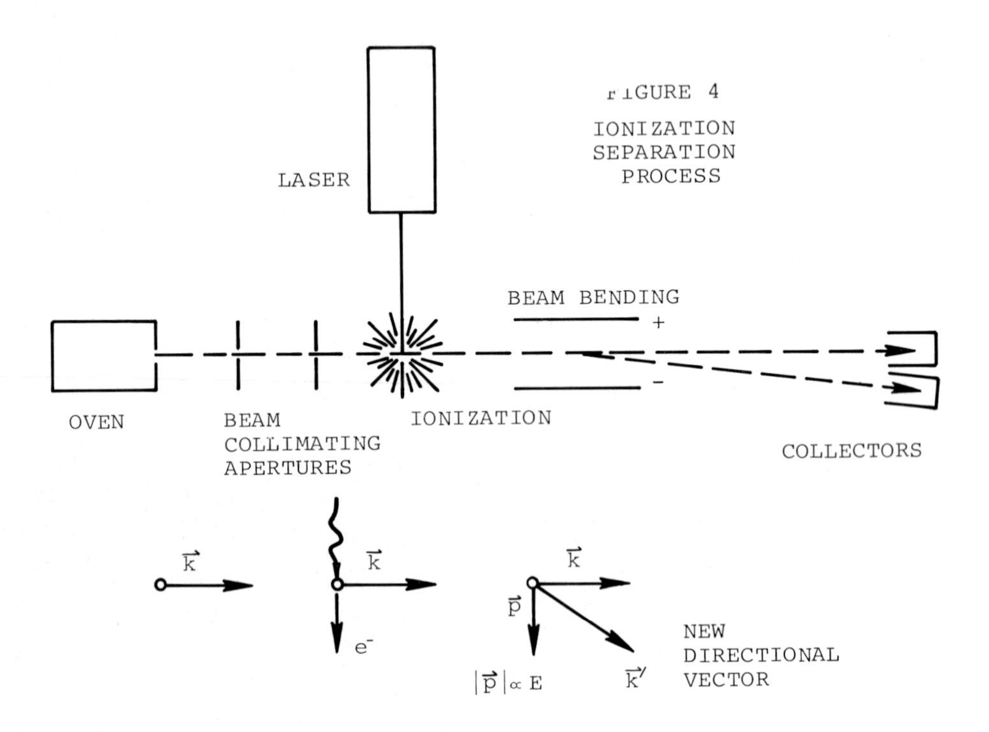

FIGURE 4
IONIZATION SEPARATION PROCESS

Ionization can be achieved through a variety of techniques which include using two, three, or more photons. Some of the various possibilities are exhibited in Figure 5. The electron, normally in its ground state, is selectively excited with one photon (the absorption cross section being large for only one isotope) and is subsequently ionized by a second photon as in case 1 or by two photons as in case 2. Case 3 shows a three-photon approach and case 4 illustrates a possible four-photon process.

Both the radiation pressure and the selective photo-ionization processes have been done using a dye laser. The radiation pressure technique has been used to enrich barium isotopes 134, 135, 136, 137 and 138. The selective photo-ionization technique has been used to separate calcium isotopes

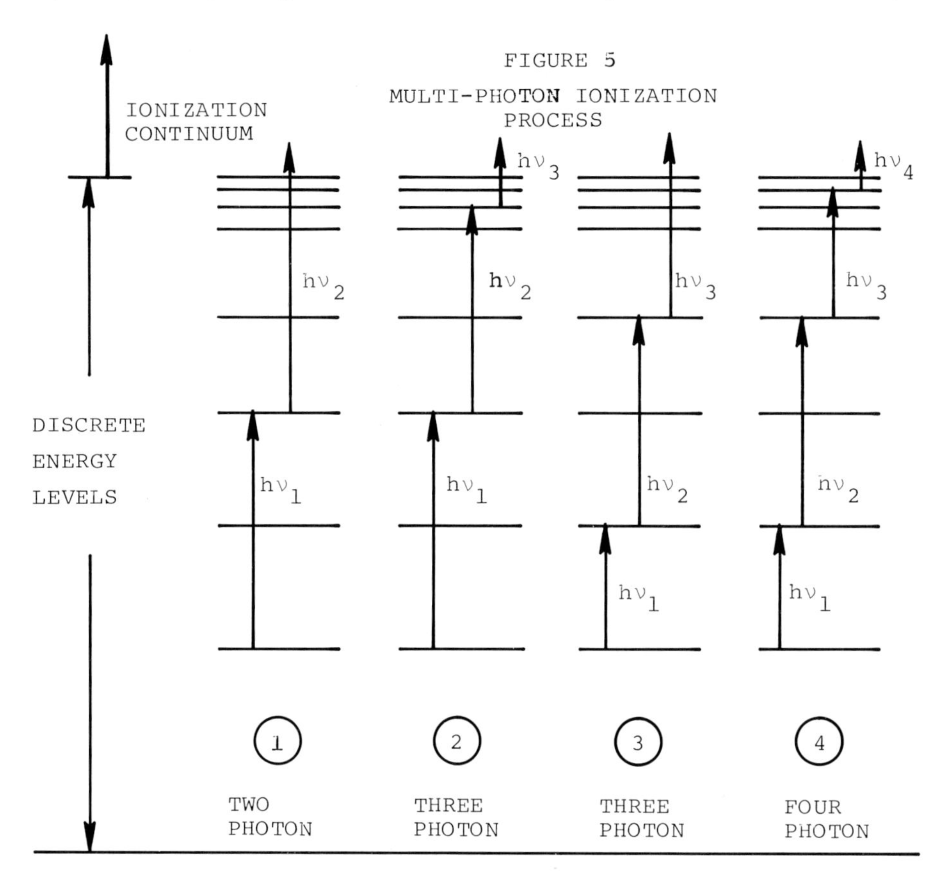

FIGURE 5
MULTI-PHOTON IONIZATION PROCESS

40 and 42, and uranium isotopes 235 and 238. Case 1, illustrated in Figure 5, has been used for the isotope enrichment of uranium. Case 3 has used a dye laser to provide the third photon where a wavelength of about 6000 angstroms was required.

Why are lasers and dye lasers required in these processes? Primarily because high power, specific wavelengths, and narrow linewidths must be used.

Figure 2 also showed molecular vapour undergoing vibrational state changes and utilizing both photoionization and photodissociation as the enrichment mechanism. In most cases, high power CW lasers are required to obtain an enriched product by these methods.

The second process of interest is that of photochemical enrichment. In this case, one starts with an isotopically mixed material and chooses a mode of isotope selection.

FIGURE 6

PHOTOCHEMICAL ISOTOPE ENRICHMENT

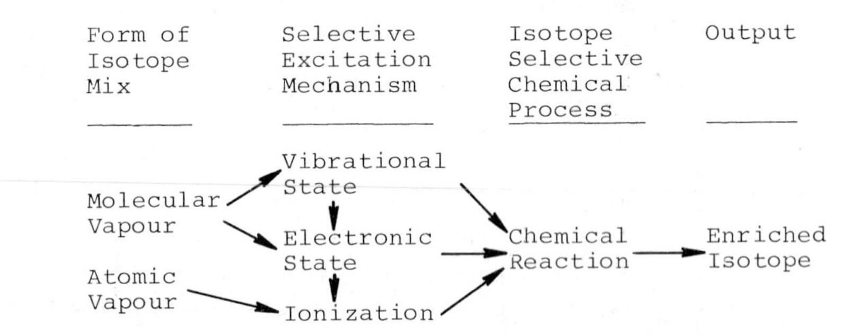

A photochemical process is applied which utilizes this mode and an isotopically enriched product is obtained. The possibilities are illustrated in Figure 6. The enrichment occurs through extraction by way of a chemical reaction rather than a physical process. A molecular vapour may be enriched utilizing vibrational, electronic, or ionization techniques which differentiate between isotopes. The chemical process is then initiated which involves the selectively excited isotopic state in the reaction to produce an end product which can then be extracted. Similar processes are utilized for separating atomic vapours.

An electronically-induced chemical reaction of thiophosgene with dieth-oxy-ethylene uses an argon ion laser and a nitrogen-excited dye laser to enrich either chlorine 33 or 35. Other techniques are being pursued for use in the enrichment of U235. Reports on this program can be expected within the year.

It appears, then, that a good argument can be made that dye lasers have contributed extensively to the development of physiochemical research. This presentation would not be complete, however, if we did not at least briefly consider what future dye laser developments appear likely and project how these developments may contribute to physiochemical research. Dr. Schaefer discussed this in greater detail at this conference.

There is interest in the frequency-doubling of dye lasers. There is a great need for a good source of tunable narrowband radiation, both pulsed and CW, in the ultra-violet (2000 to 3500 Å). The existence of such a product would most certainly enhance many efforts for isotope enrichment.

Mode-locked dye lasers, capable of generating frequency-tunable picosecond pulses have been reported by D.J. Bradley, Physics Department, Imperial College, London, and will be reported on in this conference by Ippen and Shank. The availability of this type of instrument will open entirely new areas of investigation. Processes taking place in the time scale of 10^{-12} seconds can be investigated. This would include vibrational and orientational relaxation of molecules in liquids, the study of primary radiationless transitions of electronically excited large molecules and many others.

In conclusion, one should expect further significant contributions to be made by new dye lasers to physiochemical research. There is much experimental work left to be done that will require the use of dye lasers.

REFERENCES

1 G. Flynn, Dept. of Chemistry, Columbia University Personal Communication.
2 T.W. Haensh, Dept. of Physics, Stanford University Personal Communication.
3 W.M. McClain, Dept. of Chemistry, Wayne State University Personal Communication.
4 B.S. Hudson, Dept. of Chemistry, Stanford University Personal Communication.

5 W.C. Lineberger, JILA, Dept. of Chemistry, University of Colorado, Personal Communication.

DISCUSSION

R. PAUGH

R.A. Keller - Do you know of any laboratory which is trying to optimise frequency doubling in CW dye laser systems?

R. Paugh - As of this date, I am not aware of any commercial laser company which is developing a frequency doubled CW dye laser system. Frequency doubling work with cw dye lasers is being done, much of it in Europe. I am aware of experiments in progress in France at the Laboratoire Aimé Cotton and in Germany in the AEG Telefunken Research Laboratories. There may be others.

C. Varma - My question is related to the study of the temperature dependence of the rate of chemical reactions by means of dye laser heating. For electronic ground state reactions dye laser radiation is not resonant with transitions in the reaction partners. The only mechanism for raising the temperature, of which I can think, is Raman scattering. Would present dye lasers cause a sufficient heating through this mechanism to give an observable effect?

R. Paugh - It is not clear that Raman scattering would cause sufficient heating to produce observable effects. However one could heat the reactants indirectly to achieve the desired temperature increase. The temperature dependence of the reaction rate, while clearly influenced by the temperature of one of the constituents of the reaction, is more strongly influenced by the overall temperature of the reaction environment and the reaction constituents. Hence one could supply heat indirectly by way of the matrix which contains the reaction constituents.

S. Kimel - By improving the dye laser to a linewidth required for isotopic selectivity, often to less that 0,001 cm^{-1}, there is a detrimental effect in that there is a decrease in the interaction cross section between the Doppler-broadened spectral linewidth and the often much narrower dye laser linewidth.

R. Paugh - You are correct in your statement. There is a trade-off between the line width required to discriminate between isotopically shifted absorption lines and the interaction cross section. Obviously one desires to reduce the linewidth of the dye laser only to the point where discrimination is obtained, but not so far as to sacrifice uncessarily reaction cross-section.

D.J. Bradley - What is the difficulty in producing a 1 GHz bandwidth CW dye laser?

R. Paugh - The difficulty is not in producing a 1 GHz bandwidth CW dye laser but in producing a CW dye laser which has sufficient longitudinal modes within that bandwith that the bandwidth is effectively being used A dye laser with a 30 cm optical cavity has modes spaced at approximatively 550mHz.That is 2 modes per GigaHertz.Ideally one would like to have substantially longer dye laser to reduce the C/2L mode spacing and thereby increase the number of modes in the 1 GHz bandwidth. The limit, of course, would be a dye laser of infinite cavity length, but that presents some technical problems.

EFFICIENT POWER AMPLIFICATION OF TUNABLE NARROW BAND VISIBLE RADIATION

P.BURLAMACCHI, R.PRATESI and R.SALIMBENI
Laboratorio Elettronica Quantistica CNR - Via Panciatichi 56/30
50127 FIRENZE,(Italy)

SUMMARY

A waveguide dye laser, realized by means of a flash lamp pumped slab cell with the appropriate thickness, has been proved to exert a high gain. In the present paper results are presented which show the possibility of efficiently amplify low power monochromatic radiation up to high energy and high repetition rate pulses. The high gain also permits easy and accurate tuning down to the GHz range in bandwidth, with the use of a single grating.

INTRODUCTION

Tunable laser in the visible spectrum are now available and their potential use in photochemistry or biology is fairly well individuated. The problem of attaining high energy or high average power of narrow band radiation is now very urgent in view of import applications,like isotope separation. Efficiency,reliability and low cost of the equipment are of primary importance. To this end we will consider the possibility to operate flashlamp pumped dye lasers in a waveguide configuration, which can be considered as alternative to conventional high energy lasers. The waveguide dye laser permits a much higher gain per pass;therefore it is suitable for high efficiency power amplification and for particular tuning schemes.

WAVEGUIDE AMPLIFIER DESIGN

One of the most important feature of organic solution dye lasers is the very high gain which can be achieved with both laser and flashlamp pumping. This is in part due to the fact that the quantum yeld for fluorescence remains near the value 1 even at large dye concentration, namely $10^{-3} \div 10^{-2}$ M .On the other hand, for an efficient power or energy extraction from the laser,the gain profile across the dye cell must be optimized. For a given transverse dimension 'd' of the cell in the direction of the pumping light, the dye concentration must be adjusted to give maximum gain at the cell axis, which corresponds to a compromize of uniform inversion and maximum utilisation of the pump intensity.

Clearly, the dye concentration cannot differ substantially from the value for which the mean absorption lengh $\bar{\alpha}^{-1}$ of the solution nearly equals the thickness of the cell: $\bar{\alpha}d \simeq 1$ .Since at high concentration the absorption coefficient is very large (α_p = 121 cm^{-1} for a 10^{-3} M solution of R6G in ethanol) it follows that to achieve high gain from an optimized cell, the cell thickness must be very small (a few tenths of millimiter, typically). This can be easily done by using planar cells, which, unlike cylindrical cells, have no limitation on the active volume and flashlamp coupling.

In the case $\alpha d \cong 1$ and at high concentration, in addition to the inhomogeneity of the gain profile, there is an important inhomogeneity of the refractive index, caused by the partial conversion of the pumping energy into heat. A planar cell pumped from both sides behaves, than, like a generalized cylindrical lens. Laser light will hence propagate bounching periodically between the glass walls and emerge from the cell in two typical beams as can be seen from Fig. 1. This waveguide character of the planar laser provides several important improvements of the laser performance, like good energy and frequency reproducibility. A detailed analysis of the principle of operation and performance of the waveguide dye lasers is reported in previous work by the authors together with extensive reference to literature [1 ÷ 5].

NARROWBAND LIGHT AMPLIFICATION

A waveguide amplifier usually emits a strong beam of broad band light due to amplification of spontaneous emission (super-radiance). The superradiance background may compete with the narrowband light to be amplified, however, due to the highly homogeneous broadening of the fluorescence spectrum, the super-radiance is almost completely quenced in the presence of the signal light.

Gain measurements of the waveguide amplifier have been performed and compared with measurements made by other authors for conventional lasers [6 ÷ 8].

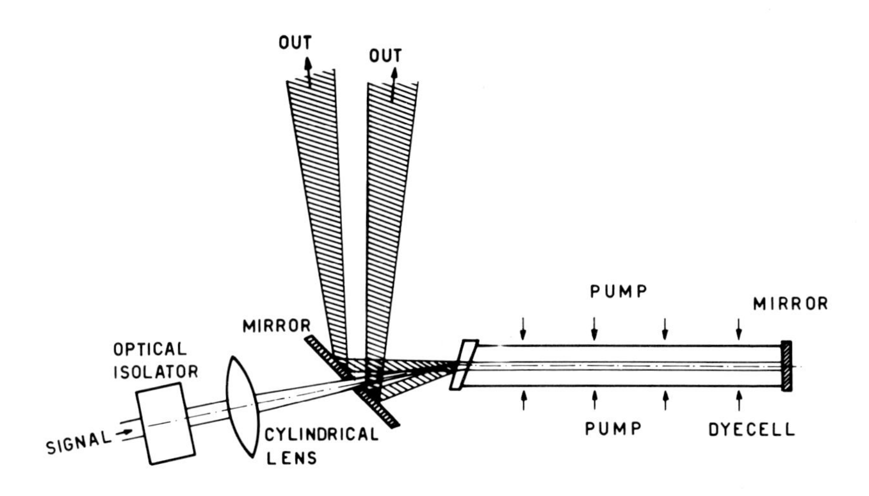

Fig. 1. Schematic drawing of a slab amplifier. A double pass scheme is illustred in this case. The optimum cell thickness is around 0.4 mm.

The schematic drawing of the oscillator-amplifier system is illustrated in Fig. 1. Rhodamin 6G at a concentration of $0.8 \cdot 10^{-3}$ M in ethanol was used for the slab amplifier.

The oscillator consisted in a conventional grating tuned dye laser pumped in synchronism with the amplifier. The pulse duration was about 2µs. The amplifier was decupled from the feedback of the output mirror of the oscillator by means of an optical isolator. The signal light was focused into the slab amplifier by means of a cylindrical lens and attenuated by calibrated neutralfilters. Single pass gain and double pass gain were determined be measuring the energy pulse of the oscillator and amplifier by means of calibrated photodiode. The superradiance background was measured by dispersing the light output be means of a grating.

The plots of Fig. 2 show the absolute energy gain for various input energy for single pass and double pass amplification.

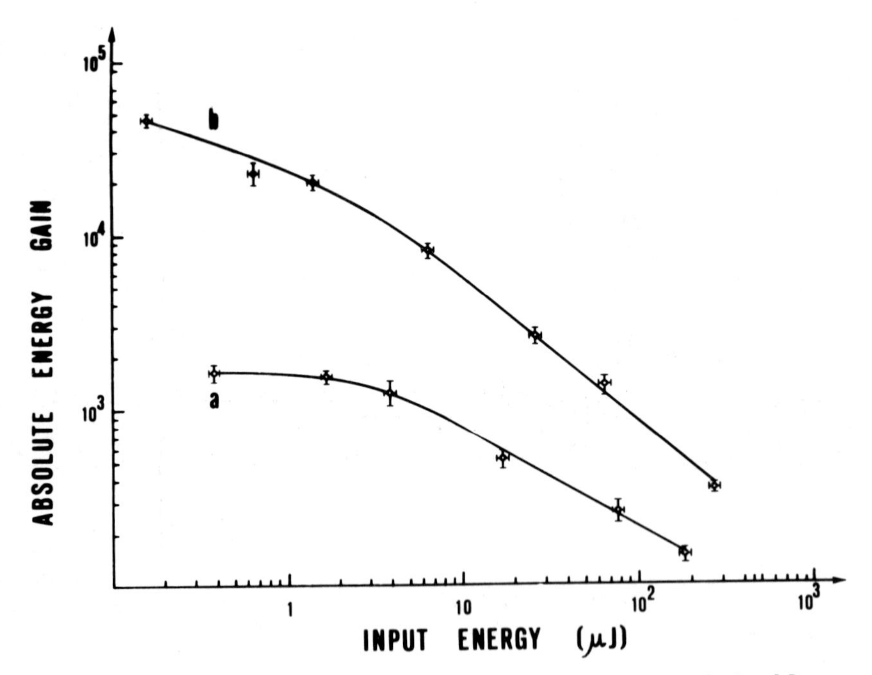

Fig. 2. Absolute energy gain for single pass and double pass amplification as a function of input energy.

As can seen the gain becomes strongly saturated at input levels of ~ 100 μJ . At this input energy the superrandiant background energy over the entire spectrum is reduced by two orders of magnitude under the signal energy.

In the case of the double pass amplifier the total energy output is of the order of 100 μJ , with an overall efficiency of the order of 10^{-3}. In absence of signal about the same output energy is obtained in the form of superradiant broad band ligth. For small signals, down to the μJ range the gain becomes less saturated and approaches the small signal gain. However the measurenment range was limited by the background fluorescence. Maximum gain approaches a factor of 10^5. This result has to be compared with the much lower gain acheaved with even more energetic pumping in conventional amplifiers [8].

TUNABLE OSCILLATORS

The slab amplifier can be easily turned into an oscillator. By terminating the slab with two end mirrors, one totally and one partially reflecting (as an example a silvered mirror and a glass output window) the amplifier may oscillate with high efficiency [2] and high energy density [4] . The oscillators have proven to be extremely reliable, insensitive to thermal fluctuations, resonably free from early termination even at high energy levels, and severe mirror misalignement may be tolerated.

The high gain demonstrated by the waveguide amplifier indicates that extremely high loss resonators may be used. This fact, together with the particular geometry of the slab, suggests the use of a slit spectrograph in substitution to one end mirror in order to obtain a tunable oscillator. Very stable narrowband output down to 1GHz in bandwidth has been obtained with this configuration. To this limit strong superradiant uncollimated ligth is present and the output energy is vely low, so that further amplification is needed in order to obtain fully saturated output. The use of Fabry-Perot interferometers in the parallel beam inside the spectrometer may reduce the bandwidth well below 100 MHz.

CONCLUSIONS

The present work tries to demonstrate that the principle of waveguide propagation in a flashlamp pumped dye laser can conveniently fulfill the problem of efficient power amplification. Gain as high as 10^5 may be achievedin a double pass amplifier, and input energy of 0.1 J is sufficient to completely saturate the amplifier. Low power level higly monocromatic oscillators may be therefore brought to high peak powers with a few amplification stages. Furthermore the waveguide appears to be scarcely affected by fluctuations of the refractive index, shock waves and thermal effects and therefore appears to be suitable for high average power operation.

Tuning possibilities are very promising owing to the fact that bandwidth in the GHz range may be acheaved by means of a single grating spectrograph in substitution of one end mirror.

REFERENCES

1. P. Burlamacchi, R. Pratesi and L. Ronchi, Appl. Opt. 14 (1975) 79
2. P. Burlamacchi and R. Pratesi, Appl. Phys. Lett. 23 (1973) 425
3. P. Burlamacchi, R. Pratesi and U. Vanni, Opt. Comm. 9 (1973) 31
4. P. Burlamacchi, R. Pratesi and R. Salimbeni, Opt. Comm. 11

(1974) 109

5. P. Burlamacchi, R. Pratesi and R. Salimbeni, Appl. Opt. 14 (1975) 1311
6. B.G. Huth , Appl. Phys. Lett. 16 (1970) 185
7. P. Flamant and H.Y. Meyer, Appl. Phys. Lett. 19 (1971) 491
8. J.B. Marling, J.G. Hawley, E.M. Liston and W.B. Grant, Appl. Opt. 13 (1974) 2317

DISCUSSION

P. BURLAMACCHI, R. PRATESI, R. SALIMBENI

D.J. Bradley - 1) What energy or power do you get at a bandwidth of 1 GHz? 2) Have you tried high repetition rate operation?

P. Burlamacchi - 1) At 1 GHz bandwidth the oscillator is very near threshold. Treshold is here determined by the slit aperture. We did not measure the energy in our preliminary experiments but we expect to be in the mw range. At 1 GHz strong off-axis superradiance exists, which is completely saturated for slit apertures which correspond to 5 - 7 GHz bandwith. 2) A high repetition rate apparatus is in progress. The wave-guide operation is very reliable and scarcely influenced by temperature and early termination so that we have reasons to be optimistic.

F.P. Schäfer - Did you observe a spread of beam divergence angle during the pulse, since from the mechanism of your waveguide laser, which integrates over the heat input, you should see that? The farfield photograph you showed was evidently a time integrated photograph.

P. Burlamacchi - This effect was indeed observed and reported in previous work (Optics Communications 9 (1973), 31).

C.B. Harris - Do you know for certain whether or not your gain is limited at high pump intensity by off-axis superradiance? Could you be limited by power broadening?

P. Burlamacchi - Off-axis superradiance can always be overcomes by a sufficiently powerful signal. Small signal gain may be limited by superradiance and power broadening but I believe that power amplification with a saturated amplifier will increase at high pump intensity.

O. de Witte - Have you an idea of the concentration of excited molecules in your cell?

P. Burlamacchi - Sorry. I have never computed this parameter for our pumping conditions and dye concentrations. We only analyzed the relative distribution as a function of the distance from the axis of the cell in various situations (Appl. Opt. 14 (1975), 79).

LASER EMISSION FROM ORGANIC CRYSTALS

N. KARL
Physikalisches Institut, Teil 3, und Kristallabor der Universität Stuttgart, D-7-Stuttgart 80, Germany F.R.

SUMMARY

In this contribution a review of recent results on stimulated emission from organic molecular crystals will be given. Two types of coherent light emission from organic molecular crystals have been described: I) Superradiance from very thin undoped and tetracene - doped anthracene crystals at low temperatures and II) true laser emission in a few distinct cavity modes from anthracene and tetracene-doped crystals such as fluorene and p-terphenyle at room temperature. In this latter case mode selection and tuning is possible. A new tuning method has been introduced, making use of a thin wedge-shaped air gap between two halves of the active laser crystal whose outer faces are forming the resonator.

1. INTRODUCTION

Many pure and doped organic crystals which are built up from molecules with π-electrons exhibit strong fluorescence in the visible and near UV with high quantum yields and high transition probabilities. Nevertheless coherent emission from organic crystals was unknown for a long time. In principle one could imagine several possibilities to be realizable: coherent emission from 1.) pure, undoped crystals in which the laser starting level is a mobile Frenkel exciton; 2.) doped crystals with a laser transition between two localized levels of the dopant. In the secondcase there are two ways for the population of the upper laser level:

a) direct absorption of the pumping light at the dopant ("guest"), or b) population by energy transfer from mobile host excitons which have been excited by the pumping light. Because of the low thermal stability of most organic crystals (small enthalpy of sublimation, low melting point) the crystals cannot be loaded with a too high energy density. Therefore low threshold four-level systems with transitions into empty vibrationally excited levels of the electronic ground state will be more realistic than three level systems which need a much higher inversion of the order of half of the lasing molecules. Similarly, undoped crystals will be more difficult candidates under most experimental conditions because the absorption coefficients for available pumping wavelengths in general will not lie at a useful intermediate value such as to avoid thermal destruction of the crystal at the one hand but nevertheless to provide sufficient inversion density at the other. It is for this reason why only superradiation from very thin crystals or from surface regions has been obtained so far for undoped crystals and why this type of emission is only obtainable at rather low temperatures, preferably at liquid helium. On the other hand there are interesting questions connected with high density Frenkel excitons in pure crystals such as the search for Bose-Einstein condensation.

For doped crystals the matrix must be sufficiently transparent for the pumping wavelength for true volume excitation to occur. The most simple way is to pump the guest directly. This is easy, because at least at not too low temperatures the absorption of organic molecules takes place in very broad spectral regions.

In any case, for reaching the laser threshold, sufficient pumping energy must be available within the fluorescence lifetime which is generally in the order of a few ns. The nitrogen laser has proved to be at the same time a very efficient, convenient and nevertheless inexpensive pumpe

source which is able to produce for a few nsec a UV (337 nm) light flux of 1 GW/cm^2 (10^{27} photons/cm^2 sec) and more. Frequency doubled, Q-switched ruby or neodymium lasers may be a suitable alternative.
A high amplification can be expected for systems with a short fluorescence lifetimes if in addition the fluorescence quantum yield is high and if losses by intersystem crossing into the triplet manifold and by triplet-triplet absorption can be neglected.

2.SUPERRADIATION FROM UNDOPED ANTHRACENE CRYSTALS

The possibility of obtaining stimulated emission from four level systems in organic crystals has been studied theoretically by Broude et.al. [1] befor the first realization of any organic laser. The high fluorescence quantum yield and the rather narrow low temperature linewidth of many organic crystals should favour stimulated emission. Broude and Sheka [2] seem to have been the first who observed a weak superlinear rise of the fluorescence intensity of the $S_1 \rightarrow S_o + 1400\ cm^{-1}$ line of anthracene crystals with the excitation intensity. They used flash-lamp excitation at 77 K. The effect recently has been studied in more detail by Galanin et.al. [3] and Benderskii et.al. [4] at liquid He, using an N_2-Laser (λ = 337,1 nm) as an excitation source. Galanin et.al. found that for excitation intensities exceeding a certain threshold ($\approx 10^{22}$ photons/ cm^2 s) by a factor of only 3 or 4 a thousandfold increase of the intensity was observed in a narrow line at 23 692 cm^{-1} †) which corresponds to the transition from the lowest singlet-S_1 excited state of the anthracene crystal to the 1405 cm^{-1} †) totally symmetric vibrational level of the electronic ground state. In the experiments the starting level turned out to be a freely moving Frenkel exciton and not a state localized at an X-trap or at an impurity. Other exciton transitions showed a sublinear rise only, which can be explained by losses due to the mutual annihilation of mobile S_1-excitons [6,7];

----

†) the precise values are taken from ref. [5]

an impurity line went into saturation [3]. The line width of the line showing superlinear increase narrowed down to $\leq 2\ cm^{-1}$ with increasing excitation, whereas the other lines showed some heat broadening. The emitting volume was in the order of 10^{-6} cm^{-3}.

From all of this it was clear that the 23692 cm^{-1} transition was superradiating by stimulated emission. The energy level diagram is given in Fig.1

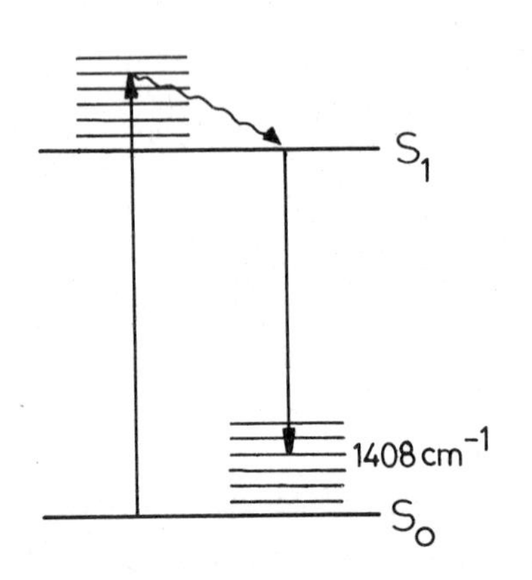

Fig.1 Energy level diagram of the anthracene superradiator and laser. In the superradiating pure crystal the corresponding S_1-level is at 25097 cm^{-1} above the ground state [5]; in the doped crystal laser it is at $\approx 25850\ cm^{-1}$.

The gain exponent has been obtained using the formula

$$\alpha = \frac{1}{\tau_o} \frac{S_1}{8\pi c\tilde{\nu}^2 \Delta\tilde{\nu} n^2} \beta \qquad (1)$$

where τ_o is the radiative lifetime, S_1 is the density of excited states, $\tilde{\nu}$ and $\Delta\tilde{\nu}$ are the position and width of the transition (in wavenumbers), n is the index of refraction and β is the fraction of the amplified band relative to the entire spectrum of the transition. With the values $\tau_o = 5 \cdot 10^{-9}$ sec, $S_1 = 2 \cdot 10^{18}\ cm^{-3}$, $\Delta\nu = 5\ cm^{-1}$, $\beta = 0{,}02 - 0{,}05$ (estimated), Galanin et.al. [3] obtained $\alpha = 1000 - 2500\ cm^{-1}$. This large value was nevertheless too small to explain the high gain observed, as the small signal light penetration depth of the N_2-laser wavelength is only 0,3 µm. Benderskii et al. [4] estimated a gain coefficient of $\approx 50\ cm^{-1}$. They showed that the larger the diameter of the excited area of the crystal platelet the higher is the amplification and the larger the reduction of the line width. This gave clear evidence that stimulation must have occurred in a layer transverse to the direction of the exciting light beam (diameter ≈ o,l - 4 mm) possibly by a zig - zag propagation guided by internal reflection.

In the early experiments the stimulated emission was more or less isotropic. Later Avanesjan et al. [8] showed that for very perfect anthracene crystals it emerged mainly from the side facets of the platelet. The threshold for stimulated emission for the best crystals was only (3-6)$\times 10^{19}$ $cm^{-2}sec^{-1}$ at 4.2 K. An amplification factor of 2 cm^{-1} was reported. The discrepancies in the amplification factors reported [3,4,8] seem to be due to different estimates of the actual S_1-exciton density (being dependent on the actual values of the nonlinear lifetime and excitation depth). An independent measurement of this quantity will be necessary. A light intensity of ≈ 10^{23} photons cm^{-2} sec^{-1} sets an upper excitation limit because of crystal damage.

In a following paper [9] Avanesjan et al. reported that the stimulated emission from thin anthracene platelets (thickness ≈ 1μm) is not homogeneously distributed over the side facets of an area of $3\cdot 10^{-5}$ cm^2, but rather, concentrated on a few channels of a width of o,l - o,5 mm. The authors estimate the internal power density to reach loo MW cm^{-2}. This power density should be sufficiently large to give rise to stimulated Raman scattering. Actually for excitatin densities exceeding 10^{22} $cm^{-2}sec^{-1}$ a line at 22292 cm^{-1} beginns to rise steeply with the excitation density. By comparison with measured coefficients of Raman scattering this line is explained to be due to Stokes - type stimulated Raman scattering of the internally generated anthracene superradiance or laser emission at the 14o5 cm^{-1} anthracene vibrational level. (It should be mentioned that very efficient stimulated Raman scattering of ruby laser light in naphthalene crystals has been found quite recently by Srivastava and Gupta [1o]).

In tetracene - doped anthracene crystals (≈10^{-4} mol/mol) of a thickness of o,8 - 1,2 μm Avanesjan et al. [11] observed that I.) beginning from an excitation intensity of $3\cdot 10^{20}cm^{-2}s^{-1}$

superratiation from a phonon side band of the tetracene $S_1 \rightarrow S_o$ transition at 2o222 cm^{-1} is obtained, narrowing from 15o cm^{-1} to 2 - 3 cm^{-1} at $I_o > 10^{22}$ $cm^{-2}sec^{-1}$; II.) starting from 10^{21} $cm^{-2}sec^{-1}$ the anthracene superratiation described above is obtained in addition. The tetracene guest superluminescence saturates with the beginning of the anthracene stimulated emission. This obviously means that the channel of anthracene - S_1 - exciton energy transfer to the tetracene guest becomes restricted by S_1 - S_1 annihilation and stimulated emission above a certain excitation density. More detailed investigations of these effects hopefully will yield some information on energy transfer processes in doped crystals.

3. LASER EMISSION FROM DOPED ORGANIC CRYSTALS

Laser emission from doped organic crystals has been obtained first by Karl [12]. In summary, stimulated emission at room temperature was observed from cleaved fluorene crystals doped with $\approx 10^{-3}$ mol/mol anthracene. It occurred from the vibrationally unexcited singlet - S_1 state into the 14o8 cm^{-1} vibrationally excited level of the singlet - S_o ground state of the anthracene molecule (see Fig.1) in the fluorene crystal,after pumping into vibrationally excited bands of the S_1 state by a focussed nitrogen laser. The output data of this four level laser were 1o W in $\leq$ 3 nsec around λ = 4o8 nm (245o3 cm^{-1}) in a few TEM_{oo} longitudinal modes extending over 1nm (6o cm^{-1}). The mode-spacing $\Delta\lambda$ was determined by the thickness of the parallel crystal platelets used (L = o,2 - 2 mm), acting at the same time as the amplifying active medium and as the resonator, according to equ. (2)

$$\frac{\Delta\lambda}{\lambda} = \frac{\lambda/2}{n_1 L} \tag{2}$$

Because of different reflectivities of the uncoated surfaces of the anisotropic crystal for light of different polarization, the emission was strongly polarized along the corresponding larger index of refraction n_1 of the optically biaxial orthorhombic crystal. The experimental arrangement used and new results obtained will be described in the next paragraphs.

3.1 Experimental arrangement and conditions for obtaining laser emission from doped organic crystals

As already mentioned in the introduction, nitrogen lasers are extremely useful pumping sources because of their characteristic property of emitting very short and intensive ultraviolet light pulses. The nitrogen laser used in the experiments of the author is shown in Fig. 2 [13].

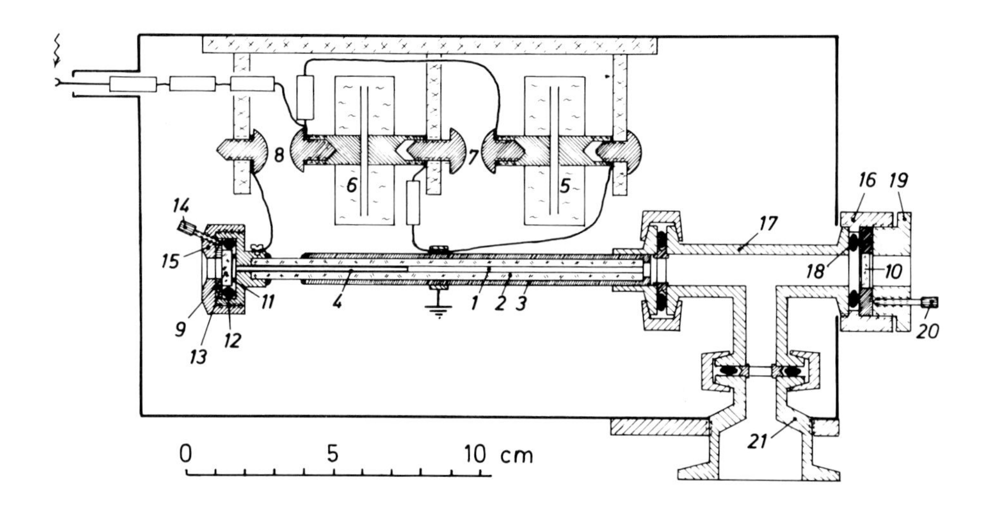

Fig. 2 Nitrogen laser used for the excitation

Briefly, it consists in a coaxial discharge through the bore (1) of a pyrex or fused silica tube (2) from a coaxial capacitor (formed by a stainless steel tubing of 2xo,1 mm (4) and the outer shield (3)) as proposed by Ericsson and Lidholt [14]. In our design the coaxial capacitor is charged by a very short high voltage pulse produced in a Marx voltage doubler circuit. For this the two ceramic capacitors (5) and (6)[x] are charged in parallel and dicharged in series. In contrast to the usual superradiating set-up, efficient optical feed back by two mirrors (9) and (1o) was introduced (bare Al or 98% dielectric at one side and a fused silica etalon at the other) effecting beam narrowing domn to 1 mrad. (A small beam divergence is a

[x]) LCC - HTD 2o kV, 5oo or 2ooo pF ("made in France")

necessary condition for obtaining high focussed power density which can exceed 1 GW cm^{-2} in a focus of 5 µm diameter with the laser described.) Further details, such as the use of commercial vacuum connectors may be seen from the figure. With (1o) as output mirror a positive voltage (1o - 15 kV) should be applied; the optimum gas pressure was between 8 and 16 torr (purified) nitrogen. The output was 2 kW in 3 - 4 nsec at a repetition rate up to 5o Hz §). In Fig. 3 the shape of the output pulse is shown.

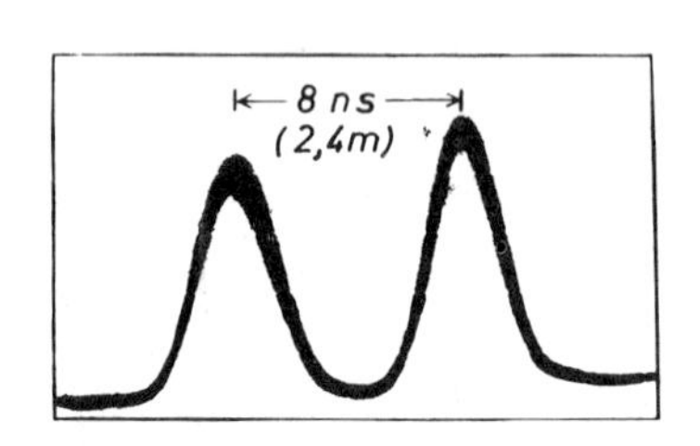

Fig.3 Shape of the nitrogen laser pulse. Time calibration is done by use of a second pulse, which is an optically delayed portion of the first one (about 1oo pulses superimposed).

The pumping arrangement is sketched in Fig.4

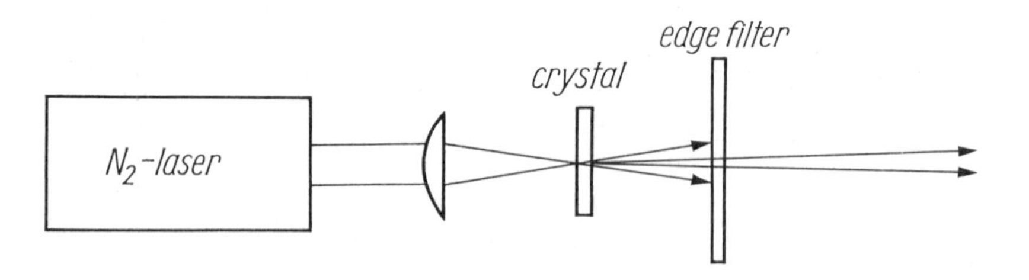

Fig.4 Experimental arrangement (after [12]).

The nitrogen laser is focussed by a f = 2,5 cm fused silica lens down to a diameter of the beam waist of 15 µm; (this number was determined by burning holes into an inconel layer of a 1o% neutral density filter). In this focus the crystals are destroyed after a few shots. Therefore they are positioned somewhat off focus for most experiments, such that the diameter of the beam is in the order of 1oo µm. The crystals, generally

§) the laser can also be operated with Ne (λ = 614,3 nm), Kr (λ = 81o,4 nm) and Xe (λ = 9o4,5 and 978,o nm) at a reduced output. The exact λ - values are taken from ref.[15]

between o,1 and 2 mm thick are mounted on a laser mirror mount carried by a xyz - micrometer translation stage. An edge filter blocks the transmitted portion of the exciting light (amounting to a few percents).

Optical feed-back was accomplished in most experiments by reflection at the uncoated crystal surfaces. Indices of refraction exceeding 2 in some cases are in favour of this simple arrangement. An external resonator is not necessary to obtain laser emission.

For the selection of suitable systems of doped crystals, the following points are important: A dopant with a high fluorescence quantum yield must be incorporated into a matrix which is sufficienly transparent for the exciting light. It turns out that a useful doping level ($\approx 10^{-3}$ mol/mol) is only obtainable for compounds of very similar size and shape. Further, crystals of optical quality with flat parallel surfaces - either natural or by cleaving - must be available; (an optical polish of the soft crystals is rather difficult). Bridgman, sublimation and solution growth has been used, starting mostly from zonerefined material.

3.2 Results

Laser emission similar to that from anthracene in fluorene has been obtained since in a few other single-crystalline matrices: anthracene in 2,3-dimethylnaphthalene, in dibenzofurane, and octahydroanthracene, and for tetracene in p-terphenyle ($\lambda \approx 530$ nm). Light pulses of ≈ 3 nsec duration with a power of 1oo W (≈ 100 MW cm^{-3})were measured in the case of anthracene in 2,3-dimethylnaphthalene, which is an especially useful matrix because it cleaves extremely well, comparable to mica. At one spot of the crystal up to a few hundred light pulses could be produced before serious thermal degradation occurred. A typical spectrum of anthracene in 2,3-dimethylnaphthalene is shown in Fig. 5. The observed fact that a central mode is lower than its neighbours is general for crystals of a certain thicknessrange. With deuterated anthracene the emission is shifted by ≈ 40 $cm^{-1}$to shorter wavelengths.

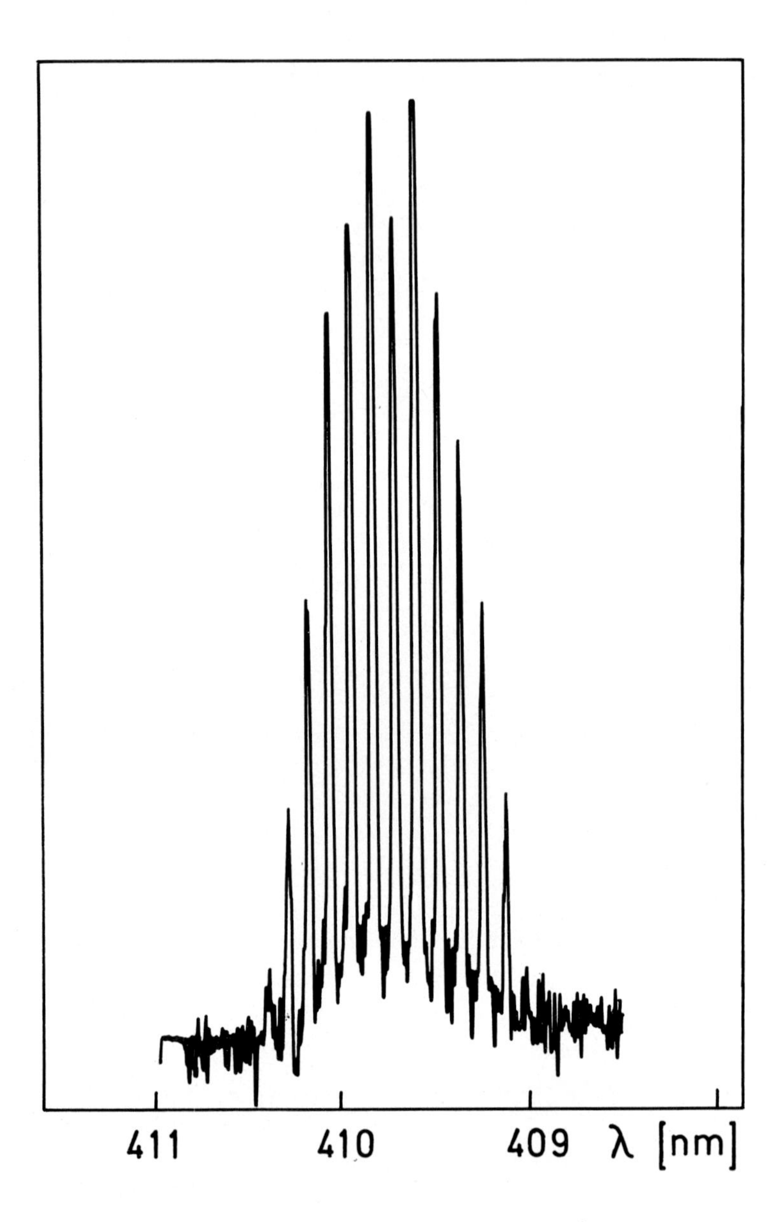

Fig.5 Spectrum of the laser emission from a 38o μm thick anthracene - doped, cleaved 2,3-dimethylnaphthalene crystal (doping level ≈ 10^{-3} mol/mol; 3oo °K)

The chemical formulae of the compounds used are listed in Fig.6

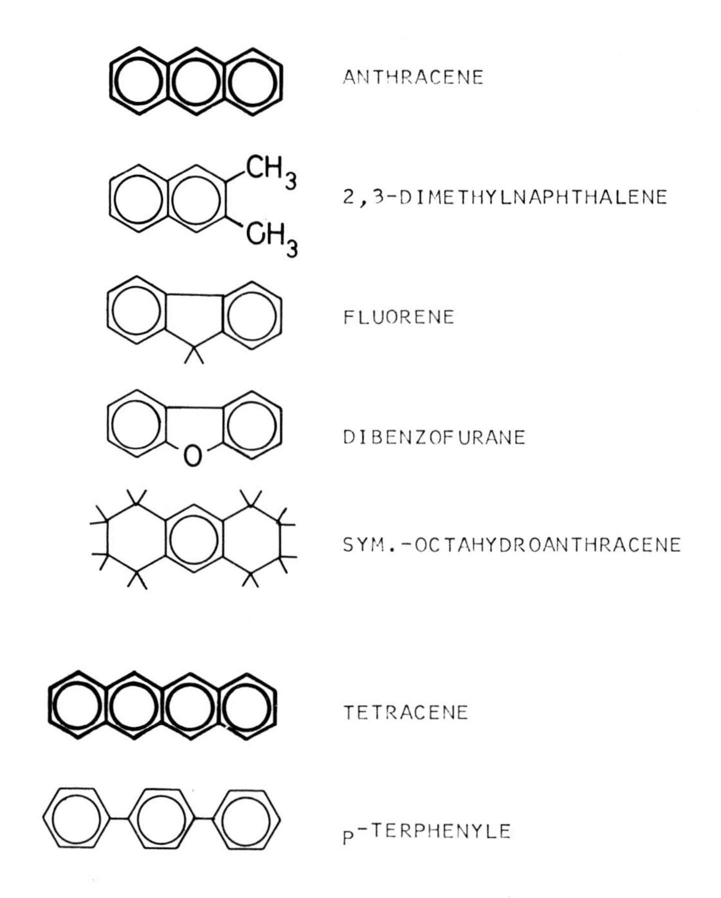

Fig.6 Compounds used. Notice the similar size of the dopant (fat) and host. Smaller π-electron systems are necessary for the host to be transparent at the pump wavelength.

Mode selection can be achieved [16] by using an additional Fabry - Perot (see Fig. 7). The simplest way to do this is to cleave a crystal partly. As long as the wedge-shaped air gap is sufficiently parallel over the crossection of the beam waist, i.e. for $2w_o \cdot tg\delta \ll \lambda$ the resonator will not seriously be distorted.

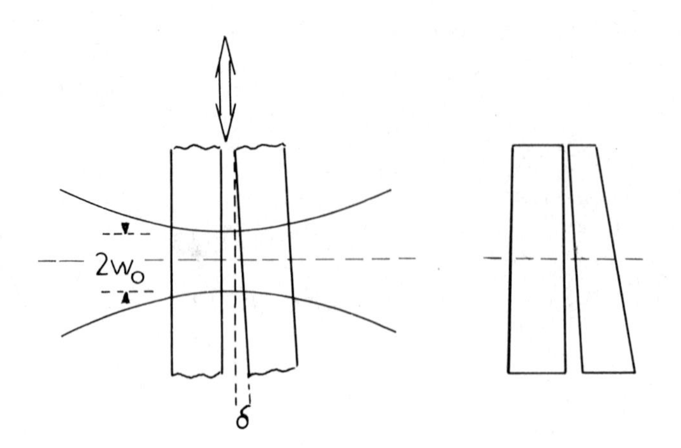

Fig.7

Folded and wedged Fabry-Perot arrangements for mode selection and tuning

Tuning is achieved [16] when the whole arrangement of Fig.7 is moved perpendicularly to the exciting beam, thus changing the effective thickness of the air gap. In this way a single mode can be tuned over the gain region. An example for single mode selection is given in Fig.8. For avoiding mode hopping during tuning the two crystal plates must be wedged too. A mathematical solution of this problem is presently under way.

By measuring gain profiles of stimulated emission at low temperatures, a high resolution coherent spectroscopy seems to be possible. Further, the inversion obtainable in the systems described may stimulate experiments designed to study coherency effects which in turn may shed light on relaxation processes in organic molecules.

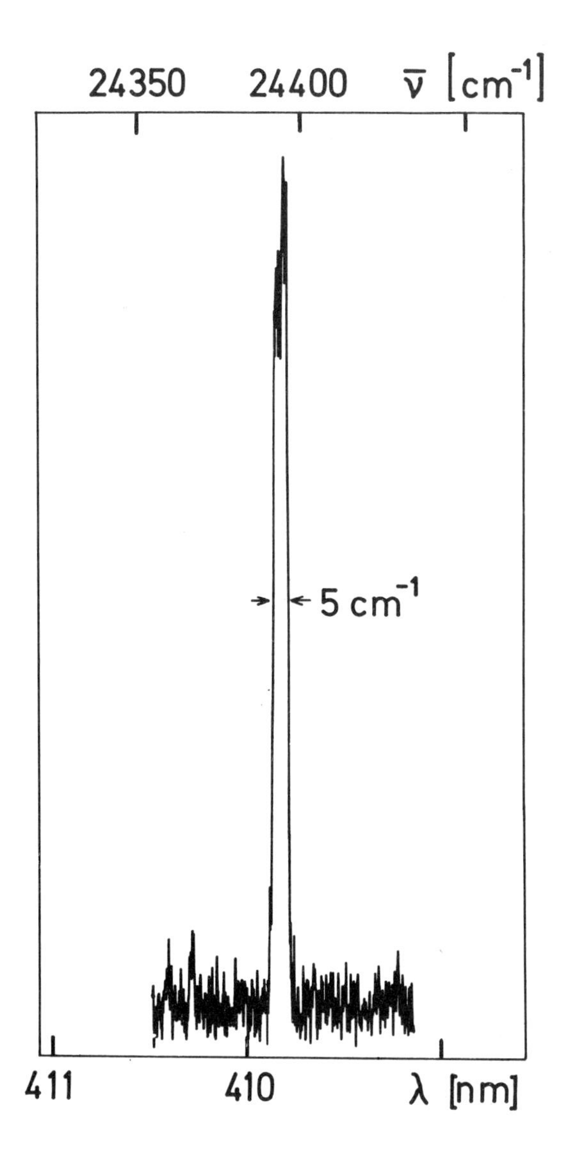

Fig.8

Single mode selection by a folded Fabry-Perot, realized by a partly cleaved 2,3-dimethylnaphthalene crystal, ≈360 μm thick, doped with anthracene.

Acknowledgements

The author whishes to express his gratitude to Prof.Dr.H.C.Wolf for his support and to A.Valera for providing the spectrum of Fig.8 . The crystals used originated in part from Prof.Dr.A. Schmillen and from Prof.Dr.K.H.Hausser (grown by H.Zimmermann); in part they were grown in the Stuttgarter Kristallabor by M.Gerdon, M.Herb, A.Huber, and W.Tuffentsammer. The present investigations are supported by the Deutsche Forschungsgemeinschaft.

REFERENCES

[1] V.L.Broude, V.S.Mashkewitch, A.F.Prihot´ho, N.F.Prokopjuk, and M.S.Soskin, Fiz.Tver.Tela 4,2976 (1962)

[2] V.L.Broude and E.F.Sheka, in Kwantovaya Electronica, p.188 "Naukova dumka" Kiev(1966)

[3] M.D.Galanin, Sh.D.Khan-Magometova, and Z.A.Tchigikova Pis´ma Zh.Exper. Teor. Fiz. 16,141 (1972)

[4] V.A.Benderskii, V.Kh.Brikenstein,V.L.Broude, and A.G. Lavrushko Pis´ma Zh.Exper.Teor.Fiz.17,472 (1973)

[5] E.Glockner and H.C.Wolf, Z.Naturforsch. 24a,943 (1969)

[6] N.A.Tolstoy and A.P.Abramov, Fiz.Tver.Tela 9,34o (1967)

[7] A.Bergman, M.Levine,and J.Jortner, Phys.Rev.Lett.18,593 (1967)

[8] O.S.Avanesjan, V.A.Benderskii, V.Kh.Brikenstein, V.L.Broude L.I.Korshunov, A.G.Lavrushko, and I.I.Tartakovskii Mol.Cryst.a.Liqu.Cryst. 29,165 (1974)

[9] H.S.Avanesjan, V.A.Benderskii, V.Kh.Brikenstein, and A.G.Lavrushko, Phys.Stat.Sol.(a)27,K77 (1975)

[10] G.P.Srivastava and S.C.Gupta, Optica acta 21, 157 (1974)

[11] H.S.Avanesjan, V.A.Benderskii, V.Kh.Brikenstein, V.L.Broude, and A.G.Lavrushko, Phys.Stat.Sol.(a)19,K121 (1973)

[12] N.Karl, Phys.Stat.Sol.(a)13, 651, (1972)

[13] N.Karl, presented at the spring meeting of the Deutsche Physikalische Gesellschaft in Kiel (1972)

[14] K.G.Ericsson and L.R.Lidholt, Appl.Optics 7, 211 (1968)

[15] D.Rosenberger, Phys.Lett. 14, 32 (1965)

[16] N.Karl, presented at the spring meeting of the Deutsche Physikalische Gesellschaft in Stuttgart (1974) under Q 15

DISCUSSION

N. KARL

C.R. Goldschmidt - It is interesting to note the similarity between your experiments with doped crystals and energy transfer dye lasers in liquid solutions. Due to triplet-triplet considerations the pure Anthracene crystal seems to be a very promising laser. Did you try to excite directly the bulk of the crystal by other light sources than the N_2 laser which is absorbed in about 1000 A.

N. Karl - We are considering such experiments. Using an intermediate dye laser (e.g. p.p'-diphenyl-stilbene in benzene) as an excitation source - pumped by a stronger nitrogen laser - seems to be a promising way.

J. Langelaar - Concerning the volume excitation of pure organic crystals did you consider 2 photon excitation for instance with a fundamental Ruby laser pulse? 2) Is there a chance to obtain stimulated emission by flash-lamp pumping of organic crystals?

N. Karl - The answer to the first question is that we have no ruby laser available. If one had sufficient power to create the inversion necessary by two photon absorption, this would be the easiest way to obtain a homogeneous excitation of larger volumes.
The second question is difficult to answer, because in spite of the very high amplification there are serious practical obstacles : first the heat dissipation makes much more difficulties than in liquids which can be circulated through the excitation volume; second it will be difficult to obtain laser-quality crystals of dimensions larger than a few mm, which means that only short amplification lengths will be realizable. If you have available a short duration high intensity flash lamp, with which laser emission can be obtained from a liquid dye laser with a fluorescence lifetime comparable to that of the guest of the doped laser-crystal considered and with an active length of say between 1 and 5 mm, then the crystal laser certainly can be operated too. The fact that the molecules are well oriented in the crystalline matrix will even lower the threshold for certain crystal orientations. In addition, levels can be much sharper in the crystal than in the liquid and the relative positions of singlets and triplets can be different.

Y. Meyer - Could you comment on the compared overall laser band width and the width of the vibronic fluorescence bands in this case? Do you observe also laser emission from the other fluorescence vibronic peaks which have nearly the same emission cross section?

N. Karl - The observed laser band width under the experimental conditions of only small feed-back by surface reflections was in the order of 1 nm, which is much smaller than the comparable (room temperature) fluorescence band width of the corresponding vibrational band. The other vibrational peaks have not been observed by us so far in stimulated emission. In fact there are appreciable differences in the fluorescence intensities of the different vibrational bands; laser emission was obtained for the strongest band. If one could introduce losses for that band by absorption or dispersive elements I am sure that the other bands will lase too, as the starting level for all of these transitions is the same. This will be extremely interesting at low temperatures, where very sharp lines are obtained already in normal fluorescence.

R.M. Hochstrasser - Would you please comment on how the biaxial nature of your host crystals influences the properties of the light propagation in the laser. 2) In regard to Langelaar's remark about two-photon absorption I might mention that the two-photon absorption of 337.2 nm radiation by 2,3-dimethyl-naphtalene is resonance enhanced due to the small (ca. 10^3 cm^{-1}) separation of the N_2 laser line from the crystal exciton absorption : we have seen extremely strong two-photon signals in this system using a N_2 laser.

N. Karl - The answer to the question is that most organic molecular crystals are of rather low symmetry. The crystal structures of these van der Waals crystals are mainly determined by the condition of "closest packing" (Kitaigorodskii) of the mostly rather anisotropic molecules. Monoclinic structures with one symmetry plane are very common. Fluorene is an example of an orthorhombic crystal. Except for the directions of the optical axis there are two distinct modes with orthogonal polarization. We always observed strongly polarized laser emission of that mode which experiences the higher index of refraction and therefore the stronger reflexion. The indices of refraction can exceed n = 2 for certain orientations, which corresponds to an amplitude reflexion coefficient of 1/3. There are interesting questions concerning the orientation of the guest transition moments relative to the polarization modes of the crystal.

Thank you for the comment concerning 2,3-dimethylnaphtalene. Perhaps this might be a suitable system for an energy transfer laser. With such systems interesting questions arise like energy transfer kinetics and temporal mode development by exciton diffusion into depleted regions.

PROPERTIES OF RARE EARTH DOPED INORGANIC GLASSES AS RELATED TO THEIR LASING ABILITIES*

R. REISFELD
Department of Inorganic & Analytical Chemistry, The Hebrew University of Jerusalem, Jerusalem, Israel

Large Nd:glass laser systems capable of delivering multikilojoule subnanosecond pulses are being designed at present for fusion applications [1]. Design of these systems requires data characterizing the dynamic spectral properties of the rare-earth-doped glasses of potential laser use.

There are a number of characteristics which distinguish glass from other solid lasers: its properties are isotropic; it can be doped at very high concentrations with excellent uniformity; the glass material affords considerable flexibility in size and shape and may be obtained in large homogeneous units of diffraction-limited optical quality.

The emission line of Nd^{3+} laser at 1.06μ arises from the $^4F_{3/2}$ $^4I_{11/2}$ transition. Transitions of a similar origin in other rare earth (R.E.) ions may provide laser transitions in the variety of wave lengths ranging from 0.27μ to 2μ. This can be achieved by selecting a proper combination of R.E. with glass host.

In the present paper, we shall discuss the basic properties needed to obtain the maximum emission from the excited levels of rare earth materials in glasses which are essential for optimum laser operation. We shall discuss the radiative transition probabilities, the multiphonon nonradiative relaxation rates which influence the quantum yields of fluorescence and the energy transfer between ions having high absorption coefficients to the emitting rare earth ions.

Because of the limitation of space, we shall discuss in this paper only the basic principles. Reference will be made to the spectral data whenever obtainable in literature. Finally, we shall present some new data on energy transfer efficiencies and probabilities.

1. RADIATIVE TRANSITION PROBABILITIES

Optical line spectra of triply ionized rare earth ions originate from

*Partially supported by U.S. Army Contract No. DAERO-75-G-029.

transitions between levels of $4f^N$ configuration. The energies of these levels arise from a combination of the Coulomb interaction among the electrons, the spin orbit coupling and the interaction with the crystalline field. Transitions between f^N levels involve no change in parity and are therefore forbidden by the Laporte rule. The observed narrow band forced electric dipole transitions become partially allowed if odd harmonies in the static or dynamic crystal field admix states of opposite parity (such as 5d) into the 4f level. This can occur statically if the rare earth ion resides in lattice sites lacking inversion symmetry. A decade ago Judd [2] and Ofelt [3] showed that calculation of electric dipole (ED) probabilities may be performed by treating the $4f^N$ and the excited opposite parity configurations as degenerate levels with a single average energy separation between them.

In this theory the electric dipole matrix element for the ρ-th component of polarization can be expressed as

$$\langle\psi_a|P_\rho|\psi_b\rangle = \sum_{q,t\ \mathrm{even}} Y(t,q,\rho)\ \langle f^N\gamma SLJJ_z \| U^{(t)}_{\rho+q} \| f^N\gamma' S' L' J' J_z'\rangle \qquad \text{(I)}$$

where the energy denominator $(E_{n'1'}-E_{4f})$av, the odd parity crystal field parameters and the radial integrals are all incorporated into the phenomenological parameters $Y(t,q,\rho)$. The number and type of components A_q^k which enter into the $Y(tq\rho)$ can be determined from group theory if the site symmetry of the R.E. ion is known. Since at most f, combined with g and d states are involved, $k \leq 7$.

In an intermediate coupling scheme the line strength S for ED transitions reduces to a simple expression containing three intensity parameters and three matrix elements given by:

$$S_{ED}(aJ:bJ') = e^2 \sum_{\lambda=2,4,6} \Omega_\lambda |\langle f^N[\gamma SL]J \| U^\lambda \| f^N[\gamma' S' L']J'\rangle|^2 \qquad \text{(II)}$$

The connection between these line strengths and spontaneous emission probability is

$$A(aJ:bJ') = \frac{64\pi^2\nu^3\chi}{3(2J+1)hc^3} S(aJ:bJ') \qquad \text{(III)}$$

in which ν is the frequency in sec^{-1}; χ is the field correction for electric dipole transition; $\chi = (n(n+2)^2)/9$ (n = refractive index).

Since experimentally the oscillator strength for various transitions may be easily obtained from absorption measurements from the ground state, the following connection between the spontaneous emission probability and the oscillator strength p permits calculation of the transition probability between any pair of degenerate levels

$$A_{ED} = \frac{8\pi^2 e^2 \nu^2 \chi}{mc^3(2J+1)} p \qquad (IV)$$

and

$$P = \frac{\sigma}{2J+1} \sum_{\lambda=2,4,6} \tau_\lambda |\langle f^N \psi J \| U^\lambda \| f^N \psi' J' \rangle|^2 \qquad (V)$$

σ being the wave number in cm^{-1}

$$\text{with } \Omega_\lambda = \frac{3h}{8\pi^2 mc} \frac{n^2}{\chi} \tau_\lambda = 9.0\times10^{-12} \frac{n^2}{\chi} \tau_\lambda \qquad (VI)$$

The Ω parameters are obtained from the absorption spectra. The matrix elements U^λ may be calculated from the reduced matrix elements of Nielson and Koster [4] and the 3j and 6j symbols [5] by the methods described by Wybourne [6].

It was shown by Riseberg and Weber [7] and Reisfeld et al. [8] that the matrix elements of the majority of R.E. ions are only slightly dependent on the environment in which they are situated, the main factor responsible for the intensity of various transitions being the three Ω parameters. The Ω parameters for the various rare earth in borate, phosphate, germanate and tellurite glasses were calculated in our laboratory. The radiative transition probabilities were obtained using their values. They may be found in a review by Reisfeld [9]. The radiative transition probabilities for Nd^{3+} in various laser glasses were obtained recently by Krupke [10]. The corresponding data for the 1.064μ line of Nd^{3+} in laser glasses obtained in [10] are presented in Table I. The induced emission cross section σ_p presented in the table is related to the radiative transition probability by

$$\sigma_p \Delta\bar{\nu} \equiv \int \sigma(\bar{\nu}) d\bar{\nu} = \frac{\lambda_p^2}{8\pi c n^2} A[(^4F_{3/2}):(^4I_{11/2})] \qquad (VII)$$

where $\lambda = 1/\bar{\nu}$; λ_p is the wavelength of the fluorescence peak (1.06μ) and $\Delta\bar{\nu}$ is the effective fluorescence linewidth determined by numerical integration of the fluorescence line shape.

The radiative quantum efficiency of a fluorescent level η is defined as

$$\eta = \frac{\Sigma A}{\Sigma A + \Sigma W} = \tau_{meas} \Sigma A \qquad (VIII)$$

where W is the nonradiative relaxation probability and τ_{meas} the measured lifetime.

Table I

Radiative properties of 3669A, S33, ED-2 and LSG-91H laser glasses

Property	3669A	S33	ED-2	LSG-91H
Calculated radiative lifetime, τ^{c}_{rad} (μs)	1000	400	372	384
Measured fluorescence lifetime, τ_1 (μs)	520	244	310	300
Calculated radiative quantum efficiencies, η_c	0.52	0.61	0.83	0.78
Measured linewidth, $\Delta\bar{\nu}_{eff}$ (cm^{-1})	296	300	300	310
Calculated induced-emission section, σ_ρ (10^{-20} cm^2)	1.2	2.8	2.9	2.6
Measured stimulated-emission cross section, σ_L (10^{-20} cm^2)		3.2	3.1	2.5
Calculated induced-emission cross section at λ = 1.064μ σ(1.064), (10^{-20} cm^2)	1.05	2.62	2.71	2.45
Small-signal gain coefficient at 1.064μ, γ, (cm^2/J)	0.056	0.139	0.145	0.130
$S_c(^4F_{3/2}, ^2G_{9/2})$, 10^{-20} cm^2	0.10	0.22	0.27	0.25
$S_c(^4F_{3/2}, ^4I_{11/2})$, 10^{-20} cm^2	0.97	2.6	2.8	2.6

2. MULTIPHONON RELAXATION

The radiative quantum efficiency of a level can be reduced from unity as a result of multiphonon and ion-ion relaxation processes. At lower R.E. concentrations only, the first process plays an important role.

The dependence of nonradiative relaxation on energy gap ΔE between two electronic levels between which the relaxation occurs is best represented by the Miyakawa-Dexter [11] theory as

$$W = W(0) \exp(-\alpha\Delta E) \tag{IX}$$

The parameter α is given by

$$\alpha = \frac{1}{\hbar\omega} [\ln\{p/g(n+1)\} - 1] \tag{X}$$

where p is the number of phonons active in the process; $\hbar\omega$ is the energy of the phonons which contribute mainly to the multiphonon process; $p = E/\hbar\omega$; n is the number of phonons excited at the temperature of the system $n_i = [\exp(\hbar\omega/kT)-1]^{-1}$ and g is the electron lattice coupling constant.

Although the exponential dependence of equation IX is an approximation, its validity has been verified for the multiphonon relaxation process between $4f^N$ states of trivalent rare earth ions in glasses by Reisfeld and Eckstein [12]. They have found that the multiphonon relaxation rates in various rare earth ions in a variety of glasses (phosphate, borate, germanate and tellurite) is determined only by the number of high energy phonons separating the energy levels. This brings us to the conclusion that W(0) and α are independent of the character of $4f^N$ states involved and depend only on the energy gap (expressed in no. of phonons). The phonon energies of the various glasses are presented in Table II.

Table II
Phonon energies of various oxidic glasses

Glass	Bond	Phonon energy (cm^{-1})
Borate	B - O	1340 - 1480
Phosphate	P - O	1200 - 1350
Germanate	Ge- O	975 - 800
Tellurite	Te- O	750 - 600

As a result of this study we can predict and control fluorescence quantum efficiencies by incorporating the ion in question in an appropriate glass host. The temperature dependence of intensities in multiphonon processes is represented by [12]:

$$W(T) = W(T_0)[1 - \exp(-\hbar\omega/kT)]^{-p} \qquad \text{(XI)}$$

In ref. [12] ln W(T) was plotted vs. ln[1 - exp(-$\hbar\omega$/kT)] for temperature dependence of Er^{3+} $^4S_{3/2}$ fluorescence. The slope of the plot corresponds to 4 phonons which equal the energy difference between $^4S_{3/2}$ and the next $^4F_{9/2}$ level in tellurite glasses in which Er^{3+} was incorporated.

3. ENERGY TRANSFER

By energy transfer in glasses we understand the phenomenon wherein excitation into the absorption bands of donor ions results in an emission band from the acceptor ions. These phenomena are observed in glasses in relatively dilute systems suggesting that long range interactions are possible. There are several interactions between the ions which are responsible for energy transfer and cooperative relaxation. The electric multipolar interaction arises from the Coulomb interaction between the electron bands of the ions.

1. The electrostatic interaction is represented by

$$H_{es} = \sum_{i,j} \frac{e^2}{K(\vec{r}_{Ai} - \vec{R} - \vec{r}_{Bj})} \tag{XII}$$

Here, r_{Ai} and r_{Bj} are the coordinate vectors of electrons i and j belonging to ions A and B, respectively; $\vec{R}$ is the nuclear separation and K is the dielectric constant. The various multipolar terms appear from a power series expansion of the denominator. This expansion was expressed by Kushida [13] in terms of tensor operators. The leading terms are the electric dipole-dipole (EDD), dipole-quadrupole (EQD), and quadrupole-quadrupole (EQQ) interaction. These have radial dependence of R^{-3}, R^{-4} and R^{-5}, respectively.

2. A second ion-ion coupling is the magnetic dipole-dipole interaction:

$$H_{MDD} = \sum_{i,j} \left[\frac{\vec{\mu}_i \cdot \vec{\mu}_j}{R^3} - \frac{3(\vec{\mu}_i R)(\vec{\mu}_j R)}{R^5} \right] \tag{XIII}$$

where $\vec{\mu}_i = \ell_i + 2\vec{s}_i$ and $(\vec{\ell}_i \vec{s}_i)$ are the orbital and spin operators for the i-th and j-th electrons of ions A and B respectively. The selection rules ΔS, ΔL, $\Delta J = 0, \pm 1$ for transitions between $4f^N$ states are again relaxed by SLJ state admixing. The MDD interaction has the same long-range R^{-3} radial dependence as the EDD interaction.

3. Exchange interaction: The matrix element of exchange interaction is of the form

$$H_{ex} = \sum_{i,j} J_{ij} \vec{S}_i \vec{S}_j \tag{XIV}$$

J_{ij} represents the isotropic or Heisenberg component of the exchange integral and is a function of R, θ and φ. S is the spin of the electron (for detailed derivation see [14]).

The probability p(exc) of the exchange interaction can be written as

$$p(exc) = \frac{2\pi}{h} \chi^2 S \tag{XV}$$

with $S = \int g_d F_d(E) g_a F_B(E) dE$ the overlap integral and $\chi = \exp(-R/L)$ where R is the interionic distance and L the average radius of donor and acceptor.

The ion-ion interactions differ in their dependence on donor-acceptor distance for various mechanisms. The radial dependence of the ion-pair transfer rate is derived from the square of the matrix element H_{da}, in general equation for interionic interaction which has the form

$$W_{da} = \frac{2\pi}{\hbar} |\langle\psi_d(2')\psi_a(1')|H_{da}|\psi_d(2)\psi_a(1)\rangle|^2 \int E_d(E)F_a(E)dE \qquad \text{(XVI)}$$

The dependence of the transfer rate on distance is R^{-6}, R^{-8}, R^{-10} for EDD, EDQ and EQQ, respectively and R^{-6} for MDD. Direct exchange involves an exponential decrease of the wave function, contained in the calculation of J_{ij}. Therefore in dilute systems it is believed to be ineffective.

A method for calculating the probability and efficiency of energy transfer between inorganic ions from the donor and acceptor luminescence intensities and lifetimes was proposed by Reisfeld et al. and is described in detail in [15].

When the absorption of the donor ion is much higher than the acceptor ion at the wavelength at which the system is excited (which is the practical case in optimum laser pumping), then the transfer efficiency η_t is given by a simple formula

$$\eta_t = 1 - \frac{\eta_d}{\eta_d^\circ} \qquad \text{(XVII)}$$

η_d being the fluorescence efficiency of the donor in presence of acceptor ion, and η_d° the fluorescence efficiency of the donor alone.

p_t, the transfer probability, may be obtained from the formula

$$p_t = \frac{1}{\tau_d}\left(\frac{\eta_d^\circ}{\eta_d} - 1\right) \qquad \text{(XVIII)}$$

τ_d being the measured lifetime of pure donor.

Table III presents the number of efficiencies and probabilities of energy transfer which were obtained in our laboratory.

In all cases presented in Table III the measured probabilities of transfer were higher by a few orders of magnitude than calculated from the formulas of resonant transfer. There may be several reasons for such deviation.

1. In many real systems d-d (donor-donor) transfer cannot be neglected. Since the donors are identical ions d-d transfer is resonant and may be more rapid than d-a transfer if the two concentrations are comparable. Excitation energy may then be able to migrate before the donors before passing to the activator.

Table III*

Efficiency and probabilities of energy transfer concentration (wt%)

Glass	Donor	Acceptor	$p \times 10^6$ sec^{-1}	η
$B_{(a)}$	Ce^{3+}	Tb^{3+}		
	0.025	1.0	2.0	0.07
	0.025	2.0	6.7	0.19
	0.025	2.5	8.7	0.23
	0.025	3.0	13.7	0.32
$B_{(b)}$	Ce^{3+}	Tm^{3+}		
	0.13	0.05	2.4	0.09
	0.13	0.15	4.8	0.17
	0.13	0.25	7.0	0.25
	0.13	0.50	11.6	0.41
	0.13	0.75	14.0	0.49
	0.13	1.00	14.9	0.52
$B_{(c)}$	Tl^{+}	Gd^{3+}		
	0.01	1.0	28.3	0.85
	0.01	3.0	38.3	0.89
	0.01	5.0	47.6	0.91
	0.01	7.0	61.7	0.93
$G_{(d)}$	Pb^{2+}	Eu^{3+}		
	1.0	1.0	0.33	0.09
	1.0	2.0	1.14	0.25
	1.0	3.0	2.57	0.41
	1.0	5.0	5.00	0.59
	1.0	7.0	7.32	0.68
$B_{(e)}$	Bi^{3+}	Eu^{3+}		
	1.0	0.5	3.45	0.50
	1.0	1.0	0.64	0.18
$G_{(e)}$	Bi^{3+}	Eu^{3+}		
	1.0	0.5	1.14	0.29
	1.0	1.0	0.72	0.20
$B_{(e)}$	Bi^{3+}	Sm^{3+}		
	1.0	0.5	-	-
	1.0	1.0	0.69	0.19
$G_{(e)}$	Bi^{3+}	Sm^{3+}		
	1.0	0.5	0.63	0.28
	1.0	1.0	2.13	0.43

B = Borate; G = Germanate.

(a) R. Reisfeld and J. Hormadaly, J. Solid State Chem., 13 (1975).
(b) R. Reisfeld and Y. Eckstein, unpublished results.
(c) R. Reisfeld and S. Morag, unpublished results.
(d) R. Reisfeld and N. Lieblich-Sofer, J. Electrochem. Soc., 121, 1338 (1974).
(e) R. Reisfeld and L. Boehm, unpublished results.

*A detailed description of fluorescence and excitation spectra and quantum efficiencies of rare-earth ions can be found in ref. [16].

A practical way of treating such a system would be an analysis of the time development of the donor luminescence. For determination of the transfer mechanism it is advantageous to vary donor and acceptor concentration [17,18].

2. Phonon-assisted energy transfer. The energy mismatch which is high because of the inhomogeneous broadening in glasses, between the levels of two rare earth ions may be matched by interaction of phonons of the host glass. An example of such study in crystals may be found in ref. [19].

3. Superexchange between the rare earth ions and the surrounding oxygens may provide an additional path of energy transfer. While all the mentioned mechanisms appear to be active additional work is needed to answer the above questions. Practically the obtained transfer probabilities may permit calculations for laser designs.

REFERENCES

1. Problems with fuel pellets for laser induced fusion. Physics Today, March 1975, p. 17.
2. B.R. Judd, Phys. Rev., 127(1962)750.
3. G.S. Ofelt, J. Chem. Phys., 37(1962)511.
4. C.W. Nielson and C.F. Koster, Spectroscopic Coefficients for the p^n, d^n and f^n Configurations. M.I.T. Press, Cambridge, Mass., 1964.
5. Rotenberg, Bivins, Metropolis and Wooten, The 3-j and 6-j Symbols, M.I.T. Press, Cambridge, Mass., 1959.
6. B.G. Wybourne, Spectroscopic Properties of Rare Earths, Interscience, New York, 1975.
7. L.A. Riseberg and M.J. Weber, in E. Wolf (Editor), Progress in Optics, Vol. XIV, North Holland Publishing Co., 1975, in press.
8. R. Reisfeld, L. Boehm, N. Lieblich and B. Barnett, Proc. Tenth Rare Earth Conf., 2(1973)1142.
9. R. Reisfeld, Structure and Bonding, 22(1975).
10. W.F. Krupke and J.B. Gruber, IEEE, J. Quantum Electronics, 10(1974) 450.
11. T. Miyakawa and D.L. Dexter, Phys. Rev. B, 1(1970)2961.
12. R. Reisfeld and Y. Eckstein, J. Chem. Phys. (1975).
13. T. Kushida, J. Phys. Soc. Japan, 34(1973)1318,1327,1334.
14. R.K. Watts, in R. DiBartolo (Editor), Optical Properties of Solids, Plenum Press, 1975.

15. R. Reisfeld, Structure and Bonding, 13(1973)53.

16. R. Reisfeld, J. Res. NBS A. Phys. Chem., 76A(1972)613.

17. M.J. Weber, Phys. Rev. B, 4(1971)2932.

18. J.-C. Bourcet and F.K. Fong, J. Chem. Phys., 60(1974)34.

19. N. Yamada, S. Shinoya and T. Kushida, J. Phys. Soc. Japan, 32(1972) 1577.

DISCUSSION

R. REISFELD

L. Burlamacchi - The τ_λ parameters which you presented depend on the strength and symmetry of the crystal-field on the ion. In most glassy networks, the crystal-field parameters for iron group ions are known to be distributed over a wide range. It is possible that in your systems τ_λ are distributed over a range rather than single value.

Mme Reisfeld - In a non-metallic substance such as a glass, there is every reason to believe in Beer's law

$$\log (I_0/I) = \ell (\varepsilon_1 c_1 + \varepsilon_2 c_2 + \varepsilon_3 c_c + \ldots\ldots)$$

It is indeed an important question whether the dispersion of properties of the cation (network modifier) sites in glasses is small and also whether a definite number (2,3,..) of distinct sites can be recognized However, in all cases, the local ε_n are definite linear combinations (weighted with the matrix elements of U^λ) of three τ_λ characterising each species n. Hence, the phenomenological τ_λ values are simply the average values weighted by the concentrations c_n and one cannot derive evidence for large or negligible dispersion of τ^n_λ of different sites just by observation of the band intensities. Jørgensen and I believe that one cannot draw extended conclusions about the symmetry of sites from intensities since the nature of the bond to the neighbour atoms seems to be much more important for the size of, in particular, τ_2 determining the hypersensitive transitions.

LASER A CHELATES D'URANYLE, ESSAI DE REALISATION

Y. MACHETEAU, A. COSTE, M. LUCE et P. RIGNY
CEN-SACLAY, DGI/SEPCP, 91190 GIF sur YVETTE

SOMMAIRE.-

Les caractéristiques spectroscopiques en absorption, fluorescence et excitation de quelques chélates d'uranyle ont été déterminées. Les déclins de fluorescence correspondants ont été mesurés à basse température. La possibilité d'obtention d'une émission stimulée avec les chélates d'uranyle est discutée, en se basant sur les caractéristiques des chélates d'Europium (B_4EuNa et B_4Eu pipéridine) qui ont donné l'effet laser.

INTRODUCTION.-

L'étude relative à la luminescence des chélates d'uranyle est encore limitée, les principaux travaux mentionnés dans la littérature sont dûs à KONONENKO et ses collaborateurs, [1, 2]. Nous avons continué ces recherches en étudiant les spectres de fluorescence et d'excitation d'un certain nombre de chélates d'uranyle, nouveaux pour la plupart :

- $UO_2(TTA)_2$ O. Phénanthroline
- $UO_2(TTA)_2$ guanine
- $UO_2(TTA)_2$ pipéridine
- $UO_2(FOD)_2$ O. Phénanthroline
- UO_2 FOD OH O. Phénanthroline

Pour chacun de ces composés nous avons déterminé quelques caractéristiques essentielles.

Enfin, l'analogie chimique entre les terres rares et les actinides laisse espérer que certains composés d'actinides pourraient présenter un effet laser. En particulier, les chélates d'uranyle, homologues de chélates de terres rares donnant des lasers bien étudiés, peuvent être intéressants.

Préparation et techniques opératoires

D'une façon générale, ces chélates sont préparés à partir d'un système biphase [3] à savoir :

- une phase aqueuse composée d'un sel d'uranyle (nitrate par exemple) en solution dans l'eau avec le chlorhydrate de la base considérée
- une phase organique correspondant à la β dicétonate:thénoyl trifluoro-acétylacétone (TTA) ou hepta fluoro 1-1-1-2-2-3 diméthyl 7-7 octanédione 4-6 (FOD), en solution dans le benzène.

L'extraction a lieu pour un pH supérieur à 8,5. Ces composés se forment uniquement lorsque le ligande (TTA ou FOD) se trouve sous la forme énolique.

Les deux ligandes précités ont pour formule :

TTA : (thiophène : H-C, C-H / H-C, S) $C - C(=O) - CH_2 - C(=O) - CF_3$

FOD : $CF_3 - CF_2 - CF_2 - C(=O) - CH_2 - C(=O) - C(CH_3)_3$

Ainsi préparés, ces chélates sont mis en solution dans un mélange éthanol-méthanol (rapport volumique éthanol/méthanol égal à trois). Un tel mélange constitue à basse température une matrice parfaitement transparente. Les spectres de luminescence ont été obtenus avec une concentration en chélate de 1.10^{-5} $M.l^{-1}$ afin d'éviter tout phénomène de réabsorption. Dans ces conditions, (basse température et faible concentration) la cinétique de photoréduction du chélate est négligeable, [4], et l'échantillon reste donc inaltéré pendant l'enregistrement d'un spectre.

Solubilité

La solubilité de ces composés dans le mélange éthanol/méthanol est faible (tableau I) mais bien suffisante pour en étudier la luminescence.

TABLEAU I

Composés	Solubilité apparente dans le mélange (éthanol-méthanol) à la température ambiante ($M. l^{-1}$)
UO_2 $(TTA)_2$ O. phénanthroline	1.10^{-3}
UO_2 $(TTA)_2$ guanine	1.10^{-3}

../

(suite)

UO_2 $(TTA)_2$ pipéridine	5.10^{-5}
UO_2 $(FOD)_2$ O. phénanthroline	5.10^{-2}
UO_2 FOD OH O.phénanthroline	5.10^{-3}

Absorption

Les spectres d'absorption (U.V visible) présentent des bandes intenses mais assez larges et peu structurées, (Fig. 1à5). Un abaissement de la température n'apporte aucune modification sensible. Ces spectres sont très semblables à ceux du ligande et de la base correspondants, sauf pour chacun d'entre eux, la bande de plus grande longueur d'onde. En effet, cette dernière ne peut être attribuée ni à la molécule chélatante ni à la molécule adduct. Ceci a déjà été observé, en particulier par BELFORD et al. [5], qui ont obtenu une bande à 3600 Å, avec les chélates bis acétylacétone UO_2(VI) bis-trifluoroacétylacétone UO_2(VI)... et bishexafluoro acétyl-acétone UO_2(VI). Dans ce cas précis, l'augmentation du nombre d'atomes de fluor dans le ligande ne fait pas varier la position de la bande. Ces auteurs pensent qu'il s'agit probablement d'une transition $n - \pi$ ou $\pi - \pi$. Cette hypothèse est corroborée par le fait que les agents chélatants apparentés et les chélates de cuivre, n'ont pas de bande d'absorption à 3600 Å, [6]. Par ailleurs, SAGER et al. [7] ont aussi observé une transition électronique de même nature ($\pi - \pi^*$) entre 2800 et 3800Å, lors d'une étude sur les chélates de terres rares (tri cétonates). Toutefois, il paraît vraisemblable que dans les chélates d'uranyle, le groupement UO_2 doit avoir un effet inductif sur les transitions du ligande.

Dans notre cas, cette transition électronique se situe entre 3558 et 3800Å, suivant le chélate considéré. La nature même des ligandes et les valeurs du coefficient d'absorption molaire (tableau II) indiquent qu'il s'agit bien d'une transition entre orbitales de type π délocalisées (transition $\pi - \pi^*$) ou d'une transition $n - \pi^*$. Les orbitales n sont localisées et de basse énergie. La disparition de cette bande d'absorption en milieu acide montre que la transition est vraisemblablement de type $n-\pi^*$. La faible intensité de ces transition ($n - \pi^*$) est due en général à une symétrie locale plus importante que la symétrie moléculaire globale. Notons aussi que cette transition masque partiellement les bandes correspondant au groupement uranyle. En solution saturée, on peut mettre en évidence les bandes d'absorp-

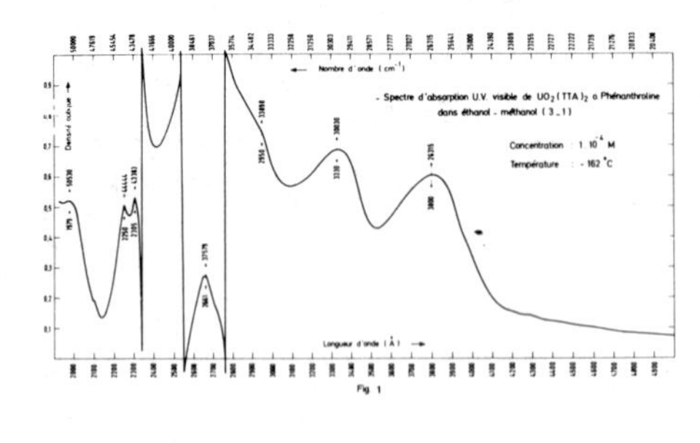

Fig. 1

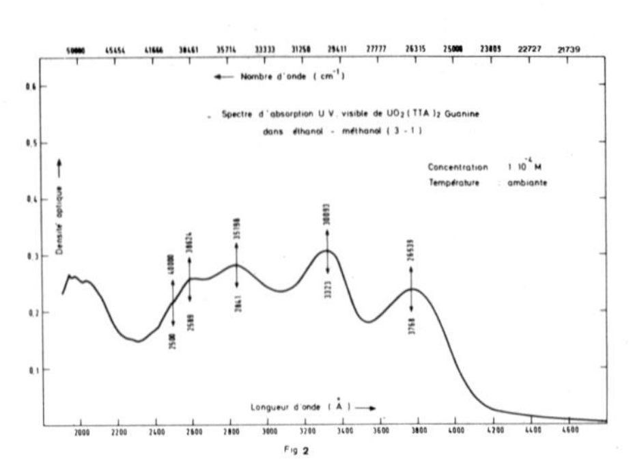

Fig. 2

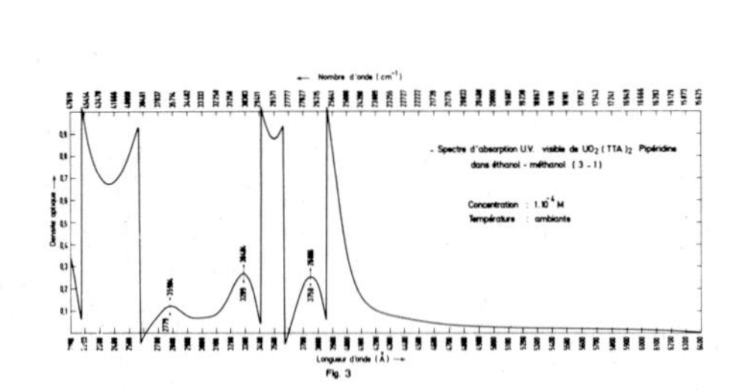

Fig. 3

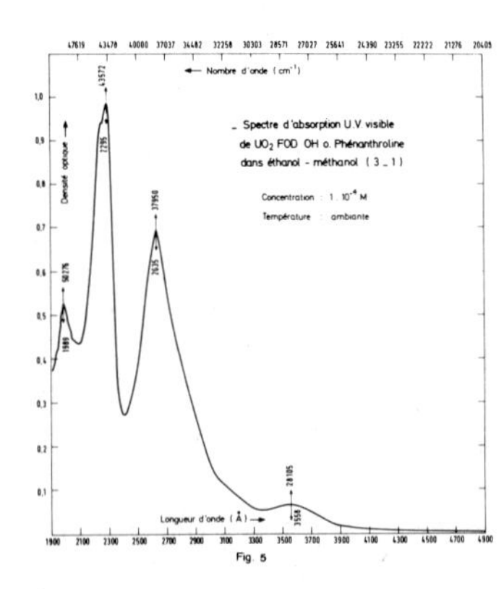

Fig. 5

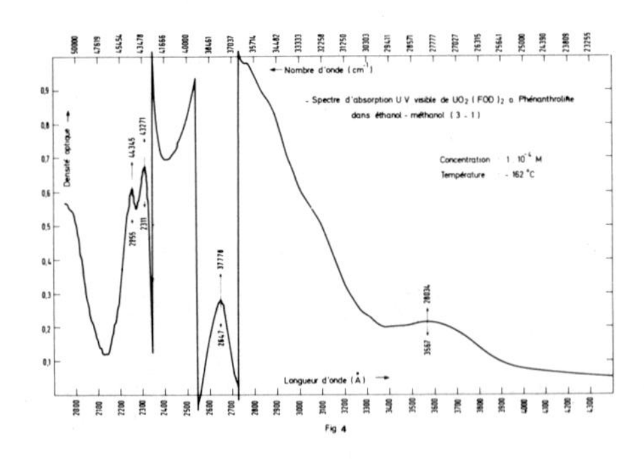

Fig. 4

tion relatives au premier état électroniquement excité.

TABLEAU II

Composés	Position de la bande correspondant à la transition $n - \pi^*$	Coefficient d'absorption molaire ($M^{-1}.cm^{-1}$)
$UO_2(TTA)_2$ O. Phén.	26 315	3 200
$UO_2(TTA)_2$ guanine	26 539	11 750
$UO_2(TTA)_2$ pip.	26 666	13 750
$UO_2(FOD)_2$ O. phén.	28 034	6 000
UO_2 FOD OH. O.phén.	28 105	3 250

A partir des données mentionnées dans le tableau II, nous pouvons calculer la force de l'oscilateur correspondant à la transition électronique $n - \pi^*$, pour chacun des composés (tableau III).

Pour cela, on utilise la formule approximative :

$$f = 4,32.10^{-9} \int \varepsilon \, d\bar{\nu}$$

avec $\varepsilon = \frac{1}{c\,l} \log. \frac{Io}{I}$

c = la concentration en mole par litre

l = l'épaisseur traversée en cm

$\bar{\nu}$ = le nombre d'onde par cm

TABLEAU III

Composés	Position de la bande (cm^{-1})	Force de l'oscillateur à la température ambiante
$UO_2(TTA)_2$ O. phén.	26 315	$4,6.10^{-2}$
$UO_2(TTA)_2$ guanine	26 539	$1,7.10^{-1}$
$UO_2(TTA)_2$ Pip.	26 666	$2,5.10^{-1}$
$UO_2(FOD)_2$ O. phén.	28 034	$1,0.10^{-1}$
UO_2 FOD OH O.phén.	28 105	$5,8.10^{-2}$

De plus, nous avons vérifié qu'un abaissement de la température ne modifie pas sensiblement la force de l'oscillateur.

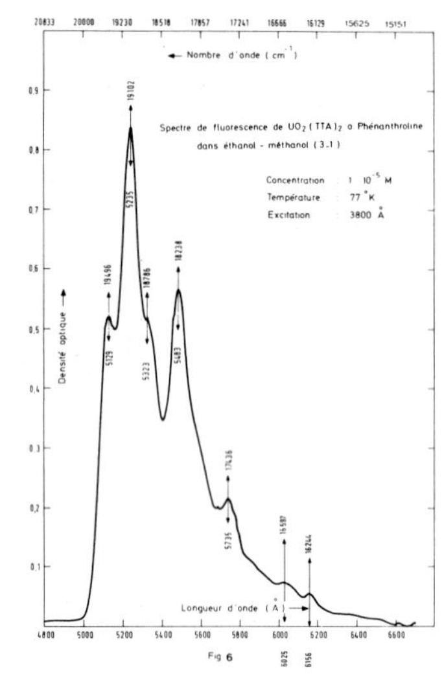

Fig 6

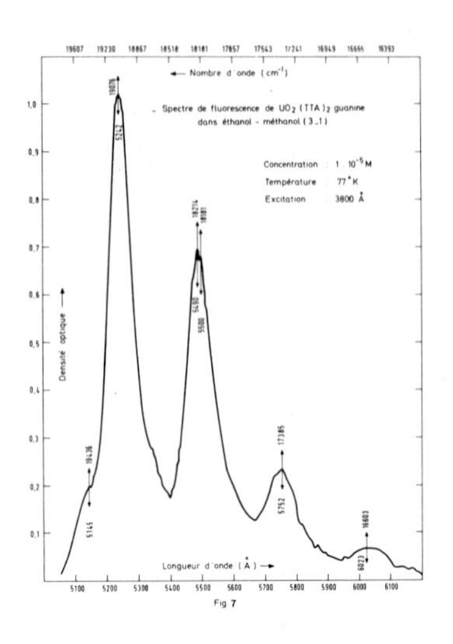

Fig 7

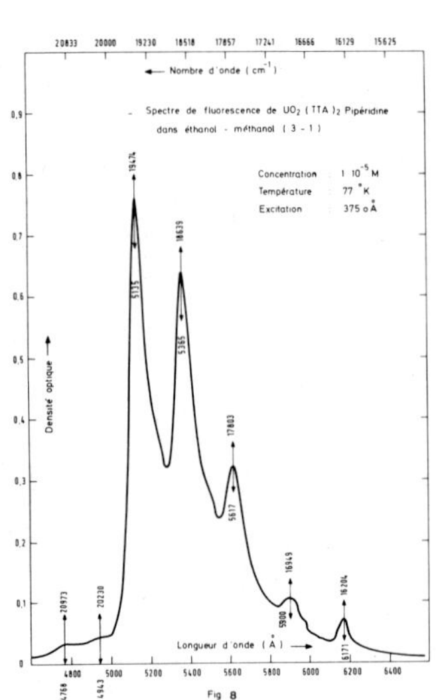

Fig 8

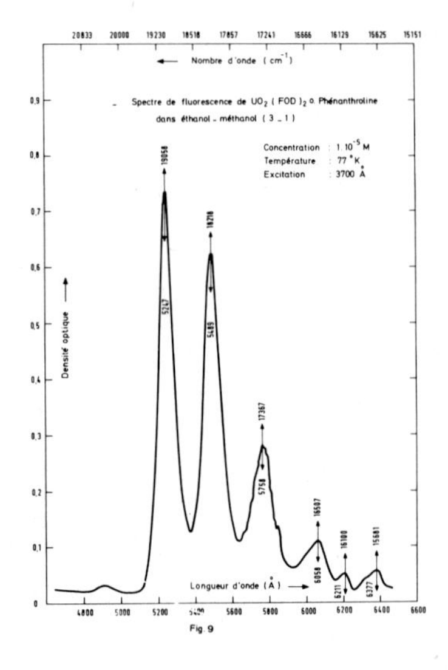

Fig. 9

Comme l'indiquent les valeurs obtenues pour la force de l'oscillateur, ces transitions n - π^* ne sont pas tout à fait permises. Ceci peut être dû principalement à une symétrie différente des fonctions d'onde des deux états électroniques, ou bien à un faible recouvrement des orbitales impliquées dans la transition.

Luminescence

Fluorescence

Afin de situer les niveaux excités de la molécule, nous avons étudié la luminescence verte que présentent à basse température et en solution ces chélates d'uranyle (Fig. 6à10).

Pour chacun d'entre eux, l'excitation était effectuée à la longueur d'onde correspondant à celle de la transition électronique n - π^*, ou bien à l'une des bandes d'absorption du ligande. Avant d'aborder l'étude des spectres de fluorescence, nous rappellerons succintement la position des niveaux d'énergie relatifs à l'ion uranyle solvaté (milieu perchlorique). En effet, nous avons vu que dans ces composés, la présence de la transition n - π^* ne permettait pas d'obtenir en totalité les vibrations du groupement

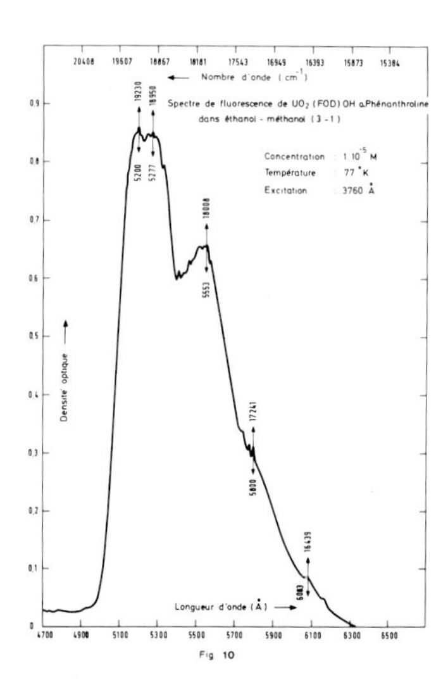

Fig 10

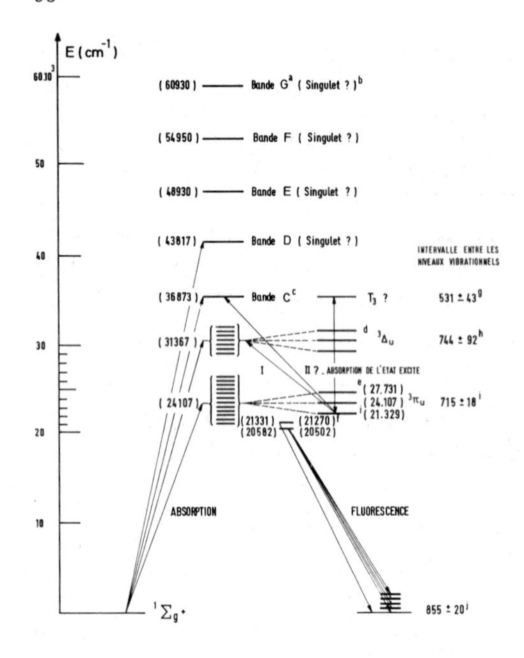

Fig.11 Niveaux d'énergie de l'ion uranyle en milieu perchlorique d'après BELL et BIGGERS [8]

a) Nomenclature des bandes U.V.

b) Multiplicité suggérée par BELL et BIGGERS.

c) La bande C ne manifeste pas de structure vibrationnelle en absorption et est considérée par les auteurs comme une composante (singulet) des séries C à G, qui sont séparées par un intervalle constant de 6015 ± 39 cm^{-1}.
Ce niveau doit être dégénéré avec un niveau triplet (appelé T_3) lequel présente une structure vibronique lorsqu'il est excité à partir de l'état triplet $^3\pi$ u. La dernière absorption indique que l'état T_3 doit être positionné à 38750 cm^{-1}, c'est-à-dire très près de la bande C à 36 873 cm^{-1}.

d-e) Les 2 groupes de niveaux centrés autour de 31 367 cm^{-1} d'une part et 24 107 cm^{-1} d'autre part, sont considérés comme des états triplets $^3\Delta$ u et $^3\pi$ u.

f) L'émission à partir du niveau situé a 21 270 cm^{-1} représenterait seulement 4,66 % de l'émission totale.

g) Donnée obtenue à partir du spectre d'absorption de photolyse flash.

h-i) Structure fine due à la vibration de valence symétrique dans l'état excité.

j) Structure fine due à la vibration de valence symétrique dans l'état fondamental.

UO_2^{2+}. La figure 11 reproduit donc les niveaux d'énergie de cet ion, [8]. D'après ce schéma, le premier état excité serait un état triplet $^3\pi_u$ dont les composantes correspondraient aux trois premières transitions électroniques , la dégénérescence étant levée par un couplage spin-orbite, [9]. De JEAGERE et al. [10]pensent que le spectre est composé d'une série de bandes vibrationnelles propres, groupées en trois transitions électroniques différentes. Le système de bandes situé entre 20 000 et 22 000cm^{-1} serait dû à une transition $^1\Sigma_g^+ \rightarrow {}^1\emptyset_g$ et celui situé entre 22500 et 27000cm^{-1}, à une transition $^1\Sigma_g^+ \rightarrow {}^1\Delta_g$ [11].

Quant aux spectres de fluorescence, nous avons remarqué que l'intensité relative des bandes et leurs longueurs d'onde restaient inchangées, quelle que soit la longueur d'onde excitatrice. Ceci démontre que les transitions responsables de la luminescence sont issues du même niveau électroniquement excité et que l'excès d'énergie apporté est dissipé sous forme de transitions non radiatives [12]. Dans l'ensemble, les spectres sont assez bien structurés pour les quatre premiers chélates; par contre, celui correspondant à UO_2FOD OH O. phén. ne l'est pas du tout. La substitution d'un ligande par un groupement hydroxyle perturbe donc de façon très nette l'environnement de l'ion uranyle. Pour chaque spectre, l'écart moyen entre les quatre bandes principales représente la fréquence de vibration symétrique de l'ion uranyle, mesurée par spectroscopie Raman et Infra-rouge, [4], (tableau IV).

TABLEAU IV

Composés	Position des bandes (cm^{-1})	$\Delta\nu$ (cm^{-1})	Ecart moyen	Vibration symétrique $\nu_1(\Sigma_g^+)$
$UO_2(TTA)_2$ O.phén.	19 102	864		
	18 238	802	835	845
	17 436	839		
	16 597			
$UO_2(TTA)_2$ guanine	19 076	862		
	18 214	829	824	840,5
	17 385	782		
	16 603			
$UO_2(TTA)_2$ Pip.	19 474	835		
	18 639	836	841	-
	17 803	854		
	16 949			
$UO_2(FOD)_2$ O.phén.	19 058	840		
	18 218	851	850	-
	17 367	860		
	16 507			

Ainsi, chaque bande s'interprète comme due à la transition du premier niveau électroniquement excité à différents niveaux de vibration de l'état fondamental. L'équidistance des bandes n'est pas rigoureuse, mais ceci vient du fait que l'ion uranyle ne vibre pas comme un oscillateur harmonique. La constante d'anhamonicité dans l'état fondamental est de l'ordre de $10cm^{-1}$ [13]. L'examen des spectres de fluorescence montre que les sous-niveaux apparaissent peu, ce qui ne permet pas de faire une identification précise, comparativement à certains sels d'uranyle. Cette absence de structure fine est due très probablement au fait que dans le cas des chélates, l'excitation de l'ion uranyle s'effectue par transfert d'énergie intramoléculaire [2]. De plus, il est assez difficile de fixer la bande de résonance (0-0) correspondant à la transition entre le premier état électroniquement excité (v'= 0) et l'état fondamental (v" = 0), car elle est mal définie. Cette bande de fluorescence de plus courte longueur l'onde doit être commune au spectre d'émission et à celui d'absorption.

A la lumière de ces résultats, le mécanisme de fluorescence peut être schématisé de la façon suivante (Fig. 12).

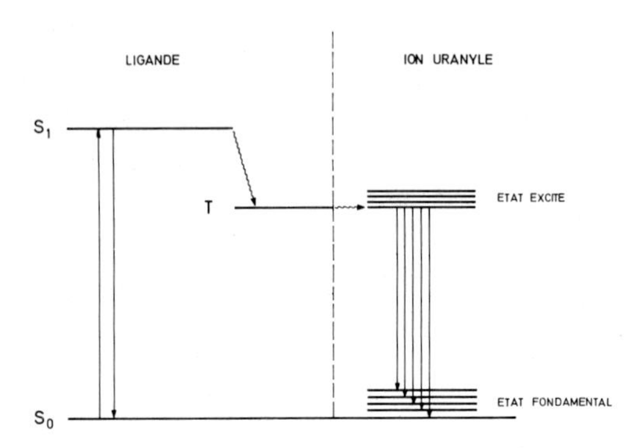

Fig. 12 – Représentation schématique du mécanisme de fluorescence des chélates d'uranyle

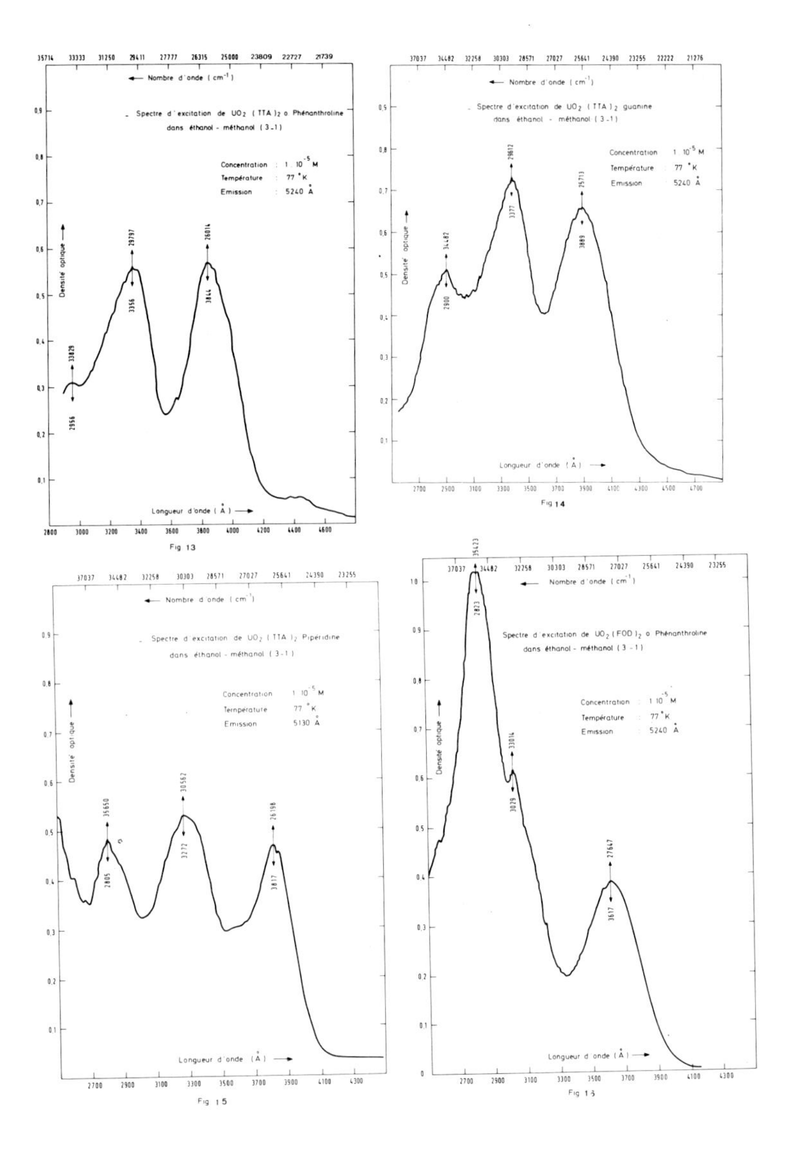

Fig 13

Fig 14

Fig 15

Fig 16

La fluorescence de l'ion uranyle sera d'autant plus intense que les transferts $S_1^* \rightarrow T_1^*$ et triplet-ion se feront dans de bonnes conditions, c'est-à-dire sans compétition avec d'autres processus non radiatifs. En particulier, pour qu'une émission stimulée puisse se produire éventuellement, il est indispensable que le niveau triplet ait sensiblement la même énergie que celle d'un niveau vibrationnel de l'ion uranyle. Dans le cas du chélate $UO_2(TTA)_2$ O.phén., la position du triplet ($\approx 21142 cm^{-1}$) est favorable, comme le montre la figure 11.

La largeur à mi-hauteur de la bande la plus intense est de cent angströms environ pour chacun des chélates à deux ligandes. Signalons que pour certains chélates de terres rares, le tétrabenzoylacétonate sodium de terbium en particulier, cette largeur de raie est de l'ordre de sept angströms seulement [14].

Enfin, signalons aussi que ces chélates en milieu alcoolique, fluorescent peu à la température ambiante, comparativement aux chélates de terres rares. Ceci peut être dû à une désactivation non radiative par l'alcool. Or, nous savons que l'uranyle fluoresce entre 5000 et 6500Å, soit un domaine énergétique compris entre 57 et 43kcal.mole^{-1}, alors que les premiers états excités de l'alcool correspondent à une énergie très supérieure. Par conséquent, une désactivation par transfert d'énergie électronique ($U^* + A \rightarrow U + A^*$) est exclue, [15] car elle mettrait en jeu un processus hautement endothermique. On peut penser que cette désactivation s'effectue selon un processus de "quenching" physique ou chimique, ce qui expliquerait l'influence très marquée de la température.

Excitation

Les spectres d'excitation (Fig. 12 à 16) comme ceux d'absorption sont peu structurés. Ils sont constitués d'une part, par les bandes du ligande, d'autre part, par celle correspondant à la transition électronique $n - \pi^*$) ; avec dans tous les cas un léger déplacement vers les faibles longueurs d'onde. Ainsi, la base ne transfère pas son énergie d'excitation à l'ion uranyle par l'intermédiaire du ligande.

Durées de vie

Les déclins de fluorescence ont été mesurés à l'aide d'un dispositif d'excitation pulsée (lampe flash, durée du pulse :$1,5.10^{-6}$s) et d'un photomultiplicateur relié à un oscilloscope. Pour effectuer des mesures sur une seule raie, nous avons intercalé un monochromateur. Le tableau V mentionne les

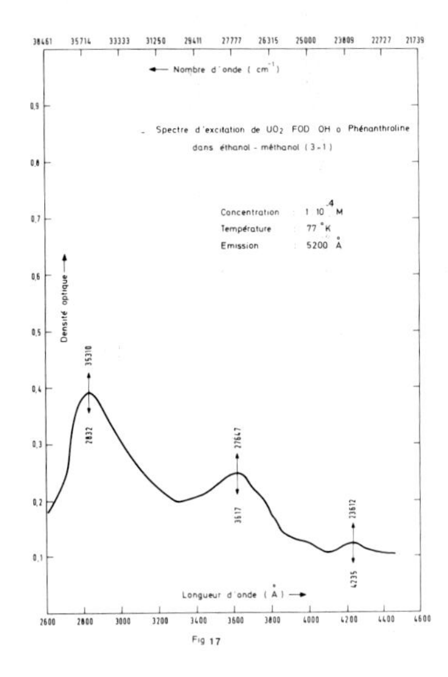

Fig 17

durées de vie mesurées sur quelques raies choisies arbitrairement.

TABLEAU V

Composés	Longueur d'onde de la raie (Å)	Durée de vie (s) Concentration: 1.10^{-5}M Température: -195°C	Intensité totale (unité arbitraire) Concentration: 1.10^{-4}M Température : -184°C
$UO_2(TTA)_2$ O.phén.	5 235 5 483	$1,2.10^{-4}$ $1,1.10^{-4}$	2,6
$UO_2(TTA)_2$ guanine	5 242	$1,1.10^{-4}$	1,6
$UO_2(TTA)_2$ pip.	5 365	$1,2.10^{-4}$	(solubilité trop faible)
$UO_2(FOD)_2$ O.phén.	5 489	$1,6.10^{-4}$	1,3
UO_2 FOD OH O.phé.	-	-	1

Les valeurs obtenues pour la durée de vie ne diffèrent pas de celles correspondant à des mesures globales effectuées par ailleurs.

Nous avons aussi mesuré les durées de vie de ces composés en solution dans l'éther et sur les poudres à la température de -195°C. Les valeurs obtenues sont pratiquement identiques à celles mentionnées dans le tableau V. Ceci laisse donc supposer qu'à cette température, la vitesse de "quenching" par les alcools est négligeable.

Dans tous les cas, les déclins de fluorescence mesurés sur une seule raie sont exponentiels.

Enfin, nous pouvons calculer la durée de vie de la transition $n - \pi^*$. En effet, l'inverse de la probabilité d'une transition ou temps de vie moyen (t_o) peut être évalué indirectement par rapport à la force de l'oscillateur, en utilisant l'expression suivante :

$$t_o = 1,5.\ \bar{\nu}_{max.}^{-2}\ .\ f^{-1}$$

avec $\bar{\nu}_{max.}$, la fréquence en cm^{-1} par rapport au maximum de la bande d'absorption. Les valeurs obtenues sont de l'ordre de 1.10^{-8} s.

Ainsi, on voit que le transfert d'énergie par l'intermédiaire de l'état triplet est à la fois rapide et efficace.

Essai de réalisation d'un laser à chélates d'uranyle

Pour effectuer ces essais, nous utilisons une cavité confocale dont le schéma de principe est représenté sur la figure 18. Les miroirs ont subi

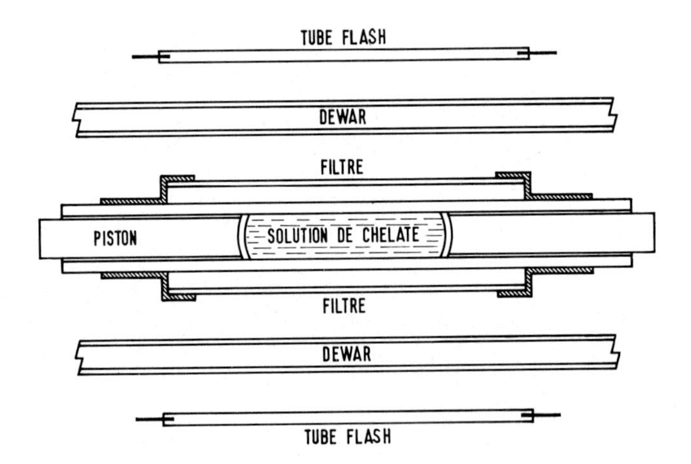

Fig. 18 _ CAVITE CONFOCALE
Les deux pistons peuvent coulisser librement

préalablement un traitement en couches multidiélectriques centrées sur la longueur d'onde de 5240 Å. La puissance électrique fournie par les flashs est de 1200 joules environ. Afin d'avoir des solutions plus concentrées, le solvant utilisé est composé de diméthylformamide (23,8%, d'éthanol (66,6%) et de propanol (9,5%).

L'inversion de population par unité de volume, nécessaire pour obtenir l'amplification optique, peut-être calculée. Nous prendrons comme exemple le chélate UO_2 $(FOD)_2$ O. phén. qui est le plus soluble des composés étudiés et nous supposerons que l'émission stimulée se produit dans des conditions sensiblement identiques à celles des chélates de terres rares (B_4EuNa ou B_4Eu pipéridine, avec B : benzoylacétone), [14, 16 et 17]. Pour effectuer ces calculs, nous utiliserons la relation suivante [18]:

$$(N_2 - g_2/g_1 N_1) = \frac{8\pi \, \nu^2 \, \tau \, \Delta\nu}{C^2 \, n} \, \alpha$$

en

- ν : la fréquence de la raie : $5,71.10^{14}$ HZ
- $\Delta\nu$: la largeur de la raie : $10,89.10^{12}$ HZ
- ζ : La durée de vie radiative d'émission spontanée de l'ion ixcité : 5.10^{-4} s
- c : la vitesse dé la lumière : 3.10^{10} $cm.s^{-1}$
- n : l'indice du milieu : 3.10^{10} à - 140°C
- α : les pertes inhérentes au milieu, en admettant que celles-ci soient dues uniquement à la diffusion Rayleigh : $1,50.10^{-2}$ $Neper.cm^{-1}$

On trouve ainsi la valeur de $9,9.10^{18}$ $ions.cm^{-3}$, soit $1,6.10^{-2} M.1^{-1}$. Il faut sans doute une concentration nettement supérieure pour obtenir l'émission stimulée. En effet, les bandes d'absorption du chélate sont très intenses et par suite, limitent la profondeur de pénétration de l'énergie de pompage. Ce phénomène se traduit en général par une diminution du rendement relatif au pompage optique et à l'inversion de population. Afin de pouvoir les comparer , nous avons rassemblé dans le tableau VI, les différentes caractéristiques des chélates d'Europium (B_4EuNa) et d'uranyle $UO_2(FOD)_2$ O. phén.

TABLEAU VI

Chélate / caractéristiques (à - 140°C)		B_4 Eu Na	$UO_2(FOD)_2$ O. phén. hypothèse d'une émission stimulée
Concentration minimale nécessaire pour obtenir l'émission stimulée	théorique	$5,5.10^{17}$ ions cm^{-3}	$9,9.10^{18}$ ions.cm^{-3}
	réelle	$6.10^{18}-1,1.10^{19}$ ions. cm^{-3}	-
Longueur d'onde de l'émission		6111Å	5247 ou 5489Å
Largeur de la raie d'émission		.6Å	≃ 100Å
Déclin de fluorescence		8.10^{-4}s	$1,6.10^{-4}$s
Durée de vie radiative		8.10^{-4}s	≃ 5.10^{-4}s
Excitation		2460Å-3190Å	2647Å-3567Å
Nombre de modes ($8\pi^2$ $\Delta\lambda/\lambda^4$)		$0,1.10^{12}$	

L'examen de ces différents paramètres laisse espérer l'obtention de l'effet laser avec les chélates d'uranyle, bien que la largeur de raie ne soit pas tellement favorable.

En effet, la condition d'oscillation d'une cavité laser fait intervenir le produit $\zeta\Delta\nu$, où ζ est la durée de vie radiative associée à la largeur spectrale de la raie considérée $\Delta\nu$. Or, si la durée de vie est relativement petite, la largeur spectrale est grande, comparativement à celle des chélates de terres rares.

D'autres facteurs non favorables peuvent aussi intervenir, comme par exemple :

- absorption transitoire à partir du niveau métastable,
- saturation du niveau triplet dû à un transfert trop lent vers l'ion uranyle
- instabilité du chélate en solution.

Il est vraisemblable que dans ces conditions une augmentation de la puissance de pompage ne pourrait pas compenser de telles pertes.

Il faut également noter que le calcul du taux d'inversion est délicat à cause de la variation du coefficient d'absorption avec la longueur d'onde.

Comme nous l'avons déjà signalé, le pompage optique est très inégal en profondeur ; de plus il dépend de la géométrie du système d'excitation, de la puissance disponible du tube flash et aussi de la durée de vie réelle du niveau métastable.

Ces premiers essais ne se traduisent pas encore par des résultats positifs. Les travaux en cours portent principalement sur la purification des produits et l'augmentation de la puissance de pompage.

CONCLUSIONS.-

Les chélates d'uranyle que nous avons étudiés présentent en solution et à basse température une fluorescence verte intense. L'excitation de l'ion uranyle s'effectue par transfert d'énergie intramoléculaire ; la base n'intervient pas dans le mécanisme de fluorescence. Les premiers essais de réalisation d'un laser à chélates d'uranyle n'ont pas encore donné de résultats positifs. Une augmentation de la puissance de pompage et une purification plus poussée des produits s'avèrent nécessaires.

REMERCIEMENTS.-

Les auteurs remercient MM. MEYER, ASTIER et CROZET du groupe de physique moléculaire de l'Ecole Polytechnique pour leur collaboration au cours des essais.

BIBLIOGRAPHIE.-

[1] L.I. KONONENKO et L.M. BURTNENKO, Russian Journal of Inorganic Chemistry (1968), 13, (1) , pp. 101-104

[2] L.I. KONONENKO, N.S. POLUEKTOV et L.M. BURTNENKO, Opt. Spek. Trosk, (1967), vol. 23, N° 4, p. 598

[3] M. LUCE, G. BESNARD et G. FOLCHER,(publication à paraître).

[4] Y. MACHETEAU, travaux non publiés.

[5] R.L. BELFORD, A.E. MARTELL et M. CALVIN, J. Inorg. Nucl. Chem. (1960), vol. 14, pp. 169-178.

[6] R.L. BELFORD, A.E. MARTELL et M. CALVIN. J. Inorg. Nucl. Chem. (1956), vol.2,pp. 11-31

[7] W.F. SAGER, N. FILIPESCU et F.A. SERAFIN, J. Phys. Chem. (1965) 69, n° 4, pp. 1092-1100.

[8] J.T. BELL, et R.E. BIGGERS, Journ. Mol. Spectroscopy (1965), 18, 247 et (1968), 25, 312-329.

[9] S.P. Mc. GLYNN et J.K. SMITH, J. Mol. Spectroscopy (1961), 6, 164

[10] S. De JAEGERE et C. GÖRLLER-WALRAND, Spectrochimica Acta, (1969), 25A, 559-568

[11] C. GORLLER-WALRAND et L.G. VANQUICKENBORNE, The Journal of Chemical Physics (1971, 54, n° 10, 4178-4185.

[12] P. PRINGSHEIM, Fluorescence and Phosphonescence, Interscience, New-York, (1951)

[13] L.H. JONES, Spectrochim. Acta, (1958), 10, 395

[14] Y.H. MEYER, Le Journal de Physique, (1966), 27, pp. 415-421

[15] R. MATSUSHIMA, J. Am. Chem. Soc. (1972), 94:17, 6010-6016.

[16] O. De WITTE et Y. MEYER, J. Chem. Phys. (1967), 64, 86

[17] Y. MEYER, R. ASTIER et J. SIMON, C.R Acad. Sc. Paris, (1964), t 259 4604

[18] YARIV et GORDON, Proc. I.E.E.E, (1963), 51, 4.

DISCUSSION

Y. MACHETEAU, A. COSTE, M. LUCE, P. RIGNY

Mme Reisfeld - I would like to comment on the origin of the low wave absorption and fluorescence. Professor Jørgensen and myself have performed lately calculation on the M.O. origin of these absorption and fluorescence and we come to the conclusion that the transitions responsible for these phenomena are the Laporte forbidden transition $\pi\mu \rightarrow \delta_g(\phi_g)$ rather than the spin forbidden transition.

Y. Macheteau - We did not try any quantitative interpretation of the origin of the fluorescence of the uranyl ion; the important discrepancies which exist in the literature as to assignments of the energy levels in this ion dissuaded us to do so.

S. Leach - La solubilité des chélates dans un mélange d'alcools méthylique et éthylique est faible. Est-ce que vous êtes certain que vous avez une bonne dispersion dans la solution rigidifiée et comment l'avez-vous testé?

Y. Macheteau - Les spectres d'absorption dépendent peu de la température et sont pratiquement les mêmes pour les solutions rigidifiées et pour les solutions fluides; d'un autre côté, les temps de déclin de la fluorescence mesurés à des concentrations de chélates allant de 10^{-5} à 10^{-3} molaire, restent invariables dans ce domaine. Nous interprétons ces deux observations comme indiquant l'absence d'effets importants d'agrégations éventuelles.

P. Rigny - The wonder with chelates is not that the uranyl chelates do not give rise to a laser effect, but rather that the rare earth chelates do give a laser effect. This can be traced to the fact that the 4f electrons are localised very close to the nucleus and thus well shielded from the environment. The present work shows that the 5f electrons are not as well shielded since they extend much further away from the metal ion.

EXCITED STATE ABSORPTION SPECTROSCOPY OF 1.1'-BINAPHTHYL RELATED TO LASING ABILITY

J. LANGELAAR
University of Amsterdam, Laboratory for Physical Chemistry,
Nieuwe Prinsengracht 126, Amsterdam, The Netherlands

INTRODUCTION

The relation between molecular structure, physical properties, and laser ability of organic molecules has become a subject of several studies [1-5], since the work of Drexhage [6]. For aromatic hydrocarbons the essential information can be obtained from spectroscopic data amoung which the excited state spectra by nanosecond laser flashphotolysis [7]. From the photophysical properties of the parent molecule it can be deduced which substitiuent has to be chosen in order to increase the laser ability as has been shown for anthracene [3-5]

Naphthalene, the next aromatic hydrocarbon of interest because it gives a fluorescence in the UV region around 330 nm, is less suitable. In contrast to anthracene, in naphthalene the transition from the fluorescent singlet state to the ground state is sign forbidden [8]. The second excited singlet state (^1L_a), which lies only 4000 cm^{-1} above the fluorescent state (^1L_b), however, has a large transition probability to the ground state. Hence, in order to obtain laser ability in naphthalene derivatives, one has to shift the ^1L_a level below the ^1L_b level to obtain the same level ordening as in anthracene.

It is well known that this situation is obtained in excimers (excited dimers) due to the interaction between an excited molecule and a ground state molecule. Naphthalene excimers can, however, only be obtained in small concentrations and are unstable at room temperature [9].
Two naphthalene moieties joined by a C-C bond, i.e. 1.1'-binaphthyl, do not possess these disadvantages, while the spectroscopic properties are similar to those of the naphthalene excimer.

THE SPECTROSCOPIC PROPERTIES OF 1.1'-BINAPHTHYL

1.1'-binaphthyl was studied by Friedel et.al. [10] and Hochstrasser

[11] in fluid solution. The absorption spectrum in fluid and rigid solution is identical to that of naphthalene, which indicates that the resonance interaction between the two moieties in the ground state is very small.

In contrast to rigid solutions the fluorescence spectrum in fluid solutions consists of a broad emission considerably shifted to the red. A change of the dihedral angle in the excited state is responsible for the large stabilisation of the fluorescent state of 1.1'-binaphthyl. A SCFMO-CI calculation as a function of the dihedral angle Θ (fig. 1) in combination with nanosecond excited state spectra indeed indicates that the geometry of 1.1'-binaphthyl in the fluorescent state in fluid solutions is nearly coplanar [12,13].

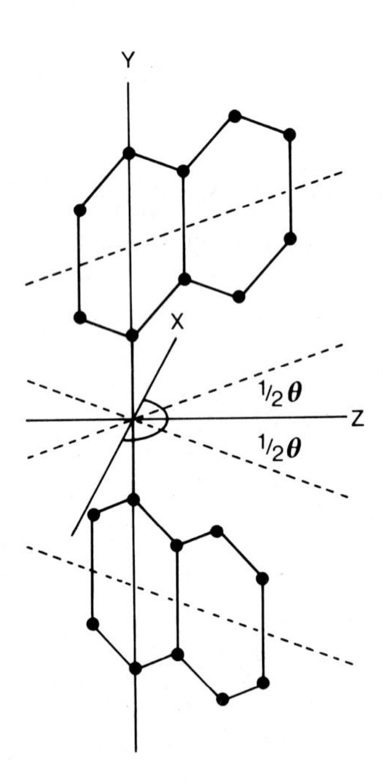

Fig.1. 1.1'-binaphthyl. The angle of twist Θ is used as a parameter for the SCF MO calculations.

The change of the level ordening and hence the change of the character of the emiting state (^1L_a-level) introduces the following properties in fluid solutions which will increase the laser ability of 1.1'-binaphthyl with respect to naphthalene:

I High fluorescence quantum yield (∅ = 0.77; $\tau_f \simeq 3.5$ ns.)

II Broad fluorescence spectrum i.e. 330 - 420 nm

III No overlap between absorption and emission spectrum due to the change in geometry after excitation (fig.2)

IV High absorption cross section at 265 nm (4 x Nd^{3+})

V No severe excited singlet and triplet quenching over the spectral region of the fluorescence [14]

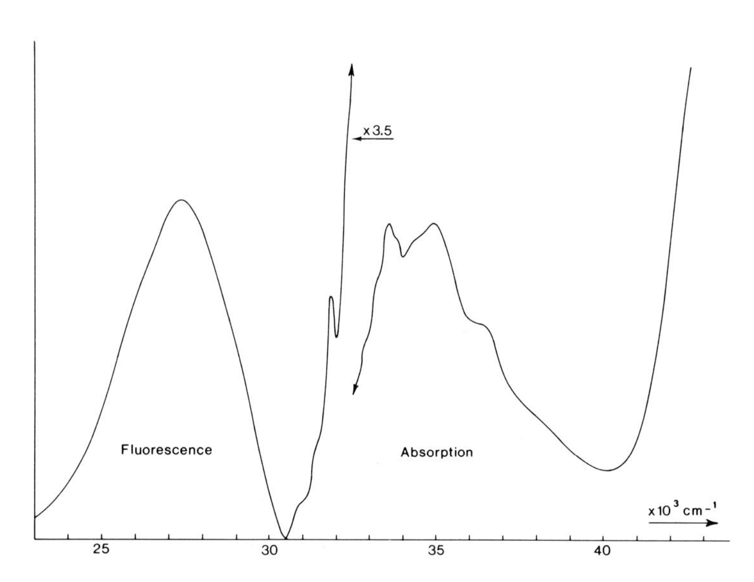

Fig.2 Absorption (77°K) and fluorescence (293°K) spectrum of 10^{-4}M 1.1'-binaphthyl in MTHF

THE LASER ABILITY OF 1.1'-BINAPHTHYL

Laser action of 1.1'-binaphthyl is expected according to the spectroscopic properties described above.

Experiments were carried out by using an oscill/ampl. Q-switched quadrupled Nd^{3+}-glass laser system [7] as excitation pump (300 KW; 20 ns at 265 nm). The dye laser consists of a dye cell, 1 cm square in cross section transversely pumped by the 4th harmonic Nd^{3+}-glass laser pulse, and two plain mirrors, one High Reflection and one High Transmission. The glass-laser pulse is focused by a cylindrical lense into a line that coincides with the optical axis of the dye laser. A 3.10^{-3} M solution of 1.1'-binaphthyl in ethanol is used as active medium. The observed laser emission around 360 nm is detected with an EG & G radiometer system.

The energy conversion efficiency for 1.1'-binaphthyl with the simple dye laser arrangement is about 0.3%; with the same configuration we observed for the well-known UV dye PBD a conversion efficiency of 0.5% and for Rhodamine 6G (also excited at 265 nm) 3%. The pulse duration (FWHM) of the 360 nm dye laser emissions was 16 ns, hence considerably shortened as compared with the pump pulse (20 ns). Due to the low repetition rate of the pump laser (1 pulse per 2 minutes) and its instability at 265 nm, the use of a more sophisticated reonator configuration with a tuning element is not very attractive. For experiments with a tuning element in the dye laser cavity a high repetition pump source i.e. a N_2-laser is desired.

In order to shift the absorption spectrum of 1.1'-binaphthyl to the red into the region of the N_2-laser wavelength (337 nm) one has to add a substituent preferably at the 4-positions. For instance 4.4'-DI(N-Butoxy) - 1.1'-binaphthyl (DBN) has a good absorption cross-section at 337 nm. Moreover, DBN has a quantum yield of $\emptyset$ = 0.97, which makes it very attractive as laser dye [15]. Unfortunately, it is not a standard available chemical, so that it has to be synthesized. Work on this is in progress.

REFERENCES

1 M.J. Weber, M. Bass, IEEE J. Quantum Electr., QE-5(1969)175

2 C. Rullière, M.M. Denariez-Roberge, Canad. J. Phys., 51(1973)418

3 J. Ferguson, A.W.H. Mau, Chem. Phys. Lett., 14(1972)245.

4 Th.G. Pavlopoulos, IEEE J. Quantum Electr., QE-9(1973)510.

5 J. Langelaar, Appl. Phys. 6(1975)61

6 K.H. Drexhage, in F.P. Schäfer (editor) Topics in Applied Physics, Vol.1 Dye Lasers, Springer-Verlag Berlin, Heidelberg, New York 1973. Chapter 4.

7 D. Bebelaar, Chem. Phys. 3(1974)205

8 P. Pariser, J. Chem. Phys. 24(1956)250

9 J.B. Aladekomo, J.B. Birks, Proc. Roy. Soc. A, 284(1965)551

10 R.A. Friedel, M. Orchin, L. Keggel, J. Am. Chem. Soc. 70(1948)199

11 R.M. Hochstrasser, Can. J. Chem. 39(1961)459

12 M.F.M. Post, J. Langelaar, J.D.W. van Voorst, Chem. Phys. Lett., 32(1975)59

13 M.F.M. Post, SCFMO-CI Calculations, in preparation

14 M.F.M. Post, unpublished results

15 I.B. Berlman, Handbook of Fluorescence Spectra of Aromatic Molecules, Academic Press, New York, 2nd ed. p.353

DISCUSSION

J. LANGELAAR

C. Rullière - What is the wavelength laser emission of binaphtyl and whatit, in the anthracene, the absorption coefficient of $S_1 \rightarrow S_n$ in the fluorescence region?

J. Langelaar - The laser wavelength of 1-1'-binaphtyl is about 360 nm. The molar extinction coefficient of a $S_1 \rightarrow S_n$ transition in the fluorescence region of anthracene, if there is any, is far below 4 000 l mol^{-1} cm^{-1} (D. Bebelaar, Chem.Phys., 3(1974), 205).

J. Joussot-Dubien - Concerning the exciplex formation using 1-1'-binaphtyl which leads to laser action, have you looked into the possibility of extending the chain which links the two naphtalene rings and relate it to laser action?

J. Langelaar - In our study, we have shown that the spectroscopic properties of naphtalene excimers and 1-1'-binaphtyl in fluid solutions is quite similar. We therefore do not expect any change by increasing the chain length between the two naphtalene moieties as far as it is concerned with their laser possibilities. Dr. Zachariasse (M.P.I. Göttingen) is studying the spectroscopy of these compounds. May be that one of his colleagues here can give more information.

H. Staerk - Work of this kind is presently in progress in our laboratory (W. Kühnle and K.A. Zachariasse). In these studies aromatic moelcules like naphtalene or pyrene are linked together by a $(CH_2)_n$ chain with n = 1...20. The most interesting outcome of the investigation of excimer formation is, e.g. in the example of pyrene, that excimer fluorescence is observed with 3, 5, 6 or more CH_2 groups but not with 4 chain elements between the moieties. Thus, a primary requirement for studying laser action is to find the number n which leads to a sterically optimal configuration with high excimer fluorescence yield and under these conditions a radiative lifetime which is appropriate for lasing action. This is also important for our studies of intramolecular heteroexcimer formation and their lasing ability e.g. : of 1-(9- anthryl)-n-(p-dimethylaniline)-propane (W. Kühnle, M. Schulz, H. Staerk, A. Weller).

M.W. Windsor - What is the equilibrium angle between the two naphtalene moieties in the ground state of 1-1'-binaphtyl? Also, what is the depth of the rotational well? We have recently made π-orbital calculation of twist angles in the crystal violet molecule.

J. Langelaar - The equilibrium angle between the two naphtalene moieties in the ground state is around 90°; most probably between 80 and 90° according to investigations in the crystalline phase.

J.D.W. van Voorst - No value for the energy of the rotational barrier can be given, since this energy is composed out of : 1) the charge in the π-electronic energy and 2) the repulsion of the hydrogen atoms. Only the first energy was of relevance for the spectroscopic properties studied and has been calculated. For the second part no experimental data are available in literature.

J. Langelaar - We only calculated difference between π electron energies and assumed that other contributions are identical in two states and thus will cancel.

M. Stockburger - Is the oscillator strength of the 1-1'-binaphtyl, in liquids where the conformation in the ground and first excited singlet states is much different, comparable with that of a strong transition from a L_a electronic state?

J. Langelaar - This is indeed the case. The values of the oscillator strengths of the strong transitions in 1-1'-binaphtyl in fluid solutions are comparable to the ^1L_a and ^1B_b transitions in naphtalene.

SUBSTITUENT EFFECTS ON THE ABILITY OF MOLECULES TO LASE

C. RULLIERE[*], J.C. RAYEZ[*], M.M. DENARIEZ-ROBERGE[+], J. JOUSSOT-DUBIEN[*]

[*]Laboratoire de Chimie Physique A - Université de Bordeaux I
33405 TALENCE (France)[ŧ]

[+]L.R.O.L. Département de Physique - Université Laval - QUEBEC 10 - (CANADA)

INTRODUCTION

Due to the extensive use of lasers in various technical areas, the search for new lasing dyes has recently reached unprecedented proportions. No less than three hundred molecules can be made to lase when pumped by means of Q-switch lasers. Although the spectral range in which these dye molecules can be made to lase covers the 11000 to 3400 Å range, only twenty of them can be made to lase continuously, and their emissive spectral range is limited between 8000 to 4200 Å. In order to extend the wavelength range in which dyes can lase, the reasons why a molecule can lase must be understood.

Until recently dyes having lasing properties were obtained on a trial and error basis. However, it has been noted by DREXHAGE [1] and PAVLOPOULOS [2] that it is possible to predict the lasing ability of molecules by considering their molecular structure.

Recent experimental data [3] now enable us to show that substitution at specific positions on the molecular frame improves the lasing ability of dye molecules. Before presenting some conclusive examples, we shall recall the principal characteristics of a lasing dye.

A laser action is essentially dependent on the emission cross-section of the molecule, which is related to the singlet state radiative lifetime and to its fluorescence quantum yield. Experimentally it has been found that the radiative lifetime must be smaller than ten nanoseconds. Consequently the fluorescence process must arise from an allowed transition. Furthermore the

---

[ŧ]Equipes de Recherches Associées au C.N.R.S. n° 167 et 312

fluorescence quantum yield should be higher than 0.5 and the light emitted by a molecule must not be absorbed by neighboring molecules in excited singlet or triplet states. Therefore the $S_1 \rightarrow S_n$ and $T_1 \rightarrow T_n$ absorption spectra must not overlap with the fluorescence spectrum.

Column 1 of fig. 1 shows the lasing characteristics of pyrene, naphthalene, trans-stilbene and trans-butadiene. It is seen that these dyes have extremely poor lasing characteristics. However in column 2 we show substituted derivatives of the same molecules from which we have obtained good lasing action. The substituent effect is seen to be the predominant factor in enabling these molecules to acquire lasing properties and this will be discussed more in detail in the following sections.

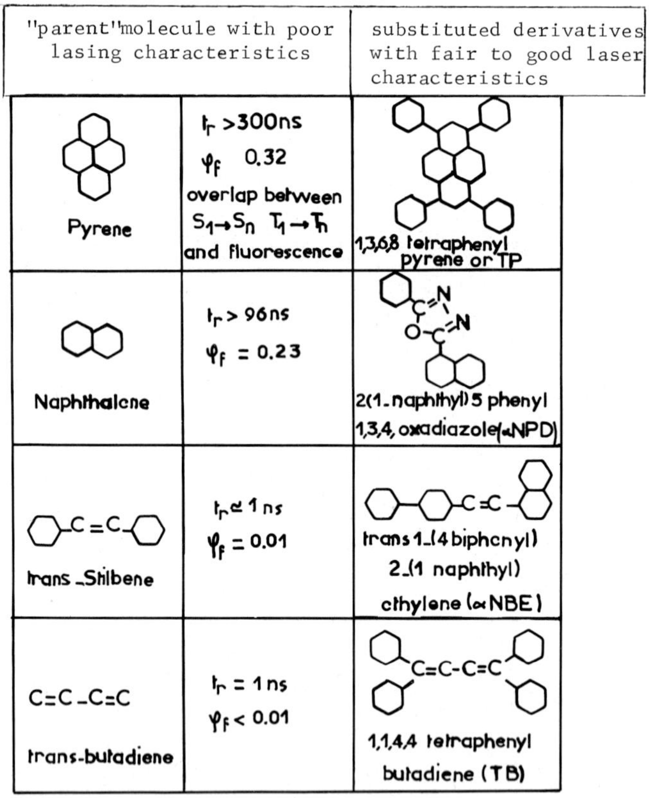

"parent" molecule with poor lasing characteristics		substituted derivatives with fair to good laser characteristics
Pyrene	t_r >300ns φ_f 0.32 overlap between $S_1 \rightarrow S_n$ $T_1 \rightarrow T_n$ and fluorescence	1,3,6,8 tetraphenyl pyrene or TP
Naphthalene	t_r > 96ns φ_f = 0.23	C=N O-C=N 2(1-naphthyl) 5 phenyl 1,3,4, oxadiazole (αNPD)
C=C trans-Stilbene	$t_r \simeq$ 1 ns φ_f = 0.01	C=C trans 1-(4 biphenyl) 2-(1 naphthyl) ethylene (αNBE)
C=C-C=C trans-butadiene	t_r = 1 ns φ_f < 0.01	C=C-C=C 1,1,4,4 tetraphenyl butadiene (TB)

Fig. 1 : Illustration of the influence of substituents on the lasing ability of molecules. Column 1 : "parent molecules" which do not lase. Column 2 : substituted derivatives which lase. t_r is the radiative lifetime of the first excited singlet state and φ_f the fluorescence quantum yield. Values taken from literature [4,5,6,7.]

I - EFFECTS OF SUBSTITUENTS ON THE TRANSITIONS INVOLVED IN LASER EMISSION

A : Pyrene

Pyrene is seen to lack all the necessary characteristics of a good lasing dye. As shown in fig. 1, its radiative lifetime, t_r, is much too long and there is extensive overlap between its $S_1 \rightarrow S_n$ and $T_1 \rightarrow T_n$ absorption spectra[7,8] and its fluorescence spectrum.

The reason for the inhibiting effect of long radiative lifetime and spectra overlap on lasing ability of pyrene can be better understood from an examination of its various transitions and associated polarizations. Fig. 2 shows the energetic and polarisation scheme of the various transitions in pyrene involved in laser action.

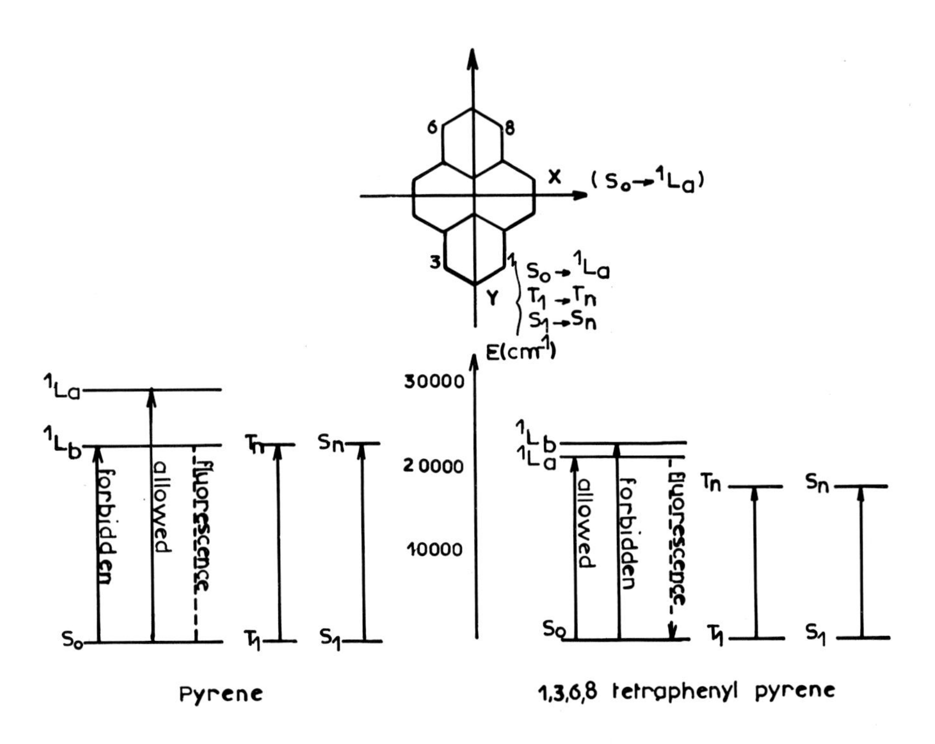

Fig. 2 : Scheme of different transitions of pyrene involved in laser action.

We see that the radiative lifetime of this molecule is long because the first excited singlet state, from which the emission occurs, is ^1L_b symmetry and thus the $^1L_b \rightarrow S_o$ transition is symmetry forbidden. We also see that the allowed transitions $S_o \rightarrow {}^1L_a$, $T_1 \rightarrow T_n$ and $S_1 \rightarrow S_n$ are polarized along the long axis (Y) of the molecule and that only the $S_o \rightarrow {}^1L_b$ transition is polarized along the short axis (X).

It is known that the addition of auxochromic substituents along a particular axis in a molecule affects preferentially the transitions polarized along that axis, whereas the transitions polarized perpendicular to that axis are practically unaffected. In pyrene, the addition of phenyl rings at the 1, 3, 6,8 positions has such an effect on the transitions polarized along the Y axis.

Examination of the fluorescence and absorption spectra of 1,3,6,8 tetraphenyl-pyrene (hereafter designated by T P) [4] shows that the $S_o \rightarrow {}^1L_a$ transition is red shifted by approximatively 4000 cm^{-1}, whereas the $S_o \rightarrow {}^1L_b$ transition is practically unaffected (see fig. 2). As a consequence the lower state is now the ^1L_a state which is just under the ^1L_b state, as shown by the small red shift (800 cm^{-1}) of the fluorescence spectrum. The fluorescence now originates from the ^1L_a state, and since the $^1L_a \rightarrow S_o$ transition is strongly allowed the radiative lifetime is much shorter (2.38 ns) than in pyrene, which favours lasing action.

At the same time the $T_1 \rightarrow T_n$ transitions of T P undergo the same effect. This is supported by both theoretical and experimental results [9] . The $^3B_{3g} \leftarrow {}^3B_{2u}+$ transition (X polarized) of pyrene is not affected by phenyl substitution, whereas $^3A_g - \leftarrow {}^3B_{2u}+$ (Y polarized) is red shifted, thus diminishing the overlap with fluorescence spectrum. We have, however, observed a new band in the blue region (23500 cm^{-1}) which overlaps with the fluorescence spectrum. Fortunately, its absorption coefficient is smaller than that of the $^3A_{g-} \leftarrow {}^3B_{2u}+$ transition of pyrene, and consequently its inhibiting effect on laser action is low when we consider the short lifetime and high fluorescence quantum yield (0.9) [4] of T P . Theoretical calculations show that the same situation is operating in the $S_1 \rightarrow S_n$ transitions [10] .

B - Naphthyl-phenyl-oxadiazole molecules

The situation in these molecules is not as simple as in the preceeding case. The $S_o \longrightarrow S_1$ and $T_1 \longrightarrow T_n$ spectra of 2-(1-naphthyl-5-phenyl 1,3,4 oxadiazole (α NPD) and 2-(2-naphthyl)-5-phenyl 1,3,4, oxadiazole (β NPD) are of a complex nature [10] . However, if we consider the NPD molecules as being composed of two chromophoric entities - a phenyl-oxadiazole (hereafter designated by PO) and a naphthyl group - , which influence each other, the spectral data of the molecules can be interpreted as a superposition of each moitie mutually perturbed. This interpretation is well supported by both theoretical and experimental results, [11] which show that the two chromophores are not coplanar, but that there is an important twist (to an angle of the order of 60°) at the bond joining the two chromophores (see fig. 3).

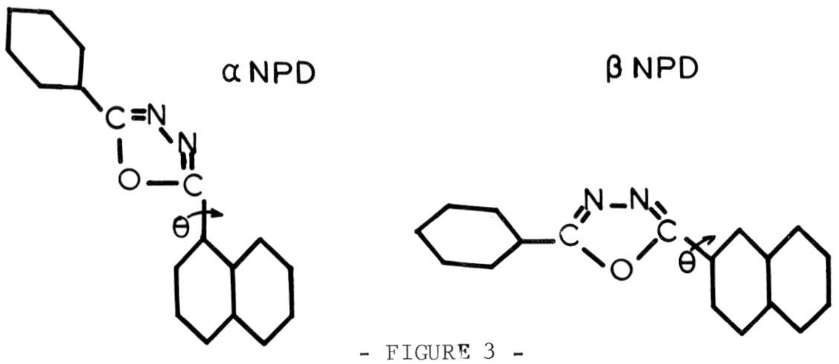

- FIGURE 3 -

According to this interpretation, the position of the PO chromophore on naphthalene is very important. In naphthalene the fluorescent state is ^1L_b and the forbidden $S_o \longrightarrow {}^1L_b$ transition ($t_r >$ 93 ns) is long axis polarized (X). Thus the perturbation induced by the PO chromophore on this transition will be less important in α NPD than in β NPD, and therefore the ^1L_b state will be energetically lower in β NPD than in α NPD. However, the perturbation of naphthalene on the first excited singlet state of PO will be more important in α NPD than in β NPD, because in β NPD the electrostatic repulsion between the nitrogen atoms of PO and the carbon atoms of naphthalene is less important, due to the distance between these atoms (see fig. 3). Theoretical calculations (P.P.P. method) show the first excited singlet state of both α and β NPD as originating from the PO chromophore and the second excited singlet state as being due to the ^1L_b state of naphthalene ; this calculation being carried for an angle of $\theta = 0°$.

The theoretically calculated energy gap ΔE between the first and second excited singlet states of α NPD and β NPD as a function of the angle θ is shown in fig. 4.

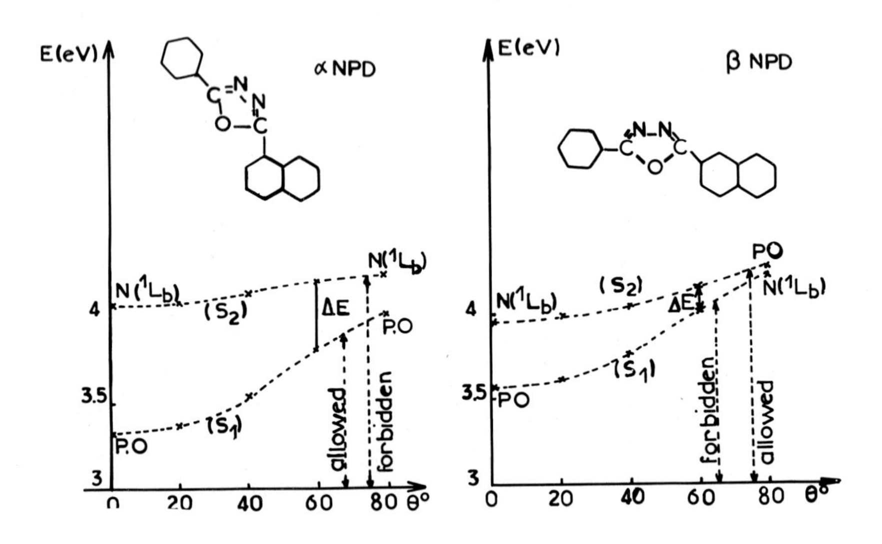

Fig. 4 Theoretical energy values of S_1 and S_2 of α NPD and β NPD as a function of twisting angle θ.

For the reasons described above, ΔE is larger in α NPD than in β NPD and at a larger angle there is even a crossing of states in the case of β NPD, which leads to a fluorescent state with ^1L_b character. Therefore the fluorescent transition $S_o \longrightarrow S_1$ will be more allowed in α NPD than in β NPD. This is well supported by the radiative lifetime values of these compounds (t_r = 2.4 ns in α NPD and t_r = 18 - 23 ns for β NPD [4,12]). Also the fact that there is no spectral overlap between the $T_1 \rightarrow T_n$ absorption bands and the fluorescence emission band [10] allows us to conclude that laser action must occur in α NPD.

II - EFFECT OF SUBSTITUENTS ON NON RADIATIVE PROCESSES : TRANS STILBENE AND TRANS BUTADIENE

The radiative lifetime of trans-stilbene and trans-butadiene are of the order of 1 ns [13,14]but unfortunately the competing non radiative photochemical processes are so fast that they strongly quench the fluorescence process, and the fluorescence quantum yield is less than 0.01 [5,6] . This is mainly due to the lack of rigidity of the molecule. It has long been known that the rigidity of a molecule must be high in order to have a high fluorescence quantum yield [1,14] . In molecules of low rigidity, electronic energy can be dissipated by rotation about single bonds [15] , such as in the case of trans-stilbene and trans-butadiene.

In trans-stilbene rotations about different bonds lead to the trans-cis photochemical isomerisation. This process is so fast that at room temperature in liquid solution the fluorescence quantum yield is 0.01 [5] . Similarly, trans-butadiene undergoes a rapid cyclization to form cyclobutene, and as a consequence trans-butadiene has no detectable fluorescence [6] .

However, as shown in fig. 5, if aromatic substituents are placed at the extremities of these molecules the non radiative losses are significantly reduced and high fluorescence quantum yields are obtained.

The role of the substituents in this case is to hinder the deformation of the molecule. This may be due to three factors. The larger size of the molecule may hinder the rotation around the different bonds. Similar results have been obtained in stilbene by simply increasing the viscosity of the solvent, thereby bringing the fluorescence quantum yield up to 0.6 [5] . Due to the introduction of the aromatic substituents the single and double bonds may lose their marked character in favor of a more delocalized π electronic system over the entire molecular frame. This has been observed, for example, in long chain polyenes where the alternating single and double bonds tend to acquire aromatic character with increasing chain length [17] . The larger size of the molecule may also hinder the formation of isomers by steric hindrance. The variation of the fluorescence quantum yields with the size of substituents,as shown in fig. 5 for stilbene derivatives, well support these hypotheses.

C=C–C=C	$\varphi_f < 1\%$
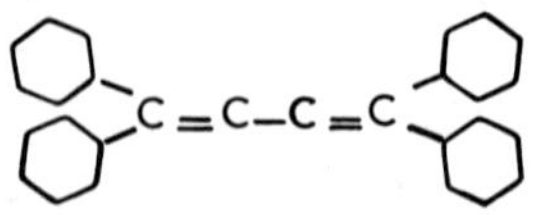 C=C–C=C	$\varphi_f = 60\%$
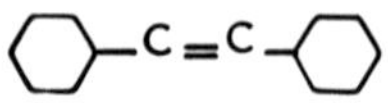 –C=C–	$\varphi_f < 1\%$
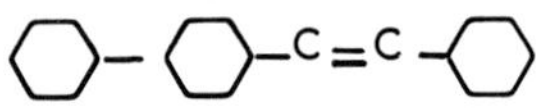 –C=C–	$\varphi_f = 60\%$
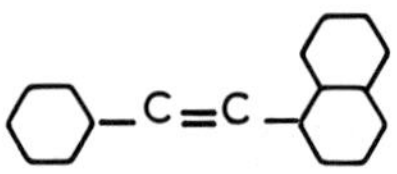 –C=C–	$\varphi_f = 80\%$
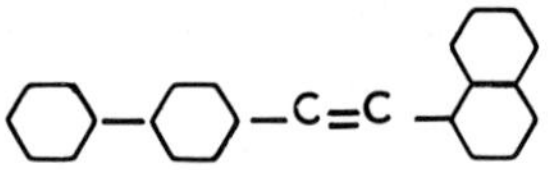 –C=C–	$\varphi_f = 95\%$

Fig. 5 : Role of substituents on the fluorescence quantum yield in butadiene and stilbene (taken from literature [4,5,16]).

III EXPERIMENTAL RESULTS

As expected, we have been able to obtain good lasing action in the substitued molecules discussed above when excited by the second harmonic of a Q-switch ruby laser (half pulse width 20 ns power = 1-5MW) The principal lasing characteristics are summarized in fig. 6. It should be noted that the lasing action in these molecules (except in α NPD) was obtained by simply using the cell walls as mirrors in the cavity, indicating a high gain.

1,3,6,8 tetraphenyl-pyrene (TP)	λ_e = 4220 to 4330 Å R = 4 %
1,1,4,4 tetraphenyl-butadiene* (TB)	λ_e = 5010 Å R = 4 %
2-(1-naphthyl)5-phenyl 1,3,4 oxadiazole (αNPD)	λ_e = 3725 and 3913 Å R = 1,5 %
trans-1(4-biphenyl)-2-(1-naphthyl) ethylene (αNBE)	λ_e = 4108 Å R ≤ 50 %

*Recent experimental results [18] show that in this compound, laser action at room temperature arises from a photoisomer of 1,1,4,4 tetraphenyl-butadiene. To obtain laser action from the normal form of 1,1,4,4 tetraphenyl-butadiene high viscosity solvent must be used. It seems then that in this compound the substituent effects are not as efficient as in other compounds.

Fig. 6: Lasing characteristics of substitued molecules discussed above
λ_e : laser wavelength emission. R : yield of laser action.

CONCLUSION

These examples have shown that it is possible, from consideration of the molecular structure, to predict lasing action in molecules. If the photophysical and photochemical characteristics of "parent" molecules are known, it is then possible to predict which substituent and which position will enhance lasing action. In this study we have limited ourselves to aromatic substituents, however several other substituents can induce similar effects on non lasing molecules. Considering the great variety of auxochromic groups available today, there is little room for doubt that chemists will be able to synthesize molecules that have been predicted to have good lasing characteristics.

REFERENCES

1. K.H. DREXHAGE - Topics in Applied Physics - F.P. SCHÄFER-SPRINGER VERLAG Berlin (1973)

2. T.G. PAVLOPOULOS - I.E.E.E. J. Quant. Elec. QE-9 (1973) 510

3. C. RULLIERE, M.M. DENARIEZ-ROBERGE. Opt. Comm. 7 (1973) 166

4. I.B. BERLMANN - Handbook of fluorescence spectra of aromatic molecules (Academic Press, New York, 1971).

5. S. SHARAFY and K.A. MUSZKAT J. Am. Chem. Soc. 93 (1971) 4119

6. R. SRINIVASAN, Adv. Photochem. 4(1966)113

7. J. LANGELAAR, J. WEGDAM-VAN BEEK, H.T. BRINKLAND and J.B.W. VAN VOORST Chem. Phys. Lett. 7 (1970)368

8. M.F.M. Post, J. LANGELAAR and B.W. VAN VOORST Chem. Phys. Lett. 10(1971) 468

9. C. RULLIERE, E.C. COLSON and P.C. ROBERGE (submitted for publication)

10. C. RULLIERE, Ph. D. Thesis. Laval University, June 1974

11. C. RULLIERE to be published.

12. P.C. ROBERGE - Private communication -

13. R.H. DYCH and D.S. Mc LURE J. Chem. Phys. 36(1972)2326

14. G.N. LEWIS and M. CALVIN Chem. Rev. 25 (1939) 272

15. G. OSTER and Y. NISHIJIMA J. Amer. Chem. Soc. 78(1956) (58)

16. J. KLUEGER, G. FISCHER, E. FISCHER, Ch. GOEDIHU and H. STEGEMEYER Chem. Phys. Lett. 8(1971) 279

17. C.A. COULSON Proc. Roy. Soc. 169 A (1939) 413 J. Chem. Phys. 7 (1939) 1069

18. C. RULLIERE, J.P. MORAND and J. JOUSSOT-DUBIEN (submitted for publication)

DISCUSSION

C. RULLIERE, J.C. RAYEZ, M.M. DENARIEZ ROBERGE, J. JOUSSOT-DUBIEN

C. Varma - The π electron spectra of pyrene and tetraphenyl-pyrene are theoretically not related in a simple manner, because the latter contains roughly twice as many π electrons as the former. Therefore I do not think one can extrapolate in a logical fashion to lasing parameters for tetraphenyl-pyrene starting from pyrene.

C. Rullière - The absorption profiles of pyrene and 1,3,6,8 tetra-phenyl-pyrene are similar; there is a difference of only 4000 cm^{-1}. (This is however sufficient to enhance lasing ability) for example in the $S_0 \rightarrow S_1$ transitions and they have similar polarization and symmetry. Also such correlations of molecular symmetry with electronic transitions have been successfully applied to benzene naphtalene, anthracene... and polyene chains, both from theoretical considerations of the symmetry of the molecule involved and experimental facts. We must note for example that pentacene has approximatively five times more π electrons than benzene.

C. Varma - All atoms have the same symmetry, still atomic spectra are widely differing.

C. Rullière - A good answer to this question should be too long.

S. Leach - Is the effect of substitution, in the case of stilbene, on the $S_1 \rightarrow S_0$ or on the $S_1 \rightarrow T_1$ non radiative transition? In the second case this might affect laser loss mechanisms of the $T^* \leftarrow T_1$ type.

C. Rullière - The effect of substituents in stilbene is to hinder the formation of cis-isomer which is the principal cause of the low quantum yield of fluorescence in this compound.

F.P. Schäfer - Another good example for the importance of the substituents in pyrene are the aminopyrenes, e.g. 1-acetylaminopyrene-3,6,8 trisulfonate, a good laser dye. I also would like to mention that we have found that several stilbenes rigidized by the introduction of saturated bridging groups have a high fluorescence quantum yield and are good laser dyes. These dyes were kindly given to us by prof. Guido Daub of the University of New Mexico, Albuquerque.

J. Langelaar - Do you have any information about the crystalline structure and the possibility of making single crystals of 1,3,6,8 tetraphenyl-pyrene? It is interesting to see if laser action can be obtained from single crystals, in order to combine the advantage of a dye laser (high gain and tunability) and of a solid state laser(high power).

C. Rullière - No, I don't have any information about this subject.

J. Faure - Absorptions which occur from the first excited singlet state may be taken into account as well as the fluorescence quantum yield. It would be interesting to measure the variation of related cross sections as a function of substituent effects.

C. Rullière - Yes. But the problem is that these compounds have relatively short singlet lifetime, of the order of 1 nanosecond. So we must use short pulse excitation that we did not have.

OPTICAL GAIN AND TRANSIENT ABSORPTIONS : THEIR CORRELATION IN ORGANIC AMPLIFIERS AND LASERS

Y.H. MEYER, P.FLAMANT, P. GACOIN, C. LOTH, R. ASTIER
Groupe de Physique Moléculaire, Ecole Polytechnique, 75230 Paris Cedex 05 (France)

SUMMARY

Amplification and attenuation of a laser beam, flash photolysis and laser action in solutions of molecules fluoresceing between 340 and 850 nm are used to study the relation between transient absorptions and optical gain. The gain coefficient (positive or negative) as a function of time is calculated in the case of non steady state populations for a linearly rising excitation intensity. By lowering oxygen concentration the triplet absorption is found to decrease with 9,10-diphenylanthracene and to increase with p-terphenyl. The correlation between gain and transient absorption is established in the case of 7-diethylamino-4-methyl coumarin. For rhodamine 6G, it is found that the triplet state does not play any rôle in the laser action at room temperature, even in an oxygen free liquid solution. With carbocyanin derivatives, laser energy can be improved by deoxygenation and by the use of N-aminohomopiperidine in dimethylsulfoxyde.

I - INTRODUCTION

In previous paper [1 - 2] some of us have studied the relation between the theoretical small signal gain and the effective saturated gain which occurs in dye amplifiers, when excited singlet and triplet state populations are in the steady state. This relation includes the case of negative gains, i.e. the case where the transient absorptions predominate over gain [2] . Experiments performed in the microsecond time scale with rhodamine 6G, fluoresceine and 2,7-dichlorofluoresceine have shown the validity of the steady state theoretical model : in air equilibrated alcoholic solutions at room temperature, the oxygen diffusion rate is such that no decay time is long enough to cause appreciable deviation from the steady state case.

When the triplet decay time is of the order of or longer than 1 μs, which is often the case with deoxygenated solutions, the triplet-

triplet absorption can be the cause of premature stopping of the laser emission, as suggested by many authors and generally accepted. However no experimental correlation between the triplet-triplet absorption and the gain in an amplifier has been looked into, as far as we know.

What has been shown hitherto is that oxygen concentration in dye solutions has frequently a strong but unpredictable effect on the laser output intensity [3]. If the laser emission is reabsorbed by a triplet-triplet transition, oxygen can be useful as a triplet quencher ; but on the other hand it can be harmful because it increases the inter-system crossing. It is well known that many more molecules lase when excited by very short pulses such as nitrogen laser pulses, than when excited by pulses of more than 0.1 μsec. This suggests that some transient process occurs with a long time constant. However transient absorption can interfere not only directly as losses at the laser wavelength but also at the exciting wavelength by screening the pumping.

In order to study the mechanism involved with different laser molecules we looked for a correlation between the gain coefficient and the transient absorption when varying an active parameter : the oxygen concentration. For a quantitative understanding of accumulation effects on attenuation due to long life transients, we measured the instantaneous gain in a single-pass amplifier ~ 1 μs after the beginning of the almost linearly rising pumping intensity from the standard dye laser flashlamp.

II - CALCULATION OF THE GAIN COEFFICIENT $\gamma(t)$

Assuming that no ground or excited singlet state reabsorption occurs the infinitesimal gain coefficient as a function of time is :

$$\gamma(t) = \sigma_e N_1(t) - \sigma_T N_T(t)$$

where σ_e and σ_T are the emission and triplet absorption cross sections respectively.

The singlet population $N_1(t)$ is given by :

$$\frac{dN_1}{dt} = P N_o(t) - (k + \sigma_e \emptyset) N_1(t)$$

where P is the pumping rate, k the singlet decay rate, $\emptyset$ the amplified photonic flux.

We have :

$$N = N_o(t) + N_1(t) + N_T(t)$$

The amplified flux $\emptyset$ is assumed to be much smaller than the saturation flux, i.e. : $\sigma_e \emptyset \ll k$

We restrict to the case where $N_T(t)$ remains negligible as compared to N. Integrating in the case where $P(t) \ll k$:

$$\frac{N_1(t)}{N} = \frac{P(t)}{k} \left[1 + \frac{1}{kt} (e^{-kt} - 1)\right]$$

A similar equation gives the population of the triplet state :

$$\frac{dN_T}{dt} = k_3 N_1(t) - k_T N_T(t)$$

Substituing $N_1(t)$ value and integrating with $P(t) = at \ll k$:

$$\frac{N_T(t)}{N} = \frac{a}{k^2} \frac{k_3}{k_T} \left[e^{-kt} \frac{k_T}{k_T - k} + kt - \frac{k}{k_T} - 1 + \frac{k^2 e^{-k_T t}}{k_T(k - k_T)} \right]$$

We have $k_T \ll k$.

Using these values of $N_1(t)$ and $N_T(t)$ in $\gamma(t)$ we find for $t \gg \frac{1}{k}$:

$$\gamma(t) = N \frac{at}{k} \left\{ \sigma_e - \sigma_T \frac{k_3}{k_T} \left[1 + \frac{1}{k_T t} (e^{-k_T t} - 1) \right] \right\} \qquad (1)$$

The gain coefficient can reach a maximum at time t_{max} :

$$t_{max} = - \frac{1}{k_T} \ln \left(1 - \frac{\sigma_e}{\sigma_T} \frac{k_T}{k_3}\right)$$

According to the values of the parameters σ_e, σ_T, k, k_3, k_T and $P(t)$ the maximum gain coefficient can be of significant value or not (figure 1).

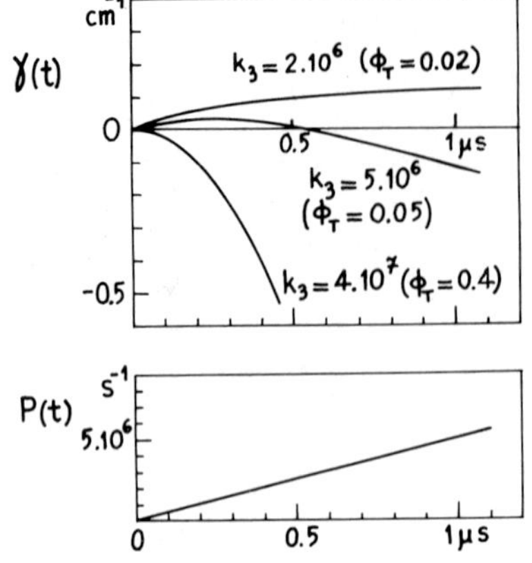

Fig.1. Gain coefficient as a function of time for a linearly rising excitation intensity in the non-steady state case, for three arbitrary values of k_3 and with :

$N = 6.10^{16} cm^{-3}$

$\sigma_T = \sigma_e = 10^{-16} cm^2$

$k_T = 2.10^6 s^{-1}$

$k_1 = 10^8 s^{-1}$

The value of σ_e can be calculated from fluorescence lifetime $\frac{1}{k}$, quantum yield Q and shape function $g(\bar{\nu})$:

$$\sigma_e = \frac{Q\ g(\bar{\nu})\ k}{8\pi c\ n^2\ \bar{\nu}^{-2}}$$

where c is light velocity and n the refractive index.

Effect of oxygen concentration on $\gamma(t)$

k_3 and k_T are little known in oxygenated solution. It is well known however that oxygen concentration has a large effect on k_T, k_3 and to a lesser extent on k, through an essentially diffusion controlled process. We shall assume that :

$$k_3 = k_3^o + q_3\ [O_2] \quad k_T = k_T^o + q_T\ [O_2] \quad k = k_1 + k_2 + k_3^o + q_3\ [O_2]$$

where k_1 is the radiative and k_2 the non radiative lowest singlet state decay rate to the ground state.

The effect of oxygen concentration on $\gamma(t)$ is then rather complex but can be studied for a given time of particular interest, such as the maximum of pumping flash at t = 1 μs. From the formula (1) it can be expected that in all cases there will be a decrease of γ for high oxygen concentrations due to the first factor with $\frac{1}{k}$ (i.e. fluorescence quenching).

III - EXPERIMENTAL

The gain was measured in an 8 to 48 cm long six-stage flash pumped amplifier (figure 2) [1][4]. The probe beam was given by a tunable dye laser pumped with a flashlamp synchronized with the amplifier flash-

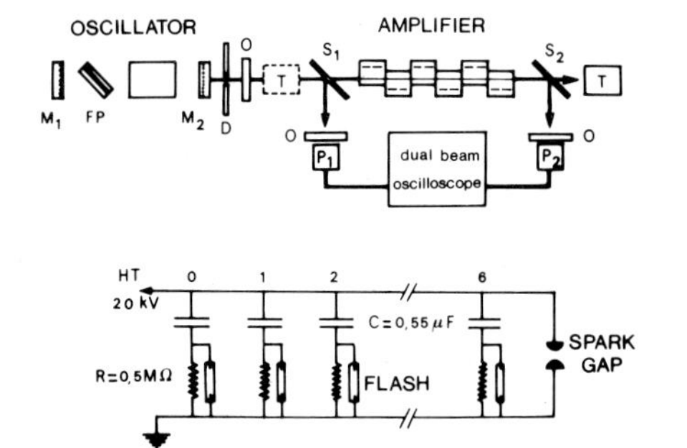

Fig.2. Amplification and attenuation measurements.

lamps..This automatically set the probe pulse at a constant delay after the beginning of the amplifier flash and at a time close to its maximum. The transient optical density (positive or negative) was deduced from the ratio of output and input pulse intensities with amplifier pumped divided by the same ratio with amplifier unpumped. A change in shape of the output pulse occurred whenever accumulation took place at low oxygen concentration. The ratio was measured on oscillograms at the input pulse maximum. Oxygen concentration was measured with a Beckman analyzer and could be adjusted in the recirculating dye solution by bubbling nitrogen, air or oxygen.

Relative fluorescence quantum yields as a function of oxygen concentration were measured in recirculating solutions with the spectrofluorimeter Cary 50-902.

Rhodamine 6G was from Eastman (10724) and purified by chromatography on alumine [5]: 7-diethylamino-4-methyl coumarin from Aldrich ; 9,10-diphenylanthracene from K & K ; p-terphenyl from Princeton Organics ; 3,3'-dimethyl-2,2'-oxatricarbocyanine iodide (DOTC), 1,3,3,1',3',3'-hexamethyl-2,2'-indotricarbocyanine iodide (HITC), 3,3'-diethyl-2,2'-thiatricarbocyanine iodide (DTTC), from Nippon-Kankoh-Shikiso ; ethanol and dimethylsulfoxyde were of spectroscopic quality from Merck (Uvasol).

IV - RESULTS

9,10 diphenylanthracene in ethanol 0.9 10^{-4}M :

The triplet absorption measured at 447 nm and after 1 μs is shown in Figure 3 as a function of oxygen concentration. For this molecule, aerated solutions correspond to the oxygen concentration least suitable for laser action. At low oxygen concentration the absorption decreases but a positive gain cannot be obtained at 447 nm (where $\sigma_e \simeq 3.10^{-17} cm^2$) even in an oxygen free solution. At the fluorescence maximum (λ= 430 nm , $\sigma_e \simeq 5.10^{-17} cm^2$) and with a slow rising excitation, laser action is not expected.

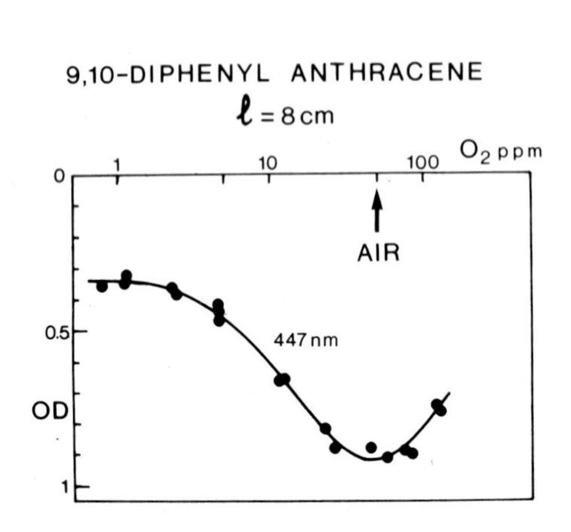

Fig.3. Transient absorption optical density as a function of oxygen concentration.

p-TERPHENYL
ℓ = 8 cm
O2 ppm
1
10
100
0
0.5
1
1.5
OD
AIR
460 nm

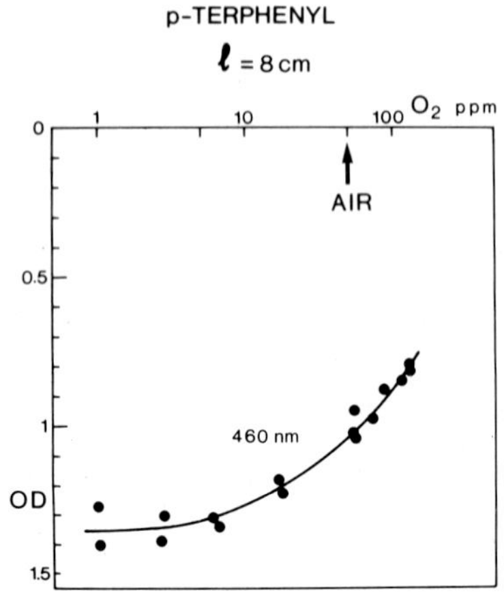

Fig.4. Transient absorption optical density as a function of oxygen concentration.

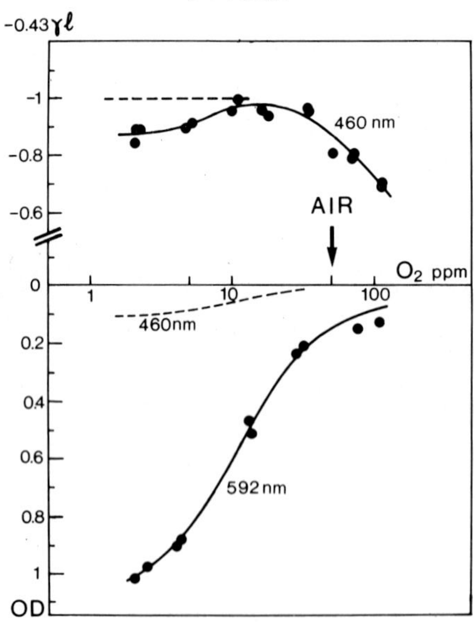

Fig.5. Correlation between transient absorption (lower curve) and gain (upper curve) for different oxygen concentrations. The lower dotted curve is the transient optical density, deduced from the effec tive measurement at 592 nm.

p-terphenyl in ethanol 0.5 10^{-4}M :

The triplet absorption measured at its maximum (460 nm) increases with low oxygen concentration (figure 4). However, at the fluorescence maximum (340 nm), the triplet absorption coefficient is more than 100 times smaller than it is at 460 nm [6] Thus from Figure 4, we can expect a triplet optical density of less than 0.01 at 340 nm. In spite of the decrease of the fluorescence yield (Table 1), high oxygen concentrations seem more favorable for laser action. p-terphenyl lases at 340 nm when pumped by a fast flash [7] or by the fourth harmonic of Nd glass laser [8]and should lase even with long flash pumping.

7-diethylamino-4-methyl coumarin in ethanol 4.10^{-4}M :

The positive gain was first measured as a function of oxygen concentration (figure 5) with the probe beam set at the maximum gain wavelength (460 nm). Transient absorption was then measured as a function of oxygen concentration with the probe beam set in a region of higher transient absorption (592 nm) where σ_e is negligible. Flash photolysis results [9][10]are used to calculate the absorption optical density at 460 nm which can be expected in this experiment from the value we measured at 592 nm. The decrease in effectively measured gain at 460 nm for low oxygen concentration corresponds to the expected value. Optimum oxygen concentration is found to be smaller by about a factor 3 than that given by air saturation. The drop in gain coefficient at higher oxygen concentration corresponds quantitatively to the decrease in fluorescence quantum yield. In spite of the probable triplet yield increase, the transient absorption remains negligible. For this molecule, the correlation between transient absorption and gain seems to be established.

Table 1

Relative fluorescence quantum yields in ethanol

Molecule	λ		Q		
	exc	obs	N_2	AIR	O_2
7-diethylamino-4-methyl coumarin	385	460	1	0.94	0.75
Rhodamine 6G.	530	570	1	0.99	0.99
p-terphenyl.	297	340	1	0.95	0.78
9,10-diphenyl-anthracene.	371	427	1	0.83	0.45

Rhodamine 6G in ethanol, concentrations from 3.10^{-6} to 10^{-4}M

A very wide transient absorption spectrum (550 - 750 nm) has been reported and assigned to the triplet [11][12], with similar extinction coefficients at 590 nm (where amplification occurs) and at 680 nm (where $\sigma_e \simeq 0$). Laser gain measurements at these wavelengths gave the following results :

a) there was no detectable transient absorption at 680 nm at usual laser concentrations and room temperature for a 48 cm path, in either oxygenated or deoxygenated solutions in ethanol.

b) the saturated gain at 590 nm was not sensitive to oxygen concentration.

These results were obtained with rhodamine 6G from Eastman without further purification. Purification by chromatography, though it did show up at least two impurities (with peak absorption maxima at 369 nm and 522 nm respectively), has no effect on measurements at 590 and 680 nm. On the other hand, older samples from various suppliers showed correlated transient absorption and gain dependence on oxygen concentration (table 2).

Table 2

Correlated impurity effects in rhodamine 6G amplifier (in ethanol 5.10^{-5}M)

	Bubbled gas	Transient absorption o.d. 680 nm	Saturated gain o.d. 590 nm
	O_2	0	- 1.52
R6G sample 1	AIR	0	- 1.52
	N_2	0	- 1.52
	O_2	+ 0.16	- 1.42
R6G sample 2	AIR	+ 0.32	- 1.32
	N_2	+ 0.83	- 0.98

Moreover the fluorescence quantum yield does not decrease by more than 1 % from nitrogen bubbling to oxygen bubbling (table 1), in sharp contrast with other molecules.

We conclude that previously reported improvements in rhodamine 6G laser output by oxygen bubbling are due, in the microsecond time scale, to quenching of transients arising from impurities rather than from the rhodamine 6G itself. Even in oxygen free solution, the triplet state which has an exceptionally low formation yield does not play any rôle in the rhodamine 6G liquid laser.

Transient absorption was also found at 440 nm with a 6.10^{-6}M solution in the spectral region where the rhodamine 6G absorption coefficient is optimum for pumping laser solutions ($\varepsilon \simeq 6000$ $\ell M^{-1} cm^{-1}$). This absorption, independent of oxygen concentration, corresponds to a steady state population (no temporal de-shaping of the transmitted pulse), and can be intense enough (o.d. = 1.35 for 48 cm) to reduce effective pumping. This is also shown by the induced absorption measured in another experiment : we observed that the optical density at 450 nm of a 1 cm cell of 10^{-4}M rhodamine 6G solution varies from 0.37 when measured with a low light flux to 0.47 with a 0.5 MW/cm^2 flux from a coumarin laser.

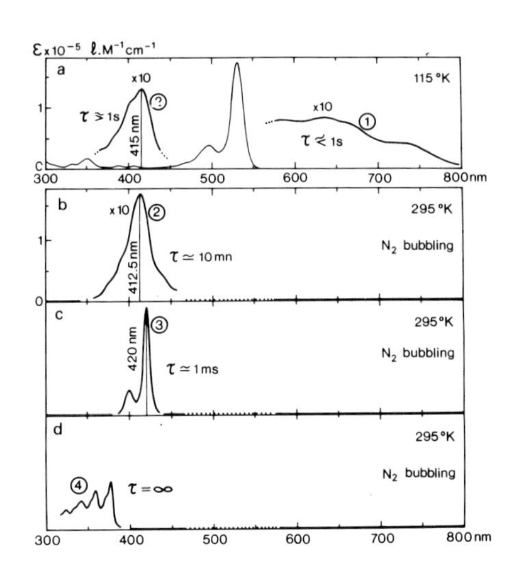

Fig.6. Rhodamine 6G flash photolysis
a) : transient spectra (curves 1 and 2) and ground state absorption at 115° K (thin line).
b) c)d) : spectra due to photoproducts.

Some additional information on long life transients was obtained from flash photolysis study of purified rhodamine 6G (figure 6) :

a) at 115° K in the mixture ethanol-methanol (3 : 1), there are two regions of transient absorption due to two distinct species : their decay times are distinct and their relative intensity depends on rhodamine 6G concentration : ε for transients in figure 6a are only indicative.

b) at room temperature in N_2 bubbled solution a photoproduct has a main band at 412.5 nm (ε_{max} = 18000 $\ell M^{-1} cm^{-1}$). It is stable in an oxygen free solution and reacts with O_2 to give back rhodamine 6G, as has been already observed [12].

c) irradiation in deoxygenated solutions gives two new spectra of related intensity : a transient spectrum with a main maximum at 420 nm and a 1 ms decay time, and a permanent spectrum with a vibrational progression beginning at 375 nm, corresponding to a species stable even in reoxygenated solution. These spectra are likely to be due to the same molecule in its ground and lowest triplet state.

From the laser gain and flash photolysis experiments, we conclude that at least three transient spectra can occur in the violet-blue region, none of which can be assigned to the rhodamine 6G triplet state. Their rôle in limiting the rhodamine 6G laser efficiency is expected to be important at high pumping intensity or for very long pulses.

Carbocyanine derivatives in dimethylsulfoxyde

The part played by transient absorption might also be significant with dyes lasing in the near IR region [13]. With a 1.4 µs rise time flashlamp excitation, a preliminary study of 1.5 10^{-4}M DOTC in DMSO, a good laser solvent [14], confirmed that oxygen concentration is of considerable importance for laser output. In air-equilibrated solution DOTC gives only little laser energy ($\simeq$ 0.2 mJ) and the duration of the output pulse is less than 0.5 µs. Bubbling nitrogen increases the peak laser intensity by a factor 20 and the pulse duration by a factor 2 (figure 7). We could not obtain laser action with 2.5 10^{-4}M HITC and 10^{-4}M DTTC in either air equilibrated or deoxygenated DMSO solution. With a 0.7 µs excitation risetime, laser action has been reported with DTTC [15].

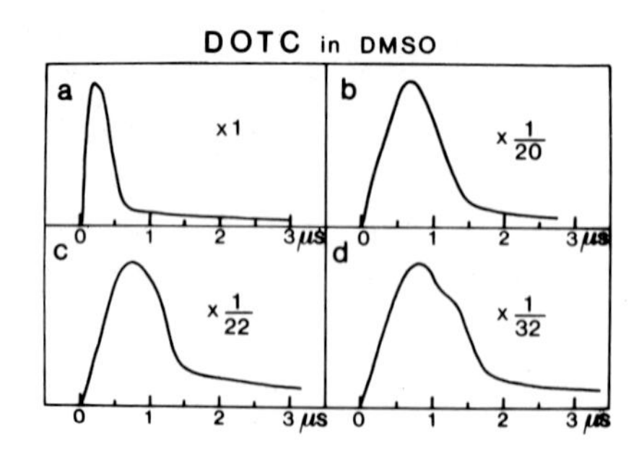

Fig.7. Laser intensity as a function of time : a) air-equilibrated solution ; b) air-equilibrated solution with addition of AHP ; c) deoxygenated solution ; d) deoxygenated solution with addition of AHP.
(conc. AHP : 0.1 cc in 150 cc)

We found that addition of N-aminohomopiperidine (AHP) allows one to obtain the laser effect with DTTC and HITC and further increases the lasing performance with DOTC (figure 7). The effect of AHP in air-equilibrated DOTC solution is at a maximum only ~ 10 mn after its addition to the solution. A subsequent decrease in laser intensity is due to chemical decomposition of the dye. A study of the rôle of AHP led to the conclusion that AHP has a double effect :

1) it removes dissolved oxygen by a chemical reaction involving DMSO.

2) independently of the presence of oxygen there is an instantaneous action of AHP similar to that of a triplet state quencher.

REFERENCES

1 P. Flamant and Y.H. Meyer, Optics Com., 7(1973)146.
2 Y.H. Meyer and P. Flamant, Optics Com., 9(1973)227.
3 B.B. Snavely and F.P. Schäfer, Phys. Letters, 28A(1969)728.
J.B. Marling, D.W. Gregg and S.J. Thomas, IEEE J. Quantum Electr., 6(1970)570.
4 P. Flamant and Y.H. Meyer, Appl. Phys. Lett., 19(1971)491.
5 J.M. Drake and R.I. Morse, Optics Com., 13(1975)109.
6 A. Bokobza, Thèse 3ème cycle, Paris, (1973).
7 H.W. Furumoto and H.L. Ceccon, IEEE J. Quantum Electr., 6(1970)262.
8 G.A. Abakumov, A.P. Simonov, V.V. Fadeev, M.A. Kasymdganov, L.A. Kharitonov and R.V. Khokhlov, Opto-Electronics, 1(1969)205.
9 D.N. Dempster, T. Morrow and M.F. Quinn, J. Photochem., 2(1973/74) 329.
10 T.G. Pavlopoulos and P.R. Hammond, J. Amer. Chem. Soc., 96(1974) 6568.
11 A.V. Buettner, B.B. Snavely and O.G. Peterson, (1969) in Molecular Luminescence, p 403, Lim E.C. (Edit.), New York ; Benjamin.
12 D.N. Dempster, T. Morrow and M.F. Quinn, J. Photochem., 2(1973/74) 243
13 A. Hirth, J. Faure and D. Lougnot, Optics Com., 8(1973)318.
14 P.P. Sorokin and J.R. Lankard, IBM J. Res. Dev., 11(1967)130.
15 J.P. Webb, G. Webster and B. Plourde, IEEE J. Quantum Electr., 11 (1975)114.

DISCUSSION

Y. MEYER , P. FLAMANT, P. GACOIN, C. LOTH, R. ASTIER

F.P. Schäfer -

From your curve for p-terphenyl showing a sharp decrease of absorption with increasing oxygen content and the numerical values for the quantum yield of fluorescence of the same molecule that showed only a slight decrease with increasing oxygen content, one would expect that p-terphenyl could be a good laser even with long flashlamp pumping when you use a higher than atmospheric pressure of O_2. Did you ever try that?

Y. Meyer - It was our conclusion that oxygen bubbling should permit laser action in p-therphenyl even with long flash pumping. We did not try higher oxygen concentration than the one given by oxygen bubbling ($\sim$ 1 atm.).

S. Kimel - Have you done similar gain studies of rhodamine at very high power and short time of excitation such as available with the nitrogen laser, where triplet formation is expected to be unimportant?

Y. Meyer - No, these experiments were concerned only with long pulse excitation.

J. Langelaar - Your conclusion that impurities may play an important role by spoiling laser gain in rhodamine 6G at slow pumping rates also holds for fast pumping rates with a N_2-laser. Extensive purification of rhodamine 6G (with maleic anhydride) gives an increase in output when pumped with a N_2-laser but moreover one can observe an increase of the tuning range.

C. Shank - You have indicated that the "purified" rhodamine 6G laser system operates independent of the oxygen concentration. How do your measurements influence the interpretation of experiments by Schäfer and Snavely who observed a strong dependence of rhodamine 6G laser pulse length with nitrogen and oxygen concentration? Also how do you explain the effects of other quenchers, e.g. cyclo-octa tetraene (COT), on the rhodamine 6G laser system?

Y. Meyer - O_2 and COT enhance laser energy in some cases. If they act as triplet quenchers it can as well be on the impurity triplet as on the rhodamine 6G triplet. Some effect on photoproducts is also probable.

F.P. Schäfer - The results of triplet quenching by oxygen in rhodamine 6G solutions obtained by Dr. Snavely and myself, which were mentioned by the previous speaker cannot be compared with the results of Dr. Meyer. The short pulses (of the order of 1 μs.) used by Dr. Meyer are by far too short to give a stationary triplet state population, while Dr. Snavely and I used pulses that were longer than 400 μs. In this latter case, the stationary triplet state population is reached even in a carefully deoxygenized solution and the laser emission consequently quenched no matter how well purified the dye sample was.

J. Faure - Dempster (Journal of Photochemistry, 2 (1973), 343) has measured the triplet-triplet absorption spectrum and extinction coefficient of rhodamine 6 G. Your results seem to be in strong contradiction with his own ones. How do you explain this discrepancy?

Y. Meyer - We confirm the presence and the ε of the photoproduct (415 nm) first found by Dempster, Morrow and Quinn. We have seen no indication of any transient band in the blue due to rhodamine 6G triplet. On the other hand, we observed a transient band at 420 nm in N_2 bubbled solution (certainly not due to R6G) which is not mentioned by these authors. At room temperature and with N_2 bubbling we observed no transients at all in the red part of the spectrum.

SOME PARAMETERS OF STIMULATED RAMAN SCATTERING RELEVANT FOR ISOTOPE ENRICHMENT

S. KIMEL and R. SCHATZBERGER
Chemistry Department, Technion-Israel Institute of Technology, Haifa, Israel

SUMMARY

Some parameters which influence the generation of stimulated Raman scattering (SRS) are discussed for the purpose of producing selective Stokes SRS from a given isotope. This is achieved by introducing into the Raman cell dye laser radiation having the frequency of the Stokes field of the isotopic species.

The principle will be illustrated with $SnCl_4$ where a minor isotopic species might be induced to generate SRS. Meanwhile, using a fluorescent amplifier we have demonstrated the appearance of SRS from a normally absent vibration in p-xylene. The molecules SCl_2 and SO_2Cl_2 are discussed as possible candidates for isotope enrichment based on the SRS technique.

INTRODUCTION

Isotope enrichment schemes which make use of lasers have in common, as the primary process, a selective excitation of one isotopic species. This is followed by some preferential photophysical or photochemical process, resulting in stable products. The secondary processes must occur before the excited isotope can decay or can transfer its energy to undesired isotope species.

Recently we have discussed the possibility of using the stimulated Raman scattering (SRS) technique as the primary process for achieving a sizeable vibrationally excited transient population of one single species in an isotopic mixture [1]. This method is probably not economical, since only a fraction of the energy of each scattered photon is usefully extracted. However, there are some notable advantages of the SRS method, compared with direct infrared laser excitation [1]. Therefore it is worthwhile to investigate various quantitative aspects of SRS.

II. PARAMETERS FOR THE GENERATION OF SRS

The parameters characterizing a laser source - frequency, bandwidth, pulse energy and duration - which may influence the effectiveness of the generation of SRS have been discussed in the literature [2]. Molecular properties of the medium, such as those causing self focusing of laser beams, are also known to be of importance. We mention here only some aspects which are relevant to our problem.

The frequency factor $(\nu_L-\nu_M)^4$, where ν_L denotes the laser frequency and ν_M a molecular vibrational eigenfrequency, favors the use of higher frequency laser sources to obtain spontaneous Raman Stokes scattering and consequently also SRS. Thus, everything else being equal, excitation by a ruby laser at λ_L=693.4 nm is preferable compared to a Nd laser at λ_L=1060 nm. In many cases frequency doubled laser sources have been used for generating SRS but here the lower intensity of the source must be considered in the overall balance, especially since nonlinear processes play an important role in SRS. No SRS seems to have been reported using a N_2 laser source at λ_L=337 nm.

When increasing ν_L the excitation approaches the first electronic absorption band ν_e of the molecule and gives rise to the preresonance Raman effect. The frequency dependence for the preresonance Raman scattering intensity I may be expressed by [3]

$$I \propto I_L(\nu_L-\nu_M)^4 e^{-\varepsilon(\nu_L)c\ell} e^{-\varepsilon(\nu_L-\nu_M)c\ell} \sum_{e,f} \frac{\nu_e\nu_f+\nu_L^2}{(\nu_e^2-\nu_L^2)(\nu_f^2-\nu_L^2)} .$$

Here I_L is the intensity of the laser source, e and f refer to virtual electronic states with eigenfrequencies ν_e and ν_f, and $\varepsilon(\nu)$ denotes the absorption coefficient at frequency ν of the medium. As a rule the SRS process is determined substantially by the values of $I/\Delta\nu$, where $\Delta\nu$ denotes the width of the Raman line. Often the relation is much more complex [4] but for mixtures of isotopic species, all having identical gain coefficients, the SRS intensities will just depend, exponentially, on the concentration of the isotope. In the limit when $\nu_L \rightarrow \nu_e$ the scattering becomes of the resonant type and enormous increases ($\gtrsim 10^6$) for the scattered intensity have been reported [3]. However, usually there is no corresponding lowering of the threshold for SRS [5]. This may be due to several factors: the resonance Raman lines are broadened [6], the absorbed incident laser light may cause fluorescence which competes with SRS and there may be direct absorption of the Stokes radiation [7].

It has been observed that certain vibrational Raman lines are preferentially enhanced by the resonance effect. Qualitatively this may be understood in terms of the strength of the vibronic coupling. For example, the totally symmetrical C-H stretching vibration ν_1 in benzene shows weaker Raman scattering than the symmetric C-C stretching vibration ν_2. However, because ν_1 is strongly coupled to the first electronic transition B_{2u} at 260 nm, the ν_1 band becomes more intense than ν_2 in the case of near-resonance [3].

While the parameters discussed so far may influence the overall intensity of SRS bands, and in some cases even of specific vibrations, it is clear that no useful isotopic selectivity can be obtained. For this we have to turn to special optical arrangements.

III. SELECTIVE AMPLIFICATION OF SRS

When a cell containing a fluorescing medium is placed in the optical path of a laser beam, in front of a Raman cell or closely behind it, the intensity of SRS lines lying in the fluorescence band may be enhanced selectively [8-10]. The fluorescent radiation serves as a broad-band amplifier for the Stokes field at frequency $\nu_L - \nu_M$ which together with the intense source field at frequency ν_L is necessary for the generation of SRS. In this way a number of SRS inactive substances - or SRS inactive vibrations - were affected to produce SRS in the spectral region of the fluorescent additive. It is natural to entend and refine this general amplification mechanism for use in isotope selective excitation by employing a tunable dye laser [11] instead of a broad-band fluorescent continuum.

Our feasibility study of producing selective SRS, commences with chlorine containing compounds. We are considering $SnCl_4$ for which the demonstrated [12] isotope selectivity results from the relative abundance of the natural isotopic composition of $SnCl_4$. Our first aim is to obtain SRS also from minor isotopic species.

Figure 1 shows our experimental arrangement which is being assembled for this purpose. A Q-switched pulse from a ruby laser (1-2 J energy and 20 nsec duration), which can be monitored by diode D_3, undergoes some mode selection by etalon E and passes through band filter F_1 in order to remove stray light from the flash lamp. The pulse is incident on beam splitter B_1.

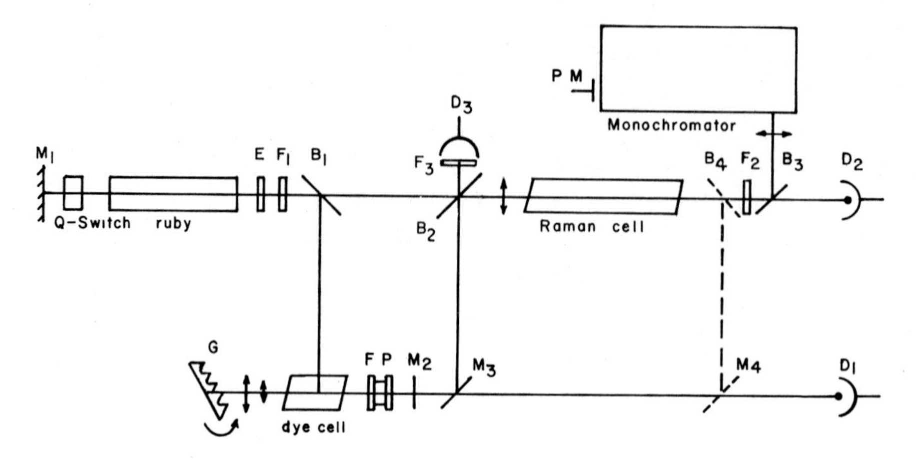

Fig.1. Experimental setup for selective enhancement of SRS

The through-going beam is focused into a 25 cm long Raman cell containing liquid $SnCl_4$. The ruby pulse deflected by B_1 excites a dye laser (DTDC iodide in acetone) in a transverse configuration. A grating G, in conjunction with beam expander and Fabry Perot etalon FP, ensures effective tunability of the dye laser in the region 710-730 nm with a spectral purity of $\lesssim 1\ cm^{-1}$ which is adequate for our purposes. Prior to each set of measurements the spectral characteristics of the dye laser are investigated with a high-resolution double monochromator (Spex Model 1402, equipped with a cooled photomultiplier, PM, type EMI 9558B) without the Raman cell in position. The dye laser intensity is monitored with diode D_1 which collects the radiation passing through mirror M_3. The pulse deflected from mirror M_3 is introduced into the Raman cell by a frequency selective beam splitter B_2; the interference filter F_3 prevents any dye radiation from reaching D_3.

The intensity of the first Stokes line of the SRS spectrum at λ=712 nm is monitored with photodiode D_2 after interference filter F_2 has removed the source radiation at λ_L=694.3 nm. Part of the SRS Stokes radiation is deflected by beam splitter B_3 into the monochromator. Figure 2(a) shows the response of $D_1(\lambda)$ and D_2 (λ). The abscis denotes the frequency $\nu(cm^{-1}) = 1/\lambda_L - 1/\lambda$ when tuning the dye laser, by changing the Fabry Perot setting in a point by point fashion, over the region λ=712-713 nm. The $D_1(\lambda)$ response is predicted to be constant over this

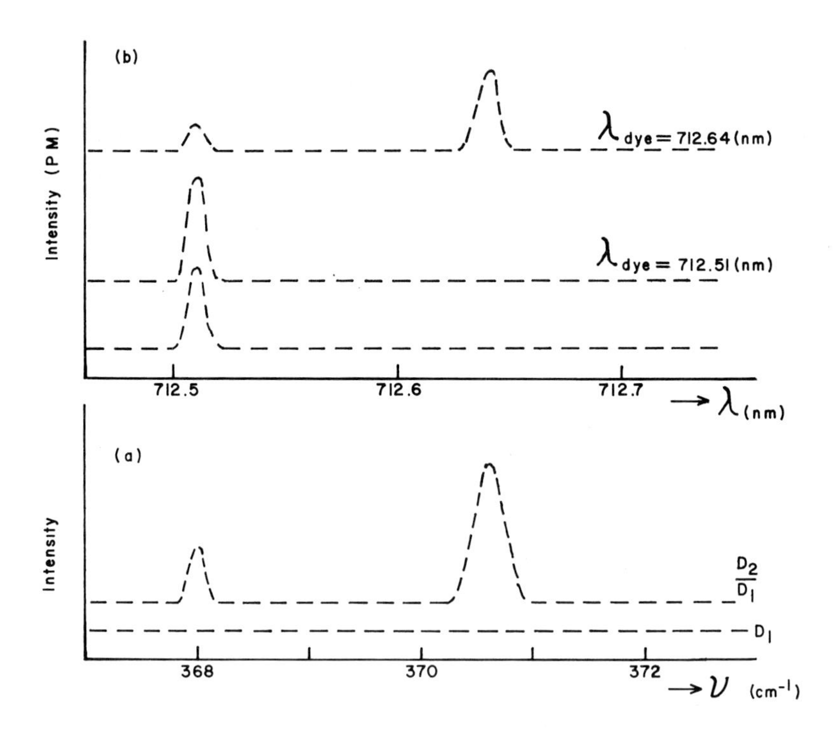

Fig.2. (a) Output of dye laser, $D_1(\lambda)$, and of $SnCl_4$ SRS, $D_2(\lambda)$, plotted as a function of $\nu = 1/\lambda_L - 1/\lambda$. (b) SRS spectrum of $SnCl_4$, obtained with and without dye laser radiation.

minute spectral interval. In fact, the $D_2(\lambda)$ and D_3 responses serve to normalise the $D_2(\lambda)$ readings for ambient intensity changes and for shot-to-shot variations. The $D_2(\lambda)/D_1(\lambda)$ response, is expected to exhibit two features:

(i) A mild enhancement when the dye laser is scanned through λ=712.51 nm; corresponding to the first Stokes shifted line associated with the vibration ν_3=368.0 cm^{-1} of $Sn^{35}Cl_3^{37}Cl$, which is the most abundant (41%) species in natural $SnCl_4$ and is predominant in SRS [12] since it has the lowest effective threshold.

(ii) a relatively large enhancement in the presence of dye laser radiation at λ=712.64 nm; corresponding to the Stokes line due to ν_3=370.6 cm^{-1} of the next abundant (32%) isotopic species, $Sn^{35}Cl_4$.

This line is normally absent in the SRS effect [12].

Figure 2(b) illustrates the intensity readings of photomultiplier PM when scanning with the monochromator. Now M_3 and B_2 are replaced by mirror M_4 and beam splitter B_4 so as to prevent dye laser radiation from entering the monochromator directly. With the dye laser tuned at λ=712.51 nm (middle curve in Fig.2(b)) there should be but small increase, if at all, in the Stokes SRS intensity compared to that without the presence of dye laser radiation (lower curve). By contrast, when the dye laser is tuned at λ=712.64 nm (upper curve) there should be a large intensification of the normally absent SRS Stokes line at that wavelength. Because of the competitive nature of the SRS process this will be accompanied by a corresponding decrease of the $Sn^{35}Cl_3{}^{37}Cl$ line at 712.51 nm.

The next stage in the research will be an investigation of the compounds SCl_2 and SO_2Cl_2 for which the regular Raman spectra have been reported [13,14] but for which, as yet, no data exist regarding SRS.

SCl_2 shows a strong Raman band at the symmetric S-Cl stretching vibration ν_1. In solid samples (77°K) fine structure is clearly resolved [13] with frequencies 505 cm^{-1} (attributable to $S^{35}Cl_2$, abundance 54%) and 498 cm^{-1} ($S^{35}Cl^{37}Cl$, abundance 36%). In the spectrum of the liquid phase the isotope effect of the ν_1 band around 519 cm^{-1} causes an asymmetrical displacement towards the low-frequency side.
Liquid SCl_2 decomposes according to

$$2\ SCl_2 \rightarrow S_2Cl_2 + Cl_2$$

If isotopically selective excitation of the ν_1 vibrational level of SCl_2 can be achieved by SRS and if the photochemical decomposition rate of the irradiated molecules is faster than the rate of all competing relaxation and energy transfer processes(which lead to isotope scrambling) then the product Cl_2 molecules will retain the isotopic identity of the initially excited parent compound. The restrictions on the rates of the competing reactions are very stringent indeed and only few laser-induced photochemical isotope enrichment reactions have been reported; one of the more recent ones being the chlorine enrichment produced by the selective (electronic) excitation of gaseous $CSCl_2$ [15]. It should be noted [1] that the typical time scale for these processes in the liquid phase necessitates the use of mode-locked laser sources.

One of the reasons for choosing SCl_2 is that there may be no need for additional optical excitation to dissociate the molecule nor for scavenger reagents to remove the dissociation fragments. SCl_2 is transparent in the spectral region $\lambda > 700$ nm; its absorbance increases monotonically from $\varepsilon \simeq 0.1$ cm^2 mole^{-1} at λ=600 nm to $\varepsilon \simeq 700$ cm^2 mole^{-1} at λ=260 nm [16]. Thus there should be no non-selective decomposition by the ruby radiation but using different excitation lasers it might, perhaps, be possible to study for this compound the influence of the resonance effect on SRS intensities.

The Raman active symmetric stretching vibration ν_3 of crystalline SO_2Cl_2 (at 77°K) shows [14] a triplet (of decreasing intensity) at 414.5 cm^{-1}, 410.5 cm^{-1} and 406.5 cm^{-1} attributable to the species $SO_2{}^{35}Cl_2$, $SO_2{}^{35}Cl^{37}Cl$ and $SO_2{}^{37}Cl_2$, respectively, which in natural SO_2Cl_2 are present in the ratio 9:6:1. An earlier Raman study [16] of the liquid phase reported lines at 408 cm^{-1} and 403 cm^{-1} with relative intensities of 13:4. Here too we assume that the ensuing one-step photochemical decomposition

$$SO_2Cl_2 \rightarrow SO_2 + Cl_2$$

will retain the isotopic identity of the selectively excited parent molecule. Again there may be no need for additional excitation nor for scavenger reagents. Similarly, the resonant Raman effect can be studied by varying the excitation laser; electronic transitions occur around 420 nm, 330 nm, 370 nm and 250 nm (strong) [16].

At present we are in the process of calibrating the equipment and testing our capability to affect the appearance and intensity of SRS in a controlled way. Figure 3 shows the SRS spectrum of p-xylene, recorded with the set-up described in Fig.1. For simplicity we have replaced the monochromator with a low-resolution spectrograph (30A/mm) and the dye laser with an 1 cm long cell containing a fluorescent medium (5×10^{-5} M cryptocyanine/ethanol) which covers the region of the low vibrational Stokes lines. It can be seen that an additional SRS Stokes line of p-xylene appears at λ=737 nm (corresponding to the symmetric C-C stretching vibration ν_5 = 830 cm^{-1}) when the fluorescent cell is placed behind the Raman cell. When reducing the distance d between the cells we observe an enhancement of the intensity of the line at λ=737 nm at the expense of the first Stokes line at λ=877 nm (corresponding to the symmetric C-H stretching vibration ν_1 = 3000 cm^{-1}), which normally is the single Stokes line

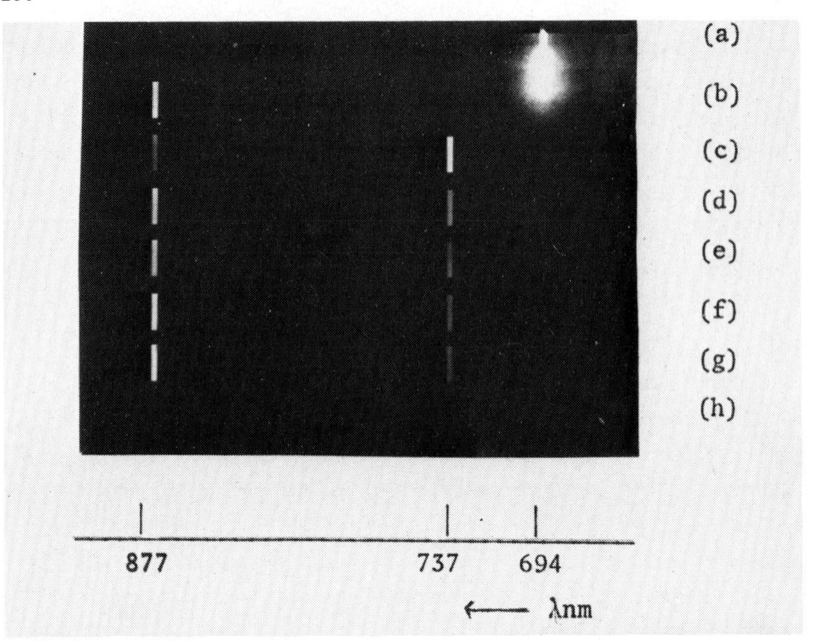

Fig.3. Observed SRS spectrum of p-xylene.

(a) Without Raman cell and without fluorescent cell (b) With p-xylene Raman cell only (c) With Raman cell and fluorescent cell, d=1 cm (d) d=5 cm (e) d=10 cm (f) d=15 cm (g) d=20 cm (h) with fluorescent cell only. In spectra (b)-(h) the ruby radiation was removed with filter F_2.

in an SRS spectrum of p-xylene.
Incidentally, in the spontaneous Raman spectrum the ν_1 line has a lower peak intensity and a broader frequency profile than the ν_{13} line [17]. This indicates that the SRS spectrum cannot be predicted simply on the basis of the characteristics of the corresponding spontaneous Raman spectrum. Clearly additional factors play a role and optimal excitation conditions must be established for each individual case [4].

In conclusion, it would seem that the problem of selectively generating SRS from one chosen isotope can, in principle, be solved. Also, the efficient use of the laser-induced non-equilibrium distribution for photochemical isotope enrichment in low-pressure gas samples has been studied in various laboratories, both experimentally and theoretically [18-20]. However, more fundamental work is needed regarding the kinetics of photochemical reactions in liquids before practical use can be made of the selective excitation potentialities of SRS for isotope enrichment schemes.

REFERENCES

1. S. Kimel, A. Ron and S. Speiser, Chem. Phys. Letters, 28 (1974) 109.
2. W. Kaiser and M. Maier, in F.T. Arecchi and F.O. Schulz-Dubois (Editors), North Holland Publ. Comp.,Amsterdam, 1972, p. 1077.
3. J. Behringer, Molecular Spectroscopy, Specialist Periodical Reports, The Chemical Society, London, 2 (1974) 100; and R.E. Hester, ibid. p. 439.
4. I.I. Kondilenko, P.A. Korotkov, V.I. Malyi and N.G. Golubeva, Optics and Spectr., 34 (1973) 271.
5. Y.S. Bobovich and A.V. Bortkevich, Soviet Physics Uspekhi, 14 (1971) 1.
6. I.I. Kondilenko, V.E. Pogovelov and Khun Khue, Optics and Spectr., 32 (1972) 337.
7. Y.S. Bobovich and A.V. Bortkevich, Optics and Spectr., 28 (1970) 56.
8. Y.S.Bobovich and A.V.Bortkevich, Optics and Spectr., 24 (1968) 238.
9. N.D. Shvedova and L.M. Sverdlov, Optics and Spectr., 30 (1971) 309.
10. H.J. Weigmann, M. Pfeiffer, A. Lau and K. Lenz, Optics Communications, 12 (1974) 231.
11. R. Frey, J. Lukasik and J. Ducuing, Chem. Phys. Letters,14 (1972) 514.
12. D.H. Rank, R.V. Wick and T.A. Wiggins, Applied Optics, 5 (1966) 131.
13. R. Savoie and J. Tremblay, Can. J. Spectr., 17 (1972) 73.
14. J. Tremblay, C. Nolin and R. Savoie, Can. J. Spectr., 18 (1973) 36.
15. M. Lamotte, J.H. Dewey, R.A. Keller and J.J. Ritter, Chem. Phys. Letters, 30 (1975) 165.
16. Gmelin, Handbuch der Anorg. Chemie (8th Ed.), 9B part 3 (1963) 1768, 1812
17. I.I. Kondilenko, P.A. Korotkov and G.S. Litvinov, Optics and Spectr., 32 (1972) 484.
18. V.S. Letokhov and A.A. Makarov, Soviet Phys., JETP, 36 (1973) 1091.
19. N.V. Karlov, Applied Optics, 13 (1974) 301.
20. Y.S. Liu, Applied Optics, 13 (1974) 2505.

DISCUSSION

S. KIMEL, R. SCHATZBERGER

M. Delhaye - 1) Do you know the rate of the recombination reactions and exchange reactions in $SnCl_4$, SCl_2 and SO_2Cl_2? 2) Did you observe the anti-Stokes side of the spectrum when exciting at low wavelength?

S. Kimel - 1) $SnCl_4$ is a rather stable molecule and is not expected to undergo photolysis easily. This molecule was chosen primarily because to our knowledge this is the only compound for which isotopic selective SRS has been reported and it would seem to be a good first choice to test our capability to achieve SRS also from a minor isotopic compound? Some thermal and photolytic decomposition and recombination rates of the molecules SCl_2 and SO_2Cl_2 have been reported in Gmelin's Handbook. By contrast, exchange reactions have not yet been studied at all.
2) In the p-xylene experiment we did not make a quantitative study of the generation of anti-Stokes scattering since we consider it to be just another loss factor which play a role in addition to mechanisms such as backward stimulated Brillouin scattering.

P. Rigny - If you did not yet perform experiments with isotopes, could you however make a guess on the selectivity of the method, i.e. the minimum distance in frequency which is required between the vibrational lines, for the selective SRS to work?

S. Kimel - The minimum distance between the same vibration of two isotopic species depends on the relative mass difference of the atoms involved and also on the symmetry of the Raman-active vibration. Typical values may range from > 1000 cm^{-1} in the case of H/D isotopes to $\simeq 0$ for the case of heavy atoms like $^{235}U/^{238}U$ in the symmetrical stretching vibration of UF_6. The chlorine containing molecules I have discussed all have isotope separations of $\simeq 2$ cm^{-1} which seems to be an optimal distance for this type of work.

C. Varma - As you have mentioned picosecond laser pulses are required for your stimalated Raman pumping scheme. Where are the bandwidths of both the laser pulse and the Raman band going to limit the selectivity?

S. Kimel - In transformlimited pulses, this does not seem to be a serious problem since we are employing vibrational isotope separations of the order of 1-2 cm^{-1}. However, at very high intensities of the incident radiation, non-linear processes such as self-focussing and self-phase modulation become important and the resulting very broad SRS is indeed suitable for selective excitation.

D.J. Bradley - I would say that ultra short laser pulses will turn out to be unsuitable for this type of work.

S. Kimel -- I agree that when considering everything (spectroscopy, photochemistry, etc...) pulses in the range of 100 picoseconds would seem to be of optimal duration for selective SRS in liquids.

LASER INDUCED PHOTOCHEMICAL ENRICHMENT OF CHLORINE ISOTOPES

M. LAMOTTE[†], H. J. DEWEY, J. J. RITTER and R. A. KELLER
National Bureau of Standards, Washington, DC 20234

SUMMARY

There has been a lot of interest displayed in laser based isotope separation schemes. This interest lies in both uranium for defense and energy needs, and in the light isotopes for scientific, medical, and environmental studies. The principal advantage of a laser based process is that, in contrast to most enrichment schemes, energy is supplied only to the isotope of interest. An overview of the laser processes with emphasis on the use of visible lasers to enhance bimolecular, photochemical reactions is presented. Data on the successful enrichment of chlorine isotopes by the photochemical reaction between thiophosgene and diethoxyethylene is used to illustrate the major points.

Selective Excitation

Every laser based isotope separation process has two principal parts: selective excitation of some isotopic species and separation of the selectively excited species from the reaction mixture. A high resolution (100,000) absorption spectrum of thiophosgene ($CSC\ell_2$) is shown in Fig. 1. Pseudo-rotational band heads for the various vibronic transitions are easily distinguished and transitions corresponding to the various chlorine isotopic molecules are clearly resolved. It is easy to see that selective excitation of an individual chlorine isotopic molecule presents no particular problem. In a more general sense, the spectral properties required for selective excitation are:

1) Line absorption spectrum - This restricts the search to relatively small molecules in the gas phase and to electronic transitions which do not lead directly to a dissociation.
2) The isotope of interest must participate in the transition in such a fashion that there is a resolvable isotope shift associated with that isotope. For example, in the molecule CrO_2Cl_2 the chromium atom lies near the center of mass and chromium isotope shifts are not observable.

---

† Department of Chemical Physics, Faculty of Sciences of Bordeaux, Talence, France. NATO Guest Worker at NBS, 1973-74.

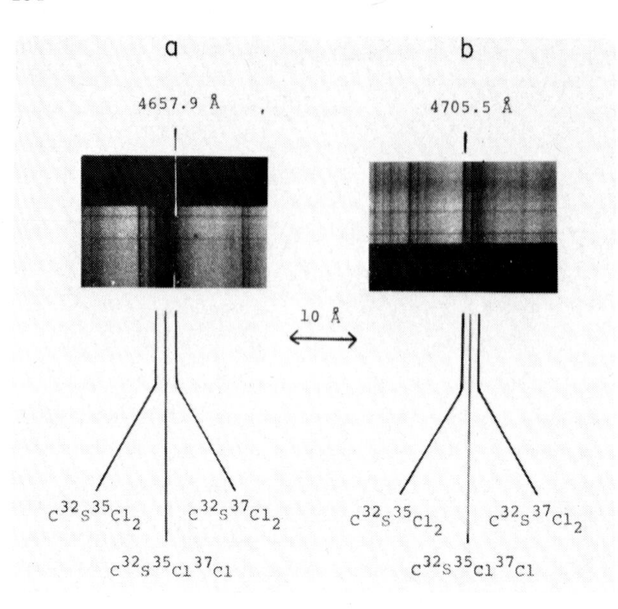

Figure 1 - Absorption spectrum of thiophosgene. Resolution ∿ 100,000. λ increases from left to right.

3) The electronic transitions should occur in the spectral region between 3500 Å and 10,000 Å. This requirement results simply from the present availability of tunable lasers. As lasers improve, the spectral limits will expand - already frequency doubling gives small powers down to 2180 Å [1] and complicated mixing techniques have been successful down to 800 Å. [2]
4) A correct assignment of the observed absorptions to the molecular transitions involved is very useful for the choice of the particular transition to irradiate.

Sometimes a low resolution spectrum misleads one as to the amount of care necessary for selective excitation. In Fig. 2 are shown a low and a high resolution absorption spectrum of the molecule ICl. In the low resolution absorption spectrum the band heads corresponding to the ^{35}Cl isotopic species are clearly visible and band heads corresponding to the ^{37}Cl species are discernable. In the high resolution spectrum the positions of the band heads are not so evident and the absorptions from both isotopes are heavily intermixed. In fact, if you consider one of the sharp dark transitions shown in Fig. 2, you will find that it is in reality composed of several molecular transitions, often from both isotopic species.

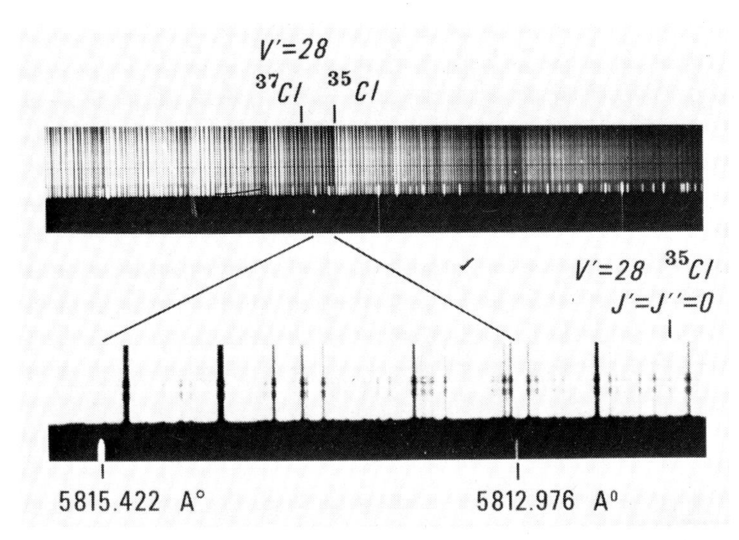

Figure 2 - A comparison of a medium resolution and a high resolution absorption spectrum of ICl. Pressure 36 torr, path length 70 cm.

Once we find a molecule which possesses the proper spectroscopic properties for selective excitation, we can then inquire about the laser properties necessary to accomplish this selective excitation. The width and stability of the laser emission must be better than the molecular Doppler width which is around 0.01 Å. This is no real problem with presently available lasers, in fact, stabilities and linewidths of 10^{-4} Å or better are available. Some skill is necessary to operate the laser at these levels. The tuning of the laser emission inside of the Doppler width of the molecular absorption can be a problem. For example, 0.005 Å out of 5000 Å represents one part in a million which is better than the resolution of almost all spectrographs. A combination of a good spectrograph (resolution > 100,000) and a measurement of the molecular absorption as a function of laser wavelength can usually get the laser to the right wavelength.

To separate reasonable amounts of material it is necessary to have high average powers. For example, a one watt laser in a process with a quantum yield of 1 will separate (enrich) a mole of material in 67 hrs. This may be fine for light isotopes for which there is not a large demand at the present time. However, this is clearly not adequate for uranium and lasers with Kw average powers are necessary for this job. This is one of the reasons why there is so much interest in developing a chemical laser in the visible region - no other pumping scheme appears to have the potential for such high output powers.

In summary, the selective excitation of particular isotopic species in many molecules is relatively straightforward and should present no real difficulties.

Separation Procedures

The second part of the isotope enrichment problem is separation of the selectively excited isotopic species from the reaction mixture. This is the real heart of the problem. Several schemes have been demonstrated to work. A process based on momentum transfer from the photon beam to an atomic beam has been successful in enriching Ba. [3] Each atom absorbs many photons coming from a particular direction and emits them isotropically. The net momentum change is sufficient to deflect the selected atoms several millimeters in a meter path. Multiphoton processes have been used to enrich, U, B, and S. [4-8] In most cases the first photon selects the proper isotopic species and the next photon leads to ionization or photo-decomposition. Some interesting multiphoton processes have recently been reported where a photo-induced cascade knocks the molecule up the vibrational ladder until, after absorbing $\sim$ 30 photons, it finally dissociates. [7-8] One of the first laser based enrichment schemes reported was the result of unimolecular predissociation in formaldehyde. [9] Almost all excited electronic levels predissociate to some extent but the problem is to find a level that predissociates at a rate fast enough to compete with normal relaxation processes but slow enough so that the level is not broadened to the extent that selective excitation is no longer possible. The above separation schemes have been primarily photophysical in nature. The main emphasis of this report is to discuss the use of chemical reactions of the selectively excited species for the separation procedure. Three photochemical reaction procedures have been reported: 1) Electronically excited thiophosgene ($CSC\ell_2$) was reacted with diethoxyethylene (DEE) and resulted in significant enrichment of chlorine isotopes. [10] 2) A reaction between vibrationally excited boron trichloride ($BC\ell_3$) and hydrogen sulfide (H_2S) produced enrichment in the boron isotopes. [11] 3) An interesting exchange reaction between electronically excited iodine monochloride (ICl) and trans-dibromoethylene (DBE) resulted in isotopically enriched cis-and trans-dichloroethylene. [12] The interest in the use of photochemical reactions for isotopic enrichment lies in their potentially high efficiency. Multiphoton processes are inherently of low efficiency because so much energy must be supplied to produce the extra photons for the separation process. The chemical reaction requires no extra energy, in fact, energy is given off by the reaction which could be recycled to further improve the efficiency.

Chemical Reaction Considerations

There are several criteria which must be considered when choosing a reaction for isotopic enrichment:

1) The reaction must proceed significantly faster in the excited state than in the ground state. This requirement usually presents no great difficulty for electronically excited states.
2) The reaction rate must compete favorably with normal relaxation processes in the molecule including energy transfer processes which result in a loss of isotopic specificity. This usually means that one out of every one to ten collisions must result in a reaction.
3) The reaction should result in a stable product. The production of radicals often results in a chain reaction and a consequential loss of specificity.
4) The product should not absorb the exciting radiation.
5) The product should be easily separable from the reaction mixture.
6) The starting material should be easily regenerable from the product. This permits a many step recycled process with continued improvement in the enrichment at each step.

Details of the Reaction of $CSC\ell_2$ with DEE

A lot of thought went into the choice of $CSC\ell_2$ and DEE as the reaction partners.+ There is not much known about bimolecular reactions when one of the reactants is in the excited state. Most of traditional photochemistry has concentrated on unimolecular processes following photoexcitation. The main exception in this area is the reaction of carbonyl compounds with substituted ethylenes about which a fair amount of information is known. Unfortunately, most carbonyl compounds do

---

+ We are grateful to Nicholas Turro of Columbia University and Paul Schaap of Wayne State University for helpful discussions during the formative stages of this problem.

not absorb in the visible region. Substitution of a sulphur atom for the oxygen atom results in compounds which do absorb visible radiation. A particularly attractive molecule of this class is $CSCl_2$. The spectrum of this molecule has a line structure and spans the visible spectral region. Vibrational assignments have been completed [13] and molecular isotope splittings are clearly resolved (see Fig. 1). Many substituted olefins were studied as possible reaction partners and DEE was found to be the most suitable.

The reaction appears to proceed as shown below.

$$Cl_2C{=}S^{*} \; + \; (EtO)HC{=}CH(OEt) \; \rightarrow \; S{=}C(Cl){-}CH(OEt){-}CH(Cl){-}OEt$$

The product is a liquid and precipitates onto the walls of the reaction vessel. The quantum yield was found to be as high as 0.6. For convenience, we decided to follow the isotopic enrichment in the unreacted starting material as the reaction proceeded. Some typical mass spectra of unreacted $CSCl_2$ are shown in Fig. 3. Inspection of this figure shows that either

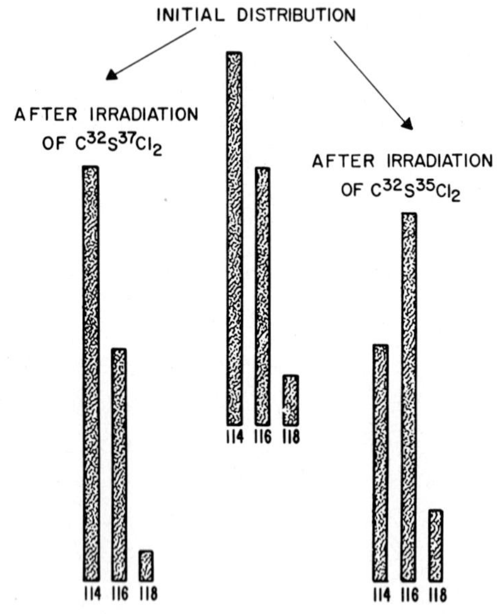

Figure 3 - Mass spectrum of unreacted $CSCl_2$ after reaction proceeded greater than 80% towards completion.

isotope can be enriched by choosing the proper irradiation wavelength. The case where $CS^{35}Cl_2$ was irradiated is particularly striking. The species $CS^{35}Cl_2$ went from a composition of 55% to 35%. The average laser power was 7 mw of which only 5% was absorbed. Milligrams of material were enriched in hours. The process could easily be scaled to produce gms of materials in hours by using a 1 watt laser and a longer cell to absorb all of the radiation.

The complicated relaxation paths available to small polyatomic molecules are listed below.

$CS^{35}Cl_2^*(^1A_2)$

$+ CS^{35}Cl^{37}Cl(^1A_1)$	$\xrightarrow{k_1}$	$CS^{35}Cl^{37}Cl^*(^1A_2) + CS^{35}Cl_2(^1A_1)$	energy transfer
+ DEE	$\xrightarrow{k_2}$	Product (35-35)	reaction
+ DEE	$\xrightarrow{k_3}$	DEE + $CS^{35}Cl_2(^1A_1)$	quenching
	$\xrightarrow{k_4}$	$CS^{35}Cl_2(^1A_1) + h\nu$	fluorescence
+ DEE	$\xrightarrow{k_5}$	$CSCl_2^*(^3A_2)$ + DEE	ISC
$+ CSCl_2(^1A_1)$	$\xrightarrow{k_6}$	$CSCl_2^*(^3A_2) + CSCl_2(^1A_1)$	ISC
$+ CSCl_2(^1A_1)$	$\xrightarrow{k_7}$	$2\ CS^{35}Cl_2\ (^1A_1)$	quenching

With the exception of the energy transfer rate constant (k_1), all of the above rate constants can be obtained by various experiments. By choosing trial values for the energy transfer rate constant it is possible to find an energy transfer rate constant which fits the amount of enrichment experimentally obtained for a particular set of reactant concentrations. However, the same rate constant does not fit the data as the ratio of DEE to $CSCl_2$ is changed. This means something is wrong with the model and we are currently trying to find the problem.

It was thought that the reaction might be proceeding via the triplet state but this hypothesis was discarded when it was found that direct irradiation of the triplet state resulted in processes which were not 1st order in the concentration of the reactants and a quantum yield of less than 0.1. Another promising hypothesis is that the reaction actually takes place from highly excited vibrational levels of the ground electronic state formed in the quenching processes outlined above. No direct experimental evidence to confirm this hypothesis exists at this time.

Impact

We mentioned above that a relatively low powered, efficient laser system will enrich a mole of isotopes in 66 hours. The current prices for enriched isotopes of several interesting light elements are given in Table 1.

Table 1

Element	Enrichment	Price/mole (K dollars)
^{37}Cl	> 95%	122.5
^{10}B	> 95%	.2
^{34}S	> 90%	290.0
^{15}N	> 95%	21.1
^{13}C	> 90%	10.0
^{18}O	> 99%	84.0

A significant reduction in the cost of the light isotopes will increase their availability and use in many areas. For example, ^{15}N and ^{13}C are being explored as tracer elements for medical applications where radioactivity must be avoided, ^{18}O is being used in a breath analysis procedure which may become as popular as medical X-rays, Cl and S could be used to study the environmental impact of certain pesticides and fertilizers, and the high neutron cross section of ^{10}B is already being used for reactor shielding and brain tumor diagnosis.

In summary, we believe that laser based, relatively efficient processes will be developed within the next several years to enrich isotopes in most of the light elements. These processes will be capable of producing tens of kilograms to hundreds of kilograms of material. The increased availability of these isotopes will increase their use in areas already explored or envisioned and more importantly in areas not anticipated at this time.

REFERENCES

1 Frequency Doubling in $KB_5O_8.4H_2O$ and $NH_4B_5O_8.4H_2O$ to 217.3 nm, C. F. Dewey, Jr., W. R. Cook, Jr., R. T. Hodgson, and J. J. Wynne, Appl. Phys. Lett. in press.

2 Nonlinear Optics in Metal Vapors, S. E. Harris, J. F. Young, D. M. Bloom, A. H. Kung, J. T. Yardley and G. C. Bjorklund, VIII International Quantum Electronics Conference (1974).

3 Separation of Isotopes by Laser Deflection of Atomic Beam I. Barium, A. F. Bernhardt, D. E. Duerre, J. R. Simpson, and L. L. Wood, Appl. Phys. Lett. 25, 617 (1974).

4 Macroscopic Separation of Uranium Isotopes by Selective Photoionization, S. A. Tuccio, R. J. Foley, J. W. Dubrin, O. Krikorian, IEEE/OSA Conference on Laser Engineering and Applications (1975).

5 Two-Photon Laser Isotope Separation of Atomic Uranium, G. S. Jones, I. Itzkan, C. T. Pike, R. H. Levy, L. Levin, IEEE/OSA Conference on Laser Engineering and Applications (1975).

6 Isotope Selective Photoionization of Calcium Using 2-Step Laser Excitation, U. Brinkmann, W. Hartig, H. Telle, H. Walther, Appl. Phys. 5, 109 (1974).

7 Isotope Selective Chemical Reactions of BCl_3 Molecule in a Strong Infrared Laser Field, R. V. Ambartzumyan, V. S. Letokhov, E. A. Rialov, and V. V. Chekalin, IEEE/OSA Conference on Laser Engineering and Applications (1975).

8 Isotopic Enrichment by Multiple Absorption of CO_2 Laser Radiation, J. L. Lyman, R. J. Jensen, J. P. Rink, C. P. Robinson, S. D. Rockwood, IEEE/OSA Conference on Laser Engineering and Applications (1975).

9 Isotope Separation by Predissociation, E. S. Yeung and C. B. Moore Appl. Phys. Lett., 21, 109 (1972).

10 Laser Induced Photochemical Enrichment of Chlorine Isotopes, M. Lamotte, H. J. Dewey, R. A. Keller, and J. J. Ritter, Chem. Phys. Lett., 30, 165 (1975).

11 CO_2 TEA Laser-Induced Photochemical Enrichment of Boron Isotopes, S. M. Freund and J. J. Ritter, Chem. Phys. Lett., 32, 255 (1975).

12 Laser Separation of Chlorine Isotopes. The Photochemical Reaction of Electronically Excited Iodine Monochloride with Halogenated Olefins, D. D. -S. Liu, S. Dotta, R. N. Zare, J. Amer. Chem. Soc. 97, 2557 (1975).

13 The 5340 Å Band System of Thiocarbonyl Chloride, J. C. D. Brand, J. H. Callomon, D. C. Moule, J. Tyrrell and T. H. Goodwin, Trans. Faraday, Soc. 61, 2365 (1965).

DISCUSSION

M. LAMOTTE, H.J. DEWEY, J.J. RITTER, R.A. KELLER

R.M. Hochstrasser - In view of your stated requirement for isotope separation that : "This restricts us to small molecules in the gas phase and to non-dissociative states" I would like to announce that David King of my laboratory has performed efficient isotope separations in a moderately large molecule in the solid state by exciting directly a predissociative state. The system was natural s-tetrazine in a benzen matrix at 4.2 k. C^{13}, N^{15} and H^2 were enriched (factors of ca. 10^4) by (what we assume is) predissociation of the singlet state into N_2 and HCN; the work is in press (JACS)

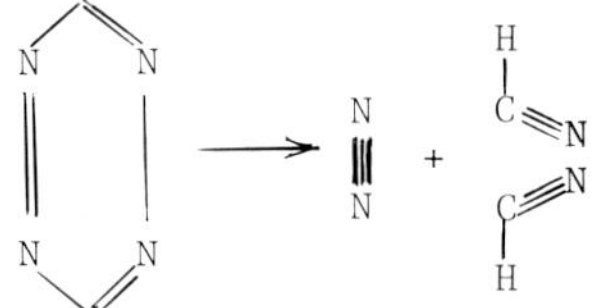

E.P. Ippen - Are the high isotope prices shown in the last slide determined by production costs or by current low demand?

R.A. Keller - The prices shown are catalog prices. I am sure that they could be reduced if a particularly large demand stimulated production.

S. Leach - Does the thiophosgene excited state rate of reactivity with diethoxyethylene depend on the vibronic level reached on laser excitation? If so, this could be affecting your result indicating that the triplet is not the active state.

R.A. Keller - We relize this possibility and experiments are in progress to determine if the reaction rate depends upon the vibronic level initially excited.

S. Kimel - Would you be able to comment on recent Russian publication in which isotope enrichment was achieved by having a discharge in a $N_2 + O_2$ mixture and obtaining isotopically enriched NO product molecules? This seems even more difficult to understand than multiphoton excitation of BCl_3 by thirty or so photons from a CO_2 laser

R.A. Keller - The Russian work was reported last year at the IEEE meeting in San Francisco. At this time the report received a lot of attention. Since the original report I have heard no further information and I have nothing to add.

C.B. Harris - You should be a little more careful in considering predissociation. If level crossing occurs before predissociation, it can still be an efficient means of separation if the level crossing does not significantly broaden the transition.

R.A. Keller - We did not intend to slight predissociation; it has been shown to be an effective process in formaldehyde. As Dr. Harris has pointed out, the predissociation may occur via an intermediate state or states (i.e. one formed by level crossing). The conclusions are the same; the level crossing must be fast enough to compete with normal relaxation processes but not too fast to broaden the transition.

J. Bastiaens - Up to now, one used the monochromaticity and the power of the laser. Are there prospects for the exploitation of the coherence charactéristics in the field of laser-induced isotopic separation?

R.A. Keller - I am sure that there must be some isotopic enrichment processes which use the coherence effects of laser. However, I know of no reported successes in this area.

DYNAMICAL INVESTIGATIONS OF ULTRAFAST VIBRATIONAL RELAXATION AND ENERGY TRANSFER IN LIQUIDS

A. LAUBEREAU AND W. KAISER
Technische Universität München, München, West Germany

Light pulses of several picosecond duration and of well-defined pulse properties are now experimentally available. We have applied these pulses for direct studies of dynamical molecular processes in liquids.

In this contribution recent investigations on vibrational relaxation processes in the electronic ground state are discussed. Our experimental techniques consist of two steps. A powerful light pulse first excites a specific vibrational mode in the liquid sample; the vibrational system is subsequently interrogated by a second weak light pulse of well-defined delay time. Different techniques are used for excitation and probing of the vibrational modes. In this paper we apply transient stimulated Raman scattering or infrared absorption for the excitation of well-defined molecular vibrations. Incoherent Raman probe scattering or fluorescence involving the first excited electronic state are used as probing mechanisms for the instantaneous excess population of the first excited vibrational state. The generation of the ultrashort pulses of visible or infrared radiation required in our experiments is discussed in a preceeding paper by Kaiser et al.

A first type of experiments is concerned with the decay and the transfer of vibrational energy in liquids. An experimental setup is depicted schematically in Fig. 1. A single intense picosecond pulse at the laser frequency $\nu_L = 9455\ cm^{-1}$ ($\lambda_L = 1.06\ \mu m$) is directed into the liquid sample and effectively excites the molecular vibration of interest via transient stimulated Raman scattering. Numerical estimates show that one molecule out of one thousand is promoted to the first excited (v = 1) state of the specific vibrational mode. When the pump pulse has passed the sample, the highly nonlinear excitation process rapidly terminates and the vibrational system relaxes freely. A second weak light pulse at the frequency

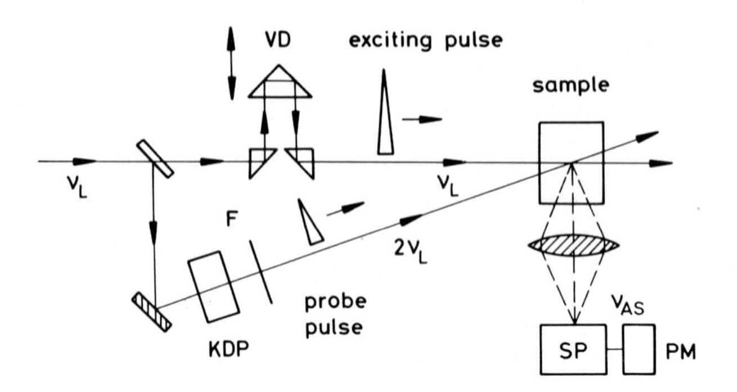

Fig.1. Experimental system for the measurement of the energy relaxation time τ' of a molecular vibration in liquids. Variable delay VD, frequency doubling crystal KDP, collecting lens CL for the incoherent probe scattering, spectrometer SP and photomultiplier PM.

$\nu_{2L} = 2\nu_L = 18910\ cm^{-1}$ ($\lambda_{2L} = 0.53\ \mu m$) is produced by second harmonic generation in a KDP crystal. This pulse enters the sample shortly after the intense pump pulse and interrogates the instantaneous degree of vibrational excitation. Incoherent (spontaneous) Raman scattering of the probing pulse produces a weak scattering signal at the anti-Stokes frequency $\nu_{AS} = \nu_{2L} + \nu_o$ which is observed at 90° scattering angle within a large solid angle of acceptance; ν_o denotes the frequency of the vibrational mode studied. This signal is a direct measure of the excited state population. Measuring the scattering signal as a function of delay time t_D between pump pulse and probing light pulse, the rise and decay of the vibrational energy stored in the (v = 1) vibrational state is directly observed.

Using this technique we have measured for the first time the energy relaxation time (population lifetime) of well-defined vibrational modes in liquids. Experimental results have been obtained on several systems. As an example Fig. 2 presents data for the mixture of CH_3CCl_3 (60 mol %) and CCl_4. The symmetric CH_3-stretching mode at $\nu_H = 2939\ cm^{-1}$ of the

CH_3CCl_3 molecule was investigated. The observed anti-Stokes scattering signal $S^{inc}(t_D)$ of the probe pulse is plotted in Fig. 2 as a function of delay time t_D. The experimental points (open circles) represent the instantaneous excess population of the first excited level of the ν_H-mode. The signal curve (broken line) rises to a maximum at $t_D \simeq 8$ psec which is delayed with respect to the maximum of the pump pulse at $t_D = 0$. This fact results from the accumulative nature of transient stimulated Raman scattering exciting the molecular vibration. For larger values of t_D the scattering signal decays exponentially. From the slope of the signal curve, the energy relaxation time τ' (population lifetime) is directly obtained. A value of $\tau' = 6.5$ psec is found for the example in Fig. 2.

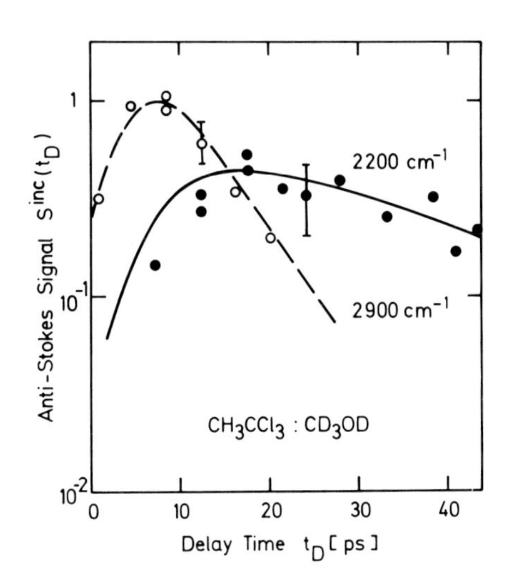

Fig.2. Measurement of the energy relaxation time τ' and of the intermolecular energy transfer in the mixture CH_3CCl_3 : CD_3OD. Open circles denote probe scattering from the CH_3-valence bond vibration of CH_3CCl_3 which is excited by the pump pulse. Full points represent scattering observed from the CD_3-valence bond vibration of CD_3OD directly indicating intermolecular transfer of vibrational energy.

It should be noted that the experimental time resolution achieved by our probing technique depends primarily on the wings and not on the duration (FWHH) of the light pulses. With our well-defined pulses of 6 psec duration our experimental system is capable of resolving relaxation times as short as 1 psec.

The question arises what physical processes determine the energy relaxation time τ'. Fig. 3 serves to illustrate the situation. The normal mode frequencies of three molecules are depicted. A polyatomic molecule such as CH_3CCl_3 possesses a large number of vibrational modes. The vibrational level ν_H at ~2900 cm^{-1} represents the CH_3-stretching mode discussed above in context with Fig. 2. This mode is excited via stimulated Raman scattering. For the pure liquid CH_3CCl_3 the CH_3-stretching mode ν_H couples to the CH_3-bending modes δ_H at approximately half the frequency. The process $\nu_H \rightarrow 2\delta_H$ is indicated by the arrows in Fig. 3. This process generates a subsequent excess population of the δ_H modes around 1400 cm^{-1} which was experimentally observed tuning the spectrometer to the corresponding anti-Stokes frequency position in our probe scattering experiment.

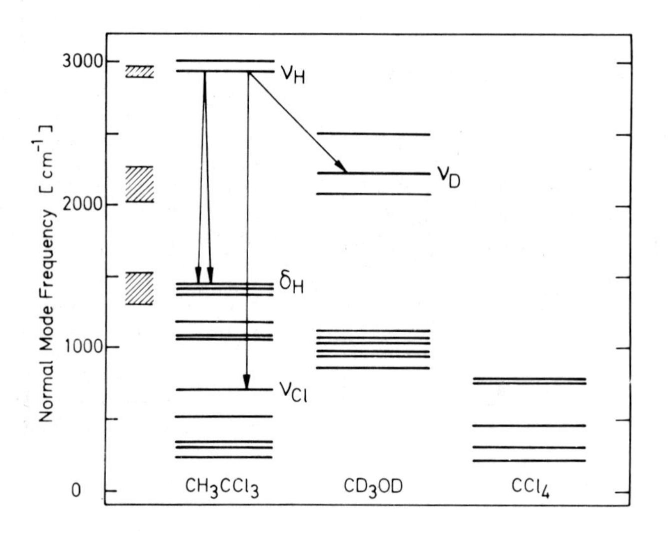

Fig.3. Frequency values of the normal vibrational modes of the liquids CH_3CCl_3, CD_3OD and CCl_4.

Direct information on intermolecular energy transfer was obtained in the mixture CH_3CCl_3 : CD_3OD. Fig. 3 shows the energy levels of the solvent CD_3OD. The CD_3-stretching vibration of this molecule at ν_D = 2227 cm^{-1} and the frequency ν_{Cl} = 713 cm^{-1} of a CCl_3-vibration of CH_3CCl_3 add up with good accuracy to the ν_H mode of CH_3CCl_3 which is experimentally excited in the mixture. An interaction between CH_3CCl_3 and CD_3OD molecules of the form $\nu_H \rightarrow \nu_D + \nu_{Cl}$ is suggested by energy resonance arguments (see arrows in Fig. 3).

Experimental data on the intermolecular energy transfer from the laser excited ν_H-vibration of CH_3CCl_3 to the ν_D-vibration of the solvent molecule are presented in Fig. 2. The data of the ν_H-vibration (open circles) of CH_3CCl_3 have been discussed above. Of special interest is the incoherent scattering signal observed at a frequency shift of $\sim 2200\ cm^{-1}$ (full points) which originates from the subsequent excess population of the ν_D-vibration of CD_3OD. The data display a delayed maximum with a slow decay of 25 ± 10 psec. This experiment gives convincing evidence of energy transfer to CD_3OD molecules. The results of Fig. 2 represent the first observation of intermolecular energy transfer in liquids. The curves in Fig. 2 are calculated from the theory of transient stimulated Raman scattering exciting the ν_H-mode (broken line) and a rate equation model for the energy transfer to the ν_D-vibration (solid curve). Knowing the values of the spontaneous cross-sections of the two vibrations, a comparison of the scattering signals yields a quantum efficiency of approximately 60% for the transfer process. Our study of the concentration dependence of the energy relaxation of CH_3CCl_3 in mixtures with CCl_4 and CD_3OD show that the $\nu_H \rightarrow 2\delta_H$ interactions have an efficiency of 40%.

A different experimental technique has been devised for dynamical investigations of molecular vibrations in highly diluted systems. This method is particularly well suited for the study of the electronic ground state of fluorescent molecules in liquid solutions. Fig. 4 illustrates schematically the transitions involved in these investigations. A powerful short pulse of infrared radiation first resonantly interacts with the vibrational system. The frequency ν_1 of an infrared pulse is properly selected in order to excite a well-defined vibrational mode by infrared absorption. 10^{15} infrared quanta are contained in the pump pulse of approximately 3 psec duration. The molecular system is subsequently interrogated by a second pulse (ν_2) of visible light. The vibrationally excited molecules make an electronic transition to a level situated close to the vibrational ground state of the first excited singlet state S_1 and produce a fluorescence emission. Molecules in the electronic ground state in levels smaller than $h\nu_1$, on the other hand, do not interact with light of frequency ν_2. The time integrated fluorescent radiation serves as a direct measure of the instantaneous vibrational excitation at frequency ν_1 when the probe pulse interacts with the excited molecules. The fluorescence signal is observed as a function of delay time t_D between the infrared excitation pulse and the probing light pulse. It should be noted that a well-defined vibrational excitation is initially produced even in large molecules with complicated vibrational spectra.

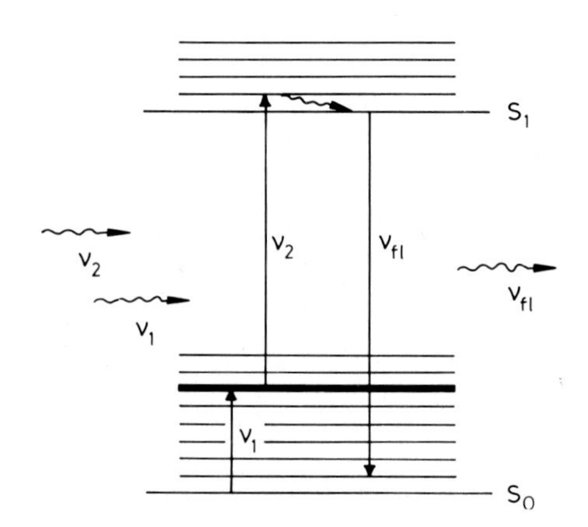

Fig.4. Schematic of the molecular energy levels and transitions during the vibrational excitation and probing process.

We have applied this novel technique to dynamic investigations of dye solutions in a large concentration range of 10^{-6} M to 10^{-3} M. As an example we discuss results on Coumarin 6 in the solvent CCl_4; the structure of the molecule is depicted on the top of Fig. 5. Two C_2H_5-groups are bonded to the conjugated ring system. The infrared transmission spectrum of Coumarin 6 around 3000 cm^{-1} is also shown in Fig. 5. Several absorption peaks are clearly resolved in the spectrum. The bands between 2865 cm^{-1} and 2970 cm^{-1} are readily interpreted as normal vibrational modes of the two

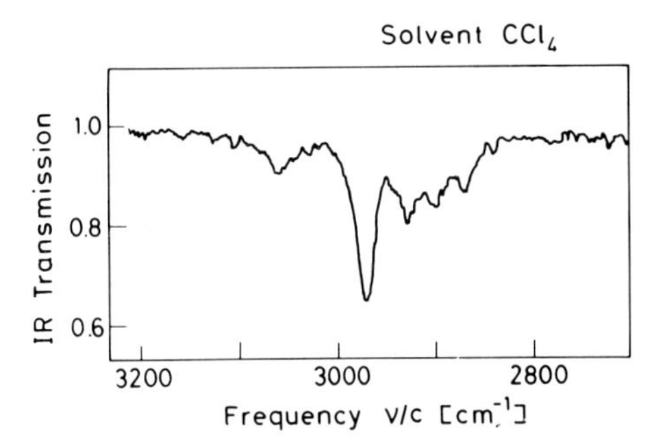

Fig.5.

Top: The Coumarin 6 molecule.

Bottom: Infrared absorption spectrum of Coumarin 6 in CCl_4 (10^{-3} M) around 3000 cm^{-1}. The line at 2970 cm^{-1} is excited in the picosecond experiment.

ethyl groups of Coumarin 6. The CH-valence bond vibrations of the ring system produce an additional absorption peak at approximately 3055 cm^{-1}. Most important for the following investigations is the maximum absorption band of the asymmetric CH_3-mode at ~2970 cm^{-1}.

In our picosecond experiment the (v = 1) state of the vibrational mode at 2970 cm^{-1} is populated by a resonant short infrared pulse. The vibrational excitation is subsequently observed as a function of time with our fluorescence technique. The probing pulse at ν_2 = 18910 cm^{-1} is prepared as follows: The primary laser pulse first passes through a nonlinear absorber cell where the pulse is shortened and, most important, the leading part of the pulse is significantly sharpened. This method effectively improves the experimental time resolution. The frequency of the pulse is then converted to the second harmonic by a KDP crystal.

Experimental results for Coumarin 6 in CCl_4 are presented in Fig. 6 for a concentration of 3×10^{-5} M.

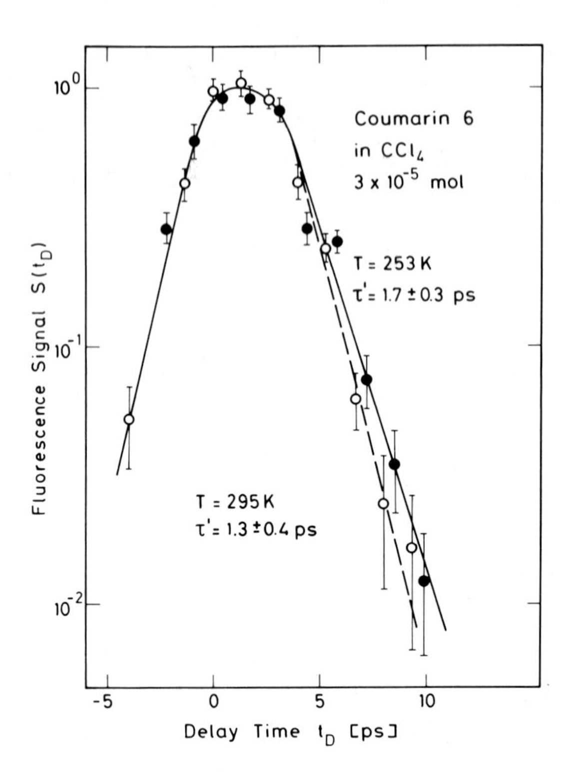

Fig.6. Ultrafast vibrational relaxation of Coumarin 6 in CCl_4 at 295 K (open circles) and 253 K (full points). The asymmetric CH_3-mode at 2970 cm^{-1} in the electronic ground state is excited and the vibrational excitation is observed as a function of time with a novel fluorescence probing technique.

The observed fluorescence signal, $S(t_D)$, initiated by the probe pulse is plotted as a function of delay time t_D for two temperatures. The time scale of a few psec should be noted. The fluorescence signal increases sharply within approximately 1 psec to a maximum value at $t_D \simeq 2$ psec. The build-up of excess population in the vibration spectrum at $\nu_1 = 2970\ cm^{-1}$ is directly seen from the rise of the signal curve. Of special interest are the decaying parts of the signal curves which extend over two orders of ten. Fast relaxation of the vibrational excitation with slightly different slopes for the two temperatures is clearly indicated by the data. At room temperature a time constant of 1.3 ± 0.4 psec is directly obtained from the exponential slope (open circle). At -20° C the energy relaxation time is measured to be τ' = 1.7 ± 0.3 psec. We have ascertained that these time constants represent true relaxation times. The liquid cell containing the Coumarin molecules was replaced by a nonlinear crystal. The intensity of the sum frequency at $\nu_1 + \nu_2$ was measured as a function of the delay time between the two pulses of frequency ν_1 and ν_2. In this way the time resolution of our system was tested and found to be better than one psec.

At the present time the energy relaxation process is not yet definitely determined. There are two possible physical mechanisms which lead to a decay of our fluorescence signal: 1) energy transfer to vibrational states of approximately equal energy with smaller absorption cross-sections to the excitedelectronic state; 2) energy decay to lower vibrational states which cannot interact with the probe pulse. Work is in progress to illucidate the relaxation process in more detail.

In conclusion it should be pointed out that different experimental techniques are now available for detailed studies of energy relaxation times and transfer processes of polyatomic molecules in liquids. Spontaneous anti-Stokes probing is well suited for high concentration systems while the fluorescence technique is particularly useful for highly diluted solutions.

This report represents a summary of several publications. For details we refer the reader to following papers:

A. Laubereau, D. von der Linde and W. Kaiser, Phys. Rev. Letters 28, 1162 (1972).

A. Laubereau, L. Kirschner and W. Kaiser, Opt. Commun. 9, 182 (1973).

A. Laubereau, G. Kehl and W. Kaiser Opt. Commun. 11, 74 (1974).

A. Seilmeier, A. Laubereau and W. Kaiser, to be published.

DISCUSSION

A. LAUBEREAU, W. KAISER

R.M. Hochstrasser - Is the 2970 cm^{-1} vibration that you populate by IR absorption one of the Franck-Condon progression forming modes of the coumarin-6 absorption and emission spectrum? Obviously it should be, in order that your interpretation be as stated; but that will depend on total energy of IR + probe pulse in relation to the positioning of vibronic levels in the fluorescence state of coumarin-G. Could you comment on this please?

A. Laubereau - The absorption and fluorescence spectra of coumarin-6 in liquid solutions look fairly broad and do not show much structure. So little experimental information is available about the absorption cross sections of individual vibrational states in electronic transitions. Theoretical arguments suggest that the CH_3 stretching mode at 1970 cm^{-1} couples satisfactorily to the incident probe pulse. It appears difficult however to draw safe conclusions from the probing process alone about the specific vibrational level we are looking at. It is the excitation process in our investigations which presents evidence that the CH_3 stretching mode at 1970 cm^{-1} is involved in the rapid vibrational relaxation process we have experimentally observed.

A.Tramer - Are you sure that the infrared light pulse excites really the v = 1 level of the C-H stretching mode and not an overtone transition involving the modes coupled to the electronic transition? Does the measured relaxation time characterize really the ν_{CH} v = 1 level?

A. Laubereau - The strong selection rules for the absorption of the infrared excitation pulse should be noted which prefer Δ v = 1 transitions. This point is readily seen from the infrared transmission spectrum of coumarin-6 I discussed in connection with one of my slides. Unfortunately I had not time enough to discuss the additional arguments we have supporting this point. For example, we also have experimental information about the absorption cross section of the vibrational excitation in the probing process. Knowing this value we can compare the absolute magnitude of the excess population observed in the probing process with calculated numbers for the vibrational population generated by the IR excitation pulse. As a result we are quite certain that the (primary) vibrational excitation produced in the experiment is an excess population of the (v = 1) CH_3-stretching mode at 1970 cm^{-1}.

M. Stockburger - The uncertainty in this experiment lies in the fact that one cannot be definitely sure that the vibrational level which is excited by an infrared pulse also couples with the electronic transition. This could be overcome when a system is formed where a clearly resolved hot band in the electronic absorption spectrum is used, which at the same time is infrared active.

A. Laubereau - Comparing with other techniques investigating vibrational relaxation of dye molecules I should like to emphasize the more detailed character of the relaxation time we are directly measuring. Certainly I agree with you that an improved knowledge about Franck-Condon factors facilitates the interpretation of the information obtained by investigations using the fluorescence probing technique.

S. Kimel - Would you be able to give absolute values for the efficiency of the SRS process to populate vibrational levels, both compared to the number of incident photons/pulse and to the number of molecules in the sample.

A. Laubereau - In our experiments we work at a moderate intensity level so that the energy conversion efficiency of stimulated Raman scattering is kept at a few percent. Starting with a laser pulse of 5 to 10 mJ energy this number corresponds to 10^{15} vibrational quanta produced in the excitation process. The excited molecules are contained in a small volume of $\sim 2 \times 10^{-4}$ cm^3. For a number density of 5.10^{21} molecules/cm^3 the occupation probability n to promote a molecule in the first excited vibrational state during the stimulated pumping process is estimated to be n 10^{-3}.

ISOTOPICALLY SELECTIVE PHOTOCHEMISTRY IN THE SOLID STATE

R. M. HOCHSTRASSER and D. S. KING

Department of Chemistry and Laboratory for Research on the Structure of Matter, University of Pennsylvania, Philadelphia, Pennsylvania 19174

1. INTRODUCTION

The laser induced photochemistry of a mixed molecular crystal system was studied at very low temperatures. The photolysis of s-tetrazine ($C_2N_4H_2$) in benzene at 1.6 K appears to be a fast unimolecular process involving the simultaneous making, and breaking of three bonds to yield N_2 and HCN quantitatively. Spectroscopically, in this temperature regime, molecular crystals have linewidths typically of 0.1 cm^{-1} to 3 cm^{-1} and zeropoint isotope shifts of 1 cm^{-1} to 10 cm^{-1}. A relatively high degree of isotopic selectivity may be achieved on excitation with a narrow band tunable dye laser leading to isotopic purification and/or the selective production of rare isotopically substituted photoproducts. The use of a large molecule lattice may well provide sufficient thermal exchange to enable the stabilization and trapping at 4.2 K or 1.6 K of hitherto unobserved reaction intermediates.

2. RESULTS

The visible absorption spectra of s-tetrazine shows two predominant states. These excited states result from an $n \rightarrow \pi^*$ electronic promotion and have been assigned as $^1B_{3u}$ and $^3B_{3u}$ states assuming D_{2h} molecular symmetry. The vibronic structures of the singlet and triplet are very similar and consist of progressions in the totally symmetric mode ν_{6a}' (ca. 700 cm^{-1}) and, more weakly, a few other vibrational levels. Tetrazine spectra in several mixed crystal systems at 4.2 K readily expose ^{13}C and ^{15}N isotope bands on the origin at moderate resolution. ^{13}C lines are, in fact, readily observable in a wide range of heteroaromatic-aromatic molecular systems at 4.2 K.

In a typical dye laser fluorescence excitation spectrum of tetrazine in benzene at 1.6 K, with ca 0.8 cm^{-1} resolution monitoring the total emission, all vibronic bands show clear evidence of isotopic structure including the presence of $C_2 N_4 DH$. The emission following excitation into the singlet manifold of a given isotopic species, $^{13}C\,^{12}C\,^{14}N_4H_2$ for instance, is unique and distinctive from the fluorescence excited by absorption into any other isotopic species manifold and gives directly the zero-point energy and ground state vibrational structure of that particular isotopic species of tetrazine. Similarly, by monitoring the emission of a single isotopic species through a monochromater with 20 μ slits, ca. 0.5 cm^{-1} resolution, the vibronic structure of the excited state is mapped out for each isotopic species of tetrazine. Figure 1 shows the energy level diagram and kinetic scheme for the photolysis of s-tetrazine in

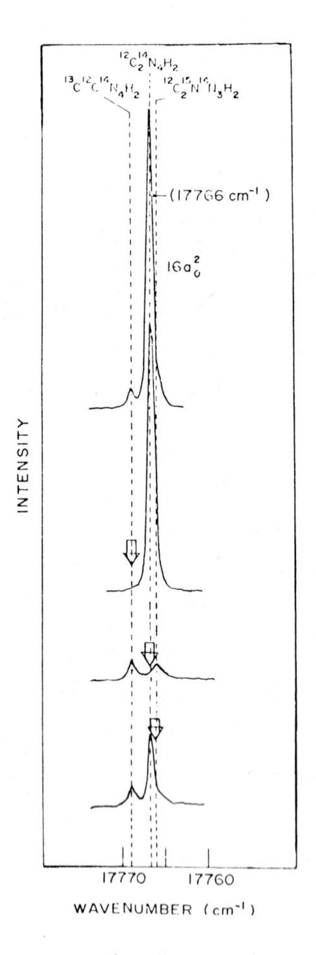

Fig. 1. S-tetrazine energy level diagram.

benzene at 4.2 K. The origin and predominent vibrational levels are shown for the ^1A_g, $^1B_{3u}(n\pi^*)$, and $^3B_{3u}(n\pi^*)$ states of $^{12}C_2\,^{14}N_4H_2$. The arrows correspond to the Franck-Condon peaks in the optical spectra, vibrational relaxation introduces no isotopic scrambling of a selective excitation, and there is no observable intersystem crossing. The observed fluorescence and phosphorescence decay rates are similar in the gas (300 K) and crystal (1.6 K) and are determined by the photochemical reaction rates k_1^p and k_3^p respectively.

The emission from both the observed states of tetrazine is weak. Much weaker and with shorter lifetimes than expected for most of the other aza substituted benzenes, and there is no observable intersystem crossing $^1B_{3u} \rightsquigarrow {}^3B_{3u}$. The $^1B_{3u}$ radiative lifetime calculated from absorption cross-section is ca. 10^{-6} sec, however the observed fluorescence lifetime is $\lesssim 5 \times 10^{-10}$ sec. The predominant process responsible for the depopulation of the excited state seems to be a unimolecular dissociation. Directly following irradiation of either the singlet or the triplet manifold of a mixed crystal of tetrazine in benzene at 1.6 K, the infrared spectrum (also at 1.6 K) shows the quantitative presence of HCN. N_2 is detected upon fractional distillation of the sample under vacuo. It is consistent with all experimental data that the photochemical reaction is the predominant mechanism of depopulating both the $^1B_{3u}$ and the $^3B_{3u}$ states. It is a single photon process occurring in both gas and condensed phase with near unit quantum yield and rate constants of $k^{(1)} \gtrsim 2 \times 10^9$ for the singlet and $k^{(3)} \sim 1 \times 10^4$ for the triplet. The ratio $k^{(3)}/k^{(1)} \sim 2 \times 10^5$ is consistent with a spin selective process, $k^{(3)}$ being determined by $k^{(1)}$ and spin-orbit coupling. This reaction may be written directly as:

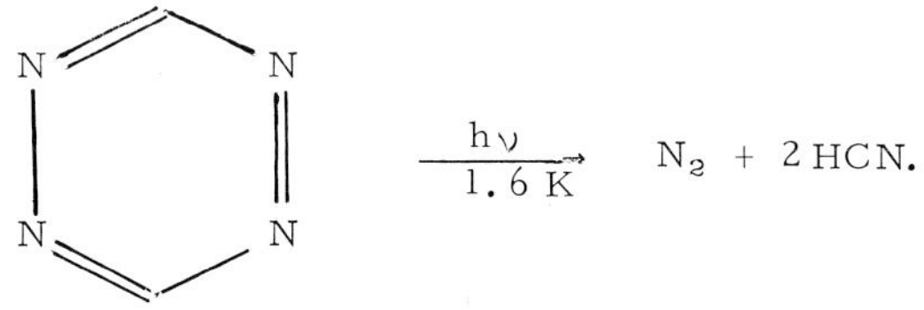

$$\text{tetrazine} \xrightarrow[1.6\ K]{h\nu} N_2 + 2\,HCN.$$

The non-appearance of an intermediate at 4.2 K could conceivably be caused by a local thermal process ulitizing the energy released in the first-step, but we expect aromatic lattices to have very high thermal conductivity at 4.2 K especially compared with diatomic and atomic lattices in which local heating might become an important factor.

The isotopic selectivity of the excitation and emission spectra of s-tetrazine in mixed crystals at 1.6 K demonstrates the negligible rate of isotopic scrambling. The relative intensities of the peaks assigned to the various isotopic species in excitation (including C_2N_4DH!) correspond to the isotopic natural abundance ratios, indicating negligible isotopic effects on the rate of the photoprocess. Selective excitation into the vibronic manifold of a specific isotopic species of tetrazine at 1.6 K enhances the photochemical reaction rate of only that particular isotopic species - all other species being unaffected. Excitation spectra obtained after irradiation of a given isotopic species conclusively show that only the excited species underwent the photochemical reaction.

Figure 2 shows the excitation spectra of the isotopically selective photodecomposition of s-tetrazine in benzene at 1.6 K. Four suitably dilute crystals showing the natural abundances of ^{13}C and ^{15}N containing tetrazine, in benzene, were placed in an optical helium dewar and cooled to 1.6 K. The $16a^2_0$ transition of the $^1B_{3u}$ $(n\pi^*)$ state of tetrazine was observed in excitation by monitoring the total fluorescence intensity as a function of exciting dye laser wavelength. The excitation and photolysis source was a narrow band (~ 0.8 cm^{-1}) tunable dye laser pumped by a 100 KW nitrogen superradiator and using fluorescein as the active medium. The transition energies for each isotopic species of tetrazine were established in the first sample. Each of the subsequent samples was irradiated at a specific isotope transition as indicated by the arrows. The respective excitation spectra of the photolyzed samples demonstrate the high degree of isotopic selectivity achieved in the decomposition reaction in this system. The situations depicted in the figure (corresponding to an enrichment of 10^4 -fold for ^{15}N and ^{13}C) were achieved by a few minutes irradiation. The selective irradiation of samples at the appropriate wavelengths produces pure samples of $^{13}C^{12}C^{14}N_4H_2$ and $^{12}C_2{}^{15}N^{14}N_3H_2$ representing enrichments in ^{13}C and ^{15}N of 10^4 fold; and 1:1 mixtures of $^{14}N^{15}N$:$^{14}N_2$, $H^{12}CN$:$H^{13}CN$, and HCN:DCN.

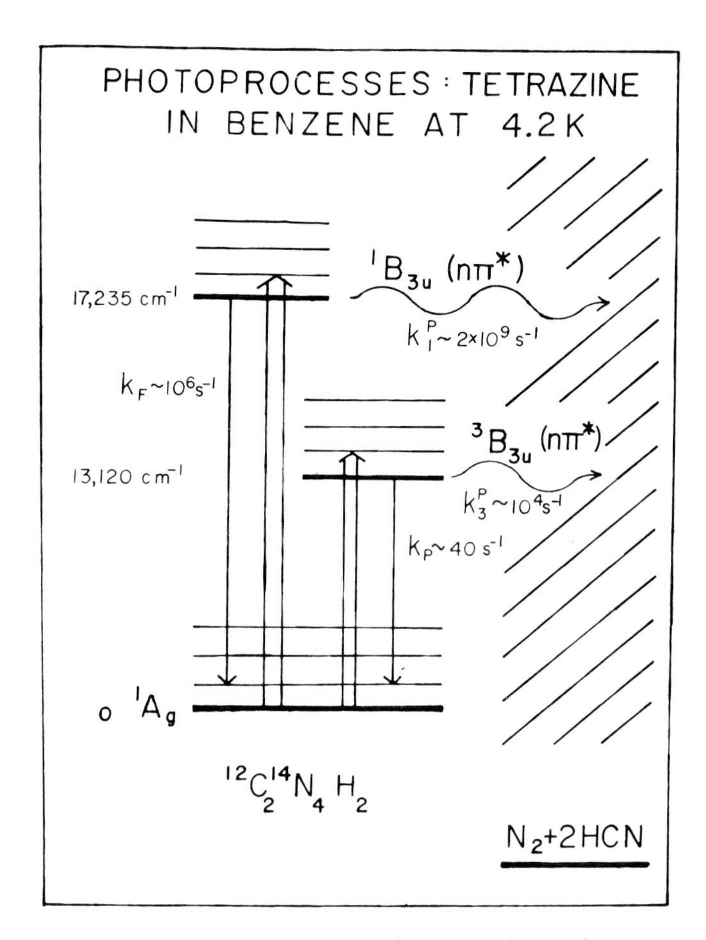

Fig. 2. Excitation spectra of s-tetrazine in benzene at 1.6 K.

3. Background Spectroscopy

Following the synthesis of s-tetrazine in 1906 by Curtius [1], and the unexpected observation of sharp line gas phase absorption spectra by Koenigsberger and Vogt [2] the time independent spectroscopy of tetrazine in the vapor [2 - 5], solutions [6, 7], and crystalline systems at low temperatures [7 - 10] have been examined. More recently, attention has been focused on the kinetics of the radiative [9 - 13] and non-radiative [13] processes of the tetrazine chromophore in various systems.

References

1. Th. Curtius, A. Darapsky, E. Müller, Ber. 40, 84 (1906).
2. J. Koenigsberger and K. Vogt, Phys. Z. 14, 1269 (1913).

3. G. H. Spencer, Jr., P. C. Cross, and K. B. Wiberg, J. Chem. Phys. 35, 1925 (1961).

4. A. J. Merer and K. K. Innes, Proc. R. Soc. London. A302, 271 (1968).

5. D. T. Livak and K. K. Innes, J. Mol. Spectry. 39, 115 (1971).

6. S. F. Mason, J. Chem. Soc., 1240 (1959).

7. M. Chowdhury and L. Goodman, J. Chem. Phys. 36, 548 (1962).

8. W. D. Sigworth and E. L. Pace, J. Chem. Phys. 54, 5379 (1971).

9. R. M. Hochstrasser and D. S. King, Chem. Phys. 5, 439 (1974).

10. J. H. Meyling, R. P. Van Der Werf and D. A. Wiersma, Chem. Phys. Letters 28, 364 (1974).

11. G. K. Vemulapalli and T. Cassen, J. Chem. Phys. 56, 5120 (1972).

12. J. R. McDonald and L. E. Brus, J. Chem. Phys. 59, 4966 (1973).

13. R. M. Hochstrasser and D. S. King, J. Am. Chem. Soc. in press.

ACKNOWLEDGMENT

This research was supported by the National Institutes of Health and by the NSF-MRL program through LRSM at the University of Pennsylvania. A fuller account is in press [J. Am. Chem. Soc.].

DETERMINATION OF DIFFUSION COEFFICIENTS OF MOLECULES IN THE TRIPLET STATE AND OF OTHER SHORT-LIVED PHOTOPRODUCTS

Ernst Georg Meyer and Bernhard Nickel
Max-Planck-Institut für biophysikalische Chemie
D-34 Göttingen

SUMMARY

A method for the determination of diffusion coefficients of molecules in the triplet state is described and exemplified by the determination of the diffusion coefficient of ^{3}anthracene* in n-hexadecane in the temperature range 20 ^{o}C ≤ t ≤ 90 ^{o}C. A general method for the determination of diffusion coefficients of photoproducts is proposed.

INTRODUCTION

In 1959 R.M. Noyes proposed a method for the determination of diffusion coefficients of reactive photoproducts [1]. The essence of the method is the spatially periodic excitation of a sample ("space intermittency") and the measurement of a second order reaction of the photoproduct of the type A + A → C. This method was later applied to the determination of the diffusion coefficient of iodine atoms in hexane [2] and in carbon tetrachloride [3]. Avakian, Merrifield, and Ern developed a similar method for the determination of diffusion coefficients of triplet excitons in molecular crystals [4,5,6]. Nickel, Nickel, and Meyer modified this method and applied it to the determination of diffusion coefficients of molecules in the triplet state in solutions [7,8,9]. The first systems investigated were ^{3}pyrene* in glycerol [8] and ^{3}anthracene* in n-hexadecane and in n-hexane [9,10]. In the present paper we report on the further development of this method [10], and we propose a new and in principle general method for the determination of diffusion coefficients of photoproducts.

SPATIALLY PERIODIC EXCITATION OF A SAMPLE AND TIME-DEPENDENCE OF THE SPATIAL DISTRIBUTION OF A PHOTOPRODUCT

A spatially periodic excitation of a sample S (Fig.1) with exactly known space-dependence is obtained with two interfering laser beams [7]. The spatial period of excitation is

$$d = \lambda/(2\,n \sin\varphi) = \lambda/(2 \sin\varphi_o) \tag{1}$$

with λ the wavelength in air, n the refractive index of the sample, 2φ the angle between the two interfering beams in the sample, and $2\varphi_o$ the corresponding angle in air. If the photoproduct is formed by a fast reaction and if the excitation is sufficiently short, then immediately after excitation at time t = 0 the initial distribution of the photoproduct is given by

$$c(x,o) = c_o\,(1 + \cos\frac{2\pi x}{d}) \tag{2}$$

(Fig.1). It is convenient to regard first the case of a stable photoproduct. The time-dependence of the spatial distribution of the photoproduct is then governed by Fick's second law

$$\partial c/\partial t = D\,\partial^2 c/\partial x^2 \quad . \tag{3}$$

A solution of (3) with boundary condition (2) is

$$c(x,t) = c_o\,(1 + e^{-\delta t} \cos\frac{2\pi x}{d}) \tag{4}$$

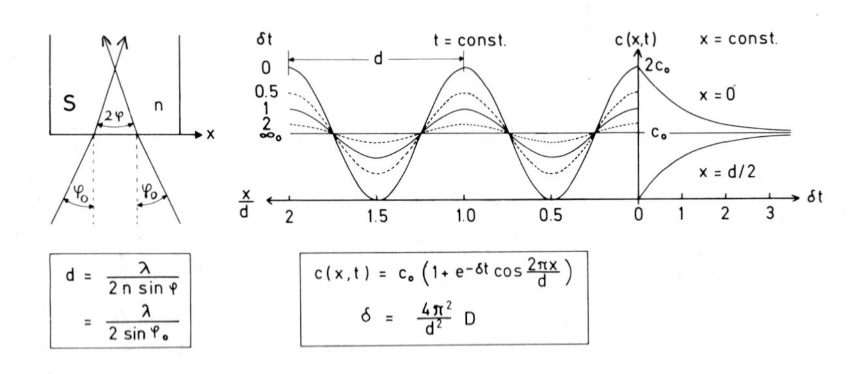

Fig. 1. Spatially periodic excitation of a sample S and space- and time-dependence c(x,t) of the concentration c of a stable photoproduct generated at time t = 0.

with the diffusion relaxation constant δ

$$\delta = (2\pi/d)^2 D \quad . \tag{5}$$

The deviation $c_o \cos(2\pi x/d)$ from the average concentration c_o at time $t = 0$ decays exponentially. In Fig.1 $c(x,t)$ is shown for constant t and for constant x.

If the photoproduct A can disappear by a first-order reaction $A \rightarrow B$ with rate constant ß and by a second-order reaction $A + A \rightarrow C$ with rate constant γ the space- and time-dependence of the concentration c of the photoproduct is governed by the partial differential equation

$$\partial c/\partial t = D\, \partial^2 c/\partial x^2 - \beta c - \gamma c^2 \quad . \tag{6}$$

A closed solution of (6) with boundary condition (2) does not exist. However, two approximate solutions of (6) turn out to be useful. For $\gamma c_o/\beta \ll 1$ the term γc^2 in (6) can be neglected and the solution of (6) is obtained simply by replacing the constant average concentration c_o in (4) by $c_o e^{-\beta t}$:

$$c(x,t) \simeq c_o e^{-\beta t} (1 + e^{-\delta t} \cos \frac{2\pi x}{d}) \quad . \tag{7}$$

It can be shown that with the less restrictive condition $q^2 = (\gamma c_o/\beta)^2 \ll 1$ the deviation Δc of $c(x,t)$ from the average concentration $\langle c(x,t) \rangle_x$ can still be represented with high accuracy by the first term of a Fourier expansion, i.e. by $\Delta c \simeq g(t) \cos \frac{2\pi x}{d}$:

$$c(x,t) \simeq c_o e^{-\beta t} \left\{ (1 - q\frac{3\beta + 4\delta}{2\beta + 4\delta}) + q\, e^{-\beta t} + q\frac{\beta}{2\beta + 4\delta} e^{-(\beta + 2\delta)t} + \left[(1 - 2q)e^{-\delta t} + 2q\, e^{-(\beta + \delta)t}\right] \cos \frac{2\pi x}{d} \right\} \quad . \tag{8}$$

DETERMINATION OF THE DIFFUSION RELAXATION CONSTANT δ

The direct measurement of the time-dependence $c(x,t)$ at a given point x is possible only if the extension of the measuring light in the direction x is small compared with the spatial excitation period d. The measured quantity could be either the light absorption by the photoproduct or the fluorescence of the photoproduct. For short-lived photoproducts in solutions d must be rather small in order that $\delta \gtrsim \beta$ and therefore in general c(x,t) is not a directly measurable quantity. In this case the diffusion relaxation constant δ can be determined from measurable quantities which are spatial average values of functions of c(x,t).

(I) <u>Determination of δ from the time-dependence of the absorption of a photoproduct measured with light having the same space-dependence as the excitation light</u>

For sake of simplicity the case of a stable photoproduct is considered. After short excitation at time t = 0 the time-dependence of the spatial distribution of the photoproduct is given by (4). The absorption of the photoproduct is measured with light having exactly the same space-dependence as the excitation light (apart from a phase diiference ϑ), i.e. the intensity I(x) of the measuring light is

$$I(x) = I_o\,(1 + \cos(\frac{2\pi x}{d} + \vartheta)) \quad . \tag{9}$$

The average light absorption ΔI is for $\Delta I \ll I$

$$\Delta I \propto I_o \frac{1}{d} \int_{x=0}^{d} c_o(1 + e^{-\delta t} \cos\frac{2\pi x}{d})(1 + \cos(\frac{2\pi x}{d} + \vartheta))\, dx \tag{10}$$

For $\vartheta = 0$ results

$$\Delta I \propto I_o c_o(1 + \frac{1}{2} e^{-\delta t}) \quad . \tag{11}$$

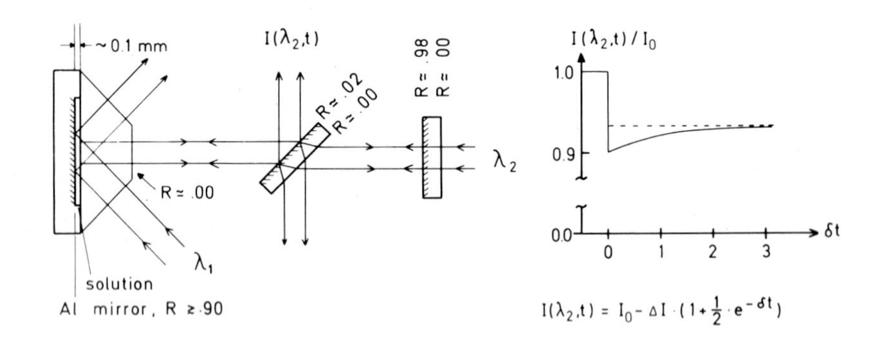

Fig. 2. Determination of δ by measuring the light absorption (λ_2) of a photoproduct with light having the same spatial periodicity and phase as the excitation light (λ_2).

The experimental realization of this method should be in general rather difficult because it would require the adjustment of five parameters: of the period d, of the orientation of the planes of constant intensity (3 parameters), and of the phase ϑ. However, at least in a special case this method should be realizable (Fig.2): Let λ_1 be the wavelength of the excitation light and λ_2 the wavelength of the measuring light with $\lambda_2 > \lambda_1$. The absorption of the photoproduct is measured with a standing light wave. The angle of incidence and thus the spatial period d of the excitation light is adjusted in order that $d_1 = d_2$. Parallelism of the planes of constant intensity and equality of phase are achieved automatically. The expected time-dependence of light absorption is shown in Fig.2. With the reflectivities R given in Fig.2 the signal ΔI is approximately 5 times higher than with a single passage of the measuring light.

(II) Determination of δ from the time-dependence of $\langle c^2(x,t)\rangle_x$. Diffusion coefficients of molecules in the triplet state

The diffusion relaxation constant δ can be determined from the time-dependence of a second-order reaction $A + A \rightarrow C$ of a photoproduct A. The spatial average rate of this reaction is $\gamma\langle c^2(x,t)\rangle_x$. For aromatic hydrocarbons in the triplet state a reaction of the type $A + A \rightarrow C$ is the mutual annihilation of triplet states T_1 which leads to a delayed fluorescence (DF):

$$T_1 + T_1 \rightarrow S_0 + S_1$$

$$S_1 \rightarrow S_0 + h\nu_{DF} \quad \text{(delayed fluorescence).}$$

The intensity I_{DF} of the delayed fluorescence is proportional to $\gamma\langle c^2(x,t)\rangle_x$. In most cases the condition $\gamma c_0/\beta \ll 1$ can be satisfied sufficiently well; from (7) then follows

$$I_{DF} \propto \gamma\langle c^2(x,t)\rangle_x \simeq \gamma c_0^{\,2}\, e^{-2\beta t}\left(1 + \tfrac{1}{2}e^{-2\delta t}\right) \; . \qquad (12)$$

In the same way for $q^2 = (\gamma c_0/\beta)^2 \ll 1$ follows from (8), retaining only terms of the order q:

$$I_{DF} \propto \gamma\langle c^2(x,t)\rangle_x \simeq \gamma c_0^{\,2}\, e^{-2\beta t}\Big\{1 + \tfrac{1}{2}e^{-2\delta t} \qquad (13)$$

$$- (\gamma c_0/\beta)\left[\frac{3\beta + 4\delta}{\beta + 2\delta}\left(1 - e^{-(\beta + 2\delta)t}\right) + 2\,e^{-2\delta t} - 2\,e^{-\beta t}\right]\Big\} \; .$$

Comparison of (13) with (12) shows that (13) additionally contains the exponential functions $e^{-3\beta t}$ and $e^{-(3\beta + 2\delta)t}$.

In practice the determination of diffusion coefficients of molecules in the triplet state consists of the following steps:

(a) Preparation of a sample with sufficiently long and constant triplet life-time ($\tau = 1/\beta \gtrsim 10$ ms, change of τ per day < 1 %).

(b) Spatially periodic excitation of a sample.

(c) Accurate measurement of decay curves of the delayed fluorescence with homogeneous and with spatially periodic excitation of a sample.

(d) Evaluation of data.

(e) Justification of the assumptions underlying the evaluation of data and verification of the consistency of the method.

Discussion will be limited here to steps (b) and (e). The smallest period of excitation for a given wavelength λ is $d = \lambda/2n$ ($2\varphi = \pi$) and corresponds to excitation with a standing light wave. This case has been described in detail [8] and will not be further discussed here. The experimental setup now being used by us for small angles 2φ is shown in Fig.3. A sample is excited in the uv and the triplet state is populated by intersystem crossing: $S_0 + h\nu \rightarrow S_1 \rightarrow T_1$. The uv beam of an argon ion laser (TEM_{00}) passes two quartz dispersion prisms (not shown in Fig.3). The wanted uv line is separated with a diaphragm. The beam is expanded with the lenses L_1 and L_2 to a diameter of 5-6 mm. Reflected beams which can cause unwanted interferences are excluded by a space filter (pinhole PH). The beam is split into two parallel beams of equal intensity by the semitransparent mirror M_1 and the mirrors M_2, M_3, and M_4. The parallelism and the coherence of the two beams can be controlled by the interference pattern displayed on the screen Scr by the beams reflected at the left surface of the prism Pr. Reproducible angles $2\varphi_0$ are obtained with a set of prisms with different angles α. Unwanted interferences of the incident beams with the beams twice internally reflected by the sample S are avoided by use of cells with the cross-section shown in Fig.3b: most of the

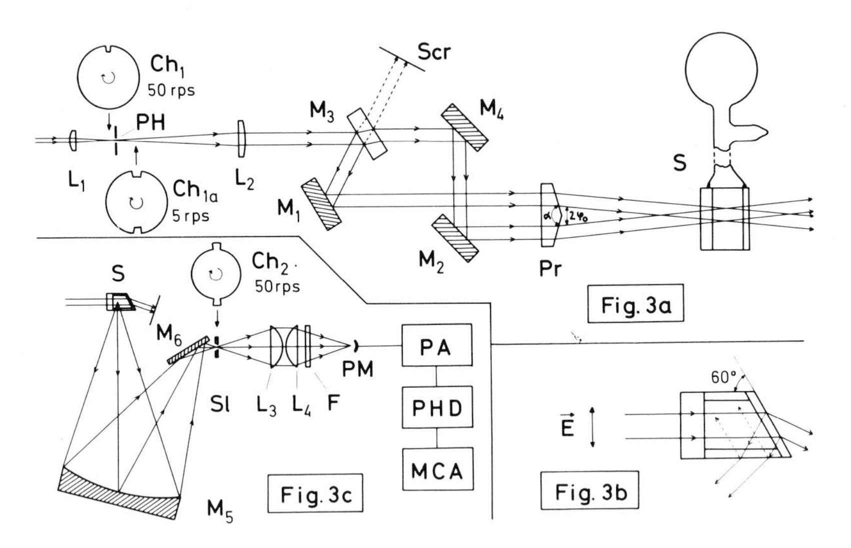

Fig. 3. Experimental setup used for small angles $2\varphi_0$.

light not absorbed by the solution can leave the cell and the interference of the incident beams with the beams twice internally reflected is negligibly small. The delayed fluorescence emitted from the region of maximum intersection of the two interfering beams is focused on to the vertical slit Sl with the toroidal mirror M_5 and the plane mirror M_6 (Fig.3c). The lenses L_3 and L_4 image the slit Sl on to the photomultiplier PM (EMI 9789 QA). The phosphorescence of the sample is absorbed by a suitable filter F. The excitation light is chopped by a fast chopper Ch_1 and a slow chopper Ch_{1a}, the fluorescence light is chopped by a fast chopper Ch_2. The excitation time is $\sim 100\ \mu s$, the time between to successive excitations is $\gtrsim 5/\beta$. The photon-counting technique is used (PA = pulse amplifier, PHD = pulse height discriminator); decay curves of the delayed fluorescence are summed up in a multichannel analyzer (MCA) working in the multiscaling mode.

(III) <u>Diffusion coefficient of ^{3}anthracene* in n-hexadecane</u>.

The consistency and the reliability of the method were tested by the measurement of the diffusion coefficient of ^{3}anthracene* in n-hexadecane. The essential results are summarized below:

(1) The consistency of the method can be verified by showing that the value obtained for the diffusion coefficient D is independent of the spatial excitation period resp. of the angle $2\varphi_0$. This independence of D on φ_0 was indeed found (concentration of anthracene $2.5 \cdot 10^{-5}$ M, λ = 351.1 nm, evaluation of data with Eq.(13)):

$\sin \varphi_0$	β/sec^{-1}	δ/sec^{-1}	$D/\mathrm{cm}^2\mathrm{sec}^{-1}$	at 25.0 °C
.014001	52.66	137.5	$5.475 \cdot 10^{-6}$	
"	51.85	138.8	$5.526 \cdot 10^{-6}$	$(5.46 \pm .08) \cdot 10^{-6}$
"	51.92	135.0	$5.376 \cdot 10^{-6}$	
.024545	53.17	409.2	$5.302 \cdot 10^{-6}$	
"	52.39	425.1	$5.501 \cdot 10^{-6}$	$(5.42 \pm .11) \cdot 10^{-6}$
"	52.58	421.8	$5.465 \cdot 10^{-6}$	

The following results were obtained with an experimental setup simpler than that described above (no space filter; beam splitting and adjustment of the angle $2\varphi_0$ with a Mach - Zehnder interferometer; use of normal fluorescence cells with square cross-section) and by evaluation of data with Eq.(12) [10].

(2) The diffusion coefficient of ^{3}anthracene* is independent of the concentration of molecules in the ground-state (this is expected for [anthracene] = 10^{-3} M - cf. [8]):

[anthr.]/M	$D/\mathrm{cm}^2\mathrm{sec}^{-1}$ at 20.6 °C	
$9.5 \cdot 10^{-6}$	$(4.88 \pm .05) \cdot 10^{-6}$	(3 values)
$2.3 \cdot 10^{-5}$	$(4.64 \pm .10) \cdot 10^{-6}$	(9 values)
$5.3 \cdot 10^{-5}$	$(4.88 \pm .10) \cdot 10^{-6}$	(2 values)

(3) The temperature-dependence of the diffusion coefficient of ^{3}anthracene* in n-hexadecane is shown in Fig.4. It can be represented by the straight line $\log(D/\mathrm{cm}^2\mathrm{sec}^{-1}) = -2.2945 - 885.7\ ^\circ\mathrm{K}/T$ with a standard deviation of 1.5 %. The slope expected from the Stokes - Einstein equation $D_{SE} = kT/6\pi\eta r_0$ is significantly different from that obtained experimentally: In the temperature range 20 °C $\leq t \leq$ 90 °C one gets $\log(T/\eta) = 5.1703 - 948.8\ ^\circ\mathrm{K}/T$ (η in cP) with a standard deviation of .8 % [11]. The dashed line in Fig.4 corresponds to r_0 = 1.37 Å

(for comparison: the average radius of an anthracene molecule is 3.5 Å).

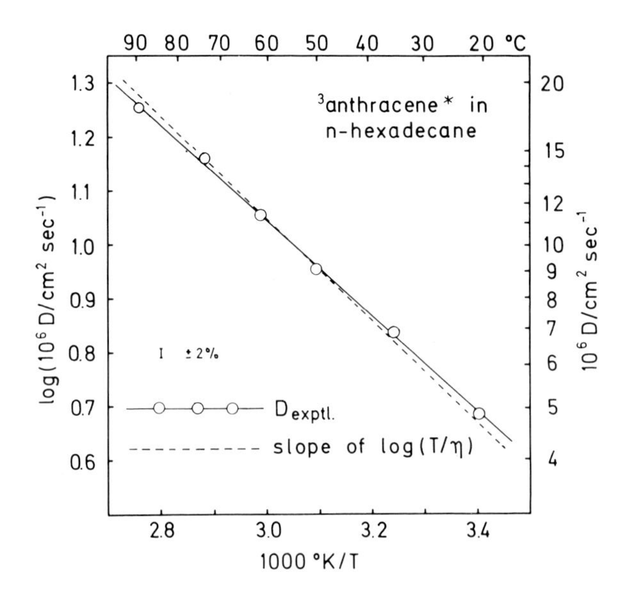

Fig. 4. Temperature-dependence of the diffusion coefficient D of ^{3}anthracene* in n-hexadecane.

CONCLUSIONS

The results reported above show that diffusion coefficients of molecules in the triplet state can be determined with a reproducibility of $\pm$ 2 % and with an estimated absolute accuracy better than $\pm$ 5 %. The range of D covered by this method is unusually large. The lower limit of D is given by excitation with a standing light wave ($d = \lambda/2n$) and by the requirement $\delta \gtrsim \beta$. Thus for ^{3}anthracene* ($\beta \gtrsim 20$ sec^{-1}; λ = 257 nm; n = 1.5) one gets $D_{min} \approx 4 \cdot 10^{-11}$ cm^{2}sec^{-1}. An upper limit of D does practically not exist. A low concentration of molecules in the ground-state is sufficient ($\sim 10^{-6}$ M). With a given sample a practically unlimited number of measurements can be made.

Acknowledgement: An argon ion laser has been placed at our disposal by the Deutsche Forschungsgemeinschaft.

REFERENCES

1 R.M. Noyes, J. Amer. chem. Soc., 81(1959)566.

2 G.A. Salmon and R.M. Noyes, J. Amer. chem. Soc., 84(1962) 672.

3 S.A. Levison and R.M. Noyes, J. Amer. chem. Soc., 86(1964) 4525.

4 P. Avakian and R.E. Merrifield, Physic. Rev. Lett., 13 (1964)541.

5 V. Ern, P. Avakian, and R.E. Merrifield, Physic. Rev., 148 (1966)862.

6 V. Ern, Physic. Rev. Lett., 22(1969)343.

7 B. Nickel, Ber. Bunsenges. physik. Chem., 76(1972)582.

8 B. Nickel and U. Nickel, Ber. Bunsenges. physik. Chem., 76 (1972)584.

9 B. Nickel and E.G. Meyer, in F. Williams (Editor), Luminescence of Crystals, Molecules, and Solutions, Plenum Publishing Corporation, New York,1973, p. 297.

10 E.G. Meyer, Diplomarbeit, Universität Göttingen, 1974.

11 Viscosity data are taken from Landolt Börnstein, 6th ed., 5th Part, Vol. a, Transport Phenomena I, (1969), p. 175 (ref. 1246).

DISCUSSION

E.G. MEYER, B. NICKEL

W. Schnabel - What is the difference of diffusion coefficients of molecules in the triplet state and in the ground state?

B. Nickel - We do not expect that for aromatic hydrocarbons large differences will be found between the diffusion coefficients of molecules in the ground state and those of molecules in the triplet state.

F.P. Schäfer - Is the diffusion coefficient of anthracene in the ground state not yet reported in the literature?

B. Nickel - The diffusion coefficient of anthracene in the ground state is known for several solvents. However, we have not yet measured the diffusion coefficient of triplet anthracene in these solvents.

CHEMICAL REACTIONS AND RELAXATION OF SELECTIVE LASER-EXCITED MOLECULES

D. Arnoldi, K. Kaufmann, M. Kneba, and J. Wolfrum
Max-Planck-Institut für Strömungsforschung, D 3400 Göttingen (G. F. R.)

SUMMARY

The effect of selective vibrational excitation of small molecules on their reactions with free atoms in the gas phase was investigated using the laser excited fluorescence method combined with a discharge-flow reactor. The molecules were chosen so that the energy of the vibrational excitation did exeed the experimentally determined Arrhenius activation energy.

For CH_3F ($\nu_1, \nu_4, 2\nu_3$) molecules, produced by excitation with CO_2-laser radiation, an upper bound of $9 \cdot 10^{10}$ cm^3/mol sec at 298 K on the rate constant for the reaction with D atoms could be deduced. This shows that vibrational energy can not effectively be used for overcoming the potential energy barrier for the $D + CH_3F \longrightarrow DF + CH_3$ reaction.

The removal of laser excited HCl (v=1) molecules was significantly accelerated by the presence of H atoms. Determination of the absolute concentration of laser excited HCl molecules show, that the observed rate constant $k_{2',b} = (3,9 \pm 1,3) \cdot 10^{12}$ cm^3/mol sec at 298 K describes the process H + HCl (v=1) $\longrightarrow$ HCl (v=0) + H (2'b).

Vibrational deactivation of HCl (v=1) molecules by Br atoms occurs with a rate of $(1,6 \pm 0,4) \cdot 10^{11}$ cm^3/mol sec at 298 K. The rate of the reaction Br + HCl $\longrightarrow$ HBr + Cl (3) increases by nearly eleven orders of magnitude to a value of k_3 (v=2) = $(8,3 \pm 3,1) \cdot 10^{11}$ cm^3/mol sec at 298 K if HCl is excited into the v=2 state. Excitation into v=2 by a HCl chemical laser occurs predominantly for $H^{35}Cl$. BrCl molecules formed in the laser induced reaction sequence show an enrichement of ^{35}Cl (90% ^{35}Cl).

INTRODUCTION

Laser light sources offer many new ways for the controlled transfer of external energy into a molecule. The coherence, collimation, high quantum flux, monochromaticity, polarization, short pulse duration and tunability of laser light now available in the infrared, visible and ultraviolett allow the preparation of molecules with an unprecedented degree of state selection. Various elementary processes such as energy transfer, radiationless transitions, unimolecular decomposition, chemical reactions and - at very high power levels - nuclear reactions can be studied in this way [1] .

In this paper, experimental investigations of the behaviour of laser-vibrationally excited gas phase molecules in the presence of reactive atoms are described. Chemical reactions between atoms and simple molecules in the gas phase could be investigated in recent years over a wide temperature range [2] . The Arrhenius parameters obtained in this way contain, however, no direct information on the efficiency of the various degrees of freedom of the molecules (relative translation, rotation, vibration, electronic excitation) for overcoming the potential energy barrier of the reactions since the rate coefficients k(T) are usually measured for reactants with a Maxwell-Boltzmann distribution of states. In order to study the effect of vibrational excitation systematically, reacting molecules must be prepared in definite vibrational states while the energy of other degrees of freedom remains unchanged. An elegant technique to excite molecules in definite vibrational states is optical pumping by laser photons. As illustrated in Fig. 1 several possibilities exist for vibrational excitation in the electronic ground state. The use of infrared lasers in resonance with transitions of infrared active molecules allows a direct excitation and detection of higher vibrational levels [3] . This is a new possibility offered by the laser radiation, since the incoherent black body light sources have a much smaller spectral density. Significant excitation of infrared inactive

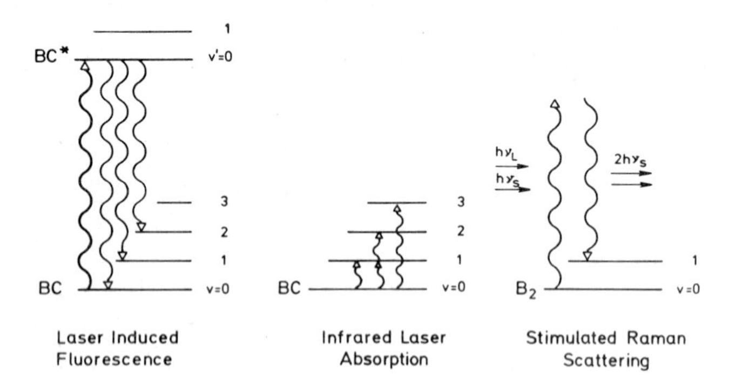

Fig. 1. Vibrational Excitation by laser Optical Pumping.

vibrations can be achieved with high energy laser pulses using the Raman effect [4, 5]. One may also use a higher electronic state of the molecule. Since the radiative lifetime of the upper electronic state is often short compared with the vibrational deactivation, one can populate higher vibrational levels of the electronic ground state by fluorescence from the upper state.

The experiments described here use excitation by infrared lasers by the coincidence between molecular laser emission lines and infrared transitions of molecules. As illustrated in Fig. 2, the molecules are chosen so that simple exothermic, thermoneutral and endothermic abstraction reactions could be investigated. In all cases the height of the potential energy barrier of the reaction is comparable to the energy of one or few vibrational quanta.

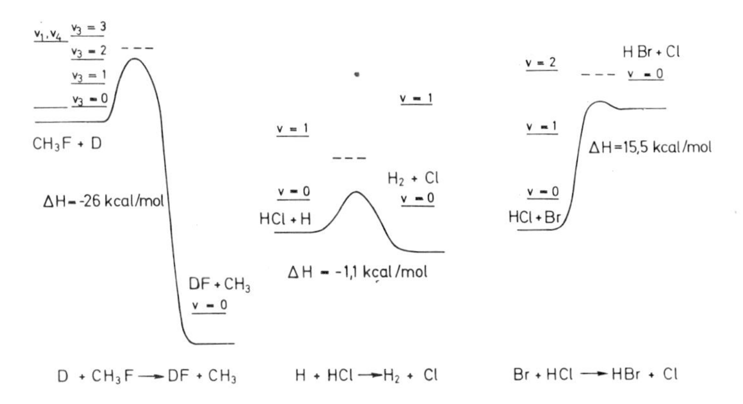

Fig. 2. Potential-energy profiles for three reactions of vibrationally excited molecules.

THE REACTION $D + CH_3F(2\nu_3; \nu_1, \nu_4) \rightarrow DF + CH_3$

A TEA-CO_2 laser, tunable to specific rotational lines in the 9, 6 and 10, 4 μm region, was employed to excite CH_3F vibrationally in the presence of D atoms. A detailed description of the experimental configuration, will be given elsewhere [6]. At 9, 55 μm a coincidence between the P(20) line of the CO_2 laser and the C-F stretching vibration (ν_3) of CH_3F exist. Flynn and coworkers [7] have measured a number of rate coefficients for the vibration - vibration energy transfer at which vibrational energy is pumped from the laser excited level into the remaining normal modes of the CH_3F molecule. In order to decouple the V-V exchange processes from the removal of vibrationally excited CH_3F molecules by interaction with the added atoms, the partial pressure of CH_3F was lowered by an order of magnitude compared to the experiments described before [7]. As shown in Fig. 3 excitation of the

ν_1, ν_4 and $2\nu_3$ level is achieved in shorter times as calculated from the

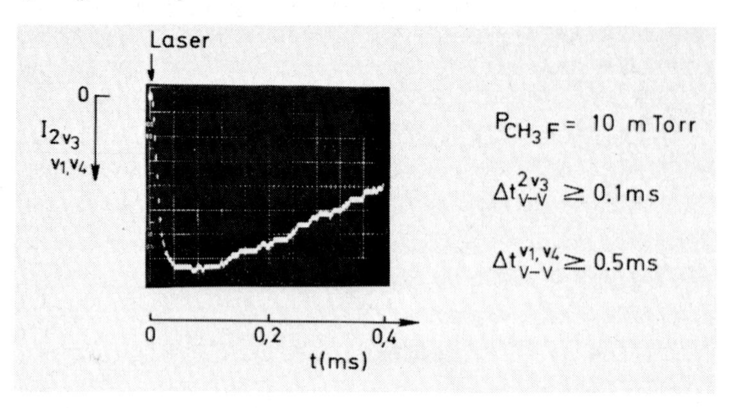

Fig. 3. $CH_3F(\nu_1, \nu_4; 2\nu_3)$ CO_2-Laser Induced Fluorescence at Low Pressures.

previously measured V -V rates [8]. The fluorescence signal results predominantly from the $\nu_1, \nu_4 \rightarrow 0$ emission at 3, 3 µm. The rise time of the signal corresponds to the response time of the detector-amplifier combination used. This was tested by observing the CO_2 (00^01) fluorescence at 4, 3 µm. A possible explanation of the observed fast population of the higher vibrational levels would be the occurence of multiquantum processes. For example, for the quantum excitation

$$CH_3F\ (\nu_3 = 0,\ J = 6) + 2\ h\nu\ (CO_2,\ P\ 20) \longrightarrow CH_3F\ (\nu_3 = 2,\ J = 7)$$

the distance between the line centers is 880 MHz [9], while the emission from the atmospheric pressure CO_2-laser is collisional broadened to 3, 5 GHz [10]. The probability of the two-quantum excitation of a vibrational overtone can significantly be increased by an intermediate level [11].

Thermal CH_3F molecules react relatively slowly with H and D atoms at room temperature [12, 13]

$$D + CH_3F \longrightarrow DF + CH_3 \qquad (1)$$

$$k_1 \approx 10^{10}\ cm^3/mol\ sec\ at\ 298\ K$$

Only a small amount of the initial atom concentration applied is therefore consumed during the passage of the mixture through the fluorescence cell. Two values (5, 2 kcal/mol [12], 9, 4 kcal/mol [13] are reported for the Arrhenius activation energy. The reason for this discrepancy is not clear yet. However, if all of the energy of the observed vibrational levels were available to overcome the potential energy barrier of reaction (1), a significant increase of $k_1(v)$ should occur. In the experiments described here, no change in the decay rate of the fluorescence from the ν_1, ν_4 and $2\nu_3$ level could be observed due to the presence of D atoms. D atoms were used instead of H atoms, since the remaining D_2 not dissociated in the microwave dischange is less efficient than H_2 in deactivating vibrationally excited CH_3F [14]. The lack of an observable change in the removal of vibrationally excited CH_3F by D atoms can be used to establish an upper bound of

$$k_1\ (\nu_1, \nu_4; 2\nu_3) \leqslant 9 \cdot 10^{10}\ cm^3/mol\ sec\ at\ 298\ K$$

From this result it can be concluded, that vibrational energy can not very effectively be used for overcoming the potential energy barrier of Reaction (1). This observation is in qualitative agreement with predictions of trajectory calculations on model potential energy surfaces. A low efficiency of vibrational energy is expected for substantially exothermic reactions, which tend to have a barrier in the reactant valley[15].

THE REACTION $H + HCl\ (v=1) \longrightarrow H_2 + Cl$

Laser Pulses generated in a HCl photochemical laser were used to excite HCl molecules vibrationally in a discharge flow reactor. A schematic diagram of the experimental apparatus is shown in Fig. 4. More details are given in ref. [16].

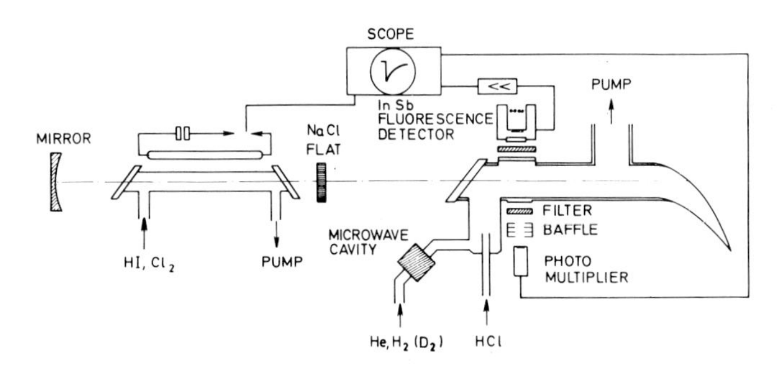

Fig. 4. Schematic of the laser-induced fluorescence discharge-flow reactor.

In contrast to the observations made for vibrationally excited CH_3F-molecules, the removal of laser excited HCl (v=1) molecules was significantly accelerated in the presence of H and D atoms.

$$k_{exp} = \frac{\Delta \ln [HCl\ (v=1)]}{[H]_0 \cdot \Delta t} = (3.9 \pm 1,3) \cdot 10^{12}\ cm^3/mol\ sec$$

As shown in Fig. 2., the energy of HCl(v=1) molecules is well above the potential energy barrier(determined from experimental data by the LEPS procedure [17] or calculated by the BEBO method [18]) of the reaction

$$H + HCl \longrightarrow H_2 + Cl \qquad (2)$$

$$k_2 = 3.7 \cdot 10^{10} \ cm^3/mol\ sec \ at\ 298\ K \quad [19]$$

Two processes can,however, contribute to the observed fast removal of HCl (v=1).

$$H + HCl\ (v=1) \xrightarrow{a} H_2 + Cl \qquad (2')$$
$$\xrightarrow{b} HCl\ (v=0) + H$$

To obtain experimental information on the relative importance of the two possible channels, the number of laser excited HCl molecules and the amount of H atoms reacted or Cl atoms formed must be known. The rate of reaction (2' a) can then be determined from the relation

$$k_{2'a} = k_{exp} \cdot \frac{\Delta\ [\ Cl\ ]}{\Delta\ [\ HCl\ (v=1)\ ]} = k_{exp} \cdot \frac{\Delta[H]}{\Delta[HCl]} \qquad (I)$$

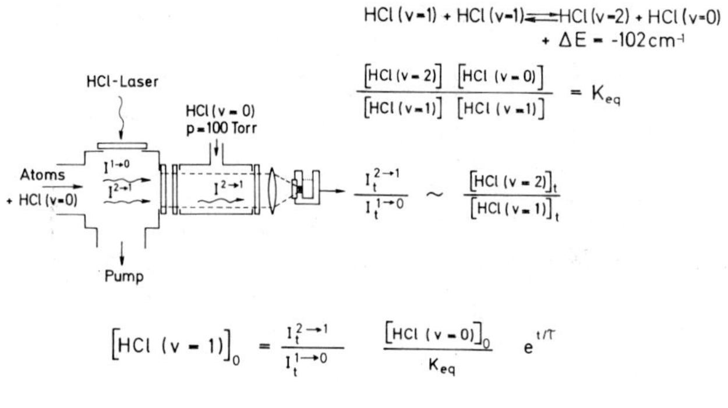

$$HCl\,(v=1) + HCl\,(v=1) \rightleftharpoons HCl\,(v=2) + HCl\,(v=0) + \Delta E = -102\,cm^{-1}$$

$$\frac{[HCl\,(v=2)]\ [HCl\,(v=0)]}{[HCl\,(v=1)]\ [HCl\,(v=1)]} = K_{eq}$$

HCl-Laser

HCl (v = 0)
p = 100 Torr

Atoms
+ HCl (v=0)

Pump

$$\frac{I_t^{2\to1}}{I_t^{1\to0}} \sim \frac{[HCl\,(v=2)]_t}{[HCl\,(v=1)]_t}$$

$$[HCl\,(v=1)]_0 = \frac{I_t^{2\to1}}{I_t^{1\to0}} \quad \frac{[HCl\,(v=0)]_0}{K_{eq}} \quad e^{t/T}$$

Fig. 5. Determination of the initial concentration of laser excited molecules.

As illustrated in Fig. 5, the amount of HCl (v=1) initially produced from the laser pulse can be obtained using the rapid equilibration between the vibrational states of HCl [20] by measuring the relative population in the levels v=1 and v=2 as function of time. With a laser energy of 4mJ in the 1→0 transition an initial excitation of

$$[HCl\ (v=1)_0] = (0,07 \pm 0,01)\ [HCl\ (v=0)]_0$$

is measured [21]. The prevalence of equilibrium between the vibrational states in the presence of H atoms was tested by comparing the decay rates for HCl (v=1) and HCl (v=2). As shown in Fig. 6, the slope of the v=2 fluorescence curve agrees closely with that of two times the value of the v=1 curve.

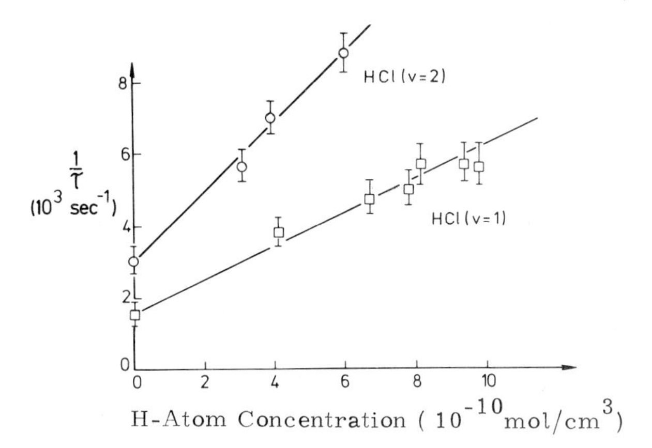

Fig. 6. Observed v=1 → 0 and v=2 → 1 fluorescence decay rates as function of the H atom concentration.

The H atom concentration could be monitored by adding a small amount of NO and measuring the intensity of the HNO* chemiluminescence. Detection of Cl atoms was possible using the emission from the Cl atom recombination glow [22]. An absolute calibration of the chemiluminescence signals was made by gas phase titration [2]. No Cl atom formation during the removal of HCl (v=1) by H atoms could be detected. A small decay of the HNO* chemiluminescence signal is observed, which can be explained by the decrease of density in the mixture due to the temperature increase after V → R, T relaxation of the HCl (v) by H atoms [21]. These experiments show, that no significant enhancement of the rate of reaction (2) due to excitation of HCl (v=0) into HCl (v=1) occurs.
An upper bound of

$$k_{2',a} \leqslant 8 \cdot 10^{10} \ cm^3/mol\ sec$$

close to the rate of HCl in the vibrational ground state could be deduced using equation (I) and a Cl atom concentration equal to our detection limit. The observed decay rate must therefore be attributed to vibrational deactivation

$$k_b \approx k_{exp} = (3,9 \pm 1,3) \cdot 10^{12} \ cm^3/mol\ sec.$$

THE REACTION Br + HCl (v=2) ⟶ HBr + Cl

Ground state Br atoms react very slowly with thermal HCl at room temperature. The rate of reaction

$$Br + HCl \longrightarrow HBr + Cl \tag{3}$$

$$k_3 \approx 2\cdot10^{1}\ cm^3/mol\ sec\ at\ 298K$$

can be calculated from the data on the reverse reaction [23]. When HCl HCl (v=1) molecules are produced in the discharge-flow reactor, the decay of the v=1 is accelerated if Br_2 is partially dissociated in the microwave discharge. The differences in the exponential decay rates with and without discharge were a linear function of the Br atom partial pressure and gave a rate constant

$$k_{exp}(v=1) = (1,6 \pm 0,4)\cdot10^{11}\ cm^3/mol\ sec\ at\ 298\ K$$

The value has been obtained recently also in another laboratory [24]. The decay of HCl (v=1) in the presence of Br atoms was accompanied by the rise of a chemiluminescence signal in the spectral region between 740 and 850 nm. This can be explained by the occurrence of the V→E process

$$HCl\ (v=1) + Br(^2P_{3/2}) \longrightarrow HCl\ (v=0) + Br(^2P_{1/2}) \tag{3*}$$

and the subsequent recombination

$$Br(^2P_{1/2}) + Br(^2P_{3/2}) \xrightarrow{M} Br_2(^3\Pi_{0^+u})$$

$$Br_2(^3\Pi_{0^+u}) \longrightarrow Br_2(^1\Sigma_{0^+g}) + h\nu.$$

A rate constant of (3*) can be calculated from the data of the reverse process ($2,6\cdot10^{10} < k_{3*} < 5,6\cdot10^{10}\ cm^3/mol\ sec$ [24]). Reaction (3) is too endothermic for HCl (v=1) to contribute to the observed decay rate.

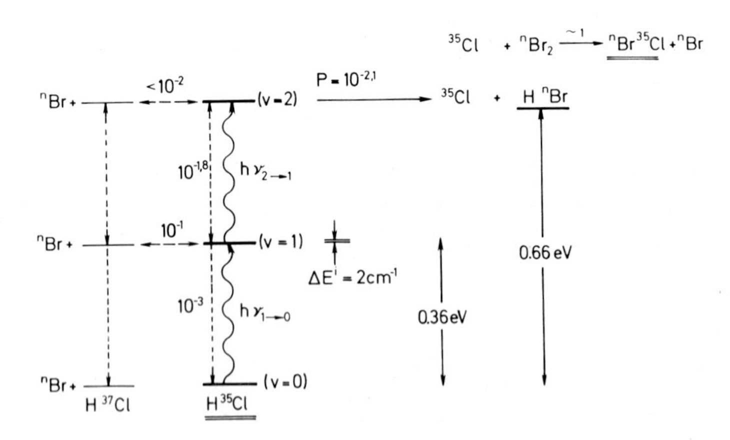

Fig. 7. Reaction of Br atoms with isotopically selective excited HCl molecules. P denotes the probability of the collision process.

Thus vibrational deactivation of HCl (v=1) can occur by V⟶R, T and V⟶E processes.
As shown in Fig. 2, reaction (3) becomes feasible energetically for the level v=2. Cl atoms formed in the reaction (3) will react rapidly with Br_2 present in the mixture to form BrCl.

$$Cl + Br_2 \longrightarrow BrCl + Br \qquad (4)$$

$$k_4 = 7 \cdot 10^{13} \ cm^3/mol\ sec \ [25]$$

To detect the BrCl formed, the laser-excited discharge-flow system was combined with a quadrupole mass spectrometer by a molecular beam sampling system. More details are given elsewhere [25]. As shown in Fig. 7, the v=2 level can initially be populated by successive absorption of two photons from the HCl laser. This occurs within 10 μsec with the photochemical laser used here and could be tested by monitoring the rise time of the laser induced 2⟶1 fluorescence. During this direct excitation of v=2 the V-V processes shown in Fig. 5 will not significantly contribute to the population of v=2. Due to the natural ^{35}Cl : ^{37}Cl = 75. 5 : 24, 5 abundance each step of vibrational excitation by the HCl chemical laser will therefore populate predominantly $H^{35}Cl$ (v). The isotopically selected Cl atom generated in (3) give predominantly $Br^{35}Cl$, which is detected by time-resolved mass-spectrometry after excitation of the mixture by the laser. As shown in Fig. 8, the enrichement for $Br^{35}Cl$ increases if higher

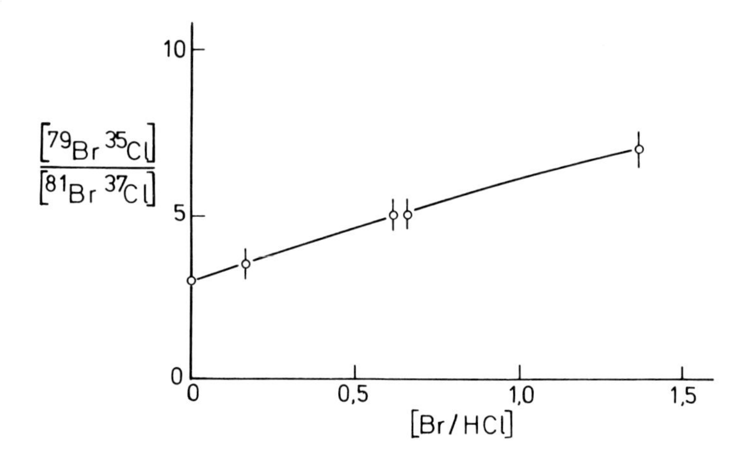

Fig. 8. Enrichement of ^{35}Cl in BrCl generated in the laser induced reaction (3) and (4) as function of the initial [Br]/[HCl] ratio.

initial [Br]/[HCl] ratios are used. This reflects the population of the v=2 level by V-V energy exchange processes, which decrease the selectivity of the vibrational excitation initially produced by the laser. As shown in Fig. 7, the process

$$H^{35}Cl\ (v=1) + H^{37}Cl\ (v=0) \longrightarrow H^{35}Cl\ (v=0) + H^{37}Cl\ (v=1)$$

requires only 10 collisions to restitute the natural isotopic abundance in HCl (v=1).
The rate constant k_3 (v=2) could be determined by measuring the decay of the HCl (v=2) concentration in the presence of Br atoms and the amount of BrCl formed under these experimental conditions according to

$$k_3(v=2) = \frac{[BrCl]_t}{[Br]_0 \int_0^t HCl\,(v=2)\,dt}$$

$$= (8,8 \pm 3,1) \cdot 10^{11} \quad cm^3/mol\ sec \quad \text{at } 298\ K.$$

Thus the rate constant of reaction (3) increases by more than 10 orders of magnitude at room temperature after excitation of ground state HCl into the second vibrational level.

ACKNOWLEDGEMENTS

The autors are grateful to Professor H. Gg. Wagner for his continuous interest and encouragement and to the Deutsche Forschungsgemeinschaft for financial support.

REFERENCES

1 C. B. Moore (ed.), Chemical and Biochemical Applications of Lasers, Academic Press, New York (1974)
2 J. Wolfrum, "Atom Reactions" in Physical Chemistry - an Advanced Treatise, ch.9, W Jost (ed), Academic Press, New York (1975).
3 L. O. Hocker, M. A. Kovacs, C. K. Rhodes, G. M. Flynn and A. Javan, Phys. Rev. Letters 17, 233 (1966)
4 J. Ducuing, C. Joffrin and J. P. Coffinet, Opt. Commun. 2, 245 (1970).
5 A. Laubereau, D. von der Linde and W. Kaiser, Phys. Rev. Lett. 28, 1162 (1972)
6 M. Kneba and J. Wolfrum (in preparation)
7 G. W. Flynn, Energy Flow in Polyatomic Molecules, ch. 6 in [1]
8 Z. Karny, A. M. Ronn, E. Weitz and G. W. Flynn, Chem. Phys. Lett. 17,347 (1972)
9 M. Betrencourt, J. Mol. Spectrosc. 47, 275 (1973)
10 T. J. Bridges, H. A. Hans and P. W. Hoff, Appl. Phys. Lett. 13, 316 (1968)
11 V. P. Chebotayev, A. L. Golger and V. S. Letokhov, Chem. Physics 7, 316 (1975)
12 W. K. Aders, D. Pangritz and H. Gg. Wagner, Ber. Bunsenges. physik. Chem. 79, 90 (1975)
13 A. A. Westenberg and N. de Haas, J. chem. Physics. 62, 3321 (1975)
14 E. Weitz and G. W. Flynn, J. chem. Physics 58, 2679 (1973)
15 J. C. Polanyi and W. H. Wong, J. chem. Physics 51, 1439 (1969)
M. H. Mok and J. C. Polanyi, J. chem. Physics 51, 1451 (1969)
16 D. Arnoldi and J. Wolfrum, Chem. Phys. Lett. 24, 234 (1974)
17 A. A. Westenberg and N. de Haas, J. chem. Physics 48, 4405 (1968)
18 H. S. Johnston, Gas Phase Reaction Rate Theory, Ronald Press New York (1966), p. 247
19 J. P. Glass, personal communication
20 B. M. Hopkins and H. L. Chen, J. chem. Physics 57, 3816 (1972)
S. R. Leone and C. B. Moore, Chem. Phys. Letters 19, 340 (1973)
21 D. Arnoldi and J. Wolfrum, in preparation
22 M. A. A. Clyne and D. H. Stedman, Trans. Faraday Soc. 64, 1816 (1968)
23 F. J. Wodarczyk and C. B. Moore, Chem. Phys. Letters 26, 484 (1974)
K. Bergmann and C. B. Moore, to be published
24 S. R. Leone, R. G. Macdonald and C. B. Moore, to be published
25 M. A. A. Clyne and W. H. Crute, J. Chem. Soc. Farad. Trans. II 68, 1377 (1972)
26 D. Arnoldi, K. Kaufmann and J. Wolfrum, Phys. Rev. Lett. (in press)

DISCUSSION

D. ARNOLDI, K. KAUFMANN, M. KNEBA, J. WOLFRUM

M.J. Pilling - Could you comment on the limit of the enhancement of the rate

$$HCl\ (v = 1) + H \rightarrow Cl + H_2$$

over that from HCl (v = 0)?

J. Wolfrum - As stated in the paper, an upper bound of $8.10^{10}\ cm^3/mol.sec.$ is obtained for the formation of H_2 and Cl in the reaction of H with HCl (v = 1).

VIBRATIONAL REDISTRIBUTION

J.D.W. van VOORST
LABORATORY FOR PHYSICAL CHEMISTRY, UNIVERSITY OF AMSTERDAM,THE NETHERLANDS

The development of pulsed lasers in spectroscopic, i.e. time resolved absorption and emission measurements, has opened a very interesting field in molecular physics. That is the problem how energy is transferred between different degrees of freedom in a single molecule or in a system of interacting molecules. Well-known examples studied are internal conversion, intersystem crossing and vibrational relaxation. A much less konown process is the vibrational redistribution as an intramolecular process, i.e. transfer of vibrational energy from one excited vibrational mode to the other vibrational modes of the molecule.

It has been found that upon excitation of pyrene vapour in the false origins (vibronically induced transitions) of the $S_1 \rightarrow S_0$ absorption band, under conditions where the molecules can be considered as being isolated during their fluorescence lifetime, the fluorescence decays non-exponentially [1,2]. The decay curves can be analyzed in terms of two exponential decaying components with characteristic times of 70 - 100 and 300 - 400 ns respectively. Excitation in the weakly allowed $S_0 \rightarrow S_1^{0-0}$ band results in a purely exponential decay with a decay-time of 430 ns. Thus it follows that in the fluorescence of pyrene the short component is only observed when the molecules are excited in a vibronically induced band, i.e., when due to an optical excitation the quantum number in an inducing mode is changed by one.

Although a large part of the thermal distribution of molecules over the vibrational levels of the ground state can be transferred to the excited state [4] as a consequence of sequence congestion effects, the absence of a short component after excitation in the $S_0 \rightarrow S_1^{0-0}$ absorption band clearly shows that sequence congestion cannot account for the fast decay.

Since in pyrene the appearance of the short component is restricted to excitation in a vibronically induced band, one has to consider separately the effect of one more quantum number of vibrational energy in an inducing mode on the radiative and non-radiative decay.[3]

As concerned with the radiative decay rate, it was shown that a change of about 24% can be considered as an upper limit.

The influence of the vibrational quantum number upon the non-radiative decay rate is most substantial for the promoting mode. Hence, in order to obtain an upper limit, it is assumed in the following that an inducing mode is also a promoting mode. When more promoting modes are allowed, the effect of an additional quantum in the inducing promoting mode will become diluted when the total non-radiative transition probability is made up. Restriction to one promoting mode that also acts as inducing mode, will give us an upper limit of the effect of a change in the vibrational quantum number upon the non-radiative decay rate.

With these restrictions, it follows that the increase in the total decay rate due to a change from $\upsilon = 0$ to $\upsilon = 1$ for an inducing mode can be estimated to be $0.24\ k^{f}(0) + k^{nr}(0)$. The values for k^{f} and k^{nr} of pyrene as determined for excitation at 27 150 cm^{-1}, are $6.2 \times 10^{5}\ s^{-1}$ and $1.7 \times 10^{6}\ s^{-1}$ respectively. This means that even in an upper limit consideration, one additional quantum of energy in an inducing mode will increase the total decay rate only by a factor of about 1.8. For higher members of a sequence of the inducing mode the increase will be even smaller.

Thus by comparing the decay rate of the short component with that of the long component after excitation in a 0 - 1 vibronic origin of the $S_0 \rightarrow S_1$ absorption band, it follows that a ratio of 3.7 to 5.2 can hardly be explained by considering only the influence of the vibrational quantum number upon the radiative decay and the non-radiative transitions like internal conversion and intersystem crossing. Therefore the fast decay rate is ascribed to a redistribution of vibrational energy from the optically excited inducing mode over all the other vibrational modes of the molecule, vibrational redistribution being the rate determining step.

Since the radiative decay rate is dependent upon the quantum number of the inducing mode, it will change upon redistribution which makes the process of redistribution observable in the fluorescence decay. This also explains why no short component is observed in the fluorescence decay of 3-methyl pyrene. As can be seen in the optical absorption spectrum the allowed part of the transition ($S_0 \rightarrow S_1^{0-0}$ band) is much larger with respect to the other bands in the $S_0 \rightarrow S_1$ absorption than in pyrene. Thus a change in one quantum of energy in an inducing mode will have almost no effect upon the radiative decay rate [5].

The value of the vibrational redistribution rate, i.e. 10^7 - 10^8, is rather small. This has important implications when one studies the behavior of excited molecules under isolated conditions in the gasphase. Suppose that after excitation certain vibrational modes become specifically populated in an internal conversion process. Then one has seriously to take into account the rate of vibrational redistribution in order to explain the experimental decay rate in terms of

1 an ensemble in which the excess vibrational energy is randomly distributed over all the vibrational modes [5]; or

2 by an ensemble in which a restricted number of vibronic states is occupied.

REFERENCES

1 C.J. Werkhoven, T. Deinum, J. Langelaar, R.P.H. Rettschnick and J.D.W. van Voorst, Chem. Phys. Letters, 11(1971)478.

2 C.J. Werkhoven, T. Deinum, J. Langelaar, R.P.H. Rettschnick and J.D.W. van Voorst, Chem. Phys. Letters, 18(1973)171.

3 C.J. Werkhoven, T. Deinum, J. Langelaar, R.P.H. Rettschnick and J.D.W. van Voorst, Chem. Phys. Letters, 30(1975)504.

4 T. Deinum, C.J. Werkhoven, J. Langelaar, R.P.H. Rettschnick and J.D.W. van Voorst, Chem. Phys. Letters, 27(1974)206.

5 C.J. Werkhoven, T. Deinum, J. Langelaar, R.P.H. Rettschnick and J.D.W. van Voorst, Chem. Phys. Letters, 32(1975)328.

DISCUSSION

J.D.W. VAN VOORST

R.M. Hochstrasser - 1) Does the fluorescence spectrum of the dilute gas change with time? 2) Is the short lifetime independent of pressure? 3) Have you considered the effect of Förster energy transfer? This would presumably change with excitation energy in the vibrational manifold of S_1. 4) Might it be worthwhile to investigate a 50:50 mixture of perproto and perdeutero pyrene to further test the isolated molecule assumption? Questions relating to longrange effects might be clarified by such an experiment.

J.D.W. Van Voorst - 1) No change in the fluorescence spectrum could be resolved at different times of the decay. 2) Measurements were done in a pressure region where the decay constant of the short component was independent of pressure, i.e. $\lesssim 5.10^{-2}$ torr. 3) Förster type energy transfer is a general and not a specific effect. All vibronic levels in the manifold will be affected in the same way. Averaged over the thermal distribution in the excited state, due to sequence congestion, still an almost exponential decay can be expected.

S. Leach - Your explanation of the pyrene results in terms of vibrational redistribution might lead to a spectrum of radiative decay rates, not just two. Did you have any indications of this?

J.D.W. Van Voorst - One quantum of vibrational energy in one specific inducing mode is redistributed isoenergetically over a number of the remaining μ modes. About seven of these modes are inducing modes. If we consider the redistributions to be random, i.e. every degenerate vibronic level has equal probability to become populated from the initially optically excited level, we end up with an ensemble in which the number of levels for which another inducing mode is populated, is percentually small. Thus for the majority of the occupied vibronic levels the radiative decay rate will be the same and a single exponential will be observed. Increase of excess vibrational energy will affect the percentage of vibronic levels in which inducing modes become populated. This effect has been studied for the redistributed long component and could be interpreted in terms of a random distribution (1)

(1) C.J. Werkhoven, T. Deinum, J. Langelaar, R.P.H. Rettschnick and J.D.W. Van Voorst, Chem.Phys. Letters, 32 (1975), 328.

M. Stockburger - Have you tried to study the two different decay domains by time-resolved spectroscopy?

J.D.W. Van Voorst - The broad and almost structure less fluorescence spectra in the two decay domains do not differ.

S. Leach - It would be interesting, in the pyrene experiment, to do time-resolved absorption spectroscopy from the excited state in order to follow the vibrational redistribution.

J.D.W. Van Voorst - Such an experiment has been performed. Dr. Langelaar can give you the details.

J. Langelaar - We indeed tried to measure time-resolved absorption spectra of pyrene vapour. Unfortunately even with a 60 cm cell we have measured at low pressures (to meet an isolated condition) only an optical density of $\leqslant 0.01$. This is insufficient to do time-resolved and spectral resolved measurements.

R.M. Hochstrasser - In your estimation of possible effects on the radiative lifetime you have always assumed Born-Oppenheimer wavefunctions. In your explanation of the short lifetime it is just a mixing of the vibrational wavefunctions within the S_1 electronic manifold that is invoked. I wish to ask why you have not considered effects due to the breakdown of the Born-Oppenheimer approwimation : you are exciting within a few thousand cm^{-1} of the 0-0 region of S_2, probably well within the interference region, where some of the vibronic states are superpositions of S_1 and S_2 Born-Oppenheimer states possibly containing a significant amount of S_2.

J.D.W. Van Voorst - Geldof, Rettschnick and Hoytink have studied this problem when they explained the absence of mirror symmetry in the S_1 absorption and emission spectra of pyrene (P.A. Geldof, R.P.A. Rettschnick, G.J. Hoytink, Chem.Phys.Letter, 10 (1971), 549. One of the bands in the progression of the S_0-S_1 absorption spectrum was ascribed to a breakdown of the Born-Oppenheimer approximation. The matrix element for the coupling turned out to be of the same order of magnitude as those for Herzberg-Teller coupling. Both the Herzberg-Teller coupling within the BO approximation and the breakdown of the BO approximation make the radiation width dependent upon the vibrational quantum number of the mode involved in the coupling. Thus for a selective number of vibronic levels the change in radiative width upon redistribution has indeed to be accounted for by a breakdown of the Born-Oppenheimer approximation. However, since the effect is of the same order of magnitude as for Herzberg-Teller coupling it will not change our estimate of the change in the radiative width.

RESOLUTION TEMPORELLE ET RESOLUTION SPATIALE EN SPECTROMETRIE RAMAN

M. DELHAYE
Service de Spectrochimie Infrarouge et Raman, C.N.R.S., Université de Lille (France)

SUMMARY

The advent of LASER has renewed the techniques of Raman Spectrometry. Recent progress are described in the present paper. Image intensifier phototubes and "multichannel" spectrometers, developped by Bridoux and coworkers enable recording of spontaneous Raman spectra in a very short time range, from microseconds with CW Lasers, to 25 picosecond with mode locked Lasers.

New devices have been studied to generate images of heterogeneous samples, using a characteristic Raman line. Laser "Mapping" and "Imaging" techniques led to Raman Microscope and Microprobe. The spatial resolution is of the order of 1 micron.

INTRODUCTION

L'apparition des sources de lumière laser a permis la renaissance et un développement considérable de la spectrométrie par effet Raman, en particulier pour l'étude des spectres de vibration de molécules ou de cristaux. L'effet Raman spontané étant un phénomène très peu intense par rapport à la diffusion sans changement de longueur d'onde, le progrès de ces techniques est précisément lié à la coexistence de sources laser intenses et de détecteurs photoélectriques excellents dans le domaine spectral où ce phénomène peut s'observer dans les meilleures conditions, c'est-à-dire dans la fenêtre comprise entre l'absorption électronique ultra-violette et l'absorption vibrationnelle infrarouge. Nous ne nous attarderons pas dans cet article aux méthodes conventionnelles bien connues, mais nous chercherons à démontrer l'intérêt de techniques de pointe qui ont été développées pour l'étude :

- d'intermédiaires réactionnels
- d'espèces chimiques instables
- d'états excités

et pour toutes sortes de problèmes physicochimiques qui demandent une étude très localisée dans le temps et dans l'espace.

De ce point de vue, on peut tout d'abord dégager les avantages essentiels des sources de lumière laser et en même temps en souligner les limitations les plus évidentes :

1° - une excellente monochromaticité, c'est-à-dire en particulier, l'absence de fond continu qui limiterait la détectivité des récepteurs

2° - la possibilité de concentrer l'énergie monochromatique émise dans le faisceau laser

- sur un très petit volume d'échantillon (la réduction des dimensions ne serait théoriquement limitée que par la longueur d'onde mais en pratique, la densité énergétique ne peut dépasser certaines valeurs fixées par la destruction thermique ou photochimique des substances)

- dans une impulsion de très courte durée (mais dont la puissance de crête doit être limitée si on veut étudier l'effet Raman "ordinaire" spontané en évitant l'apparition des phénomènes stimulés beaucoup plus intenses).

HISTORIQUE

La figure 1 représente schématiquement l'évolution des résultats obtenus dans ce domaine au cours des 25 dernières années. La durée d'enregistrement d'un spectre Raman était de l'ordre de plusieurs heures vers 1950 à l'époque des lampes à mercure. Cette durée avait pû être abaissée à une fraction de seconde vers 1960 grâce à l'amélioration des monochromateurs et à l'emploi de photomultiplicateurs de haute qualité. Ensuite le développement des sources laser et le perfectionnement de techniques permettant l'enregistrement simultané d'un grand nombre d'éléments spectraux ont permis de descendre régulièrement dans l'échelle des temps pour atteindre la microseconde vers 1969, la nanoseconde à peu près à la même époque, puis récemment le domaine des picosecondes. Ce diagramme fait apparaître à partir de 1960 notre souci de mener de front le développement des techniques dites "monocanal" et des techniques dites "multicanaux" (1) (2) (3).

Fig. 1. L'évolution de la Résolution Temporelle en Spectrométrie RAMAN

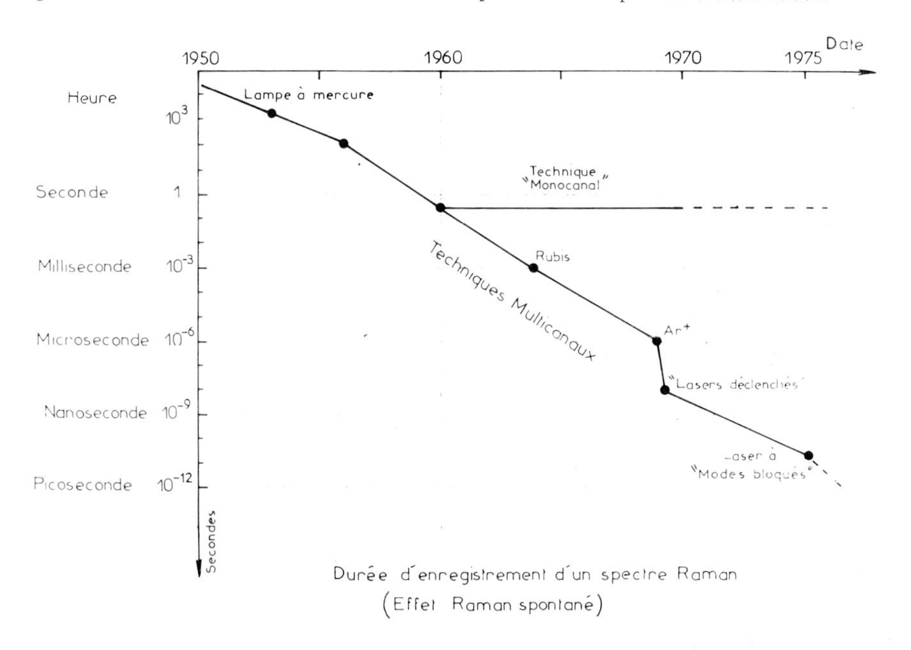

SPECTROMETRIE "MONOCANAL"

Le terme "spectrométrie monocanal" désigne l'ensemble des techniques qui, pour l'analyse d'un spectre optique, emploient un monochromateur suivi d'un unique récepteur de rayonnement (photomultiplicateur). Le résultat de la mesure est alors un diagramme à deux dimensions : en ordonnée l'intensité lumineuse, en abscisse soit le nombre d'onde correspondant au balayage du spectre par le monochromateur, soit si l'instrument fonctionne à longueur d'onde fixe, l'évolution temporelle pour un élément spectral c'est-à-dire une raie. La résolution temporelle peut, lorsque l'analyse porte sur un seul élément spectral, atteindre quelques picosecondes par l'emploi des lasers pulsés à modes bloqués. Par contre, si l'étude spectrale envisagée exige la connaissance de l'évolution au cours du temps de différentes espèces chimiques présentes dans un milieu réactionnel, il est nécessaire d'enregistrer des spectres Raman complets et non pas une seule raie spectrale. Le processus d'exploration des spectres par balayage successif des différents éléments spectraux devient alors très défavorable. Il en est de même si l'on cherche à tirer profit de la bonne résolution spatiale que permet la focalisation du faisceau laser sur l'échantillon pour déterminer la distribution d'une espèce.

SPECTROMETRIE MULTICANAUX

Les techniques de spectrométrie Raman "multicanaux" ont été étudiées et développées depuis 1960 par Bridoux. Elles consistent à analyser d'un seul coup l'ensemble de l'image spectrale grâce à un détecteur de rayonnement photoélectrique comportant un très grand nombre de canaux d'informations indépendants et fonctionnant simultanément.

Le résultat d'une telle analyse peut se représenter schématiquement comme sur la figure 2 par un diagramme à 3 dimensions : sur l'axe OZ l'intensité spectrale mesurée en chaque point, sur l'axe OY le nombre d'ondes caractérisant les vibrations moléculaires, les N canaux indépendants analysés par le récepteur définissant la résolution spectrale dans le domaine considéré, enfin sur l'axe OX une coordonnée spatiale analysée en N' canaux indépendants par le récepteur. Le plus souvent cet axe OX coïncide avec la direction de la fente d'entrée du spectromètre c'est-à-dire avec la direction des raies spectrales. Les N' canaux disponibles selon OX peuvent être employés de plusieurs façons :

- soit pour résoudre spatialement des régions différentes de l'échantillon

- soit pour résoudre dans le temps l'évolution d'un phénomène grâce à une déflexion des images à vitesse connue selon OX.

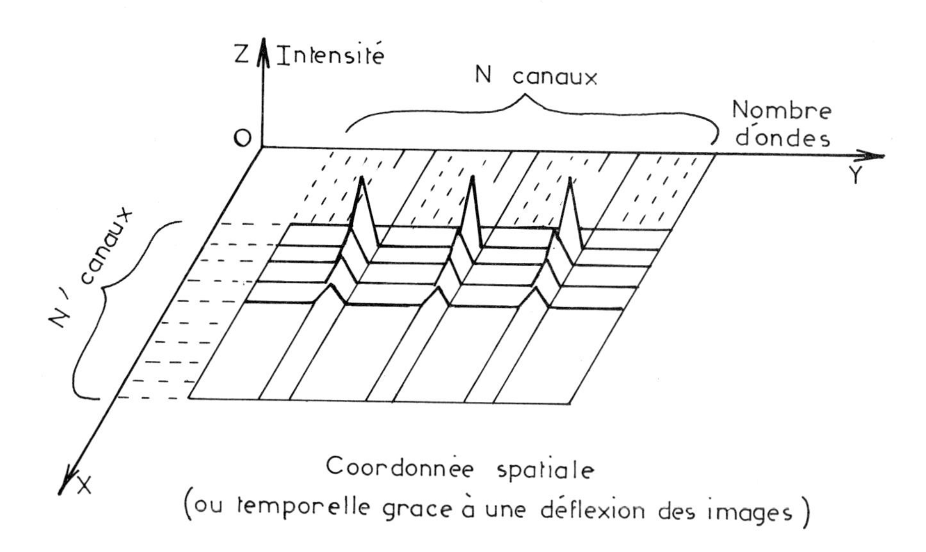

Fig. 2. Le spectromètre "Multicanaux" enregistre un diagramme à 3 dimensions.

L'ensemble des informations contenues dans une telle image spectrale doit être enregistrée dans une "mémoire". (Cible accumulatrice de charges électrostatiques, mémoire magnétique, mémoire digitale, mémoire photographique).

L'emploi d'une mémoire à très grande capacité permet d'envisager l'enregistrement d'une série d'images successives qui seront ensuite analysées lentement et transférées sur un autre support. La figure 3 précise comment sont réalisées ces différentes fonctions dans les instruments conçus par Bridoux. Un laser fonctionnant, soit en continu, soit en impulsion, excite l'effet Raman dans l'échantillon de la même manière que dans un instrument conventionnel. La lumière diffusée est projetée sur la fente d'entrée d'un spectromètre comportant un ou plusieurs étages à réseaux à grande ouverture, construit de manière à réduire la lumière parasite à un taux extrêmement faible. Les considérations développées ci-dessus montrent que le système doit, en outre, être stigmatique. A la sortie de ce spectromètre, l'image spectrale est focalisée sur la photocathode d'un tube intensificateur d'images. L'accélération et la focalisation des photoélectrons produits par chacun des éléments de surface de cette photocathode fournit une image électronique du spectre Raman, visualisée par un écran fluorescent. La mise en série de plusieurs étages intensificateurs permet d'atteindre un gain important (de 10^4 à 10^6). Cela signifie que, en moyenne, pour un photon frappant la photocathode, sont émis à la sortie de 10^4 à 10^6 photons sur un élément de surface correspondant. L'image spectrale ainsi intensifiée est transférée, soit par un système de fibres optiques, soit par une optique conventionnelle sur un récepteur secondaire capable de mettre en mémoire l'information. Plusieurs types de récepteurs secondaires ont été expérimentés, depuis le film photographique ou cinématographique, jusqu'au système de télévision pour très bas niveau comportant, soit une cible électrostatique (tube S.E.C), soit un dispositif de transfert de l'information sur ruban magnétique, disque magnétique ou mémoire d'ordinateur. La lecture ultérieure du signal mis en mémoire peut donner lieu à toutes sortes de traitements de l'information.

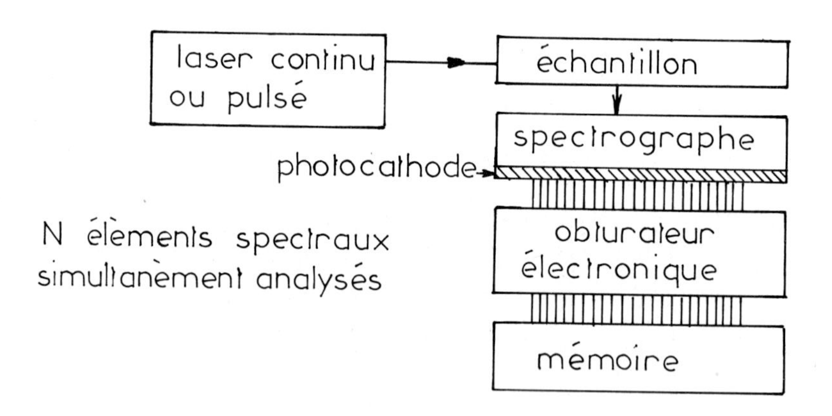

Fig. 3. Schéma d'un spectromètre Raman électrono-optique "Multicanaux"

Quelques remarques générales résumant les propriétés d'un tel ensemble méritent d'être signalées :

- Les qualités de la photocathode du tube intensificateur sont comparables à celles d'un bon photomultiplicateur. A cet égard, l'enregistrement d'un signal mesurable dans la mémoire à partir d'un seul électron initialement émis par la photocathode est théoriquement possible. On constate en pratique que les étages intensificateurs introduisent souvent des bruits parasites qui rendent difficile le fonctionnement en "compteur de photons" par intégration sur des temps longs. Par contre, ce récepteur est idéal pour l'enregistrement de l'ensemble des données d'un phénomène de courte durée. D'autre part, la "dynamique" du système est limitée non pas par la photocathode mais par les mémoires utilisées.

L'avantage le plus significatif réside dans le très grand nombre d'éléments d'informations qui peuvent être analysés simultanément. Des tubes intensificateurs d'usage courant correspondent à un nombre de canaux d'informations N x N' (Fig. 2) de l'ordre de 10^5.

Certains tubes intensificateurs comportent un étage obturateur ou un étage déflecteur ultra-rapide. Puisque le système est capable d'intégrer les informations au niveau de la mémoire, plusieurs modes de fonctionnement peuvent être envisagés :

- Laser fonctionnant en continu, image fixe
- Laser fonctionnant en impulsion, image fixe
- Déflection rapide de l'image spectrale.

Quel que soit le mode de fonctionnement, l'énergie nécessaire à l'enregistrement d'un <u>spectre Raman complet</u> par ces techniques est de l'ordre de <u>10 microjoules à 10 millijoules</u> au niveau de l'excitation de l'échantillon. Dans le cas d'un laser à Argon ionisé fonctionnant en continu avec une puissance de quelques watts, il suffit donc de quelques microsecondes. Dans le cas de lasers pulsés (Rubis, YAG ou Laser à colorant), une faible fraction de l'énergie prélevée sur <u>une seule impulsion</u>

d'une durée allant de la nanoseconde à la picoseconde est largement suffisante. Si l'on utilise un obturateur, le bruit de fond dû à l'émission thermique des photocathodes pendant des temps aussi courts, est tout à fait négligeable. Par contre, le bruit "de photons" est le même que celui qu'on observerait sur un enregistrement lent avec le même nombre de photons.

RESOLUTION TEMPORELLE

Les figures 4, 5 et 6 montrent quelques exemples de résultats obtenus par la technique "multicanaux" par Bridoux et ses collaborateurs. Les expériences les plus récentes, réalisées en collaboration avec C. Reiss, ont démontré la possibilité d'enregistrer des spectres Raman spontanés dans le domaine des picosecondes. L'énergie disponible à partir d'un laser à mode bloqué est très supérieure à l'énergie nécessaire. De grandes précautions doivent être prises pour éviter l'apparition de phénomènes beaucoup plus intenses que l'effet Raman spontané qui sont l'effet Raman stimulé ou les phénomènes de claquage dans le milieu matériel constituant l'échantillon. (5).

Même pendant des temps aussi courts, la résolution spatiale disponible selon l'axe OX permet de localiser avec une bonne précision les différentes régions d'échantillons projetées le long de la fente et illuminées par le faisceau laser. Le stigmatisme du spectromètre assure une correspondance point par point entre chaque élément de volume de l'échantillon et chacun des N' canaux répartis le long de la fente d'entrée.

Si l'on ne désire pas tirer parti de cette résolution spatiale, une déflexion rapide des images selon la direction OX peut être envisagée. La durée de l'impulsion laser n'est donc pas le seul paramètre qui définit la résolution temporelle de l'instrument. Des tubes photoélectriques dits "à balayage de fente" permettent de réaliser une déflexion électrostatique ultra-rapide des images. Des expériences actuellement menées avec de tels tubes ont cependant montré une dégradation importante de la résolution spatiale. C'est pourquoi nous préférons

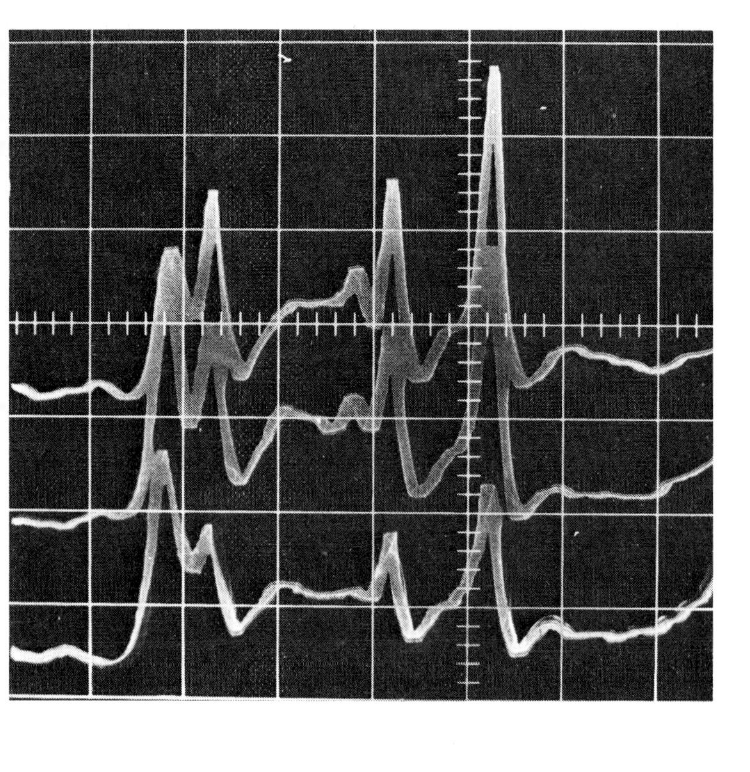

Fig. 4. Spectre Raman de mélanges de composés aromatiques, excités par une impulsion de Laser à Rubis (600 μs)

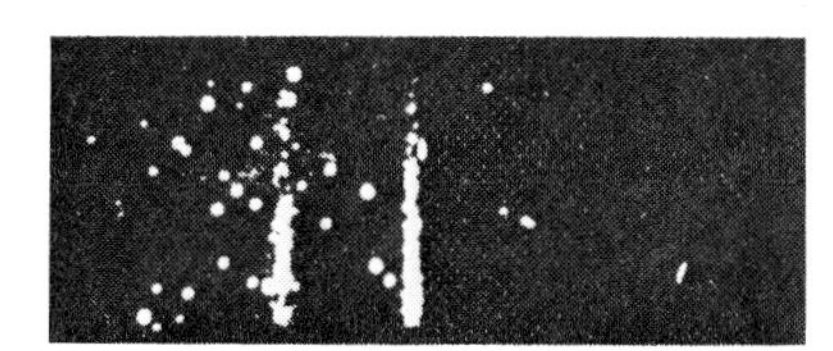

Fig. 5. Une partie du spectre Raman spontané du Méthanol excité par une impulsion de 25 picosecondes (λ exc = 5300 Å Energie = 200 microjoules)

Fig. 6. Spectre Raman "ordinaire" du Trimethylbenzène 1, 3, 5
1 impulsion de 25 picosecondes
Laser YAG doublé - 5300 Å
Energie 1 millijoule
Région 3000 cm^{-1}

employer un dispositif de déflexion optique des images. Lorsqu'il s'agit de temps très courts, il est facile de réaliser des lignes à retard optique produisant des réflexions multiples du faisceau laser qui permettent à partir d'une impulsion laser unique, d'exciter l'échantillon par un train d'impulsions successives. Ces impulsions décalées dans le temps peuvent, soit exciter la même région de l'échantillon, soit illuminer des régions différentes. La figure 7 représente l'un des schémas optiques utilisés. La figure 8 montre un exemple d'application de ces techniques dans lequel un laser à Néodyme (YAG) à mode bloqué comportant un doubleur et un quadrupleur de fréquence fournit 2 impulsions de 25 picosecondes. La première, ultraviolette, produit une excitation photochimique de l'échantillon, la seconde, retardée d'une quantité connue, par passage dans la ligne à retard optique, produit l'analyse par effet Raman.

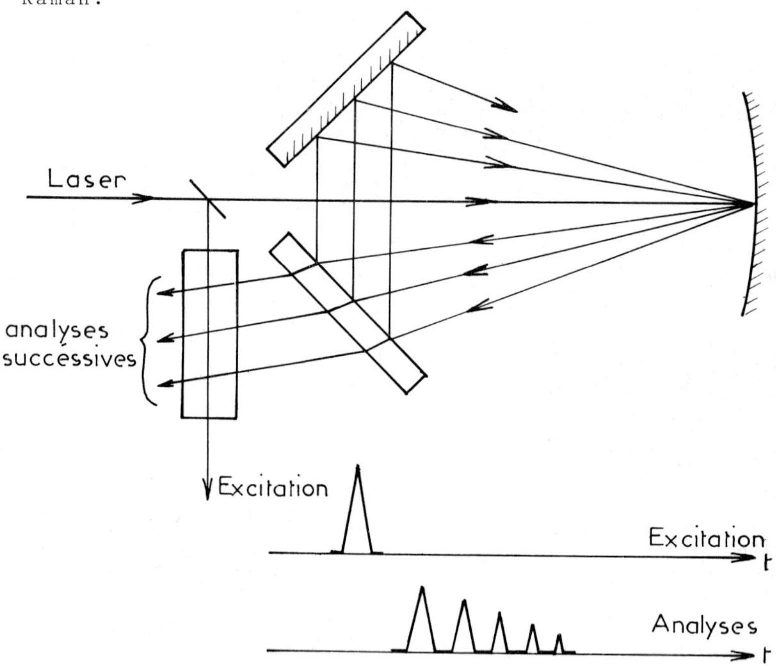

Fig. 7. Exemple de ligne à retard à passages multiples pour l'excitation de spectres Raman par un train d'impulsions.

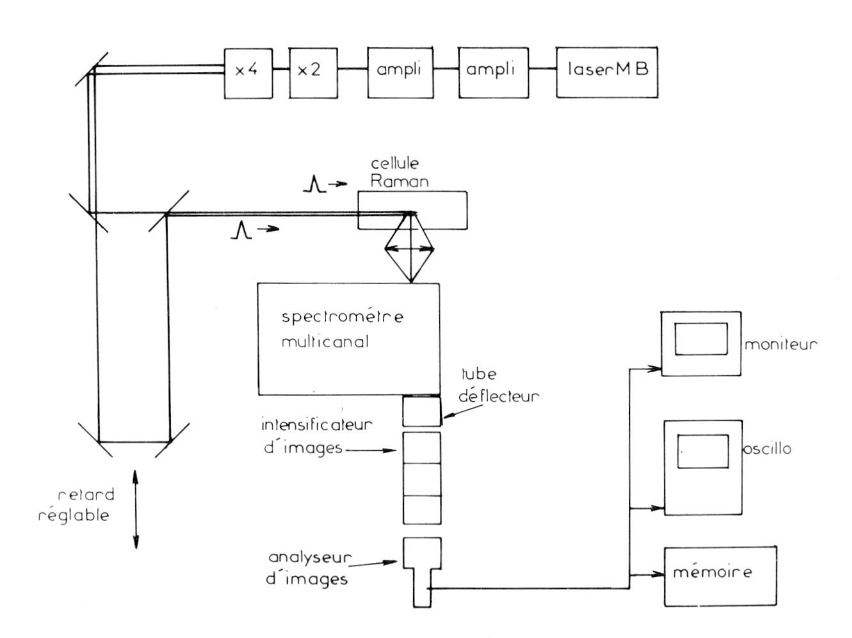

Fig. 8. Schéma général d'un montage de Spectrométrie Raman dans le domaine 10^{-11} à 10^{-8} seconde (M. Bridoux et C. Reiss)
A. Deffontaine

RESOLUTION SPATIALE : Microanalyse et Microscopie par effet Raman

La possibilité d'enregistrer des spectres Raman à partir d'un très petit volume d'échantillon excité par laser a donné lieu à d'intéressantes applications. Dans ce domaine, on peut également, soit faire appel à une technique "Monocanal", soit employer un récepteur "Multicanaux".

- La technique "Monocanal" conventionnelle consiste à illuminer un microéchantillon par le faisceau laser focalisé, grâce à une optique à grande ouverture, sur une surface qui n'est théoriquement limitée que par la diffraction.

La puissance maximale admissible dans ces conditions doit rester très inférieure au milliwatt car il est indispensable d'éviter la destruction thermique de l'échantillon. La très faible intensité du spectre Raman impose alors un enregistrement très lent.

- L'emploi d'un système à "Multicanaux" offre des avantages décisifs en microanalyse :

1) Si l'on désire enregistrer le spectre Raman d'un très petit volume d'échantillon, l'enregistrement simultané de N éléments spectraux revient, à durée égale, à améliorer le rapport Signal/Bruit de $\sqrt{N}$.

2) Si l'on désire identifier une des espèces chimiques présentes dans l'échantillon et la localiser, il est possible d'examiner simultanément un grand nombre de "points". Cela permet d'établir une "carte" de la distribution de cette espèce à la surface de l'échantillon, avec une résolution spatiale qui n'est limitée que par la longueur d'onde du rayonnement diffusé soit une fraction de micron.

La figure 9 représente le schéma d'un "microscope - micro-
sur ce principe (4).

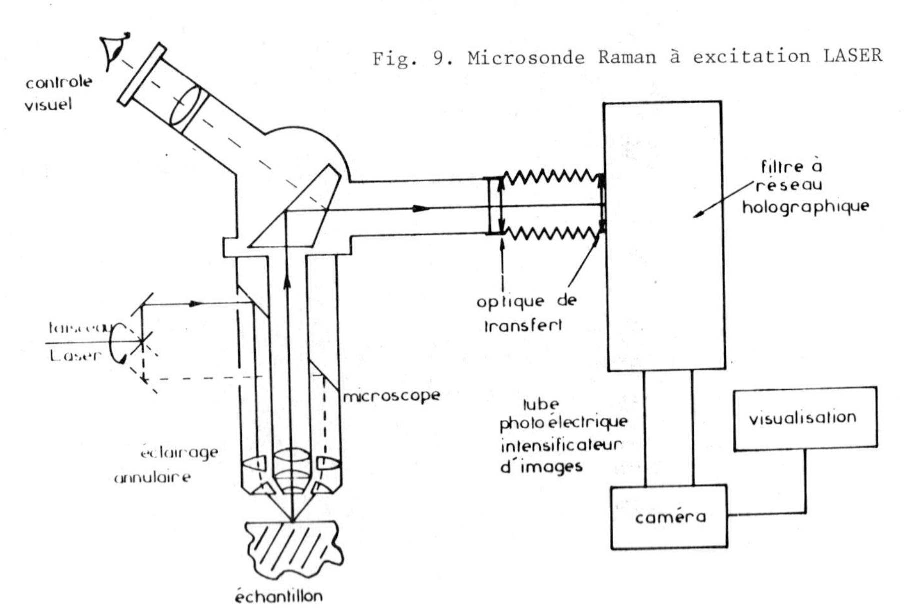

Fig. 9. Microsonde Raman à excitation LASER

La surface observée de l'échantillon est illuminée dans son ensemble par le faisceau laser, ce qui permet d'atteindre des puissances de quelques dizaines de milliwatts sans endommager les matériaux. L'image agrandie de cette surface, produite par une optique classique de microscope, est projetée sur la photocathode d'un tube intensificateur d'images, à travers un filtre optique que l'on cale sur une raie Raman caractéristique de l'un des composés présents dans l'échantillon.

Ce filtre optique doit être réalisé de manière à réduire le taux de lumière parasite à une valeur extrêmement faible, car il est nécessaire de former des images grâce à une radiation Raman très peu intense, située dans le spectre au voisinage immédiat de la radiation laser diffusée sans changement de longueur d'onde. L'emploi de réseaux holographiques concaves (Jobin et Yvon) a permis de résoudre ce problème de façon satisfaisante.

La figure 10 montre un exemple de micrographies obtenues avec cet instrument. Il s'agit d'un mélange de microcristaux d'oxyde de Titane et de nitrate d'argent déposés sans aucune préparation sur une lamelle de verre. L'image de l'échantillon obtenue en lumière blanche ne permet pas de différencier ces deux composants. Par contre, les images enregistrées en excitant avec un Laser à Argon ionisé, et en réglant le réseau pour isoler la radiation correspondant à une raie Raman caractéristique, soit de TiO_2, soit de NO_3^-, permettent d'identifier séparément chacun des cristaux. Cet exemple est tiré d'une étude de réactions photochimiques en phase solide.

Il faut souligner que de telles images sont enregistrées en un temps de l'ordre de la seconde, ce qui permet l'observation de préparations évoluant au cours de réactions chimiques ou électrochimiques.

La résolution spatiale atteinte au cours de ces expériences correspond au pouvoir séparateur du microscope optique, soit de l'ordre du micron.

IMAGE OBTENUE EN LUMIERE BLANCHE

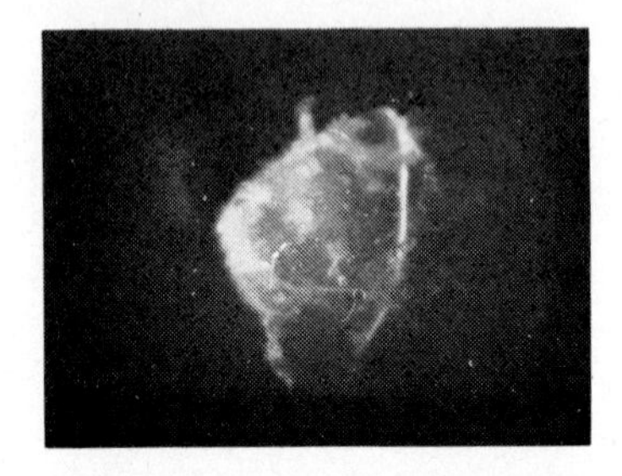

IMAGE OBTENUE DANS UNE RAIE CARACTERISTIQUE DE TiO_2 (612 cm^{-1} A_{1g})

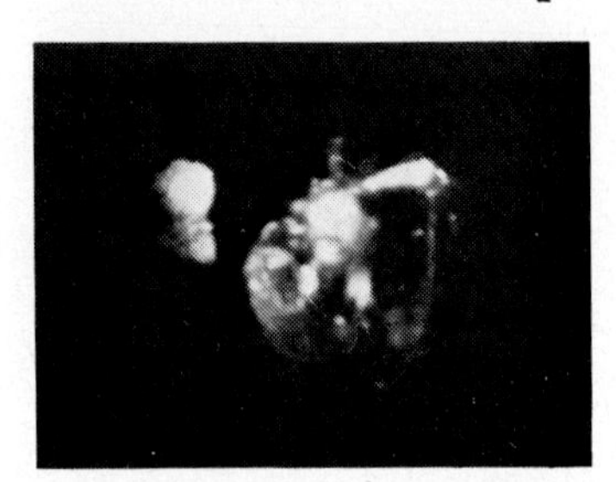

IMAGE OBTENUE DANS UNE RAIE CARACTERISTIQUE DE NO_3^- (1045 cm^{-1} ν_1)

MELANGE $TiO_2 + AgNO_3$

Fig. 10. Micrographies d'échantillons hétérogènes obtenues par "Microanalyse Raman-Laser"

CONCLUSION

Les progrès récents des techniques de spectrométrie Raman permettent de tirer parti des qualités exceptionnelles des sources de lumière LASER. Parmi les aspects nouveaux qu'offrent ces techniques, la possibilité d'observer des spectres de vibration avec une excellente résolution temporelle et spatiale permet d'une part, d'aborder l'étude d'espèces à durée de vie très brève, et d'autre part d'envisager la microanalyse d'échantillons hétérogènes.

REFERENCES

1 M. Delhaye, Appl. Optics, 7 (1968) 2195

2 M. Bridoux et M. Delhaye, Nouv. Rev. d'Optique Appliquée 1 (1970) 23

3 M. Bridoux, A. Chapput, M. Delhaye, H. Tourbez et F. Wallart in "Laser Raman Gas Diagnostic" Plenum Press (New York) M. Lapp C.M. Penney edit.

4 M. Delhaye et P. Dhamelincourt, Jnal of Raman Spec. 3 (1975) 33

5 M. Bridoux, C. Reiss et A. Deffontaine (à paraître)

DISCUSSION

M. DELHAYE

J. Faure - Dans le cas de la résolution spatiale, comment obtient-on les renseignements à l'échelle moléculaire?

M. Delhaye - Le microscope Raman permet d'établir une image monochromatique, prise dans une seule raie de vibration caractéristique d'une espèce polyatomique. Il fournit donc la distribution de cette espèce à la surface d'un échantillon hétérogène. Le même instrument peut fonctionner en micro-analyseur pour enregistrer le spectre Raman complet d'une très petite région choisie dans cette image.

C. Varma - Your review of the various techniques in Raman spectroscopy does not pay attention to the inverse Raman effect as an analytical tool. It may have certain advantages in short time scale recording of Raman spectra. The measurements will not depend critically on stray light, therefore allowing the use of a single grating spectrograph, which will increase the sensitivity. Can you comment on this point?

M. Delhaye - The inverse Raman effect is also studied by the authors cited in reference, by using the techniques above described. It seems however that the analytical applications are not exactly the same as those of spontaneous Raman because inverse Raman effect is only observed under high density of laser power in the sample, where non-linear effects are noticeable.

M. Stockburger - Have you already used the picosecond systems in order to follow a chemical reaction by spontaneous Raman scattering?

M. Delhaye - Not at the present time, because the picosecond spontaneous Raman experiments are very recent. Chemical reactions have been studied in the time range from second to microsecond.

B. Alpert - Le Raman permettant une cartographie spatiale, pensez-vous qu'il soit possible de situer les inclusions dans les membranes ou les nerfs? Cette technique permettrait-elle de pouvoir répondre expérimentalement à certaines théories de "portes" laissant passer sélectivement certains ions métalliques?

M. Delhaye - La micro-analyse et la cartographie par effet Raman présentent certainement beaucoup d'intérêt pour l'étude de problèmes biologiques. La principale difficulté dans ce domaine provient de la présence de composés fluorescents dans de nombreux échantillons biologiques.

A LASER TEMPERATURE JUMP AND PHOTOLYSIS APPARATUS WITH REPETITIVE TUNABLE EXCITATION

I. GIANNINI
Laboratori Ricerche di Base, Snamprogetti S.p.A., Monterotondo (Rome), Italy.

SUMMARY

We describe an apparatus which can be used to study small liquid and solid samples by perturbation techniques both as temperature jump and pulsed photolysis. We use a repetitive pulsed laser source tunable in the visible and near IR. Analyzer beam coming from a mercury arc lamp crosses the laser beam at 15 degrees in the sample. Fast electronics with sampling and averaging techniques allow the detection of less than 0.1‰ OD changes with 10 ns time resolution. The higher temperature jump ($\Delta T \sim 1 \div 2°C$) is reached in water solutions by absorption of the 1.36 μ - Neodymium line -. The performances of the apparatus are shown with the most important results already obtained in enzymatic systems.

INTRODUCTION

The relevance of relaxation techniques in kinetic studies is generally well known and was emphasized in many reviews papers [1, 2, 3] . On the other hand powerful and ductile pulsed light sources are now available, allowing the realization of high sensitivity experimental apparatus.
We describe in the following a new instrument for relaxation measurements where the main perturbation source is a tunable and pulsed laser system, which can be utilized alternatively

as a temperature jump or a flash photolysis apparatus [4]. As main features, this apparatus joins some different aspects of the existing laser-T-jump [5, 6, 7, 8] and flash photolysis [9, 10], instruments, i.e. the near infrared (IR) heating of temperature jump operation [5] to the repetitive and averaging techniques already used by Witt and coworkers for flash photolysis operation [9].

EXPERIMENTAL

A schematic diagram of the apparatus is shown in Fig. 1.

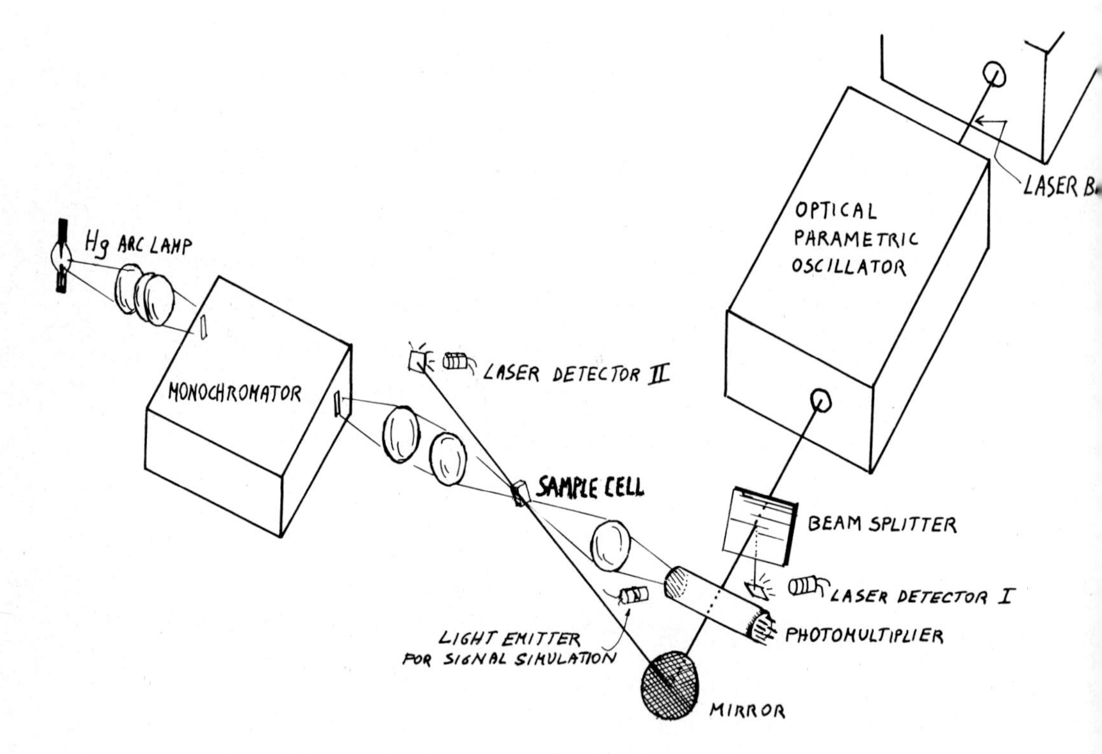

Fig. 1 - Schematic diagram of the laser-T-jump apparatus.

The laser system was the Nd:YAG Laser mod.1000 of Chromatix, which is coupled with a tunable Optical Parametric Oscillator (OPO) (Chromatix mod.1020). A complete description of the system can be found elsewhere [11] . The wavelength of the output radiation has some fixed powerful lines (the shortest at 973 nm, the longest at 1.36 μ) and a tunable region ranging from 600 nm to $\sim 3\,\mu$. The laser pulses have a duration of 100 ns and the repetition rate generally used in 75 Hz. The energy of each pulse varies from 0.1 to 3 mJ. The highest power is obtained at the laser output at 1.32 $\div$ 1.36 μ by an high transmission (50%) output mirror. At this wavelength water absorption is good ($\sim 2\ cm^{-1}$) and we can obtain temperature rise of 1 $\div$ 2°C for each pulse by focusing the laser beam on our sample.

On the area of the cell where the laser hits we have the monochromatic image of an high pressure Hg arc lamp with a very small spot dimension (HBO 100W/2 Osram). The light from the lamp goes through an high aperture quartz optics and a small monochromator (SPEX Minimate 1 : 4 aperture). We can work with a spot image of about 0.1 mm in max size. By this way the volume of the observed reaction is about 10^{-5} ml, i.e. extremely small. This condition limits the upper observable time by diffusion effects to about 1 ms. The lower observable time is limited to some fraction of the laser pulse width (i.e. some 10^{-8} sec); in fact we can do an accurate waveshape measurement of laser pulse, and reconstruct the signal by computer deconvolution.

The laser pulse is monitored by different detectors: an ultra fast silicon pin-photodiode is used in the visible region (Hp 4220); while in the near IR, where no fast detectors are available, either Germanium (OAP 12 Philips) or In As (Barnes A 100) photovoltaic detectors are used, coupled to an integrating amplifier device. The monitor beam crosses the laser beam at about 15°. The cell geometry can be varied: we have alternatively used quartz cell disposed at Brewster angle as in Fig. 1; or 1 mm. diameter capillary with flowing solutions.

Accurate thermostating of the cell is obtained in some different heat exchanger. The flowing geometry is needed when the apparatus is working as T-Jump. The total volume of the sample can be confined in this case to ~1.5 ml.
The absorbancy change following the laser pulse is measured by a Philips XP 1113 photomultiplier equipped with a wide band (~ 100 MHz) preamplifier; the photomultiplier output can be switched from anode to the 4th dynode.
A fluorescence emission monitor is realized with fiberglass light pipe, orthogonal to laser beam. The electronic signal processing follow the block diagram of Fig.2 and is only partially realized with commercial circuitery.

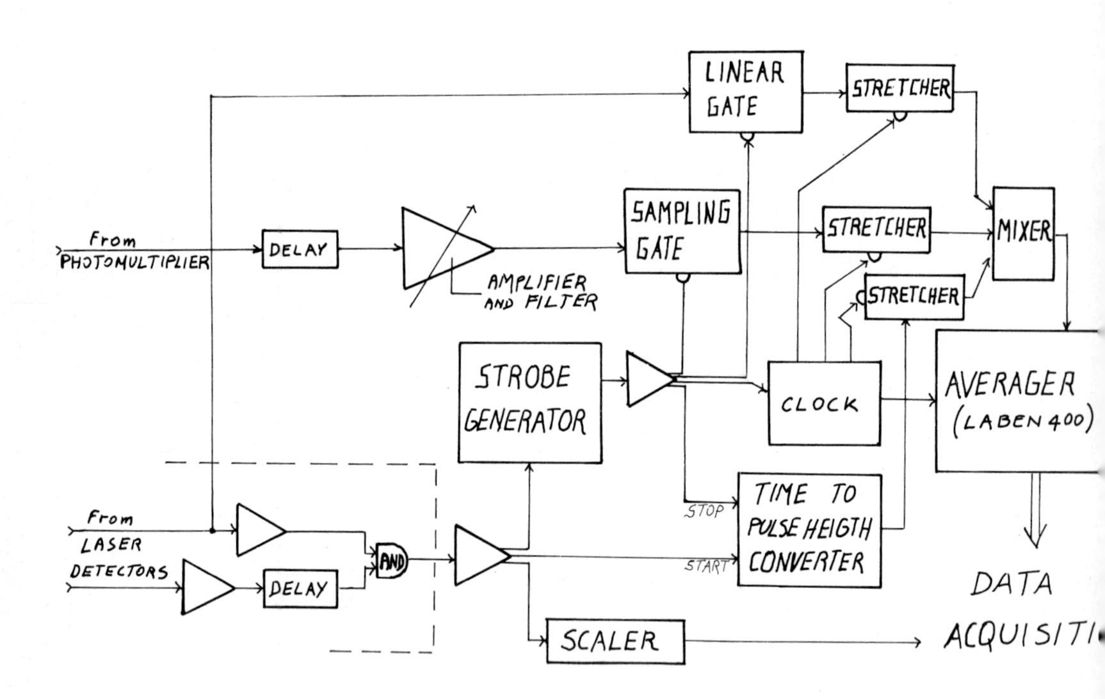

Fig. 2 - Electronics block diagram.

Fast trigger is obtained from laser detectors; the signal from photomultiplier is amplified, filtered and then analysed in the fast sampling gate, the strobe generator commands the sampling signal sequence. The signal, the laser monitor, the time delay generated by the strobe are first memorized in analogic memories (capacitors in "stretcher" circuits) and then sequentially digitalized and memorized in a multichannel analyzer (Laben 400), with averager. We usually can sum over some thousands of signals due to the high repetition rate of the laser. Output from the multichannel was displayed on the scope and punched on paper tape. Computer processing gives normalized data, plots, etc.

SOME EXPERIMENTAL RESULTS

A computer plotter trace of $\sim$1000 samples of a temperature jump operated experiment is shown in Fig. 3a. A phenol red $2 \cdot 10^{-4}$M pH 8. in Tris NCl 0.1 M is used in a flow geometry. The change in pH is monitored at $\lambda = 430$ nm. The resulting trace is only a little slower than the heating time.
Both photochemical perturbation and fast heating pulse can be successfully used in biological samples. The laser technique allows the detection of very fast conformational changes or proton transfer reactions in the 10^{-5} sec to 10^{-8} sec time region. The most important advantages with respect to other techniques available in the same time range (ultrasounds, dielectric relaxation, etc.) are due to the selectivity both of excitation and of detection of the recorded events, like the possibility to select specific probes related to catalytic activity.
Fig. 3b shows the plot coming from a photolysis experiment now under way [12]. A fast photoisomerization is detected in the complex of Human Carbonic Anhydrase B with a chromophoric azosulfonamide (Neoprontosil). The Quantum Yield of this effect is of the order of few part per thousand but the signal

averaging system allows the registration of a trace with S/N ~ 20 in only 5 minutes. The first relaxation effect shown in Fig. 3b has a relaxation time of 800 ns (two more and slower relaxation are observed in this system) 12 .
The laser operated at $\lambda = 532$ nm and only 0.2 mJ per pulse in order to avoid heating; the geometry was quartz cell at Brewster angle; the monitor was at $\lambda = 405$ nm corresponding to the second singlet transition of the dye. The relaxation measured in Fig. 3b could be interpreted as fast proton transfer at the ionizable group of the active site assential for catalytic activity [13] ; the measured rate largely exceeds the diffusion controlled limit, thus supporting the importance of intramolecular proton transfer.
Another photolysis experiment on enzyme-inhibitor complexes is shown in Fig. 3c. Here the α-chymotripsin bounded dye Biebrich Scarlet [14] (BS) is directly ionized by laser pulse; experimental conditions are similar to the previous experiment. Then we measure the recombination rate as a function of different parameters (pH, ionic strength, etc.) allowing a quantitative evaluation of the influence of the ionization of the active site groups (like the essential hystidil residue) on the proton affinity of the substrate analogue [15].

ACKNOWLEDGEMENTS

The realisation of this instrument should been impossible without the collaboration of R. Colilli.
We are indebted to Dr. L. Micheli and P. Grasselli for the essential contributions they have done during the work for their thesis at University of Rome.

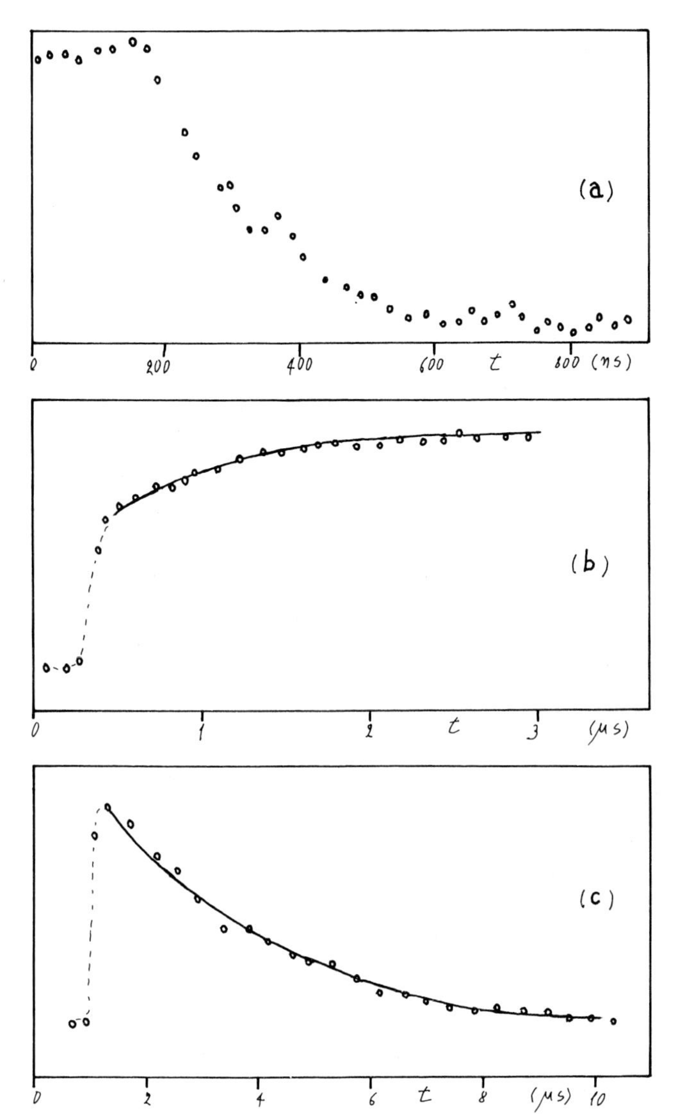

Fig.3 a: T-Jump operation; the laser is working at 1.36 μ with special output mirror. A flowing solution in a capillary is heated 1,2°C per pulse. The sample was a fast reacting mixture of Phenol Red $2 \cdot 10^{-4}$M and Tris HCl 0.1M pH 8. Absorbance change follows with small delay (20 ns) the heating effect. b: Flash photolysis operation. A photoisomerization effect is observed in a carbonic anhydrase azosulfonamide complex. The fast relaxation has 800 ns time. The equilibrium was reached at longer times with two more relaxation effects [12] . c: The proton recombination effect following the photodissociation of chymotripsin-BS complex (see text).

REFERENCES

1 M. Eigen and L. de Maeyer, in S.L.Friess, E.S.Lewis and A. Weissberger (Editors), Technique of Organic Chemistry, Vol. 8, N° 2, Wiley Interscience, New York, 1964, p.895.

2 G.G. Hammes, Adv. in Protein Chem., 23(1968)1.

3 G.G.Hammes and P.R.Schimmel, in P.D.Boyer (Editor), The Enzymes, Vol. 2, Academic Press, New York, 1970, p.39.

4 I. Giannini, in E. Wyn Jones (Editor), Chemical and Biological Applications of Relaxation Spectrometry, 1975, in press.

5 J.U. Beitz, G.W. Flynn, P.H. Turner and N. Sutin, J. Am. Chem. Soc. 92(1970)4130.
D.H. Turner, G.W. Flynn, N. Sutin and J.U. Beitz, J. Am. Chem. Soc., 94 (1972)1554.

6 E.M. Eyring and B.C. Bennion, Ann. Rev. Phys. Chem., 18 (1967)129.

7 H. Hoffman, E. Yeager and J. Stuher, Rev. Sci. Instr., 39 (1968) 649.

8 E.F. Caldin, J.E. Crooks and P.H. Robinson, J. Phys. E: Sci. Instr. 4(1971)165.

9 H. Ruppel and H.T. Witt, in K. Kustin (Editor), Methods in Enzymology, Vol. 16: Fast Reactions, Academic Press, New York, 1969, p. 316.

10 B. Alpert, R.Banerjee and L. Lindquist, Proc. Nat. Acad. Sci. U.S.A., 71(1974)558.

11 S.E. Harris, Proc. IEEE, 57(1969)2096.
R.W. Wallace and S.E. Harris, Appl. Phys. Letters, 15(1969) 111.

12 I. Giannini and G. Sodini, in E. Wyn Jones (Editor), Chemical and Biochemical Applications of Relaxation Spectrometry, 1975, in press.

13 S. Lindskog, in P.D. Boyer (Editor), The Enzymes, Vol. 5

Academic Press, New York, 1971, p. 587.

14 A.N. Glazer, J. Biol. Chem.,242(1967)4528.

15 I. Giannini and P. Grasselli, to be published.

DISCUSSION

I. GIANNINI

S. Ameen - I am not sure about the functioning of the apparatus shown because there are many complicated technical problem involved in using Nd^{+3}-laser for obtaining the temperature-jump. As far as I know 1.36μ wavelength pulses of sufficient energy and very short duration pulses are difficult to obtain and it remains problem yet to be solved for temperature-jump experiments by this method. The detection system shown is a bit difficult to realize for chemical relaxation studies. The electronics and rise-times of operational amplifiers and photomultiplier are crucial for nanosecond measurements.

I. Giannini - The 1.36 μ line of Nd-YAG laserwas obtained many years ago and is well stable in the Chromatix commercial system; the time duration is not so short (~ 150 ns) as it has been shown. For the detection system we can say that many photomultipliers are well working at the ns time scale and obviously we do not use integrated circuits in our preamplifier.

SUBNANOSECOND LASER T-JUMP

C. REISS
Institut du Radium, Section de Biologie, Orsay, France.

Lasers have proven very helpful for investigating photochemical reactions. Giant laser pulses could also be used to study kinetics of non-photochemical reactions ; the energy carried by such pulses could be brought to suddenly change some thermodynamic parameter governing a reaction equilibrium (temperature, pressure, pH, etc). The subsequent return to a new equilibrium of the reaction partners can yield informations on reaction pathways, intermediates, rate constants, etc, according to the principles of chemical relaxation, first developped by EIGEN.

Among the perturbation methods used, the most popular has been the T-jump. The temperature of a conductive solution of reacting species is suddenly rised by Joule heating, induced by a capacitor discharge. Although this method proved to be very successful in many instances, it is beset with limitations. The jump-speed does not exceed a few microseconds, the medium to perturbate has to be conductive to electricity, restrincting in practice the application to water solutions containing a fair amount of salt ; also, the T-jump generating device has to be located in close vicinity to the solution, which results in constrains on the reaction cell design.

Production of T-jumps through the absorbtion of a laser light pulse was attempted right at the beginning of the laser age. The principle is simple : the absorbtion of an electromagnetic energy results in heating. Quantitatively, this can be seen from the LAMBERT law, which states that

$$I(x) = I(o) \exp(-2.3\, D \cdot x)$$

where I(x) is the intensity of the light at depth x in the medium of absorbance D at the radiation wave length, I(o) the incoming intensity. Similarly, the temperature increase at depth x is

$$\Delta T(x) = \Delta T(o) \exp(-2.3\, D \cdot x)$$

where $\Delta T(o)$ is the temperature increase at the sample surface. If C is the heat capacity of the medium, $\Delta T(o) = 1/C \cdot \frac{\delta I(o)}{\delta x}$

so that $$\Delta T(x) = \frac{2.3\, DE}{C} \exp(-2.3\, D \cdot x)$$

where E is the energy density of the radiation.

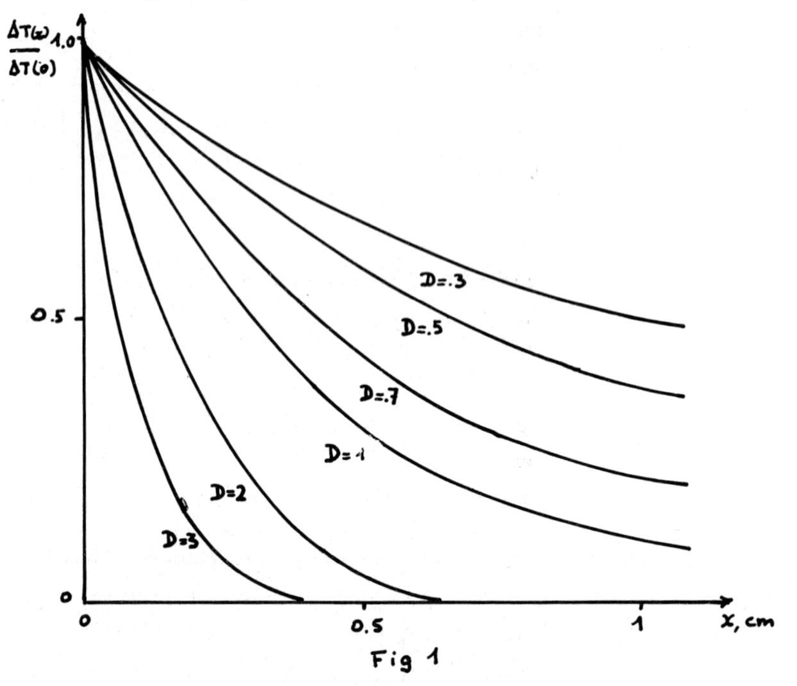

Fig 1

Before describing how laser pulse energy can in practice be transfered into heat, let us point out several observations derived from equation (1). The optical density enters $\Delta T(x)$ both through $\Delta T(o)$ and the exponential decay gradient, as can be seen in figure (1). Thus, although a high D value would be of interest since it would give rise to high $\Delta T(o)$ it would also generate a steep temperature gradient, which should be avoided both for theoretical reasons (the method of chemical relaxation applies for small perturbations) and practical considerations, that is unwanted side-effects, like the generation of shock-waves for instance. Optimal values of D for T-jumps would be in the unity range.

In contrast to the behaviour of D, the pulse energy density E enters $\Delta T(x)$ only through $\Delta T(o)$; one could then consider to increase E in order to generate more sizeable T-jumps. However, since the T-jump is to be generated in small time intervals, $\Delta\tau$, this implies enhanced power densities, $E/\Delta\tau = P$. The Maxwell equations show that the electric field V (V/cm) associated with a pulse of power P (W/cm^2) is related to P through

$$\log V = 0.5 \log P + 1.45$$

If a pulse of 0.2 J lasting for 20 ps is focused on 0.1 cm^2, P = 10^{11} W/cm^2 and V $\sim 10^7$ V/cm, which is equivalent to the electric field at an angstrom from a proton. Similarly, the magnetic field H (gauss) is given by

$$\log H = 0.5 \log P - 1.05$$

so that H $\sim 10^4$ gauss ; the radiation pressure π (atm) related to P through $\log \pi = \log P - 9$
amounts to 10^2 atm.

Although these parameters, which oscillate at some 10^{15} cycles/sec may be interesting for perturbating chemical equilibrium, they may introduce unwanted by-effects upon T-jumping, if they exceed some threshold value.

Laser heating can be achieved directly or indirectly, as to whether the energy is transfered to one of the reaction partners or to some intermediate relay. The later method was extensively investigated by Czerlinski (1), Caldin (2), Eyring (3) ; they used relay-dyes, absorbing at the laser wavelength and subsequently releasing the stored energy to the reaction mixture as heat. However, as shown above, the dye molecules have to be present in small amounts only, so as to keep D at reasonable levels. On the other hand, the power handling capability of a dye molecule is low, since once in the excited state after having absorbed a light quantum, the bleached dye molecules cannot absorb another quantum before having returned to their ground state. The energy transfer from pico- or nanosecond laser pulses to the reaction mixture have thus a poor yield. Also, the T-jump speed is governed by the relaxation time of the dye, which may last microseconds, so that the benefit of laser T-jumping is lost when relay dyes are used.

In direct laser heating, absorbtion of the radiation should not occur by one of the reacting species, but rather by the surrounding solvent-water in the present case- so as to prevent photoreaction among the former. This rises the question of producing a light pulse with selected wavelength , so that the solvent has an absorbance of ~ 1, the solutes being transparent. Dye lasers are tunable over wide ranges, but do not yield energies high enough for T-jumps. Molecular IR lasers (CO_2, CO) can be fitted with tuning devices (spin-flip, frequency multipliers), but presently they hardly deliver subnanosecond pulses, as do mode locked, solid state lasers.

Among the later, Nd^{3+} glass or YAG lase at 1.06 μ . This wavelength can be changed very efficiently both by frequency multiplication in non linear crystals, and (or) by stimulated Raman, which is highly effective with subnanosecond pulses carrying high power levels. Only very few chemicals -and specially biochemicals- absorb in the 1-2 μ region, whereas water absorbs there very efficiently (fig. 2).

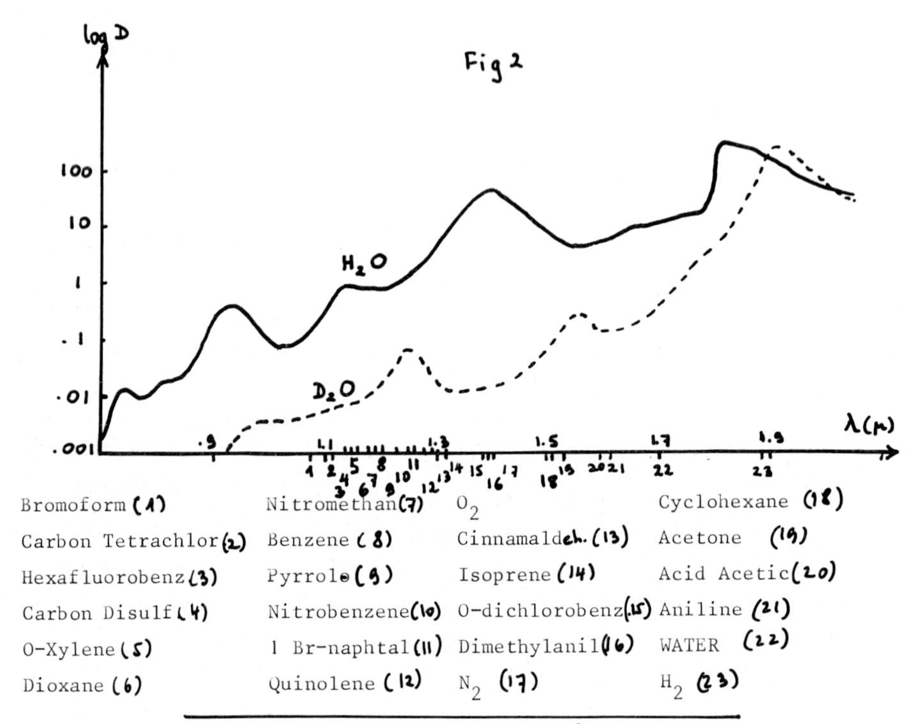

The water absorbtion wavelength can be easily reached by stimulated Raman scattering (SRS) induced by a Nd^{3+} laser pulse (1.06 μ) in various medium (fig. 2). Additional criteria to be observed upon selecting a scattering medium concern the scattering yield, proportional to the scattering cross-section of the molecule, and the ability to reabsorb the scattered light. Attention should be payed to stimulated Brillouin scattering (SBS), as its wavelength differs usually only by a few tenth of a wave number from the laser wave number, so that backward travelling SBS may reenter the laser chain and damage it. SBS is avoided if picosecond laser pulses are used, since the time needed to built up SBS is usually in the nanosecond range. One could also use devices closing the laser chain to backward travelling radiations (Faraday cell, λ/4 plate combined with a dielectric polarizer, etc...).

Among the liquids listed on fig. 2, nitrogen and benzene have very high SRS yields, as shown in table (1). They have been efficiently used for T-jump experiments in water. SRS in liquid nitrogen has been used previously by Beitz and coll (4) for heating water with a Nd^{3+} laser pulse lasting 40 ns ; their yield was ~ 20 %, whereas with picosecond pulses the SRS yield is increased fourfold. The values of the over-all heating, $\overline{\Delta T}$, measured with a platinum foil resistor, are also indicated on table 1.

A practical T-jump experiment has been performed on the reaction equilibrium.

$$I_2 + I^- \rightleftarrows I_3^-$$

a system already investigated by Turner and coll (5).

The reaction mixture contained 10^{-2} M ClO_3H, 10^{-2} M KI, 10^{-4} M I_2 in 50 % H_2O-D_2O. The amount of I_3^- was monitored through its absorbtion at 353 nm, which coincides with the third harmonic of the 1.06 μ fundamental. The experimental arrangement is shown on fig. 3.

MEDIUM	C_6H_6 (liq.)	N_2 (liq)
SHIFT	991 cm^{-1}	2326 cm^{-1}
S.R. λ	1.18 μ	1.41 μ
ABSORB. (H_2O)	0.8 cm^{-1}	50% H_2O-D_2O 4.5 cm^{-1}
$\overline{\Delta T}$ Calc.	0.9°C L=1cm	8.0°C L=1cm
$\overline{\Delta T}$ Meas.	0.8°C L=1cm	6.9°C L=1cm
ΔT(o) calc.	1.8°C	10.3°C
OPT. YIELD	>80 %	>80 %
ENERG. YIELD	89 %	86 %

Io = 0.1 cal. focused on 0.1 cm^2

$$\overline{\Delta T} = \frac{\Delta T(0)}{L}\int_0^L \exp(-2.3\ D\ x)\ dx$$

TAB. 1

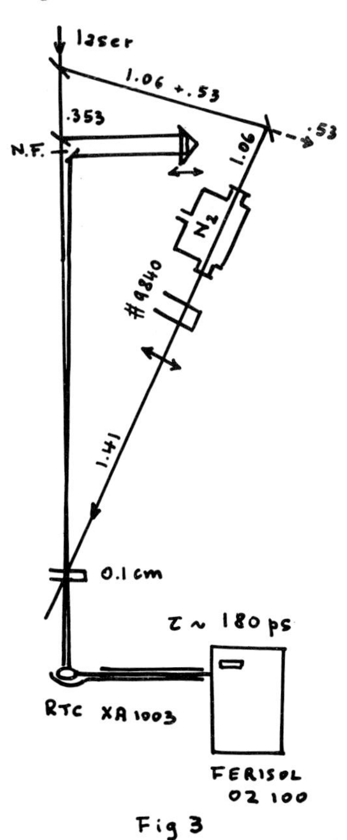

Fig 3

The measured relaxation time is 16 ± 5 ns, in good agreement with τ = 14 ns extrapolated from Turner's study (5).

Fast chemical relaxation studies demand, in addition to fast perturbation, fast analysis of the reacting species. In the case of macromolecules, and specially those of biological interest, informations on elementary reaction pathes (transconformation, H bond reactions, electron or proton transfers, etc...) are of prime importance, if the over-all changes of these particles are to be understood. Thus, the analysis should yield as many specific informations, as possible. Multichanel spectroscopy is idealy suited to fullfill these requirements, specially if the fast electronooptic methods, developped recently in the laboratory for Chemical Spectroscopy, of the Lille University, are applied. Prof. Delhaye, in an other talk at this meeting, reports on picosecond Raman Spectroscopy. Let us just mention preliminary, multichanel absorbtion spectroscopy in the visible range. The principle consists of generating a continuum of light through self-phase modulation of the laser beam focused in some medium (here water).

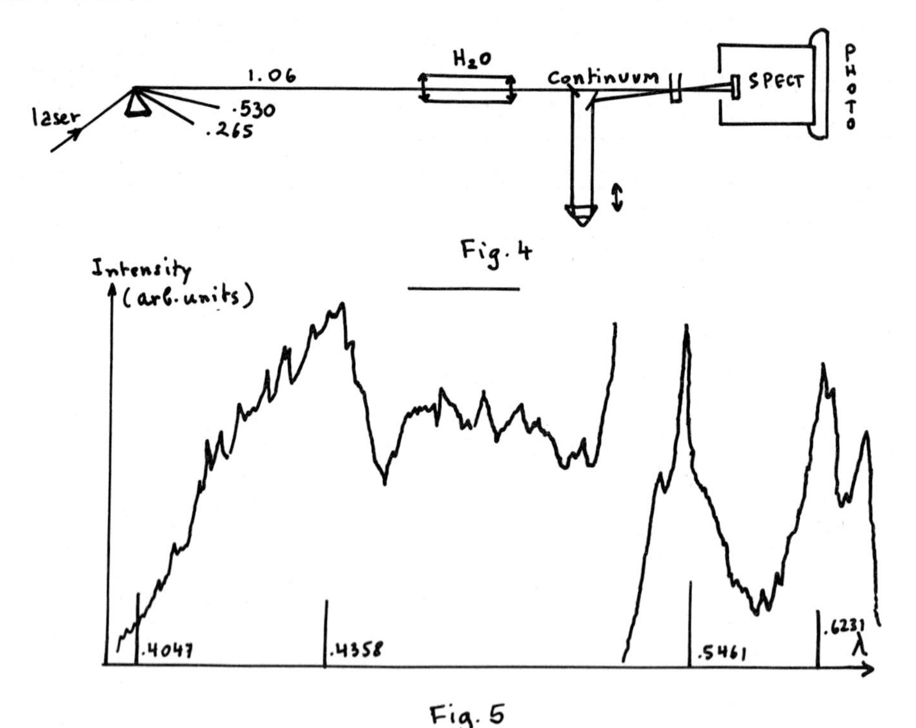

Fig. 4

Fig. 5

The powerful continuum so generated can be split into several parts with appropriate delays, so that they can travel through the absorbing sample at different time intervals before and after a perturbation. After dispersion in a spectrographe, the various spectra can be recorded on a photographic plate (see fig. 4 for an experimental arrangement). Simultaneous registration of a non-absorbed, dispersed continuum allows, with the aid of a computer treatment, to substract the emission spectrum of the continuum from the various spectra registered. Fig. 5 shows the microdensitometer tracing of such an emission spectra from water, using a 5.10^{15} W/cm^2 Nd^{3+} laser pulse.

The laser used in the present experiments has been built by QUANTEL, France. Thanks are due to the staff of QUANTEL for many helpfull advices. In the analysis part of this work, efficient collaboration by Prof. M. BRIDOUX and Dr. A. DEFFONTAINE (Lille University) and Dr. J.G. VILLAR (Laboratoire de Photosynthèse, Gif) is acknowledged. This work was supported by grants n° 72.7.0496 and 74.7.0373 from D.G.R.S.T.

(1) H. STAERK, G. CZERLINSKI, Nature 1965, 205, 63.

(2) E.F. CALDIN, J.E. CROOKS, B.H. ROBINSON, J. Phys. E. Scientif. Instr. 1971, 4, 165.

(3) E.M. EYRING, B.C. BENNION, Annu. Rev. Phys. Chem., 1967, 18, 129.

(4) J.V. BEITZ, G.W. FLYNN, D.H. TURNER, N. SUTIN, J. Amer. Chem. Soc., 1970, 92, 4130.

(5) D.H. TURNER, G.W. FLYNN, N. SUTIN, J.V. BEITZ, J. Amer. Chem. Soc., 1972, 94, 1554.

DISCUSSION

C. REISS

H. Staerk - How do you escape the development of Brillouin scattering in your Raman cell which could shatter your sample or more severe could ruin your laser? Are you below the critical pulse width of 1-2 nsec. in order to prevent this dangerous back-scattering?

C. Reiss - Yes, because, as I have mentioned, I have used 25 ps laser pulses to escape the Brillouin scattering.

S. Ameen - 1) The spectrum of H_2O and D_2O shows that the SRS of the list of solvent given would not work because the $\frac{d\delta}{d\Omega}$ and Raman gain are small. Besides, these liquids do not generate just a single wavelength of Stokes' frequency but a series of frequencies (Stokes and anti-Stokes both, plus other competing scattering processes). Thus detection and optical filtering is difficult to obtain. Only useful wavelength could be obtained by CH_3CN for heating aqueous solutions.
2) SRS of pressurised H_2 gas was used (Reference J. Amer.Chem.Soc., 97 (1975), 1590) which has comparable properties to that of N. Sukin's use of liquid N_2 for construction of laser-temperature jump spectrophotometer . Besides mixture of gases at varying high pressures can yield other useful IR frequencies for temperature-jump methods. The threshold for Brillouin scattering is probably crucial and when small energy of heating-laser pulse and small volumes are used this problem does not arise. This can be achieved by proper optical filtering and use of coaxial probe beam to heating laser pulse and a very fast detection system with good S/N.
3) An intracavity laser-temperature-jump was reported by B.D. Sykes, J.H. Balds, E.B. Priestley, J. Am.Chem.Soc., 97 (1975), 1681, which is useful in μsec range for chemical relaxation studies and yields up to 10°C temperature jumps.
4) The temperature-jump cells of various types can also partly overcome the detection of fast transient species (see S. Ameen, Ph.D.Dissertation, Max-Planck-Institut für Biophysikalische Chemie, Göttinge, 1975.
5) The stimulated Raman scattering in Er^{+3}-Yb^{+3} system can be used which is reported by V.P.Grapontsev et al. JETP Letters, 18, 1973, 251. This yields a wavelength 1.54 μ (5joules/30 nsec. pulses) as described in the above mentioned dissertation.
6) The stimulated Raman scattering in pressurized CH_4 gas yields 1.53 μ, can be used for construction of laser-temperature-jump spectrophotometer (SRS of CH_4 with Nd^{+3} glass laser).
7) Reverse Raman scattering of H_2 gas with ruby laser yields 0.97 μ .
8) Since the pulses after SRS experiments in H_2 or other gases are sharpened, the starting laser source (Nd^{+3}-glass or ruby) can be so selected that it has very short high power pulses to start with and thus even picosecond heating pulses can be generated. This could be achieved by reducing cavity length or amplification methods. So the SRS in gases can still be useful for laser vibrational heating for chemical relaxation studies. (See above said dissertation).

C. Reiss - 1) Not all the solvents shown have 80% transfer yield. But such yields are not necessary to generate sizeable T-jumps in water.

Higher order SRS lines are not troublesome because most will be also absorbed by water and will thus contribute to the heating process.
2) H_2 gas does not seem to us to be a suitable SR frequency shifter for water, because both ordinary and heavy water have high absorbance values at the generated 1.89 μ wavelength, so that very abrupt T profiles in the sample will be generated. My own experience with N_2 gas in the nanosecond range is that copious stimulated Brillouin scattering occurs. This can be fully avoided with picosecond pulses, for the Brillouin scattering will be unable to build up on subnanosecond pulses.
3) I agree and I would like to point out that this method yields homogeneous heating of the sample.
4)?
5) We have thought of using this kind of lasers for directly generating T-jumps in water. But since we have to use a laser also for generating light for analysis of the chemical reaction under study, we preferred to use a convential one.
6) Yes.
7) Yes and in our experience also forward SRS.
8) We have preferred to use a picosecond laser to begin with.

B. Alpert - L'énergie électromagnétique est absorbée par le solvant et ne peut produire des réactions photochimiques du soluté. Mais dans le cas des molécules biologiques où le solvant (eau en particulier) est adsorbé en surface, ne peut-il se produire des "points chauds" tout autour de ces molécules biologiques? Cet effet pourrait être encore plus indésirable qu'une réaction photochimique très atténuée.

C. Reiss - Un tel évènement est possible mais je ne dispose d'aucun élément d'information concernant le spectre d'absorption de l'eau de solvatation (pourquoi devrait-il être différent de celui de l'eau ordinaire?). D'autre part, si un tel effet existait réellement, on aurait enfin un moyen d'aborder l'étude de l'eau d'adsorption de molécules biologiques dont rien n'est connu à ce jour.

R.A. Keller - Have you considered the possibility of using a relay dye excited by a two photon process to set up a periodic temperature gradient?

C. Reiss - No. As mentioned, relay dyes are not efficient for heating reaction mixtures. Your suggestion may be used for generating temperature profiles with periodic spatial structure.

J. Joussot-Dubien - Comment se compare votre méthode avec celle proposée récemment dans une publication du JACS 97, (1975), 1689?

C. Reiss - Cette méthode permet des sauts de température dans le domaine des microsecondes et non des nanosecondes. Elle a par contre l'avantage d'un chauffage uniforme de l'échantillon.

MISE EN EVIDENCE EXPERIMENTALE D'UN DICHROISME INDUIT PAR LES ONDES LASER ET SES PREMIERES APPLICATIONS EN BIOPHYSIQUE

A. PLANNER* et J.R. LALANNE
Université de Bordeaux I - Centre de Recherches Paul Pascal, C.N.R.S., 33405 - TALENCE (France)

RESUME

Nous avons observé une augmentation de l'absorption moléculaire de l'anthracène, de l'acridine et du triphénylène, en solution dans le cyclohexane induite par un faisceau laser déclenché focalisé. Nous avons attribué cette observation à un dichroïsme perpendiculaire induit par le champ électrique optique de l'onde laser. Dans les mêmes conditions expérimentales, l'anthracène dilué en solution aqueuse liposomique ne conduit à aucun changement d'absorption détectable. Cette absence d'effet peut être attribué à une diminution de la mobilité orientationnelle de l'anthracène due à une rigidité plus marquée du milieu environnant. Ces variations d'absorption ne sont pas observées lors de l'application de champs électriques statiques impulsionnels d'intensité inférieure à $5\ 10^{6}\ V.M^{-1}$.

Par contre, les premiers résultats d'une étude en fonction du temps et de l'intensité du champ statique impulsionnel appliqué concernant une suspension colloïdale de graphitate d'ammonium dans le D.M.S.O. sont donnés.

INTRODUCTION THEORIQUE

L'orientation des molécules anisotropes par les champs électriques est cause de quelques effets électro optiques bien connus : effet Kerr [1] [2], génération du premier harmonique dans les milieux isotropes [3], changement de l'état de polarisation de la lumière diffusée [4][7], dichroïsme induit [5] etc... Ces effets conduisent à des résultats intéressants dans le domaine des propriétés électromagnétiques statiques et dynamiques des molécules.

C'est ainsi que l'application d'un champ électrique sur une assemblée de particules conduit généralement à une absorption anisotrope d'un faisceau lumineux polarisé linéairement. Ce dichroïsme linéaire fournit des renseignements importants sur l'orientation des moments de transition et

* Adresse permanente : Institut de Physique, Université A. Mickiewicz, POZNAN (Pologne).

l'identification des transitions électroniques dans les molécules.

Nous montrons ici qu'il peut également être induit par le champ électrique d'une onde laser intense et donner des indications sur la cinétique des phénomènes orientationnels dans les liquides.

Considérons un échantillon liquide, constitué de molécules à symétrie de révolution non polaires (molécules de Langevin [6]) absorbantes à la longueur d'onde d'analyse λA d'un faisceau lumineux se propageant suivant Ox. Ce liquide est, de plus, soumis à l'action d'un champ orientateur électrique ou optique polarisé suivant Oz. La polarisation de l'onde d'analyse sera notée ∥ (suivant Oz) ou ⊥ (suivant Oy). Dans ces conditions, et en supposant le moment de transition responsable de l'absorption dans le plan perpendiculaire à l'axe de symétrie moléculaire, le calcul montre qu'il existe des variations relatives d'intensité transmise $\frac{I^{\lambda A}}{I^{\lambda A}}$ données par :

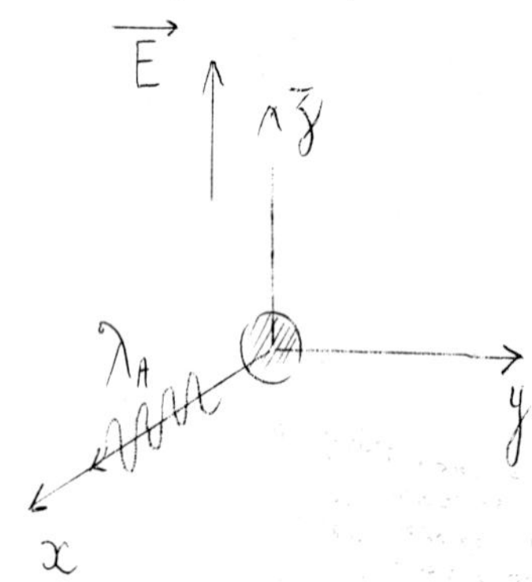

$$\frac{\Delta I_{\perp}^{\lambda A}}{I_{\perp}^{\lambda A}} = - 2.3 \; d \; \frac{1}{4} \; (1 - 3L_2) \qquad (1)$$

$$\frac{\Delta I_{\parallel}^{\lambda A}}{I_{\parallel}^{\lambda A}} = - 2.3 \; d \; \frac{1}{2} \; (3L_2 - 1) \qquad (1 \text{ bis})$$

où d est la densité optique de l'échantillon, L_2 la fonction de Langevin d'ordre 2 en cos θ (θ : angle entre le champ orientateur et l'axe de symétrie de la molécule) et $< E^2 >$ la densité moyenne énergétique.

Dans l'approximation des champs faibles, ces équations s'approximent par :

$$\frac{\Delta I_{\perp}^{\lambda A}}{I_{\perp}^{\lambda A}} \# \frac{2.3 \; d \; \gamma \; < E^2 >}{30 \; kT} \qquad (2)$$

$$\frac{\Delta I_{\parallel}^{\lambda A}}{I_{\parallel}^{\lambda A}} \# \frac{2.3 \; d \; \gamma \; < E^2 >}{15 \; kT} \qquad (2 \text{ bis})$$

où γ est l'anisotropie de la polarisabilité moléculaire du premier ordre.

On remarque immédiatement que, pour une même intensité du champ excitateur :

$$\frac{\Delta I_{\parallel}^{\lambda A} / I_{\parallel}^{\lambda A}}{\Delta I_{\perp}^{\lambda A} / I_{\perp}^{\lambda A}} = -2 \qquad (3)$$

formule qui constitue une généralisation de la loi bien connue d'Havelock [8].

Un calcul numérique, effectué à partir des équations (2) et (2 bis) montre que cet effet peut être observé dans le cas de macromolécules excitées par les champs électriques statiques conventionnels. Par exemple, des particules possédant une anisotropie de l'ordre de 10^{-20} M^3 (dimensions de l'ordre du micron) conduisent à $\frac{\Delta I_{\perp}}{I_{\perp}} \sim 0.1$ pour $d \sim 1$ et $< E^2 > \sim 15$ $J.M^{-3}$ c'est-à-dire $E \sim 2\ 10^6$ $V.M^{-1}$.

Par contre, il sera très difficile à détecter pour les molécules habituelles ou il faudra utiliser des champs de l'ordre de 10^9 à 10^{10} $V.M^{-1}$.

Seuls les champs électriques laser, appliqués brièvement, pourront être supportés sans dommage par les échantillons irradiés.

Nous avons étudié successivement les deux cas de l'excitation électrique pulsée et de l'irradiation optique déclenchée.

LES EXPERIENCES

Dichroïsme induit par champ optique (DO)

Notre expérience recherche les éventuelles corrélations entre les effets d'orientation moléculaire et les variations d'absorption résultantes. L'orientation est induite par une onde laser déclenchée intense, de longueur d'onde λ_I et les variations d'absorption sont détectées au moyen du premier harmonique de cette onde (longueur d'onde λ_A). Les fluctuations d'intensité des ondes laser déclenchées, leur polarisation, leur directivité et leur brièveté se prêtent bien à une telle réalisation. Notre dispositif est illustré par la figure 2.

La seule mesure effectuée correspond à une polarisation perpendiculaire des ondes inductrice et d'analyse. Elle fournit, pour chaque tir laser, l'intensité de l'onde d'orientation et celles (avant et après l'échantillon) de l'onde de mesure.

Dichroïsme induit par champ électrique impulsionnel (DE)

Il s'agit d'un montage mis en oeuvre pour vérifier la théorie donnée

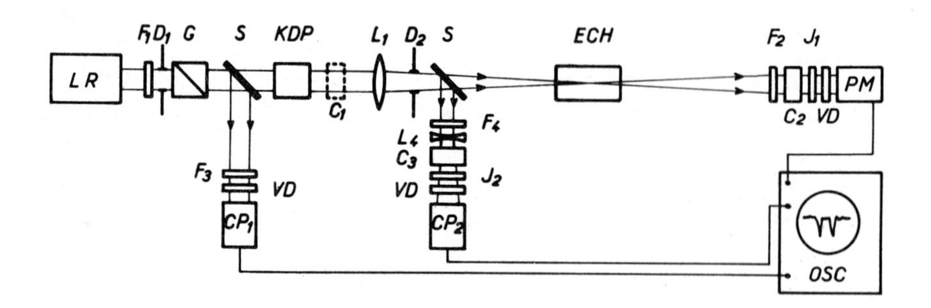

Fig. 2. Dispositif expérimental de mesure du D.O.

LR : laser à rubis déclenché, λ_I = 0,6942 μ, λ_A = 0,3471 μ, puissance moyenne 10 Mw x cm^{-2}, durée de l'impulsion ∿ 30 ns ; OSC : oscilloscope bicanon Fairchild modèle 777 ; pM : photomultiplicateur la Radiotechnique type XP 1002 ; CP_1 et CP_2 : photocellule CSF Thomson type CPA 1143 ; ECH : échantillon l = 1 et 3.4 cm ; KDP : monocristal orthophosphate monopotassique ; C_1, C_2, C_3 : cuves $CuSO_4$; D_1, D_2 : diaphragmes Ø = 0.6 cm ; F_1 : verre Schott RG 630 ; F_2, F_3, F_4 : verre Schott NG 4 et NG 11 ; G : prisme de Glan ; J_1, J_2 : filtres interférentiels 0,3471 μ ; L_1 : lentille f = 1m ou f = 25 cm ; S : séparatrice.

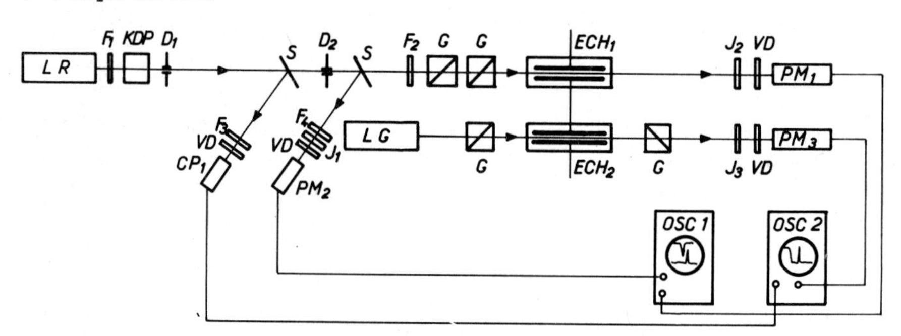

Fig. 3. Dispositif expérimental de mesure du D.E.

Les légendes sont identiques à celles données figure 2. J_3 : filtre interférentiel 0,6328 μ ; L.G : laser à gaz He-Ne Spectra Physics modèle 132, 2m watt, λ = 0,6328 μ ; ECH 1 : échantillon de mesure dans cuve Kerr, longueur des électrodes 2,5 et 29,5 cm, écartement respectif des électrodes 0,6 et 0,5 cm ; ECH 2 : échantillon CS_2 de référence, longueur des électrodes 29,5 cm, écartement 0,5 cm ; OSC_1 : oscilloscope Fairchild modèle 777 ; OSC_2 : oscilloscope Tektronix modèle 7904 ; D_1 et D_2 : diaphragmes Ø : 1.5 mm.

ci-dessus et mesurer en outre, la cinétique du phénomène décrit, mesure impossible dans le cas DO. L'orientation est induite par un champ électrique pulsé, pouvant atteindre 5 10^6 $V.M^{-1}$, montant en 200 ns et durant plusieurs dizaines de microsecondes. Ses effets sur l'absorption sont analysés par une onde laser déclenchée à 0,3472 µ pouvant être continuement retardée par rapport à l'instant d'application du champ inducteur dans l'intervalle 0,2 µs - 1,8 µs. La figure 3 présente notre réalisation.

Chaque mesure collectionne quatre informations grâce aux deux oscilloscopes : amplitude du champ E dans les échantillons et instant de mesure du dichroïsme (OSC_2) ; intensité de l'onde d'analyse avant et après l'échantillon (OSC_1).

RESULTATS

DO : Nous avons mesuré $\frac{\Delta I_{\perp}^{\lambda A}}{I_{\perp}^{\lambda A}}$ pour des solutions d'anthracène, d'acridine et de triphénytène dans le cyclohexane. La variation absolue $\Delta I_{\perp}^{\lambda A}$ est définie par :

$$\Delta I_{\perp}^{\lambda A} = I_{2\perp}^{\lambda A} - (I_{2\perp}^{\lambda A})^{*}$$

où $I_{2\perp}^{\lambda A}$ et $(I_{2\perp}^{\lambda A})^{*}$ définissent les intensités transmises respectivement sans onde inductrice rouge et avec onde inductrice rouge. Les figures 4 et 5 traduisent certains de nos résultats. ($I_{1\perp}^{\lambda A}$ est l'intensité d'analyse avant la traversée de l'échantillon).

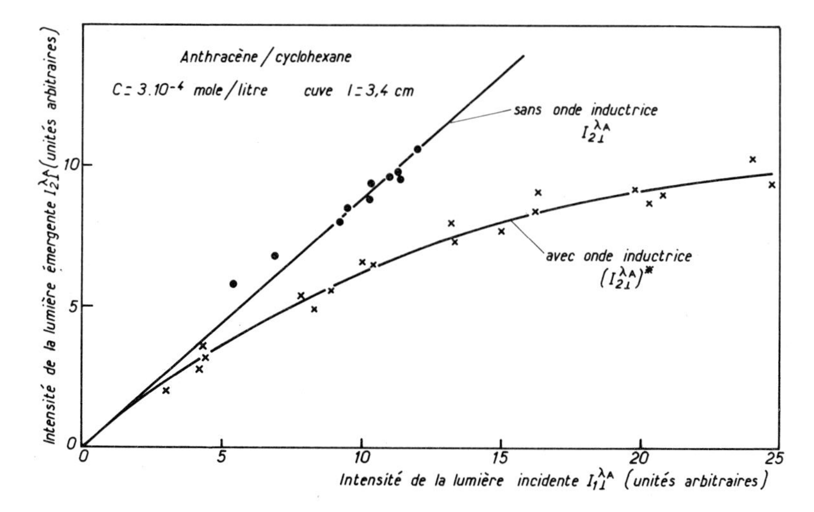

Fig. 4. Variations de la transmission à 0,3471 µ d'une solution 3 10^{-4} mole $litre^{-1}$ d'anthracène dans le cyclohexane soumise à l'action orientatrice d'un faisceau laser déclenché 0,6942 µ.

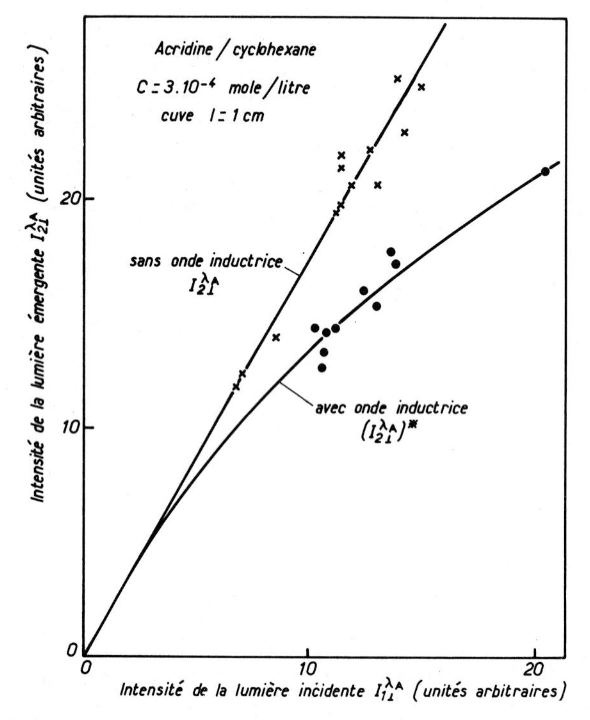

Fig. 5. Variations de la transmission à 0,3471 µ d'une solution 3 10^{-4} mole litre^{-1} d'acridine dans le cyclohexane, soumise à l'action orientatrice d'un faisceau laser déclenché 0,6942 µ.

On constate un effet dont l'amplitude est très supérieure aux fluctuations expérimentales. Une tentative de détection effectuée sur une solution aqueuse liposomique contenant de l'anthracène, de même densité optique que la solution cyclohexanique, s'est avérée vaine.

Nous avons contrôlé les points suivants :

- cet effet n'apparaît qu'aux très hautes densités énergétiques du faisceau inducteur.

- cet effet disparaît quand nous remplaçons nos échantillons par des solutions de benzène dans le cyclohexane de même constante de Kerr mais n'ayant aucune absorption à λA. Cela élimine toute possibilité d'effets parasites conduisant à des manifestations de même type, par exemple induits par l'autoconvergence des ondes et tous les effets stimulés qu'elle engendre.

- cet effet varie comme d en accord avec la relation (1).

- cet effet varie comme $< E^2 >$ en accord avec l'approximation (2). En effet, nous avons :

$$I_{2\perp}^{\lambda A} = k_1 \ I_{1\perp}^{\lambda A} \qquad (4)$$

et $$I_{1\perp}^{\lambda A} \neq k_2 < E^2 >^{\alpha} \qquad (5)$$

où α est un coefficient souvent inférieur à la valeur théorique 2 et qui sera déterminé expérimentalement. Il vient alors, si l'expression (2) est vérifiée :

$$\mathrm{Ln}\, \frac{\Delta I_{\perp}^{\lambda A}}{I_{2\perp}^{\lambda A}} = \frac{1}{\alpha}\, \mathrm{Ln}\, I_{1\perp}^{\lambda A} + k \qquad (6)$$

C'est une représentation linéaire de pente $\frac{1}{\alpha} = 0.6$

La figure 6 traduit les résultats obtenus dans le cas des solutions d'anthracène et d'acridine. L'accord avec les résultats précédents est très convenable.

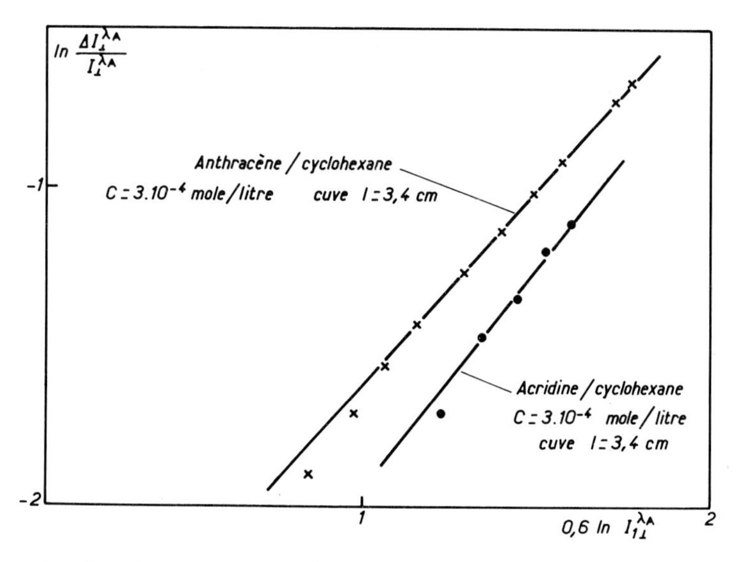

Fig. 6. Vérification expérimentale de la relation (6) pour des solutions d'anthracène et d'acridine dans le cyclohexane.

-<u>DE</u> : Aucun effet n'a été détecté dans le cas des solutions précédentes et ce, pour des champs pouvant atteindre 5 10^6 $V.M^{-1}$. D'autre part, la conductivité importante des solutions aqueuses liposomiques s'est traduite par une diminution catastrophique du champ électrique. Aucun signal de dichroïsme n'a été, dans ces conditions, enregistré. Par contre, nous avons obtenu des résultats positifs dans le cas d'une solution de graphitate d'ammonium dans le diméthyl sufoxyde (DMSO) de concentration approximative 0.1 $g.l^{-1}$.

La figure 7 traduit les résultats obtenus.

Ils sont en bon accord avec les équations (1) et (1 bis). D'autre part, la relation (3) est bien vérifiée pour le seul point de mesure en géométrie perpendiculaire.

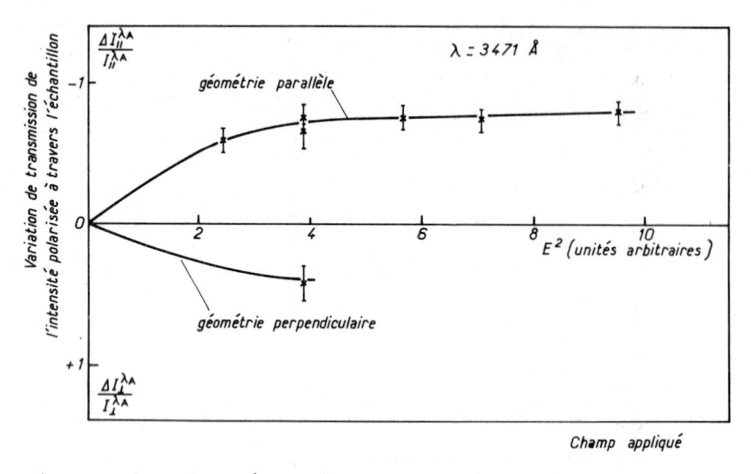

Fig. 7. Dichroïsme électrique du graphitate d'ammonium en solution dans le DMSO.

Nous avons également enregistré la cinétique d'établissement du dichroïsme illustrée par la figure 8.

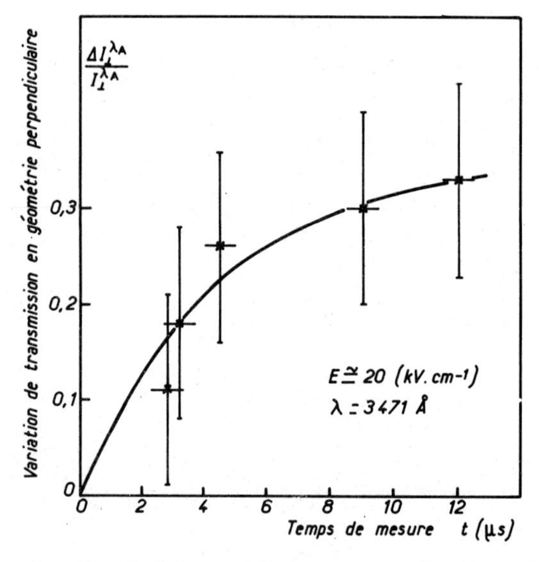

Fig. 8. Cinétique d'établissement du dichroïsme de la solution de graphitate d'ammonium dans le DMSO.

Le temps caractéristique observé est de l'ordre de 5 µs.

DISCUSSION

De nombreux travaux rendent compte de l'existence de variations de transmission d'échantillons liquides soumis à une onde laser intense absorbée [9] [10] ou non absorbée [11] [12]. Cependant, aucun de ces effets ne semble directement lié à l'orientation moléculaire. Nos résultats DO sont cohérents avec les propriétés électro optiques des molécules étudiées : polarisabilité maximale suivant l'axe moléculaire longitudinal, moment de transition responsable de l'absorption observée polarisé linéairement suivant l'axe transversal. Malgré le fait que l'approximation d'une molécule à symétrie de révolution soit très exagérée dans le cas des trois chromophores étudiés l'équation (2) conduit à une variation $|\frac{\Delta I_{\perp}^{\lambda A}}{I_{\perp}^{\lambda A}}|$ de l'ordre de 0,1 (et donc comparable à l'amplitude observée) pour la valeur $\gamma = 20\ 10^{-30}\ M^3$ et en supposant l'énergie de notre laser concentrée dans une tache de diamètre 30 µ ($< E^2 > \sim 10^8$ J x M^{-3} ; $E \sim 4\ 10^9$ $V.M^{-1}$.

L'absence de signal dans la solution d'anthracène, inclus au sein des chaînes lipidiques d'une dispersion de liposomes vésiculaires de lécithine naturelle dans l'eau, s'explique probablement par une cinétique d'orientation beaucoup plus lente, liée à la rigidité plus marquée de l'environnement du chromophore. Ce fait a été illustré par notre expérience DE sur le graphitate d'ammonium. Celui-ci s'est avéré constituer un échantillon bien choisi et adapté à la vérification de notre présentation théorique. Des ondes optiques de durée temporelle plus importante (par exemple des ondes laser relaxées suffisamment amplifiées) seront prochainement expérimentées sur les solutions liposomiques.

Le dichroïsme optique induit par les ondes laser peut conduire à la détermination du signe de l'anisotropie de la polarisabilité moléculaire d'ordre un, difficile à obtenir par les méthodes traditionnelles qui en révèlent le carré. Comme l'effet Kerr optique et outre l'apport "statique" précédemment cité, il peut fournir des informations "dynamiques" en cinétique d'orientation moléculaire.

Cependant, il présente sur lui l'avantage de sélectionner une espèce chimique et se rapproche ainsi des techniques spectroscopiques.

REMERCIEMENTS

Nous tenons à remercier M. BOTHOREL qui nous a constamment aidé dans la réalisation de ce travail ; M. LUSSAN et Mme FAVEDE qui ont eu la patience de préparer pour nous les solutions liposomiques ; et M. DUVERT, du laboratoire de Cytologie qui a examiné certains de nos échantillons en microscopie électronique.

REFERENCES

1 J. Kerr, Phil. Mag. et J. Sci., 50 (1875) 337.

2 G. Mayer et E. Gires, C.R. Acad. Sci., 258 (1964) 2039.

3 R.W. Terhune, P.D. Maker, C.M. Savage, Phys. Rev. Lett., 8 (1962) 404.

4 C. Wippler, J. Chim. Phys., 53 (1956) 328.

5 C. Bergholm et Y. Bjornstahl, Physik Z., 21 (1920) 137.

6 P. Langevin, Radium, 7 (1910) 249.

7 S. Kielich, Acta Phys. Polonica, 33 (1968) 89.

8 T.H. Havelock, Proc. Roy. Soc., 80 A (1908) 28.

9 Yu.M. Gryaznov, O.L. Lebedev et A.A. Chastov, Opt. Spectrosc. (USSR) 24 (1968) 126.

10 A. Zunger et K. Bar-Eli, IEEE J. Quantum Electron., (part I) QE-3 (1967) 358. (Part II) QE 10 (1974) 29.

11 M. Paillette, C.R. Acad. Sci., Paris, 264 B (26 juin 1967).

12 M.W. Dowley, K.B. Eisenthal et W.L. Peticolas, Phys. Rev. Lett., 18 (1967) 531.

DISCUSSION

A. PLANNER, J. LALANNE

C. Varma - Optical field induced dichroism requires in general power densities of the inducing laser pulse of at least 1 GW/cm^2. At these high power densities one should certainly take two photon absorption into account in your anthracene experiment. Can you estimate the contribution of such transitions to the observed dichroism?
A proper analysis of the optical field induced dichroism should account for the non-uniformity of the electric field over the beam cross-section. This, I think, will be impossible unless the laser operates in a well defined transverse mode.

A. Planner - The two photon absorption at 6943 Å by anthracene molecules is able to modify the intensity of the harmonic light crossing the sample simultaneously. However our experiments with anthracene in aqueous solution of liposomes (at the same optimal density) show that this effect is not important. In these last conditions we have not observed any modification of the intensity of the harmonic light crossing the sample in presence of ruby light.

MESURE DE TRES FAIBLES VITESSES PAR VELOCIMETRIE LASER : APPLICATION A L'ETUDE D'UN FLUIDE CONVECTIF

M. DUBOIS et P. BERGÉ
C.E.N. Saclay - Service de Physique du Solide et de Résonance Magnétique - B.P. N° 2 - 91190 GIF SUR YVETTE - France

RESUME

Le principe de la vélocimétrie laser par détection hétérodyne de l'effet Doppler est brièvement rappelé. Nous discutons d'une variante intéressante de cette méthode : la vélocimétrie laser avec franges dont nous décrivons une adaptation pour l'étude de la vitesse locale d'un fluide soumis à la convection de type Rayleigh-Bénard. Les résultats ainsi obtenus montrent un plein accord avec les prédictions de la "théorie linéaire".

I - EFFET DOPPLER - DETECTION PAR HETERODYNAGE OPTIQUE

Dès la découverte des lasers continus à gaz on a songé à mettre à profit leur très grand monochromatisme pour effectuer des mesures de vitesses [1]. Ces mesures sont basées sur l'effet Doppler et sur la détection de ce dernier par hétérodynage optique.

I-1 - Effet Doppler

Considérons une particule diffusante animée d'une vitesse constante V. Eclairons cette particule à l'aide d'un faisceau lumineux de fréquence ν_o . Recueillons la lumière diffusée par cette particule dans une direction formant un angle θ avec celle du faisceau incident. L'effet Doppler se manifeste par un changement $\Delta\nu$ de la fréquence diffusée par rapport à celle de la lumière incidente

$$\Delta\nu = \nu_o \frac{2V}{c} \cos\alpha \sin\theta/2 \qquad (1)$$

où c est la vitesse de la lumière de fréquence ν_o et α est défini figure 1.

EFFET DOPPLER: PRINCIPE DE DETECTION

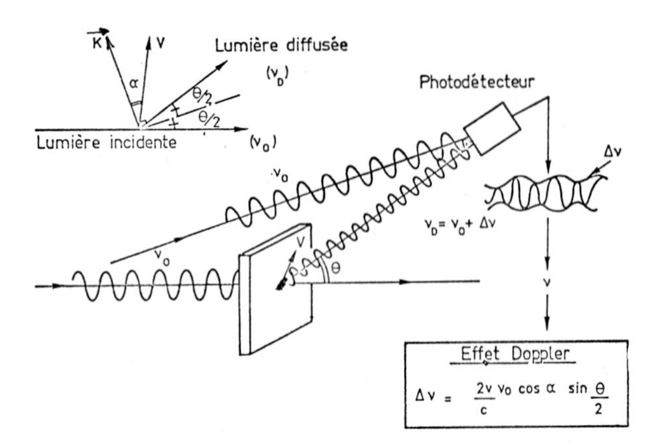

Fig. 1 - Schéma de principe correspondant à la mesure de vitesse par effet Doppler

I-2 - La méthode d'hétérodynage optique

Si $V = 1$ cm s^{-1}, l'application de la formule (1) nous apprend que $\Delta\nu$ sera au maximum de l'ordre de quelques kilohertz.

Il est parfaitement exclu de mettre en évidence un tel changement par des méthodes de spectrométrie optique classique. Les interféromètres de Perot-Fabry les plus performants ayant une résolution relative de l'ordre de 10^{-8}, ils peuvent détecter un changement de fréquence optique de l'ordre de 10^6 hertz [2]. La seule méthode praticable est celle de "l'hétérodynage optique par battements de photons" [3][4][5].

Le principe consiste à "mélanger" sur un détecteur sensible à l'intensité lumineuse le faisceau diffusé et une partie du faisceau incident. Pour que le "mélange" soit efficace, il faut que les deux faisceaux puissent interférer avec le maximum d'efficacité : ils doivent être rigoureusement colinéaires, polarisés dans le même plan, et avoir des longueurs de parcours optiques voisines. Dans ces conditions les franges formées sont dynamiques : les faisceaux battent et la fréquence caractéristique du battement, se traduisant par une variation temporelle de l'intensité reçue par le détecteur, est $\Delta\nu$, différence des fréquences optiques des deux faisceaux.

$\Delta\nu$ est facilement détecté et mesuré par analyse spectrale des fluctuations du photocourant du détecteur. De la connaissance de $\Delta\nu$ on remonte immédiatement d'après (1) à $V\cos\alpha$, projection de la vitesse de la particule sur $\vec{K}$ (Fig.1). On trouvera des détails sur la technique d'hétérodynage optique en [4] [6] par exemple et des exemples d'application sont cités en [4] [7] [8].

I-3 - Une application au domaine de la biologie

Une application de la technique décrite ci-dessus a été proposée [9] pour l'étude de la mobilité des microorganismes vivants et a été faite avec succès dans divers cas [10] [11] [12]. L'objet de la présente communication n'est pas de rentrer dans le détail de la méthode [13], ni des résultats que l'on peut obtenir [14]. Citons simplement une illustration marquante de ses possibilités. Les spermatozoïdes ont, dans le liquide séminal, un mouvement complexe et omnidirectionnel, ce qui donne lieu à un spectre de battements hétérodyne situé autour de la fréquence nulle. A partir de ce spectre on peut déduire très rapidement les caractéristiques objectives de la mobilité des spermatozoïdes et en particulier la vitesse moyenne des spermatozoïdes et la courbe de répartition des vitesses autour de cette vitesse moyenne. Quand les spermatozoïdes pénètrent dans le mucus cervical, leur mouvement n'est plus désordonné comme ci-dessus, mais au contraire la structure particulière du mucus cervical entraîne une orientation apparente de leur mouvement. La technique de diffusion de la lumière et de la détection hétérodyne du spectre nous a permis de montrer que le spectre n'était plus large et centré autour de la fréquence nulle comme dans le cas de l'étude du sperme, mais au contraire était constitué par une raie unique située à une fréquence bien définie, reliée très simplement à la grandeur de la vitesse et mettant en évidence le mouvement d'ensemble cohérent des spermatozoïdes dans le mucus cervical [15], voir figure 2. Ce type d'expérience est d'un grand intérêt dans l'étude du fonctionnement et des anomalies du processus de reproduction de l'espèce humaine en particulier.

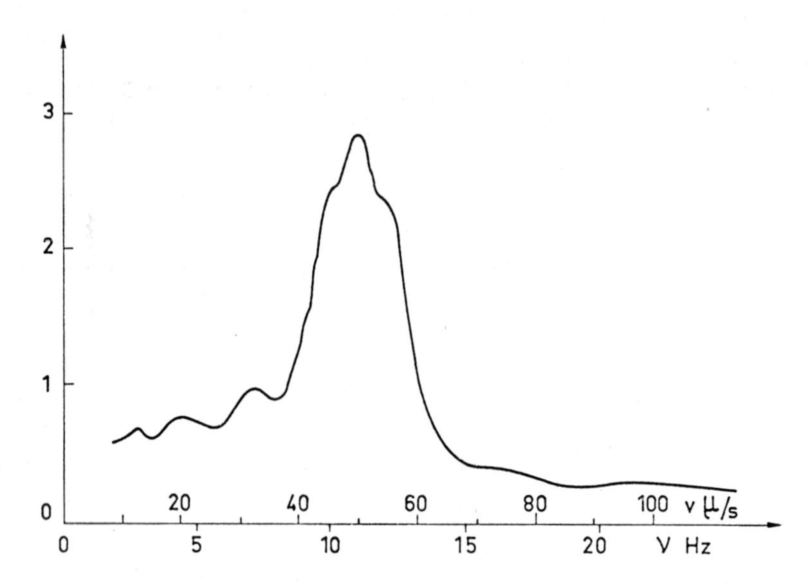

Fig. 2 - Spectre de la lumière diffusée par des spermatozoïdes humains en progression dans le mucus cervical

Quittons le domaine de la biologie pour voir un autre aspect de la vélocimétrie laser : une très intéressante variante de la vélocimétrie par hétérodynage optique consiste dans le vélocimètre laser à franges.

II - LE VELOCIMETRE A FRANGES

II-1 - Principe

Le principe en est très simple. Quand deux faisceaux cohérents - par exemple issus du même laser - se croisent, il se forme au point de croisement P un système de franges d'interférences. Ainsi dans le volume commun aux deux faisceaux l'intensité de la lumière est modulée spatialement. Les maxima d'intensité consistent en des plans parallèles équidistants et perpendiculaires à la bissectrice extérieure $\vec{K}$ de l'angle θ formé par les deux faisceaux. La distance entre les plans parallèles (interfrange) est :

$$i = \frac{\lambda}{2 \sin \theta/2}$$

où λ, la longueur d'onde lumineuse, est mesurée dans le même milieu que θ. Si une particule diffusante traverse à vitesse constante $\vec{V}$ le volume commun aux deux faisceaux, elle passe successivement dans des zones brillantes et obscures ; il s'ensuit une modulation temporelle de l'intensité diffusée à une fréquence $f = \frac{V \cos \alpha}{i} \equiv \frac{VK}{i}$ où α est l'angle entre $\vec{V}$ et $\vec{K}$. Il est très simple de vérifier que f est identique à $\Delta\nu$ obtenu dans la méthode d'hétérodynage optique. Ceci n'est pas une coïncidence fortuite : en réalité les deux méthodes sont équivalentes et on peut considérer la modulation de l'intensité diffusée dans la deuxième méthode comme le battement des fréquences diffusées par l'un et par l'autre des deux faisceaux formant les franges et ayant subi chacun l'effet Doppler. Pour des considérations plus détaillées sur cette méthode voir [16] [17].

II-2 - Montage expérimental

On peut voir figure 3 un schéma du montage optique fonctionnant sur le principe des franges. Le faisceau sortant

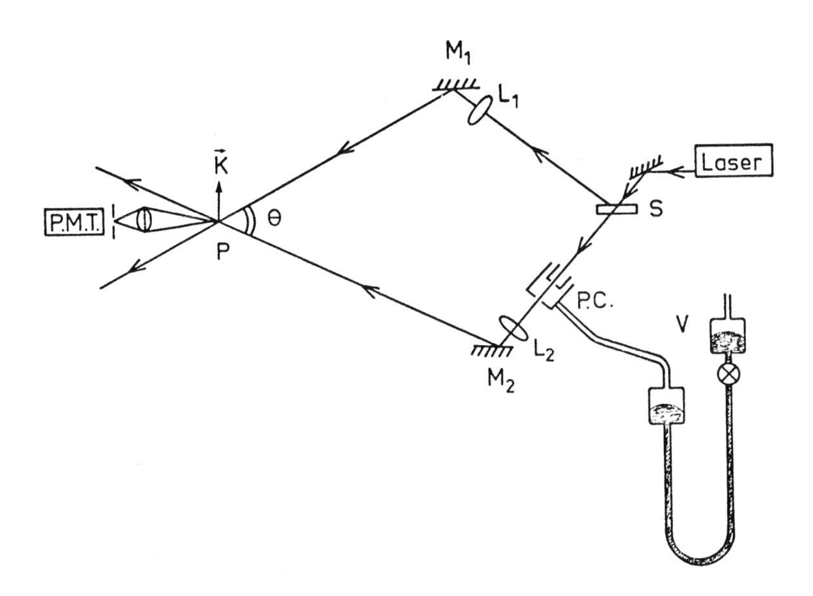

Fig. 3 - Schéma de principe du vélocimètre à franges

d'un laser HeNe de 5 mW de puissance est divisé en deux parties d'égale intensité par une lame séparatrice S. Les deux faisceaux sont envoyés sur deux miroirs réglables M_1 et M_2 qui permettent de faire varier l'angle θ dans un large domaine. Les faisceaux sont focalisés en leur point de croisement P par deux lentilles identiques L_1 et L_2. L'image du point P est formée sur un trou d'épingle placé devant devant un photomultiplicateur PMT détectant la lumière diffusée. Pour être complet un vélocimètre optique doit être capable de déterminer le sens de la vitesse. Pour ce faire, une cellule à trajet optique variable PC est insérée sur l'un des deux faisceaux et remplie d'huile lentement et régulièrement. L'augmentation continue du trajet optique qui en résulte provoque un décalage continu du système de franges dans une direction bien déterminée. La comparaison des fréquences de modulation de la lumière diffusée avec et sans variation du trajet optique permet de connaître immédiatement le sens de la vitesse des particules diffusantes. La détection des modulations du photocourant du PMT est effectuée grâce à un analyseur d'onde "en temps réel". Avec $\theta \simeq 30°$ dans un fluide d'indice n = 1,4, on a $i = 0{,}8\ \mu m$. Si l'analyseur peut détecter des fréquences de modulation de 1 Hz, on peut mesurer des vitesses V de l'ordre de 1 $\mu m\ s^{-1}$. Nous allons décrire les résultats obtenus par application de cette méthode à l'étude du phénomène de convection.

III - ETUDE DE LA VITESSE DE CONVECTION (PROBLEME DE RAYLEIGH-BENARD)

III-1 - Brève description de l'instabilité de Rayleigh-Bénard

Considérons une mince couche horizontale de fluide au repos limitée en haut et en bas par des parois rigides et supposons ce fluide pur et isotrope ; soumettons-le à un gradiant thermique constant ΔT dirigé parallèlement aux forces de pesanteur; c'est à dire que nous chauffons le fluide par dessous. Le fluide froid, plus dense, étant situé au-dessus du fluide chaud, moins dense, nous avons ainsi créé dans le fluide une situation apparemment instable. En fait, nous verrons qu'il est nécessaire de dépasser un certain gradient critique ΔT_c

pour déclencher l'instabilité. Le fluide cesse alors d'être au repos et des mouvements de convection thermique prennent place. Un des aspects les plus fascinants de la convection thermique - en dehors de son seuil critique - est la périodicité remarquable des lignes de courant hydrodynamiques : ces dernières constituent des structures géométriques parfaitement régulières.

III-2 - Bref rappel théorique

De nombreux calculs ont été élaborés pour décrire l'état d'un tel système d'instabilité, depuis Lord Rayleigh [18] en 1916 jusqu'à nos jours [19][20], où des théories plus sophistiquées prennent place. Pour une revue détaillée du problème de Rayleigh-Bénard voir [21]. Tant que le fluide reste dans un état voisin de celui du seuil de la convection, il suffit d'avoir recours aux équations linéarisées du système. Sans entrer dans les détails - que l'on peut trouver en particulier en [22] - nous en donnons ici les bases que sont les équations habituelles de l'hydrodynamique :

$$\operatorname{div} V = 0$$ équation de continuité

$$\frac{dV}{dt} + V \operatorname{grad} V = F - \frac{1}{\rho} \operatorname{grad} P + \nu \nabla^2 V$$

équation de Navier Stokes

et l'équation de diffusion de la chaleur

$$\frac{\partial T}{\partial t} + V \operatorname{grad} T = K \nabla^2 T$$

où les notations ont les significations usuelles. Ces relations sont écrites en supposant que le fluide est incompressible et compte tenu des approximations dites de Boussinesq. Celles-ci supposent que les caractéristiques physiques du fluide peuvent être considérées comme constantes en regard des variations locales de température (mise à part la variation de densité). Par ailleurs, ces équations sont écrites en prenant comme variables - non les paramètres physiques du système qui correspondent à un état d'équilibre - mais en prenant les variables dites de perturbation, soient la vitesse et la perturbation de température. Cette dernière θ représente l'écart entre la température locale du fluide en convection et celle qu'il aurait en absence de convection où les isothermes seraient représentées par des plans horizontaux. De plus, seuls les termes du 1er ordre en V et Θ sont considérés. Toutes ces simplifications aboutissent aux équations dites linéarisées :

$$\left(\frac{\partial}{\partial t} - \nu \nabla^2\right) \nabla^2 V_z = g \alpha \left(\frac{\partial^2 \Theta}{\partial X^2} + \frac{\partial^2 \Theta}{\partial Y^2}\right)$$

$$\left(\frac{\partial}{\partial t} - K \nabla^2\right) \Theta = - \left(\frac{\partial T}{\partial z}\right) V_z$$

Les solutions stables de telles équations peuvent être examinées en fonction de modes spatiaux de longueur d'onde Λ (ou de nombre d'onde $a = \frac{2\pi d}{\Lambda}$). Les équations précédentes peuvent être alors réécrites sous la forme symétrique suivante

$$\left(\frac{\partial}{\partial Z^2} - a^2\right)^3 V_z - R\, a^2 V_z = 0$$

$$\left(\frac{\partial}{\partial Z^2} - a^2\right)^3 \Theta - R\, a^2\, \Theta = 0$$

où R est le nombre de Rayleigh

$$R = \frac{\alpha g d^3 \Delta T}{\nu K}$$

α est le coefficient de dilatation volumique du fluide

ν est la viscosité cinématique

$K = \frac{\lambda}{\rho C_p}$ est la diffusitivité thermique, où λ est la conductibilité thermique, ρ est la densité et C_p la chaleur spécifique.

Connaissant les conditions aux limites de l'état du fluide, les solutions analytiques de V_z peuvent être calculées. Par ailleurs il existe une relation entre les paramètres R et a. La valeur minimum de $R = f(a)$ donne la valeur critique R_c au-delà de laquelle le fluide est en convection. Cette valeur détermine ΔT_c critique et a_c critique. Avec des parois horizontales rigides et bonnes conductrices de la chaleur

$$R_c = 1707 \qquad a_c = 3,117$$

soit $$\Lambda_c = \frac{2\pi}{a_c} = 2,016\ d$$

Les prévisions de la théorie linéaire portent donc sur les valeurs R_c et Λ_c attendues dans une couche d'extension infinie. Dans le cas pratique des géométries finies - caractérisées par leur rapport d'aspect $\frac{d}{L}$ où L représente l'extension horizontale maximum de la couche - DAVIS [23] a montré théoriquement que R_c et Λ_c ont pratiquement la même valeur qu'en milieu infini dès que $\frac{d}{L} \lesssim \frac{1}{6}$. Ceci a été confirmé expérimentalement par STORK et MÜLLER [24]. Par ailleurs c'est la forme de la géométrie horizontale de la couche qui détermine celle des structures convectives ; en particulier dans le cas qui nous intéresse - couche fluide limitée par une boîte à base rectangulaire - la structure convective consiste en un système de rouleaux parallèles au petit côté du rectangle (voir figure 4).

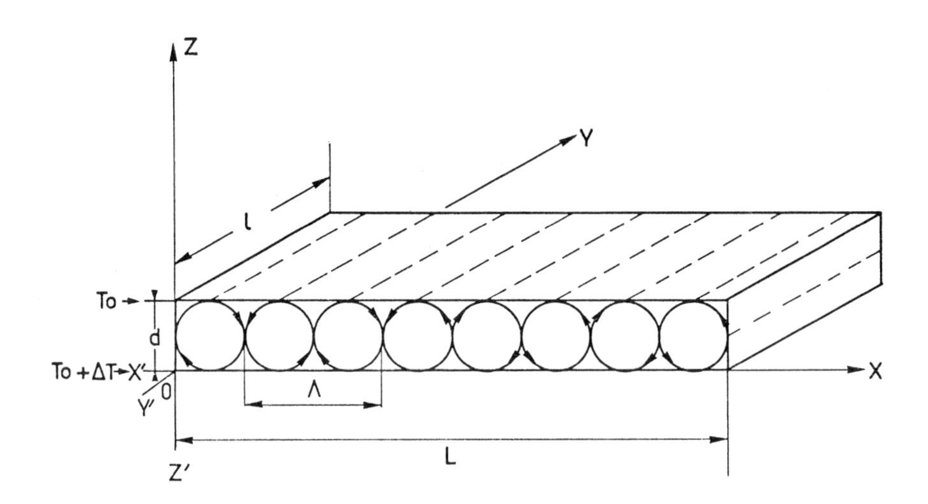

Fig. 4 - Disposition schématique des rouleaux convectifs dans une cellule rectangulaire

III-3 - Cellule expérimentale et conditions de mesure

Nous ne discuterons ici que des résultats obtenus dans le type de cellule représenté figure 5. La couche de fluide - huile aux silicones de viscosité 1 stocke - a 1 cm de hauteur et une extension horizontale suivant un rectangle de 10 × 3 cm. Elle est limitée en haut et en bas par deux plaques de cuivre et un cadre de plexiglass poli maintient l'huile latéralement. La température de chacune des plaques de cuivre est régulée à partir de circulations d'eau provenant de bains thermostatés. La stabilité et la précision de la mesure de la différence de température entre les deux plaques de cuivre est de $\pm 10^{-2}$ °C. La cellule est fixée à 3 platines de translation qui permettent de la déplacer par rapport au point P de mesure selon les 3 axes X Y et Z.

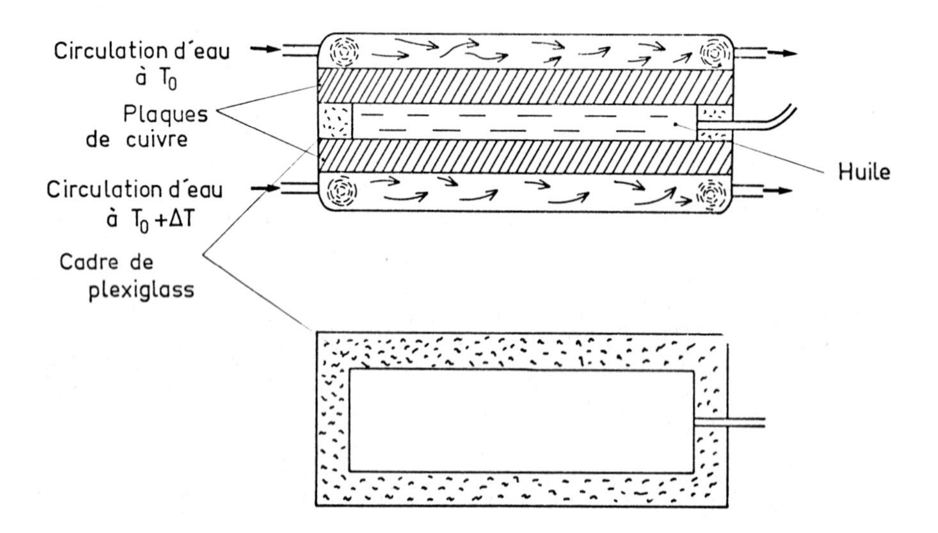

Fig. 5 - Schéma de principe de la cellule expérimentale

Deux géométries de mesure ont été adoptées ici. Pour mesurer la composante horizontale V_X de la vitesse convective, le plan des deux faisceaux laser est placé horizontalement, tel que $V_K = V_X$. L'angle θ dans l'huile est de 31° ; i = 0,825 μm s^{-1}. Dans le cas de la mesure de la composante verticale V_Z, le plan des deux faisceaux est vertical ; (ici $V_K = V_Z$) avec un angle θ dans l'huile de 21° et i = 1,75 μm s^{-1}.

Dans les deux cas les faisceaux laser traversent les

grands côtés du cadre de plexiglass (poli optique) et la bissectrice de l'angle θ est toujours parallèle à l'axe Y. La lumière diffusée est également recueillie à travers le cadre de plexiglass ; elle provient principalement de sphères de latex de diamètre 3 μm, qui ont été ajoutées en très faible quantité à l'huile étudiée.

Parallèlement à la détermination du champ de vitesses de la couche fluide, nous pouvons "visualiser" la structure convective par la modulation d'intensité que subit un large faisceau parallèle qui la traverse. En effet, les filets convectifs descendants ayant une température inférieure à la température moyenne correspondent à des zones d'indice de réfraction plus élevé. Les faisceaux lumineux y sont alors légèrement réfractés provoquant un effet de focalisation à une certaine distance de la couche. Ainsi on peut voir figure 6 la périodicité remarquable de la structure convective. L'éclairement est ici parallèle à Y ; la figure correspond à une projection de la structure sur un plan X O Z où les lignes verticales brillantes correspondent aux courants descendants.

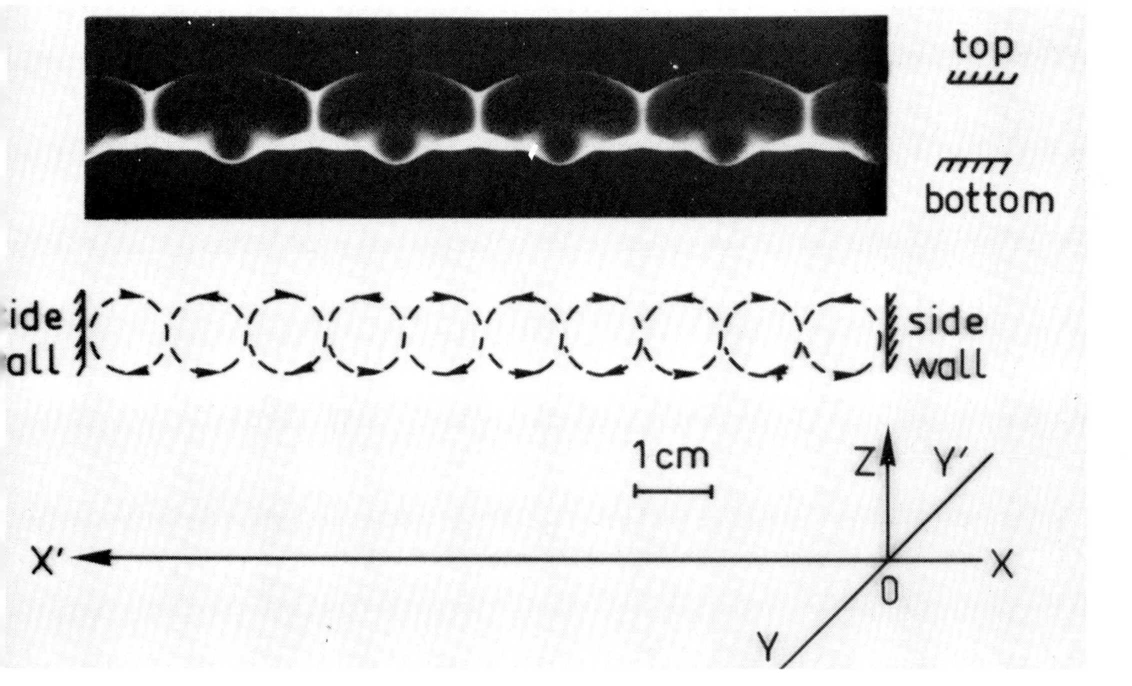

Fig. 6 - Photographie de la structure convective projetée sur le plan X O Z. Les lignes verticales brillantes correspondent aux filets fluides descendants.

III-4 - Le démarrage de la convection

Les mesures de vitesse du fluide nous fournissent une méthode très sensible de détermination du seuil convectif : on peut voir par exemple (figure 7) la variation de la composante V_x de la vitesse en fonction de la différence de température appliquée au fluide. Cette courbe illustre parfaitement le caractère critique du seuil, ainsi que la sensibilité de la détection de ce dernier, ici $\pm 10^{-2}$ °C. De la valeur ΔT_c ainsi déterminée on peut déduire R_c

$R_c = 1600 \pm 100$, l'erreur étant essentiellement due à l'incertitude avec laquelle sont connues les constantes physiques de l'huile.

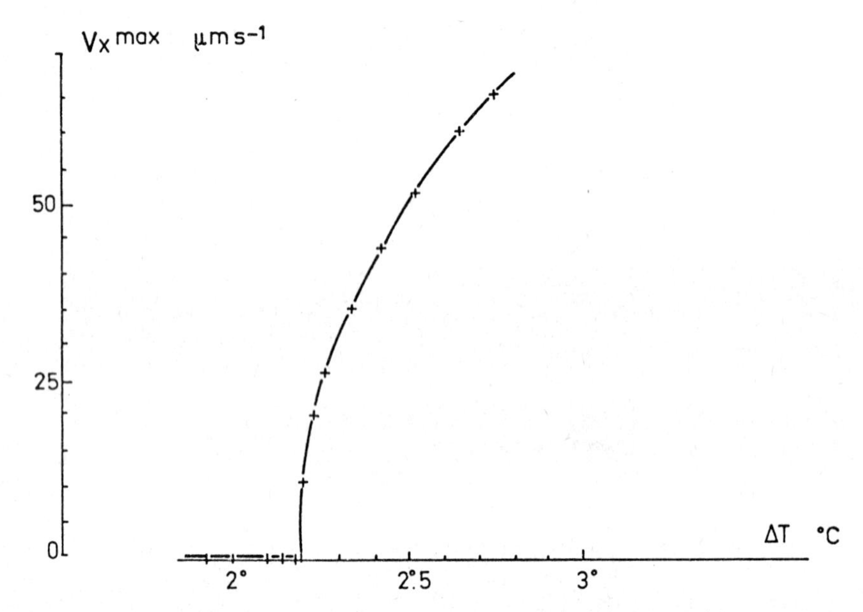

Fig. 7 - Variation de la vitesse convective avec l'écart de température entre les deux plaques.

III-5 - Le champ de vitesse convectif

Toutes les mesures rapportées ici correspondent à l'état stationnaire du régime convectif pour une valeur de $\varepsilon = \frac{R - R_c}{R_c}$ donnée.

Les mesures de vitesse permettent en premier lieu de déterminer la structure périodique du mouvement de convection dans toute la cellule.

La variation de V avec Y confirme quantitativement la nature des rouleaux de convection parallèles au petit côté de la cellule. En effet (Figure 8), la variation de $V_{X\ MAX}$ (maximisée par rapport à X et Z) en fonction de Y montre l'indépendance de la vitesse avec Y à l'exclusion des effets de bord.

La variation de V_X avec X dans toute la cellule montre bien d'une part la périodicité remarquable des rouleaux et d'autre part la constance de la grandeur de la vitesse maximum positive ou négative d'un rouleau à l'autre.

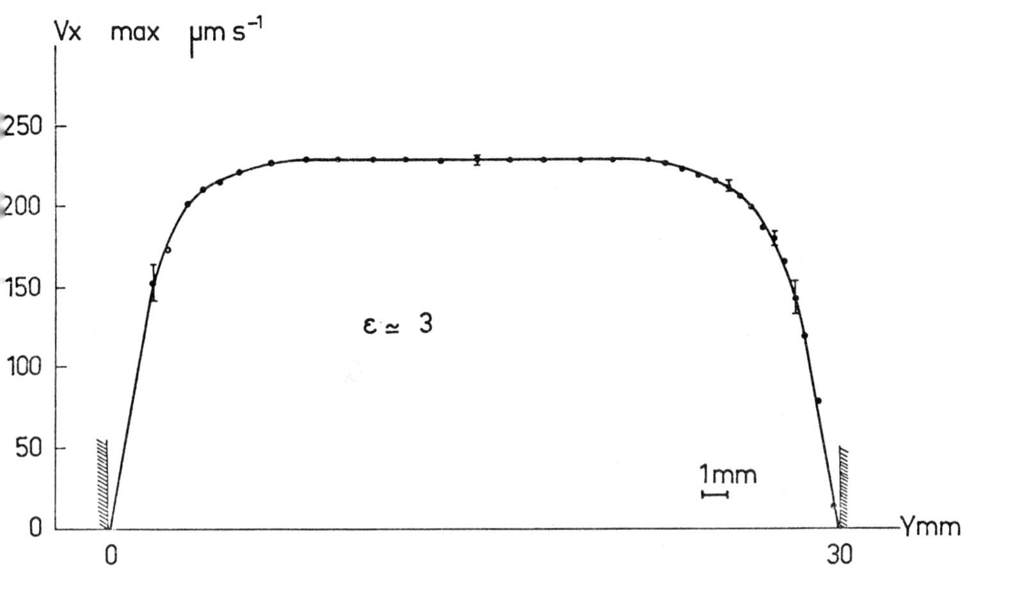

Fig. 8 - Variation spatiale de la vitesse convective en fonction de Y

Sur la figure 9, V_X est mesuré à Z = 0,2 d (V_X est maximisé par rapport à Z).

La variation de V_X avec Z (fig. 10) donne une vitesse nulle aux parois horizontales et à mi-hauteur de la cellule. Par ailleurs il y a changement du sens de la vitesse entre la

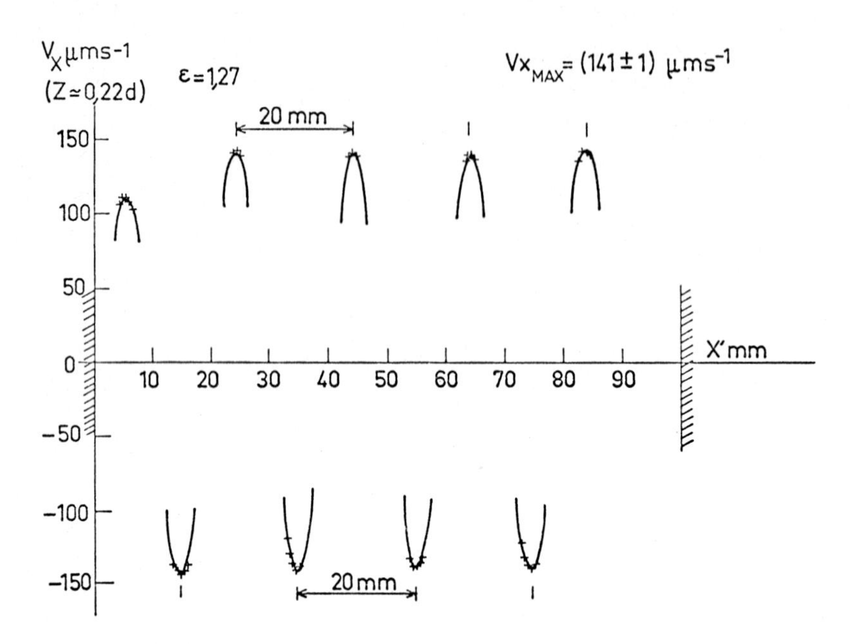

Fig. 9 - Variation spatiale de la vitesse convective en fonction de X

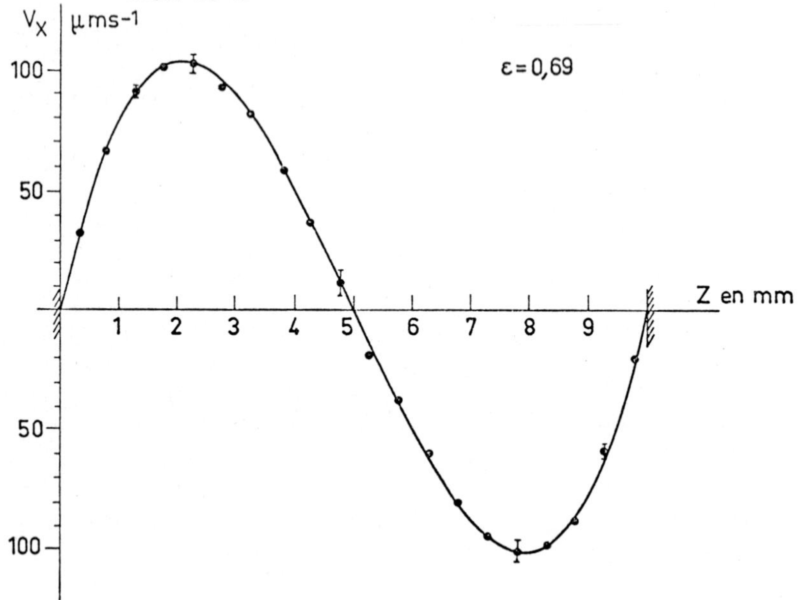

Fig. 10 - Variation spatiale de la composante horizontale de la vitesse en fonction de la hauteur Z dans la couche fluide

moitié supérieure et inférieure de la couche fluide.

Ces trois comportements de la vitesse permettent de conclure sans aucune ambigüité que le système de structure convective est bien selon l'arrangement attendu figure 4.

Etudions de plus près la dépendance de V_X et de V_Z en fonction de X. On peut voir (fig. 11) la comparaison entre la dépendance expérimentale de $V_{X\ MAX\ Z}$ et $V_{Z\ MAX\ Z}$ en fonction de X et des sinusoïdes calculées (trait plein). On peut noter un accord excellent, à nos erreurs expérimentales près. C'est dire que la variation de la vitesse dans la cellule est bien décrite par un mode de Fourier unique. Cette situation prévaut jusqu'à des valeurs de ε de l'ordre 3 (au-delà une distorsion harmonique croissante se manifeste, mais nous n'aborderons pas ce domaine plus complexe ici).

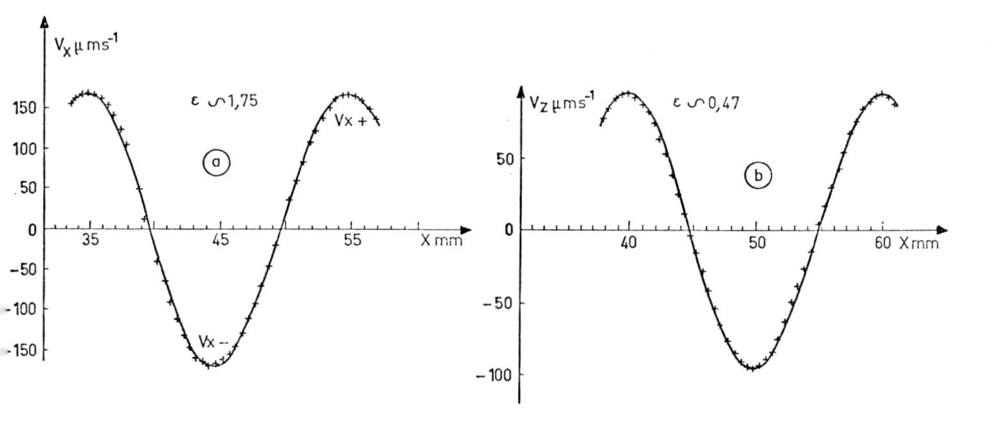

Fig. 11 - Variation des composantes horizontales V_X max Z et verticales V_Z max Z de la vitesse en fonction de X

Par ailleurs, la théorie linéaire [25][26] permet de prévoir exactement le profil V_X max X = f(Z). C'est ce profil qui est représenté en trait plein sur la figure 10. On peut noter ici aussi un accord remarquable entre nos résultats expérimentaux et les prévisions théoriques.

III-6 - Variation de la vitesse convective avec ε

Nous avons déjà vu (fig. 7) que la vitesse convective augmente avec l'écart de température appliquée. Landau [28] prévoit que la vitesse de convection doit varier comme $\varepsilon^{0,5}$. On peut voir (fig. 12 et 13) la variation de $V_{X\ MAX}$ et de $V_{Z\ MAX}$ avec ε sur des diagrammes logarithmiques. On peut noter un excellent accord avec une dépendance en $\varepsilon^{0.5}$.(Notons que cette dépendance en $\varepsilon^{0.5}$ n'est obtenue que si les conductivités des parois horizontales limitant le fluide sont élevées. Avec des parois en verre nous avons obtenu des dépendances telles que $V \propto \varepsilon^{0.59}$ [27]).

III-7 - Comparaison des grandeurs mesurées et calculées pour V

La théorie linéaire [25] [26] permet de prévoir la grandeur des vitesses $V_{X\ MAX}$ et $V_{Z\ MAX}$ pour $\varepsilon = 1$ par exemple. Pour l'huile aux silicones utilisée et avec une cellule de d = 1 cm on attend théoriquement

$$(V_{Z\ MAX})_{\varepsilon=1} = 135 \quad \mu m\ s^{-1}$$

$$(V_{X\ MAX})_{\varepsilon=1} = 133 \quad \mu m\ s^{-1}$$

alors que les expériences reportées fig. 12 et 13 donnent comme résultats

$$(V_{Z\ MAX})_{\varepsilon=1} = 140 \pm 10 \quad \mu m\ s^{-1}$$

$$(V_{X\ MAX})_{\varepsilon=1} = 132 \pm 4 \quad \mu m\ s^{-1}$$

en parfait accord avec les valeurs théoriquement attendues.

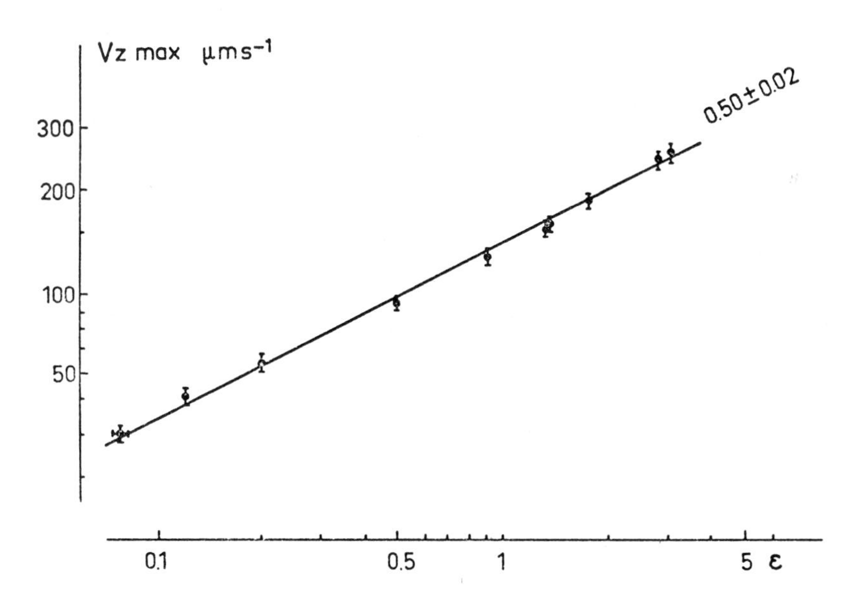

Fig. 12 - Variation thermique de la composante verticale $V_{Z\ MAX}$

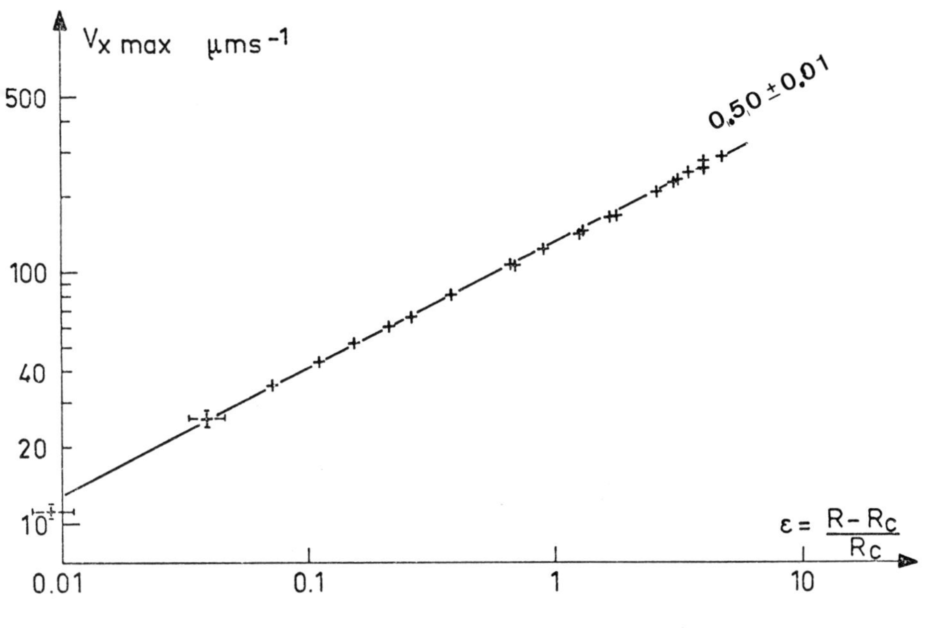

Fig. 13 - Variation thermique de la composante horizontale $V_{X\ MAX}$

CONCLUSION

Les résultats obtenus ci-dessus montrent clairement que la technique de vélocimétrie laser constitue un excellent moyen d'étude des propriétés locales de l'état convectif. Pour résumer, nous dirons que dans un large domaine en ε toutes les prédictions de la théorie linéaire sont parfaitement vérifiées. Un mode de Fourier unique de longueur d'onde constante existe dans toute la couche fluide, aussi bien pour la composante horizontale que verticale de la vitesse. De plus, un excellent accord existe entre les prévisions théoriques et les valeurs expérimentales, aussi bien pour le profil de vitesse $V_X = f(Z)$ que pour la grandeur des vitesses convectives. Enfin, et sous réserve que les parois horizontales soient bonnes conductrices de la chaleur, la dépendance de la vitesse en $\varepsilon^{0,5}$ prévue par Landau se trouve bien vérifiée.

BIBLIOGRAPHIE

1 Y. YEH, H.Z. CUMMINS
Appl. Phys. Lett. 4 176 (1964)

2 D. BEYSENS
Rev. de Phys. Appl. 8 175 (1973)

3 T. FORRESTER
J. Opt. Soc. Am. 51 253 (1961)

4 H.Z. CUMMINS, N. KNABLE, Y. YEH
Phys. Rev. Letters 12 150 (1964)

5 J.B. LASTOVKA et G.B. BENEDEK
Phys. Rev. Letters 17 1039 (1966)

6 M. ADAM, A. HAMELIN, P. BERGÉ
Optical Acta 16 N° 3 337 (1969)

7 N.A. CLARK, J.H. LUNACEK and G. BENEDEK
Am. J. of Phys. 38 5 575 (1970)

8 M. DUBOIS, P. BERGÉ et C. LAJ
Chem. Phys. Letters Vol 6 N° 3 227 (1970)

9 P. BERGÉ, B. VOLOCHINE, R. BILLARD, A. HAMELIN
C.R. Acad. Sc. Paris 265 889 (1967)

10 M. ADAM, A. HAMELIN, P. BERGÉ, M. GOFFAUX
Ann. Biol. Anim. Bioch. Biophys. 9 651 (1969)

11 J.P. BOON, R. NOSSAL, S.H. CHEN
Biophys. J. 14 847 (1974)

12 C. ASCOLI, M. BARBI, C. FREDIANI, D. PETRACCHI,
C. TRIMARCO
C.N.R. Pisa (1972)

13 P. BERGÉ, M. DUBOIS
Rev. de Phys. Appl. 8 89 (1973)

14 M. DUBOIS, P. JOUANNET, P. BERGÉ, B. VOLOCHINE, C. SERRES
G. DAVID
Annales de Phys. Biol. et Médicales Tome I (1974)

15 M. DUBOIS, P. JOUANNET, P. BERGÉ, G. DAVID
Nature 252 Dec. 20/27 (1974)

16 S.S. PENNER, T. JERSKEY
Ann. Rev. Fluid Mech. 5 9 (1973)

17 L. LADING
Appl. Optics 10 N° 8 1943 (1971)

18 Lord RAYLEIGH
Phil. Mag. and J. Science 32 529 (1916)

19 "Non Equilibrium Thermodynamics"
The University of Chicago Press, edited by Donnely, Herman, Prigogine (1966)

20 E.L. KOSCHMIEDER
Adv. Chem. Phys. 26 177 (1973)

21 M.G. VELARDE
"Hydrodynamics", Les Houches 1973, edited by R. BALIAN (Gordon and Breach, New York 1974)

22 S. CHENDRASEKHAR
"Hydrodynamic and hydromagnetic Stability"
(Clarendon, Oxford, England, 1961)

23 S.H. DAVIS
J. Fluid Mech. 30 465 (1967)

24 H. STORK, O. MULLER
J. FLUID Mech. 54 599 (1972)

25 A. SCHLÜTER, D. LORTZ, F. BUSSE
J. Fluid Mech. 23 129 (1965)

26 C. NORMAND, Y. POMEAU
A paraître

27 P. BERGÉ, M. DUBOIS
Phys. Rev. Lett. 32 1041 (1974)

28 L.D. LANDAU, E.M. LIFSHITZ
"Fluid Mechanics", Pergamon Press. London (1959)

DISCUSSION

M. DUBOIS, P. BERGE

P. Rigny - Est-ce que la largeur du pic de distribution des vitesses est toujours due à des limitations instrumentales, ou peut-on atteindre les mouvements browniens moléculaires?

Mme Dubois - Dans le cas de mesures effectuées sur l'huile aux silicones, l'intensité diffusée était due à des sphères de latex de 3 μm de diamètre, ajoutées en très faible quantité à cette huile. La largeur du spectre de diffusion, due au mouvement brownien de telles sphères, est très faible car leur coefficient de diffusion D est lui-même très faible, étant donné la forte viscosité (1 stoke) du milieu dans lequel elles baignent. Par ailleurs, l'observation se fait avec un angle θ petit (largeur spectrale proportionnelle à $\sin^2 \theta/2$). Par contre, dans le cas d'expériences réalisées dans l'eau (viscosité 10^{-2} stoke) et pour des angles θ plus importants, la largeur due au mouvement brownien pourrait ne pas être négligeable.

R. Bonneau - Puisque la vitesse varie selon l'axe des rouleaux pour être nulle aux deux extrémités (par effet de bord), quelle est la zone qui est observée lors de l'étude des variations de vitesse perpendiculairement à l'axe des rouleaux? S'agit-il d'une zone proche des extrémités, située au centre ou une intégration sur toute la longueur d'un rouleau?

Mme Dubois - Lorsque le comportement des composants de la vitesse est étudié dans un rouleau ou sur une longueur d'onde, le point de mesure (croisement des faisceaux) est situé au milieu de la cellule ($Y \sim 1/2$ si ℓ est la largeur de la cellule). Par ailleurs, le photodétecteur ne voit que le point de croisement des faisceaux et la mesure est très locale.

R. Mohan - In the last paper of Dubois et al., the $V_2 \propto (\frac{R - R_C}{R_C})^{0,66}$ was given. Could you explain why the critical exponent is given as 0.5 now?

Mme Dubois - Les expériences reportées dans la publication, P. Bergé, M. Dubois, Phys.Rev.Letters, 32 (1974), 1041, ont été faites avec des conditions aux limites différentes de celles rapportées ici. Le fluide était alors confiné par des parois horizontales en verre, qui n'assurent pas la condition $\theta = 0$ aux parois, avec la précision et l'uniformité que donnent les parois de cuivre. Dans ce cas, en effet, l'exposant trouvé pour le comportement de ν avec ε est 0,59. inexpliqué théoriquement à ce jour.

S. Ameen - It is a very interesting method which can be used to study probably the chemical oscillating reactions or oscillations in chemical reactions, as well as stabilities and structure in dissipative processes.

LASER FLASH PHOTOLYSIS STUDIES OF MACROMOLECULAR MOTIONS BY USE OF THE LIGHT SCATTERING DETECTION METHOD

G.BECK, G.DOBROWOLSKI, J.KIWI and W.SCHNABEL
Hahn-Meitner-Institut für Kernforschung Berlin GmbH, Bereich Strahlenchemie, 1 Berlin 39

SUMMARY

A technique was developed for investigating the diffusion of the fragments of macromolecules in solution generated by scissions of main chain bonds. It is based on irradiating macromolecules in solution with flashes of frequency doubled light (λ = 347.1 nm) from a ruby laser (Korad) and following the time dependence of the decrease of light scattering intensity (LSI) which is a consequence of main chain rupture. Polymers investigated so far include polyphenylvinylketone (PPVK) and copolymers of phenylvinylketone (PVK) and styrene (St). In benzene solution of PPVK LSI decreased with a half life $\tau_{1/2}$(LS) between 15 and 30 μs depending on the initial molecular weight of the polymer. Quenchers like naphthalene and biphenyl decreased the extent of chain scissioning without influencing $\tau_{1/2}$(LS). It was therefore concluded that the time for splitting a bond is much shorter than $\tau_{1/2}$(LS). Since the scissioning process involves excited triplet states further evidence for this could be obtained from triplet lifetime measurements using the optical absorption detection method. Triplet lifetimes of 100 ns (PPVK) and up to 200 ns (St-PVK-copolymers) were found. Thus it is evident that $\tau_{1/2}$(LS) corresponds to a diffusion process. The weak dependence of $\tau_{1/2}$(LS) on the initial molecular weight ($\tau_{1/2}(LS) \propto M^{0.2}$) indicates that the observed diffusion process does not correspond to translational diffusion. Disentanglement diffusion may be distinguished from translational diffusion.

INTRODUCTION

Macromolecules consisting of flexible linear chains exist in dilute solution in the conformation of random coils. When the chains are ruptured irreversibly the parts become independent and tend to separate. Since they are entangled disentanglement has to take place in order to accomplish total separation. It was the intention of this work to gain knowledge concerning the disentanglement process which is considered to be a special mode of macromolecular motion. In order to be able to observe this process by time resolved light scattering measurements certain requirements have to be considered. The most important one concerns the fact that the chemical act of bond scissioning must be much more rapid than the time the fragments need to get disentangled. Furthermore, main chain scissioning should not be counteracted by intermolecular crosslinking. Therefore, the photolytic main chain degradation of certain polymers which are scissioned with a rather large quantum yield appeared to be a suitable technique in combination with the use of a laser producing flashes of rather small half width. The requirement concerning the time of bond scissioning had to be carefully checked. In the following investigations on polyphenylvinylketone (PPVK) and copolymers of phenylvinylketone (PVK) and styrene (St) will be reported. The photolytic main chain rupture of PPVK is known to occur via excited triplet states according to a Norrish type II mechanism [1-3]. Quantum yields of about 0.25 are reported in the literature.

EXPERIMENTAL

a) Apparatus

The instrumental equipment used has been described in detail elsewhere [4]. Therefore, only a general descripton of apparatus will be given here.

A frequency doubled ruby laser (Korad) was used as photolytic light source. The analyzing light was provided by an argon ion laser. Its 514.5 nm line passed through the sample cell at right angle with respect to the photolyzing light. Light scattered by the polymer solution at an angle of 90° was directed to a photomultiplier. The signal was displayed on a storage oscilloscope. The time resolution during light scattering measurements was about 1 μs and during optical absorption measurements about 3 ns.

b) Material

Synthesis and characterization of PPVK samples have been described earlier [4]. Copolymers of PVK and St were prepared by irradiating monomer mixtures with ^{60}Co-γ-rays at room temperature. The polymer samples were reprecipitated several times from benzene solution with methanol and dried at the vacuum line (10^{-5} Torr). Table 1 contains a list of CP samples used. Polyacetaldehyde was prepared by the so-called freezing polymerization method and stabilized by end capping with acetate groups [17-19].

Table 1

Characterization of copolymer samples

Polymer Sample	PVK content mol %	number average molecular weight $\overline{M}_n \times 10^{-6}$	triplet halflife $\tau_{1/2}(T)$ (ns)
CP-St-PVK-0.6	0.6	0.51	150
CP-St-PVK-3.5	3.5	0.65	120
CP-St-PVK-9.7	9.7	0.72	95
CP-St-PVK-13.6	13.6	0.15	75
CP-St-PVK-18.2	18.2	0.39	72
homo PPVK	100.0	0.46	70

c) Evaluation of Data

Based on the Debye equation expressions were derived [4] for the dependence of measured signal voltages U on the degree of degradation α (number of scissions per base unit). Hereby an initial random molecular weight distribution of the polymer was assumed.

$$\frac{U_o - U_\infty}{U_\infty - U_L} = \frac{n_{w,o}\,\alpha}{4BcP_{\vartheta} \; mn_{w,o} + 2} \qquad (1)$$

The subscripts o and ∞ denote the state of the system before the flash and a relatively long time after the flash. By assuming that the scissioning process follows first order kinetics one obtains:

$$\ln \frac{\alpha_{\infty}-\alpha_{t}}{\alpha_{\infty}} = \ln \frac{(U_{\infty}-U_{L})^{-1} - (U_{t}-U_{L})^{-1}}{(U_{\infty}-U_{L})^{-1} - (U_{o}-U_{L})^{-1}} = -kt \qquad (2)$$

U_L: signal voltage due to LSI of the solvent, B: second virial coefficient, C: polymer concentration, P_{ϑ}: particle scattering factor, m: molecular weight of base unit, $n_{w,o}$: weight average degree of polymerization.

RESULTS

a) PPVK and Copolymers of Styrene and Phenylvinylketone

PPVK and CP-St-PVK were irradiated in benzene solution. Fig. 1 shows a typical oscilloscope trace and a plot of these data according to equ. 2. The halflife amounts in this case to about 12 μs. As expected

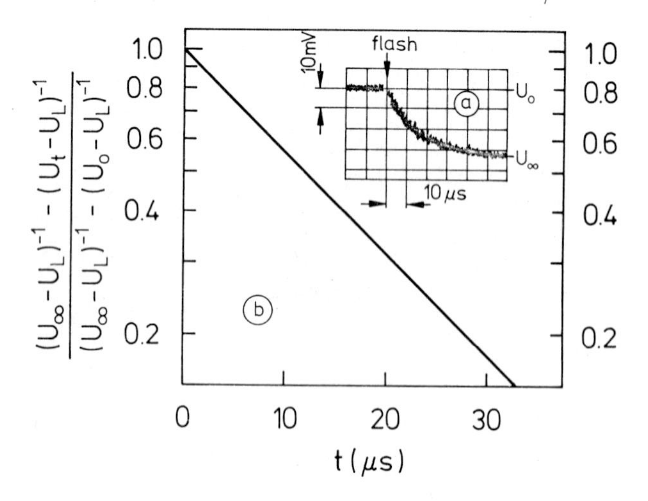

Fig. 1 - (a) Oscilloscope trace after a 25 ns flash demonstrating the time dependence of the decrease of light scattering intensity at $\vartheta = 90^{o}$. CP-St-PVK-9.7 in benzene, c = 1.6 x 10^{-3} g/ml. (b) The decay of the signal of (a) plotted according to equ. 2. Absorbed dose per flash: 9 x 10^{14} photons.

triplet quenchers like naphthalene and biphenyl inhibited degradation. However, while the degree of degradation decreased with increasing quencher concentration the halflife $\tau_{1/2}$(LS) was not influenced. It was, therefore, concluded that the observed LSI change is not due to the chemical act of bond scissioning but corresponds to a diffusion process. In order to obtain additional evidence for this assumption, triplet lifetimes were measured by following the decay of the triplet-triplet absorption spectrum formed during the flash. It was found [5] that the triplet halflife $\tau_{1/2}(T)$ of homo PPVK amounts to about 70 ns. $\tau_{1/2}(T)$ of the copolymers increased with decreasing content of PVK, a finding which is discussed in detail elsewhere [6]. Results are shown in the last column of Table 1. The fact that the change of LSI occurs with halflifes about 100 times slower than $\tau_{1/2}(T)$ confirms the above mentioned assumption of $\tau_{1/2}(LSI)$ corresponding to diffusion. In order to obtain information on the question whether or not we are dealing with a pure translational diffusion process $\tau_{1/2}(LS)$ was measured as a function of the initial molecular weight of the polymer. For this purpose different homo PPVK samples were used. The results yielded [4] a weak dependence of τ (LS) on $n_{w,o}$: $\tau_{1/2} \propto n_{w,o}^{0.2}$. Based on the assumption that the coil fragments separate by translatory diffusion a much stronger dependency was estimated: $\tau_{1/2}(LS) \propto n_{w,o}^{1.5}$ [4]. Therefore, it may be concluded that the disentanglement process observed during our investigation is to be discriminated from pure translatory diffusion. This finding was corroborated recently in our laboratory by results obtained with polyisobutylene dissolved in hydrocarbons [7]. The scissioning of bonds was accomplished in that case by irradiating the polymer solution with 15 MeV electron impulses from a linear accelerator. In the latter case it was found that $\tau_{1/2}(LS)$ decreased with increasing polymer concentration.

An analogous effect of polymer concentration was found during flash photolysis of styrene-co-PVK samples. (For experimental reasons it was not possible to carry out runs with homo PPVK in a sufficiently broad concentration range). Fig. 2 shows typical results. As the plot of the reciprocal halflife versus polymer concentration yields a straight line an analogy to translatory diffusion seems to exist. It has been reported [8, 9] that the translatory diffusion coefficient D_t depends on polymer concentration c according to equ. 3:

$$D_t = D_{t,o}(1 + k_D c + \ldots) \qquad (3)$$

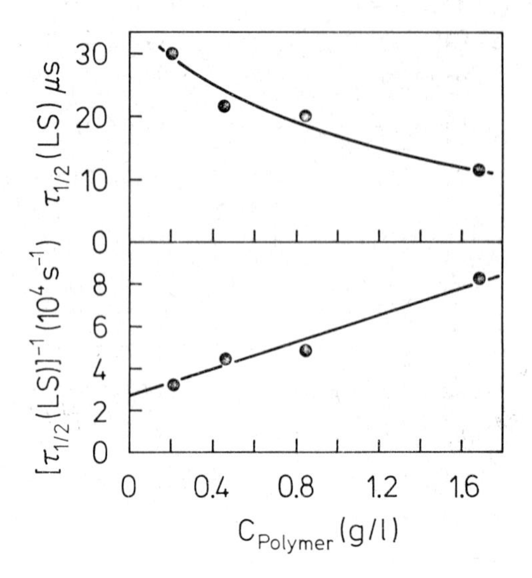

Fig. 2 - $\tau_{1/2}$(LS) as a function of polymer concentration, CP-St-PVP-9.7 in benzene

where $k_D = 2BM - k_f - \bar{v}$, k_f: friction coefficient and $\bar{v}$: specific volume of polymer. According to the sign of k_D D_t is expected to decrease or increase, respectively. Evidence for D_t increasing with increasing polymer concentration has been provided for various polymers of relatively high molecular weight [9-14]. Since diffusion coefficients are considered to be proportional to reciprocal relaxation times our results show that a similar relationship holds for disentanglement diffusion as for translation diffusion.

b) Polyacetaldehyde - an Example for Sensitized Photodegradation

The ability of a polymer to absorb light of the wavelength produced by the laser is a prerequisite for investigations like the ones described above. Unfortunately, the number of polymers fulfilling this condition is limited. It was therefore tried to use sensitizers for initiating main chain scissons. First experiments were carried out with polyacetaldehyde dissolved in acetone or benzene solution containing 10^{-2}M benzophenone [15]. The sensitized photodegradation of this polymer under the influence of stationary irradiation was reported recently [16] to occur via a free radical mechanism. As shown in Fig. 3, the light scattering intensity decreases after the flash in a similar manner as observed with PPVK in benzene. The half-life amounts to about 25 μs.

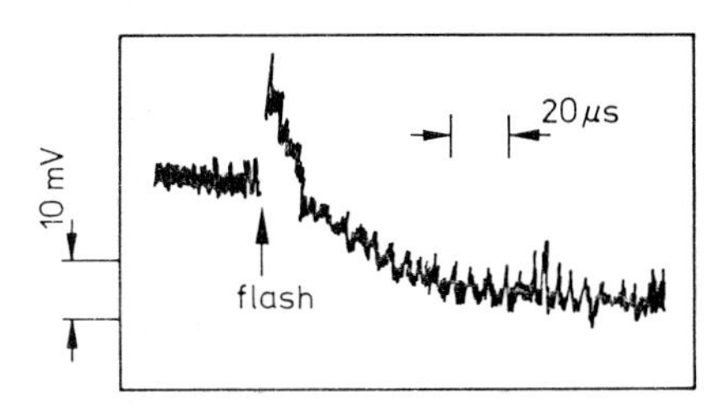

Fig. 3 - Oscilloscope trace after a 25 ns flash. Polyacetaldehyde in acetone (10 g/1). [Benzophenone]: 10^{-2}M.

CONCLUDING REMARKS

It has been shown above that it is possible under certain circumstances to observe disentanglement processes of linear flexible polymers in solution using the light scattering detection method in combination with laser flash photolysis. It was demonstrated that disentanglement diffusion may be distinguished from pure translational diffusion. In case the polymer does not absorb light of a certain wavelength main chain scissions may be initiated by a sensitizer. If the lifetime of intermediates is longer than the diffusion relaxation time the kinetics of the chemical step during the process of main chain scission may be investigated by our method. An example for this was found at the main chain degradation of polymethylmethacrylate in acetone solution [4].

REFERENCES

1 F.J.Golemba and J.E.Guillet, Macromolecules 5 (1972) 212

2 I.Lukač, P.Hradlovič, J.Manasek and D.Belluš, J.Polym.Sci.A1, 9 (1971) 69

3 C.David, W.Demarteau and G.Geuskens, Europ.Polym.J.6 (1970)1405

4 G.Beck, J.Kiwi, D.Lindenau and W.Schnabel, Europ.Polym.J. 10 (1974) 1069

5 G.Beck, G.Dobrowolski, J.Kiwi and W.Schnabel, Macromolecules 8 (1975) 9

6 J.Kiwi and W.Schnabel, Macromolecules, in press

7 G.Beck, D.Lindenau and W.Schnabel, Europ.PolymerJ., in press

8 P.J.Flory, and W.R.Krigbaum, J.chem.Phys. 18 (1950) 1086

9 L.Mandelkern and P.J.Flory, J.chem.Phys. 19 (1951) 984

10 W.N.Vanderkooi, M.W.Long and R.A.Mock, J.Polymer Sci. 56 (1962) 57

11 J.Bisschops, J.Polym.Sci. 17 (1955) 81

12 R.M.Secor, A.I.Ch.E.J. 11 (1965) 452

13 T.A.King, A.Knox and J.D.McAdam, Polymer 14 (1973) 293

14 T.A.King, A.Knox, W.I.Lee and J.D.McAdam, Polymer 14 (1973) 151

15 G.Dobrowolski and W.Schnabel, unpublished results

16 D.G.Marsh, Intern.Symp.Degradation and Stabilization of Polymers, Brussels (1974), paper 28

17 M.Letort, Compt.Rend. 202 (1936) 767

18 M.S.Travers, J.Faraday Soc. 32 (1936) 246

19 O.Vogl, J.Polym.Sci. 2 (1964) 4591

DISCUSSION

G. BECK, G. DOBROWOLSKI, J. KIWI, W. SCHNABEL

R. Mohan - After the flash photolysis, the broken molecule has to bend and stretch - internal or breathing modes of brownian motion - to disentangle itself. This is what is to be observed by intensity light scattering.
Do you get a single exponential decay of intensity with time? Have temperature dependent experiments been done?

W. Schnabel - The separation of fragments may be considered as a concerted action of motions of the different fragments. We can observe by our method with the systems described the motion of the whole fragments. Information on segmental motion, in other words, on micro-brownian motion, can be obtained by other methods in order to find out how disentanglement depends on parameters like chemical composition, chain flexibility, solvent quality, etc... At the present state the amount of data is insufficient in order to draw general conclusions. Naturally, measurements of the temperature dependence of $\tau_{1/2}$(LS) should provide interesting information. However, such measurements have not yet been carried out. Concerning the question whether a single exponential decay of light scattering intensity has been observed I would like to state that this was the case during the work with PPVK and styrene-co-PVK, where the chemical act of chain rupture is much faster than the separation of fragments.

M. Magat - Je pense que la réponse la plus directe qui pourrait être donnée à la question du Dr.Mohan proviendra des expériences utilisant des solvants très différents. D'autre part, je pense que votre méthode pourrait fournir des résultats très intéressants sur des polymères biologiques, protéines et acides nucléiques.

W. Schnabel - It is planned to extend our experiments to several other polymers and also to use different solvents. Actually interesting results have been obtained already with polyisobutene dissolved in different hydrocarbons. Main chain scissions were accomplished in this case by irradiating polymer solutions with pulses of 15 MeV electrons. By adding propanol to heptane solutions we found a significant increase of $\tau_{1/2}$(LS) with increasing fraction of propanol. This result indicates that the disentanglement process is strongly influenced by coil density. (The coil density is increased by the addition of alcohol). Concerning the other part of your comment, I agree that our method should be applied to biopolymers. Experiments with nucleic acids are already in preparation.

C. Varma - In your experiment a large number of polymer molecules will have to dissociate in order to create a sufficient transient change in refractive index. The small quantum efficiency for the dissociation imposes the use of laser excitation pulses with a large energy content. With increasing power density of the excitation pulse other effects such as heating through Raman scattering may contribute to the transient changes in refractive index. Can you comment on this with respect to the power density of your excitation pulse?

W. Schnabel - The quantum yield of chain scissioning of PPVK is rather high ($\emptyset(S)$ = 0.2 to 0.4). For the copolymers ot is somewhat less. The temperature increase caused by the energy absorption during the flash is negligible. At the highest intensity applied $5x10^{16}$ photons per flash passed the cell. 20% of them were absorbed. This corresponds to about $1.4x10^{-3}$ cal if one assumes complete conversion of quanta to heat. A temperature increase of about 0.02 degree results. Since we worked in general at much lower intensities, the temperature increase we were dealing with was even less.

SUB-PICOSECOND SPECTROSCOPY WITH A MODE-LOCKED CW DYE LASER

E. P. Ippen and C. V. Shank
Bell Telephone Laboratories, Holmdel, New Jersey 07733 (USA)

Until recently, the only lasers available for picosecond optical studies have been flashlamp-pumped devices which produce a short train of picosecond pulses with each flash. The variation of pulse duration within each train often requires the selection and experimental use of single pulses. With the development of the passively mode-locked CW dye laser [1,2], ultra-short pulses have become available at high repetition rates and pulse durations have been shortened into the sub-picosecond range. In this paper we first present the results of recent experiments which have accurately characterized the pulse output of this system. The pulses are found to be asymmetric and frequency chirped, and they can be compressed in time to a duration of about 0.3 psec. We also describe techniques which utilize these pulses to study ultra-fast molecular response times and provide illustration with several particular experiments.

The passively mode-locked cw dye laser used in this work is shown in Fig. 1. The gain medium, Rhodamine 6G in a free-flowing stream of ethylene glycol, is excited with approximately 5 watts of 5145Å argon laser light focused to a 15μ diameter spot. Near one end of the resonator a second stream contains the saturable absorber. An acousto-optic deflector at the other end provides output pulses at a rate of 10^5 per sec. Greatly improved stability of short pulse mode-locking has been achieved by the use of a saturable absorber mixture of a slow response dye (DODCI) and a fast dye (Malachite Green). With this combination the laser can be operated well above threshold. Best pulses are produced near λ = 6150Å which is the absorption peak of the Malachite Green.

Autocorrelation measurements of the output pulses are made with the experimental arrangement shown in Fig. 2. This particular scheme [3,4] is advantageous in that one obtains a background-free trace of the pulse autocorrelation function.

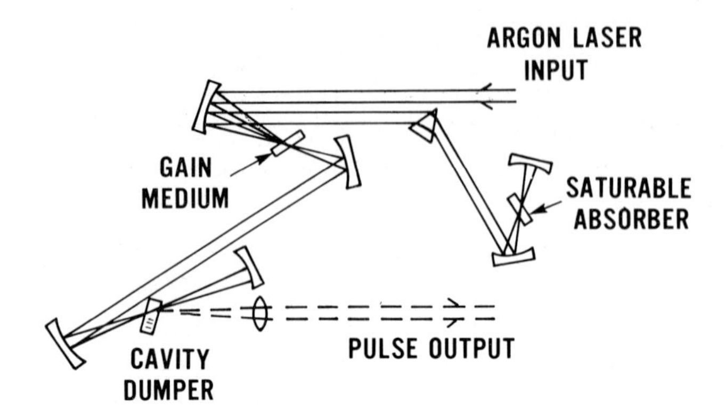

Fig. 1. Experimental configuration of the cw dye laser picosecond pulse generator.

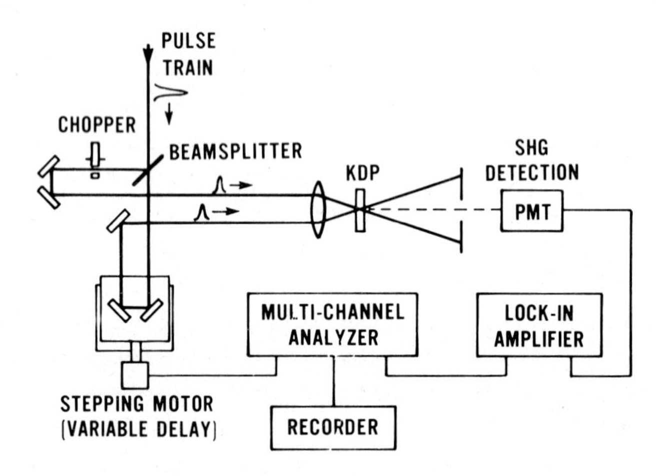

Fig. 2. Technique for optical pulse autocorrelation measurements.

The KDP crystal is only 100μ thick to provide a resolution of better than 10^{-13} sec. A stepping motor varies the time delay and indexes a multi-channel analyzer in which data may be accumulated over a number of repetitive delay scans. The uppermost curve in Fig. 3 is a typical autocorrelation trace of the laser output pulses. From this trace one infers that the pulses have an FWHM of about 1 psec and are characterized by exponential behavior in the wings.

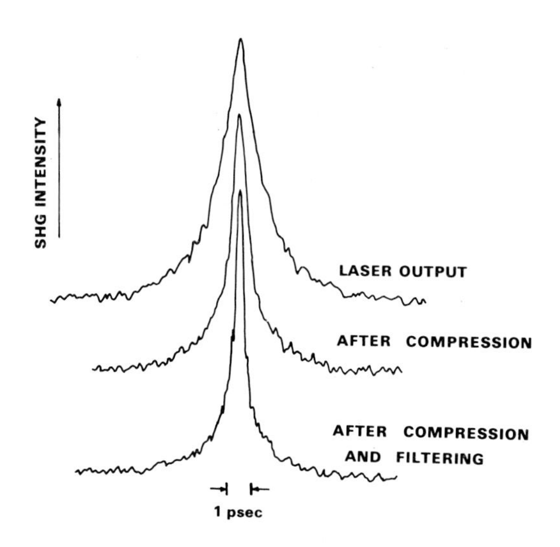

Fig. 3. Experimental autocorrelation curves of laser output pulses (upper trace), grating-pair compressed pulses (middle), and compressed pulses after spectral filtering (bottom).

By passing these pulses through a grating pair [5], ruled at 1800 lines/mm and with a slant separation of about 10 cm, one can achieve the temporal compression indicated by the middle trace in Fig. 3. Even further pulse narrowing can then be effected by passing these compressed pulses through a spectral filter which has a bandwidth of about 15Å and is tuned to select the long wavelength portion of the pulse spectrum. The resulting autocorrelation function is shown at the bottom of Fig. 3. It has a half-width of less than 0.5 psec and indicates a pulsewidth of about 0.3 psec. These ultra-short pulses have been used to probe the shape and dynamic spectral behavior of the laser output pulses. This is accomplished with arrangement shown in Fig. 4.

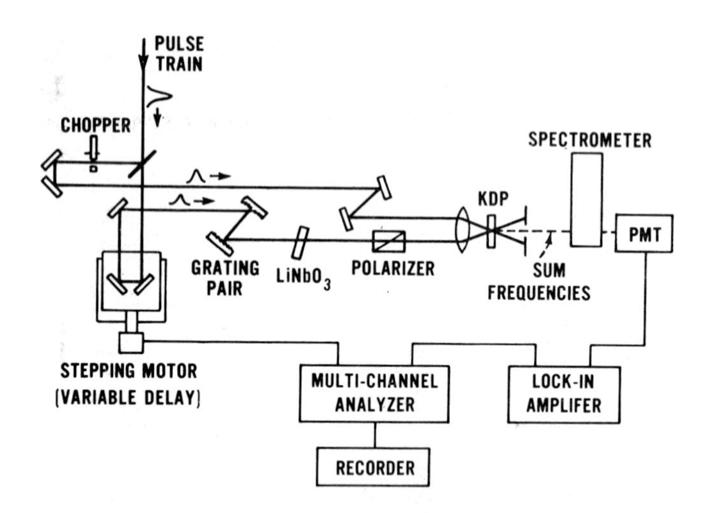

Fig. 4. Experimental arrangement for dynamic spectroscopy of laser output pulses.

The shape of the laser pulses is shown in Fig. 5(a). A temporal asymmetry characterized by a fast leading edge and a longer trailing edge is apparent. By selecting particular sum-frequency components for different correlation traces we get the results in Fig. 5(b) and (c) which reveal the pulse chirp. The long wavelength components are located in the leading part of the pulse. The shorter wavelengths follow and are less rapidly swept in frequency. This nonlinearity in the chirp explains the observation above that the compressed pulses can be further shortened by filtering out the less compressible shorter wavelengths.

Several other experiments illustrate the utility of these ultra-short pulses for making time-resolved measurements. The peak power of several kilowatts in the pulses makes it possible to operate an optical Kerr shutter [6,7] at high repetition rates. The temporal behavior of such a shutter is

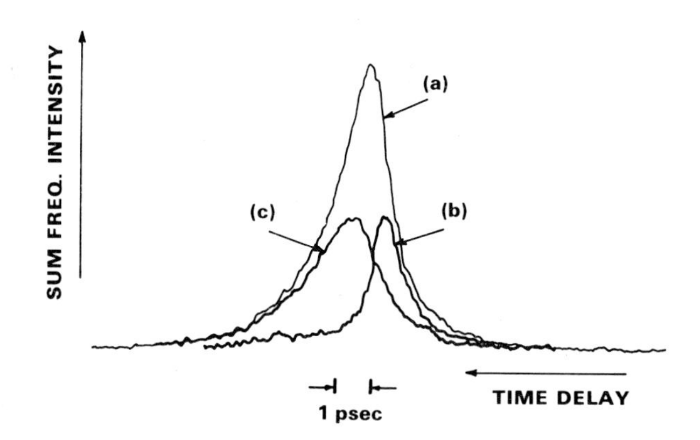

Fig. 5. Dynamic spectrogram of laser pulses showing (a) pulse envelope, (b) long wavelength components and (c) short wavelength components.

shown in Fig. 6 and clearly reveals a 2.1 psec orientational time of the CS_2 shutter medium. An advantage of repetitive operation is that the Kerr shutter may also be operated in a biased mode [7] where the incremental transmission is linearly proportional to the optical pulse power. The Kerr gate relies on an induced birefringence, but molecular orientation may also be studied in systems where one induces an anisotropic absorption change with a linearly polarized optical pulse. [8] Extremely small anisotropies can be observed with the signal averaging one achieves at high repetition rates.

A study of ultra-fast absorption recovery in the triphenylmethane dye malachite green has recently been performed with the experimental arrangement shown in Fig. 7. The observed 2.1 psec response of this dye in methanol is shown in Fig. 8. Previous work [9,10] has demonstrated that recovery times and quantum efficiencies of triphenylmethane dyes are

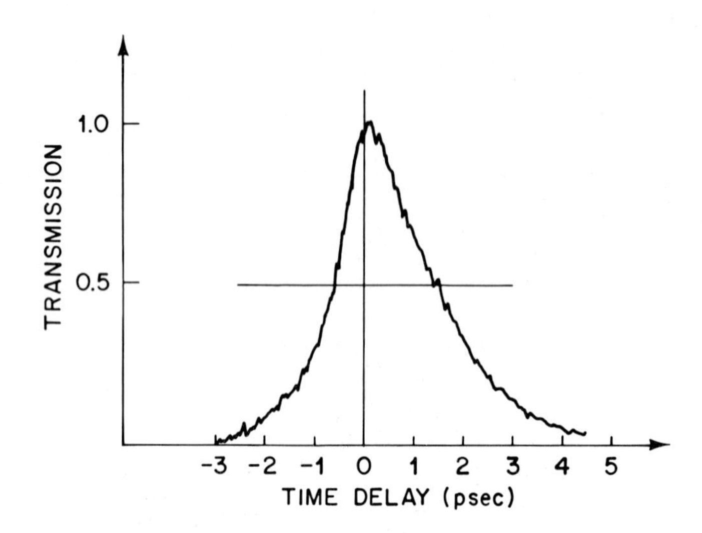

Fig. 6. Transmission response of a CS_2 optical Kerr shutter.

strongly but nonlinearly dependent on the viscosity of the solvent. It has also been suggested that this nonlinear dependence should evidence itself in non-exponential temporal behavior [9,10]. We have found, however, that recovery remains exponential in time over a wide range of solvent viscosity. Complete recovery of the dye takes considerably longer than the fluorescent state lifetime inferred from quantum efficiency data [9]. Solvent viscosity was varied by using several different glycerol-water mixtures as well as methanol and ethylene glycol. The results are plotted as solid points in Fig. 9. For the higher viscosities, two separate exponential time constants become evident experimentally. Points from a typical experimental trace are plotted logarithmically in Fig. 10. The longer time constant has the viscosity dependence above. We also plot in Fig. 9, as open circles, the initial, shorter time constant. Interestingly, these points are in good agreement with fluorescent state lifetimes inferred from the quantum efficiency data of Förster and Hoffmann [9].

Absorption recovery measurements may also be used to obtain a measure of vibrational relaxation in molecules. The

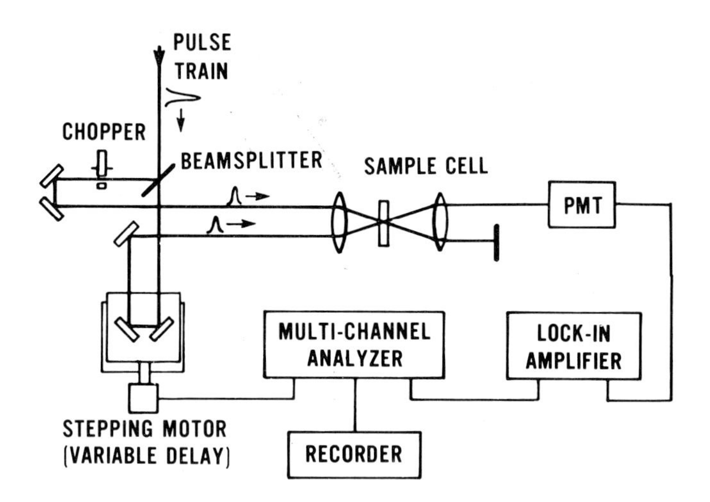

Fig. 7. Experimental arrangement for absorption recovery measurements.

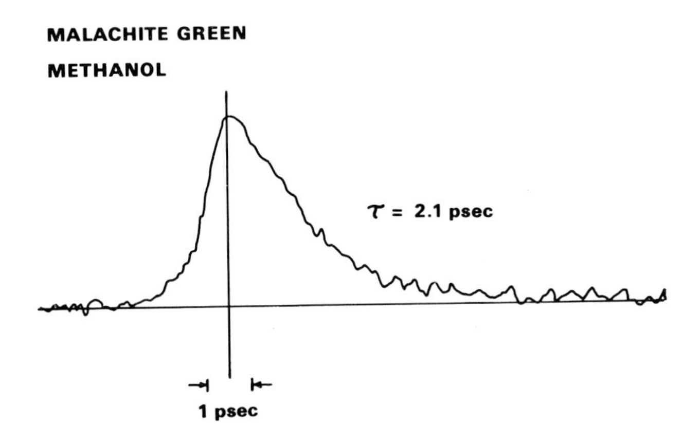

Fig. 8. Picosecond response of malachite green.

curves plotted in Fig. 11 serve to illustrate the way in which recovery data may be analyzed. The uppermost curve is the total absorption change observed in a methanol solution of cresyl violet perchlorate. This observed signal can be

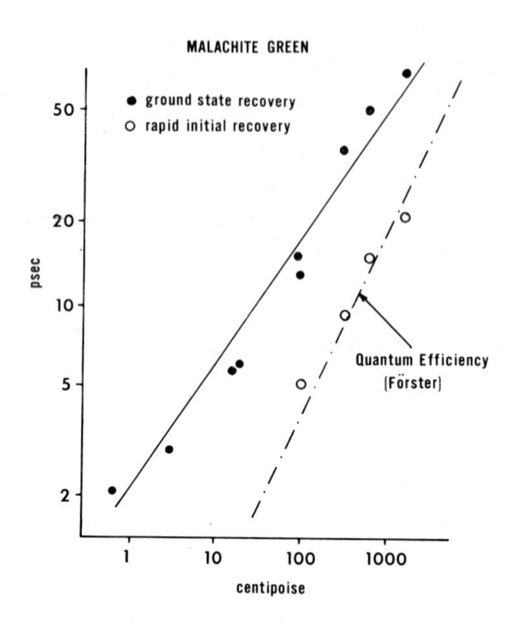

Fig. 9. Measured lifetimes of malachite green vs. solvent viscosity.

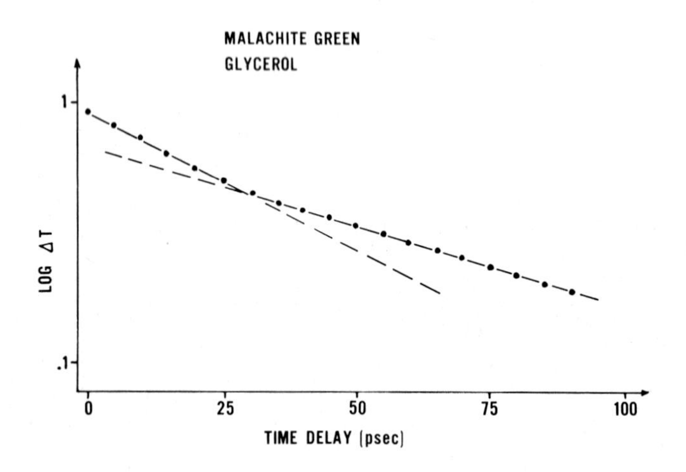

Fig. 10. Two component response of malachite green in glycerol.

separated into several different components. The pulse integrating effect of a very long excited state lifetime can be computed directly from pulse autocorrelation measurements. An additional, unavoidable, coherent beam coupling term is combined with this integral to produce the "slow" response shown in the middle of Fig. 11. Direct subtraction of this from the experimental trace gives the "fast" response shown. The resulting curve yields a 2.8 psec exponential time constant. Further experiments are required to separate the excited state and ground state contributions to this relaxation.

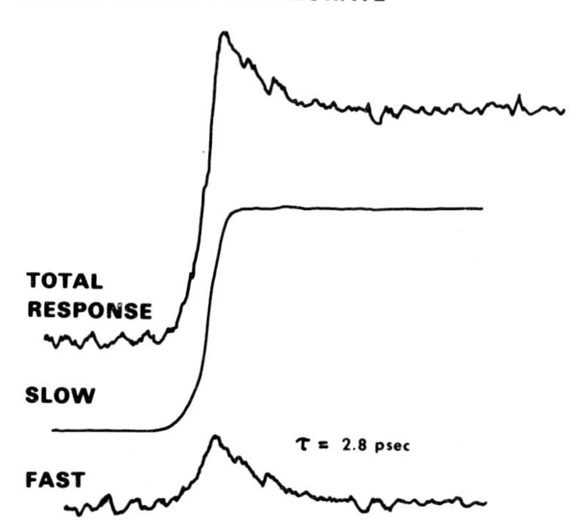

Fig. 11. Experimental separation of slow (electronic) and fast (vibrational) response of cresyl violet perchlorate.

REFERENCES

1 E. P. Ippen, C. V. Shank and A. Dienes, Appl. Phys. Lett., 21(1972)348.
2 C. V. Shank and E. P. Ippen, Appl. Phys. Lett., 24(1974)373.
3 M. Maier, W. Kaiser, and J. A. Giordmaine, Phys. Rev., 177 (1969)177.
4 H. A. Haus, C. V. Shank, and E. P. Ippen, Opt. Commun., to be published.

5 E. B. Treacy, IEEE J. Quant. Electron., QE-5(1969)454.
6 M. A. Duguay and J. W. Hansen, Appl. Phys. Lett., 15(1969) 192.
7 E. P. Ippen and C. V. Shank, Appl. Phys. Lett., 26(1975)92.
8 C. V. Shank and E. P. Ippen, Appl. Phys. Lett., 26(1975)62.
9 Th. Förster and G. Hoffmann, Z. Physik Chem. N.F., 75(1971) 63.
10 D. Magde and M. W. Windsor, Chem. Phys. Lett., 24(1974)144.

DISCUSSION

E.P. IPPEN, C.V. SHANK

R.M. Hochstrasser - What effect does molecular reorientation have on the ground state recovery data you have shown for cresyl violet perchlorate?

E.P. Ippen - We have studied molecular reorientation in separate experiments where we observe only the decay rate of the optically induced dichroism. The measured orientational times are in the range 50-100 psec in methanol and are of course longer in higher viscosity solvents. Since they are much slower than the fast responses we are observing here, they can be included in our modeling of the "slow" response.

O. de Witte - Did you look at the effect of viscosity on the rapid decay component in the absorption of cresyl violet perchlorate to see if there are any cross relaxation effects?

E.P. Ippen - We found very little change with viscosity and got almost the same fast response in glycerol that we got in methanol. We do plan to investigate this further at lower temperatures and in more solid matrices.

M.W. Windsor - As you know, we have published a study of crystal violet (CV), also a triphenylmethane dye and a close relative of malachite green, using a train of picosecond pulses (D. Magde and M.W. Windsor, Chem.Phys.Letters 24, 144(1974). We have since extended this work using single picosecond pulses. We have measured decay times via both decay of excited state absorption and rate of ground state repopulation, as a function of solvent viscosity. In brief, we find a sigmoidal dependence of lifetime on viscosity. In the region below 0.1 poise the lifetime is approximately constant at about 10 ps. Above 103 poise the lifetime is again constant at about 1 ns. In between, the lifetime depends upon the 1/3 power of the viscosity. We have developed a theoretical model that accounts reasonably well for these observations. At high viscosities, thermal rotational diffusion of the phenyl rings from the equilibrium angle ($\sim$35°) in the excited state to large angles ($\sim$40°) for which the S_1S_0 energy gap is smaller, thus giving very fast internal conversion, is rate-controlling. At low viscosities, the excited molecule effectively sees a wide range of rotational angles and the internal conversion step ($\sim 10^{11}$ to 10^{12} sec $^{-1}$) becomes rate-controlling.

E.P. Ippen - Our results with malachite green are somewhat at variance with this since at the lower viscosities (down to. 006 poise) we see no limiting of the recovery rate. Over the range .006 poise to 14 poise the recovery varies more like the 1/2 power of the viscosity than like the 1/3 power you describe. To get this dependence we were careful to separate out the initial more rapid response and to measure the exponential time constant for complete recovery.
The decidedly exponential behavior that we observe provides an additional constraint on the theoretical model. How one reconciles simple exponential recovery with the nonlinear dependence on viscosity is still an open question.

S. Ameen - In picosecond time-resolved Spectroscopic Techniques, I was wondering if one can control the amplitude of the pulses as well as the pulse width simultaneously? In other words can one vary the pulse-width while keeping the amplitude or intensity constant? or vice versa? If this can be achieved this can be a very useful and significant technique to apply laser technology to the studies of "Information Quantum Mechanics" which are being developed by Hellstrom and Kennedy and others. Besides I think one can be able to obtain knowledge of the "Storage of Chemical Information in Macromolecules" and also study what I call the "Information Capacity of Chemical Bond".

E.P. Ippen - With this system it is certainly easy to vary the pulse amplitude in a controlled, reliable manner by attenuation of the laser output. The best way to vary pulse-width without losing coherence would be to broaden the pulses by spectral filtering. It should not be too difficult to compensate for the corresponding amplitude change.

O. Svelto - The experiments you have shown only refer to measurements of the T_1 relaxation time. By also making measurements in the frequency domain (with your probe pulse) you could also measure the homogeneous (T_2) relaxation time in liquids. Can you actually do these experiments with your apparatus?

E.P. Ippen - At the present time we are limited to experiments which utilize probe and pump pulses of the same frequency. In the future we hope to amplify these pulses and obtain probe pulses at other frequencies by conventional nonlinear optical conversion techniques.

G. Mourou - We have mentioned so far T_1, longitudinal decay, and T_2, cross-relaxation decay. I have to say that we should also be aware of T_3 which is spectral cross relaxation diffusion time. This is just a comment.

M. Clerc - Could you please give us the order of magnitude of the smallest optical density variation measurable by this technique.

E.P. Ippen - We are easily able to work with density changes of about 10^{-4}. I want to emphasize, however, that these sub-picosecond pulses are sufficiently intense to produce strong saturation effects in dye molecules. For careful response time measurements one works with small density changes in order to assure linear behavior.

PICOSECOND LASER STUDIES OF PHOTOEXCITED ORGANIC SYSTEMS*

R. M. HOCHSTRASSER and A. C. NELSON
Department of Chemistry and Laboratory for Research on the Structure of Matter, University of Pennsylvania, Philadelphia, Pennsylvania 19174

1. INTRODUCTION

At first it is perhaps useful to discuss picosecond pulse techniques in relationship to other methods of studying optically induced fast processes. The simplest approach - and one that does not require a laser - is to make spectroscopic studies of the lineshape corresponding to the spectroscopic transition producing the transient. If the lifetime of the transient is ca. 5 ps, the homogeneous spectral width will be ca. 1 cm^{-1}. Such homogeneous widths are easily measureable spectroscopically in gases of small molecules where the spectral congestion is minimal. In these cases the inhomogeneous contribution (Döppler effect) to the linewidth is negligible. In solids at low temperature the inhomogeneous contribution to the linewidths is usually in the neighborhood of 1 cm^{-1}, so it is just feasible to learn about such transients using optical spectroscopy [1-3]. In solids at higher temperatures optical spectra are homogeneously broadened because many phonon states are occupied. This thermal broadening of electronic transitions is often of the order of kT and so even ultrafast processes occurring in 25 fs are barely detectable by optical spectroscopy of solids at 300k. Evidently pulsed laser techniques are essential to learning about the time behavior of such systems. In liquids and glasses there is an additional substantial inhomogeneous contribution to the linewidth and again pulsed laser techniques are needed to expose transient properties. However in these cases the pulsed laser can be used to study the dynamics of both the inhomogeneous distribution and the molecular properties. The processes I will describe in this presentation are all occurring at 300k and mainly in solutions. The homogeneous effects will be emphasized but it should be remembered that there are certainly inhomogeneous characteristic expected as well. The inhomogeneous effects are sensitive to the pumping pulse duration, spectral width and peak power.

There are important effects resulting from the relative energy width of the laser pulse and the homogeneous energy broadening (Γ) relating to the lifetime of the state to be optically pumped. When the pulse is energetically broad compared with Γ the system is prepared in a non-stationary state whose characteristics (e.g. Γ) can be investigated once the pulse has passed by the sample. Any light emitted by the sample also has a characteristic decay that is a property of the homogeneous width - that is, a molecular property. On the otherhand when the energy width of the light pulse is narrower than the width of the resonance the scattered light carries no energy or time information about the resonant state. Of course if the incident and scattered light pulses are polarized, information regarding the symmetry of the resonant state is obtainable. During the period that the light pulse is at the sample there will be light emitted but it will be in form of a pulse having approximately the width of the laser pulse [4]. During that same period it will be possible to observe transient absorptions using a second pulse but the time characteristics will only contain information about the correlation of the two laser pulses. This is quite analogous to the process of off-resonance two-photon absorption.

2. Radiationless Processes in Benzophenone and other Ketones

Studies of the spectrum of crystals and mixed crystals of benzophenone at low temperature [2] have indicated that the singlet excited states reached by near uv excitation have homogeneous widths in the range 0.8 to 1.3 cm^{-1}. It was presumed that these widths relate to electronic and low temperature vibrational relaxation processes occurring in the range 6.6 to 2.5 ps. At 300k in crystals, mixed crystals or solutions these spectra become extremely diffuse - i.e. much wider than the laser pulse. An additional cause of diffuseness in this case is that the potential barriers for internal rotations of the benzene rings are expected to be very low, such that molecules having a large range of quite different nuclear configurations are excited by the pulse. Since the ground and excited state torsional constants are different, this results in broad electronic spectra due to congestion. Excitation with a picosecond pulse therefore produces a range of transients and our solution measurements refer to that distribution.

Table I summarizes our results obtained for some aromatic ketones.

Table I

Triplet build-up times (530 nm) for aromatic ketones*

Molecule	Solvent	k^{-1}
$(C_6H_5)_2CO$	n-heptane	8 ± 1
	p-dioxane	10 ± 1
	C_2H_5OH	12 ± 2
	C_6H_6	23 ± 5
	C_6D_6	23 ± 5
	DDE	10
$(C_6D_5)_2CO$	C_6H_6	20 ± 5
xanthone	C_6H_6	8 ± 1
4-phenyl-$(C_6H_5)_2CO$	C_6H_6	6 ± 1

*From [5, 7] and previously unpublished surveys (by R. W. Anderson, G. W. Scott and H. Lutz) in this laboratory.

The experiments have been described previously [5-7]: Briefly, a ca. 7 ps pulse of 353 nm radiation ($\frac{1}{3}$ x 1060) pumps the distribution which is probed subsequently by a 530 nm pulse. The k of Table I refers to the rate constant obtained by assuming our data refer to (a) absorption of a gaussian 353 nm pulse by the singlet state S_1 (b) decay of S_1 into a 530 nm absorbing state (T), this process being describable by a single rate constant k. Obviously the actual processes occurring are more complex than this but at present our data hardly merit a more detailed treatment.

2.1 Solvent Effects:

At present there is no compelling reason to assume that the molecular electronic transition shifts are causing the solvent effect on k. Thus, until much more data is available, it seems more natural to assume that the solvents are assisting the electronic vibrational relaxation to different extents. In other words some of the excess energy released in the radiationless process is directly transferred to solvent - solute motion, resulting in a change in the orientation of the benzophenone region of the solution as a consequence of the radiationless transition $S_1 \rightarrow T$. This effect is not unexpected for processes involving the coupling of a discrete state to a sparse manifold [8].

Benzophenone in dibromodiphenylether (DDE) forms a homogeneous mixed crystal [9], and so that entry in Table I is not expected to include contributions from the inhomogenities discussed above. The mixed crystals used in this experiment were close to 1% in benzophenone so the quality of the material was not perfect. As far as we know, this is the first picosecond transient absorption experiment to be performed on a molecular crystal. This is a system in which the temperature dependence of the radiationless processes could be readily studied.

2.2 Transient Spectra:

A number of the assumptions from our earlier work may be tested by studying the development of transients at wavelengths other than 530 nm. Previously the time development of the ketone transients were measured at a few selected wavelengths obtained by stimulated Raman scattering of the 530 nm probe beam. These measurements gave results that were consistent with the only transient contributing to the 530 nm signal being the ketone triplet state, but a complete spectrum of these systems following 353 nm pumping was clearly needed. Another feature of our earlier work was that the laser oscillator contained no mode selecting device so the transverse structure of the beam was quite non-uniform and we found it was necessary to use an efficient diffuser to obtain reproducible measurements. Finally the spectral width of the pumping pulse ($\sim$ 130 cm^{-1}) was not ideally suited toprobing an inhomogeneous distribution. These are among the many considerations which led us to construct a transform limited Nd/glass system producing single TEMOO pulses having about the same energy as our earlier multimode spectrally broad pulses (viz., ca. 50 mJ at 1060 nm, 7 mJ at 530 nm, and 1-2 mJ at 353 nm). The transient spectra were photographed using either multimode or TEMOO mode selection by using the continuum generated by focusing a single pulse into CCl_4. The system was designed to minimize the chirp in the continuum pulse at the sample at less than half the pulsewidth. A main feature of our method is the presence of a good spectroscopic slit at the sample cell:

This slit limits the lateral width of the uv and continuum pulses enabling an image of the uv irradiated part of the sample to ce centered on the image of the continuum strip at the focal plane of the spectrograph. The spectra obtained are similar in appearance to those recently described by Magde and Windsor [10]. A great advantage of this technique of forming the focal plane image is that it is easy to search for new absorption spectra, since any spectra that exist appear as a dark line in the center of the continuum and that line only appears when the system is pumped with uv.

The spectrum of benzophenone in benzene is shown in Figure 1 photographed at very early times. The spectral region covered ranged

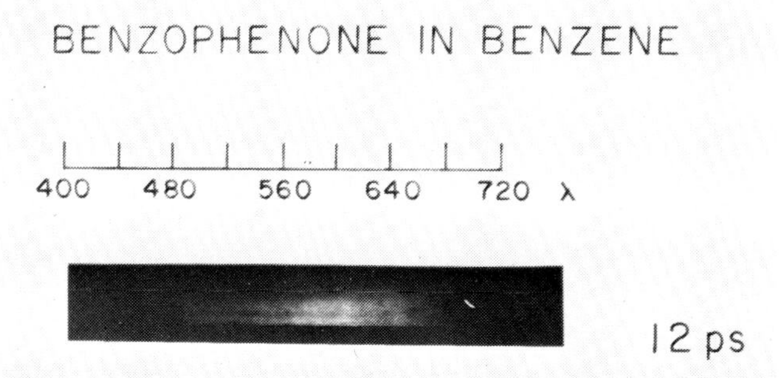

Fig. 1. Time resolved absorption spectra of benzophenone in benzene: The spectra are photographed using Kodak 103a-F plates. The continuum from CCl_4 is shown bordering the absorption streak. The absorption streak is absent in circumstances (i) The time is prior to - 6 ps (ii) The system is not pumped with 353 nm.

from ca. 430 nm to 670 nm., and it is apparent from the photographs that there are no substantial time dependent changes in the nature of the spectrum. We previously made careful studies of the kinetics at 530 nm and there is no indication from our spectra that the kinetics are different at other wavelengths. We conclude therefore that our previous experiments and the present ones in the range 430 - 670 nm represent triplet build-up, as assumed. It has not been possible to reconcile these observations with the earlier reports of the disappearance of a singlet transient following near uv excitation [11]. If there is a singlet spectrum in the range 430 - 670 nm it must be quite similar to but weaker than the triplet spectrum. The fact that the triplet spectrum was not observed to change with time is consistent with the radiationless transition being $^1n\pi^* \sim\!\sim\!> {}^3n\pi^*$ without any appreciable intermediate state (e.g. $^3\pi\pi^*$) being involved. This result was expected on the basis of earlier work [2, 12] and from an elementary density of states argument. Since the $^3\pi\pi^*$ state is much closer to the excitation energy than is the $^3n\pi^*$ state, the density of $^3\pi\pi^*$ levels at the excitation energy is expected to be comparatively low. The higher vibronic levels of $^3\pi\pi^*$ will in any case be mixed in the interference region and the $^3n\pi$ will in any case be mixed in the interference region and the $^3\pi\pi^*$ amplitude will be small in any one of these mixed levels. Because the T-T absorption of benzophenone is so diffuse around 530 nm, there would be very little spectral difference expected from triplet molecules that were in quasi-equilibrium with the solution, and those that still contained some excess vibrational energy.

2.3 Other Ketones:

Table I gives triplet build-up times for other aromatic ketones and in each case the k^{-1} is less than 10 ps.

The spectra of xanthone in benzene and ethanol are shown in Figure 2. These spectra change only their intensity with time between 0 and 100 ps, and it is apparent that they are triplet - triplet absorption spectra. The spectra have a different appearance to the benzophenone T-T absorption. Microdensitometer tracings show that the absorption in ethanol at wavelengths shorter than 600 nm is considerably suppressed relative to that in

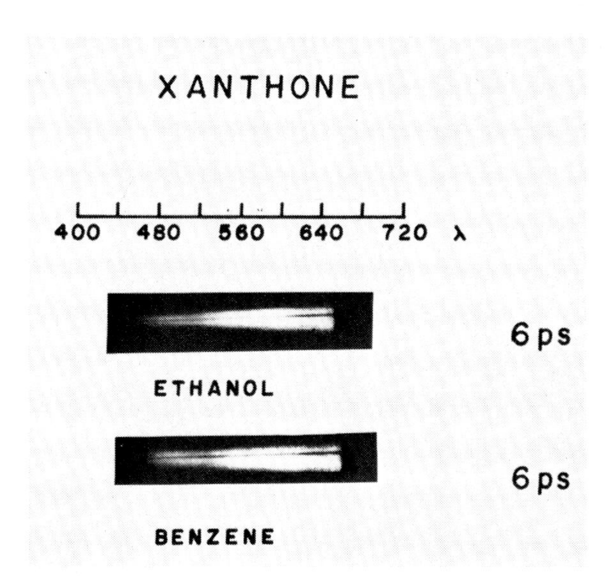

Fig. 2. Time resolved absorption spectra of xanthone in benzene and ethanol. For more details see caption to Figure 1.

benzene. It is possible that T_1 is a different type of state in the two solvents. There are experiments in rigid media suggesting that the lowest triplet state of xanthone is different in polar and non-polar media [13]. It would be useful to have two accurate photometric spectra of these T - T absorptions in order to better expose any differences that might exist both at short and long times (i. e. nanoseconds).

The phosphorescence spectrum and lifetime for 4-phenylbenzophenone are indicative of a $\pi\pi^*$ (biphenyl-like) lowest triplet state [14], and therefore the appearance spectrum of the T - T absorption is of some interest. The appearance of the spectrum around t=o is shown on Figure 3. The spectrum does not appear to evolve substantially at later times. The T-T

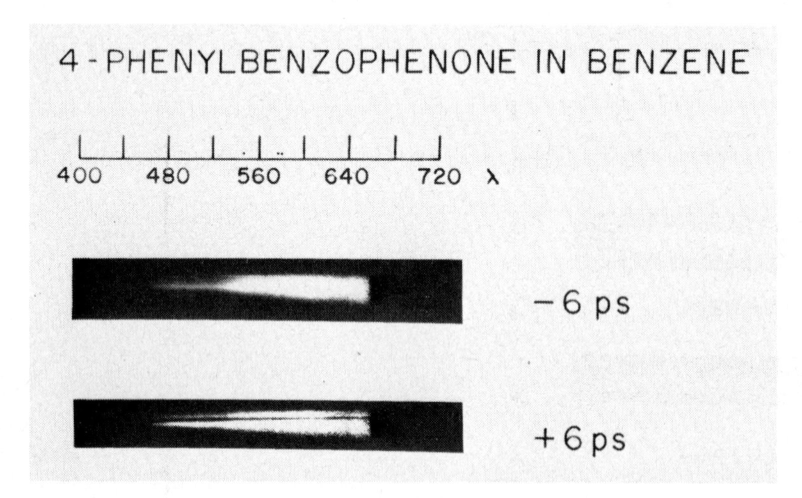

Fig. 3. Time resolved absorption spectra of 4-phenylbenzophenone in benzene. For more details, see caption to Figure 1.

absorption simply grows in with $k^{-1} = 6$ ps. This behavior is to be contrasted with that displayed by the system with biphenyl replaced by an α - naphthyl group (see below) where there is clearly a transient. Note that the T - T absorption is not similar to that of benzophenone and this is consistent with an essentially biphenyl-like lowest triplet state.

3. Energy Transfer Involving Benzophenone:

Benzophenone molecules in their triplet states transfer energy to neighboring acceptors such as naphthalene and piperylene in a few picosecond [6]. Even when the benzophenone is chemically connected to the acceptor the 530 nm transient absorption was detectable [15]. In these experiments it was assumed that the 530 nm absorption corresponded to the triplet state, however if energy transfer from T was essentially instantaneous the 530 nm transient could conceivably be the singlet state in this case. The spectrum of benzophenone in trans - piperylene shown in Figure 4 has the same appearance as the persistent triplet spectrum of benzophenone in benzene. Thus the earlier assumptions [5, 14] are strengthened. For the case benzophenone in piperylene the transient

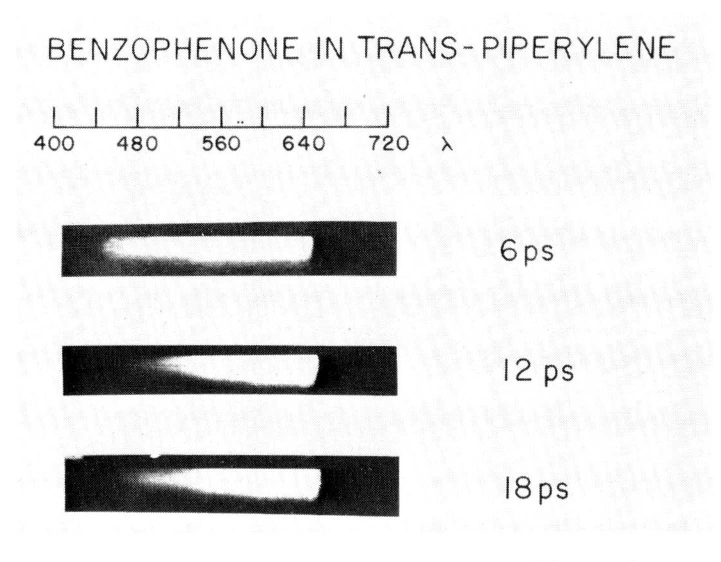

Fig. 4. Time resolved absorption spectra of benzophenone in trans - piperylene. For more detail see caption to Figure 1.

absorption has disappeared after 40 - 50 ps, and there is just a short region between 0 and 12 ps when the spectrum is clearly visible.

Figure 5 shows the spectra we have obtained for the double molecule having benzophenone connected to naphthalene through one methylene bridge [15]. In this case the transient absorption at 530 nm was most intense at ca. 10 ps and decayed away with a $k^{-1} = 10$ ps [15], and it can be seen that the spectra are again uniform with time. These spectra exist only for a very brief time and are quite useful for locating those to t = o on the time axis of the experiment.

In these experiments the benzophenone molecule is very likely left in an excited vibrational state following energy transfer [6]. For the case benzophenone-piperylene we have calculated the vibrational energy distribution in the product ground state benzophenone using known Franck-Condon factors for torsional and carbonyl stretch modes. The final distribution depends on whether the vibrational relaxation is comparable or rapid compared with the energy transfer, thus if the actual distribution could be determined experimentally the vibrational relaxation details would be exposed. We have made considerable effort (with R. W. Anderson, Jr.) to detect anti-Stokes Raman scattering from these systems close to t = o but so far the experiments have not been successful. Our calculations show that energy transfer from benzophenone to piperylene should occur respectively 3, 5, and 7 times faster from levels containing 1, 2, and 3 carbonyl stretch quanta than from the zeropoint level of T_1. Some other experiments, however, suggest that vibrational relaxation for such molecules in solution is extremely rapid (see below) and that very likely the required information is already lost before the system can be adequately probed using picosecond pulses.

4. Energy Transfer Involving Nitronaphthalene: An estimate of vibrational energy redistribution

The nitronaphthalenes undergo picosecond radiationless transitions following excitations of the lowest energy singlet states [6, 7] to form ultimately naphthalene - like triplet states. In the case of 1-nitronaphthalene k^{-1} was 12 ± 2 ps in benzene solution [7]. For 1-nitronaphthalene in piperylene the triplet lifetime was 680 ps. We interpreted this as being caused by a slow thermally assisted energy transfer - the thermal assistance being needed because the lowest accessible triplet state of piperylene is ca. 1000 cm^{-1} above the lowest triplet of

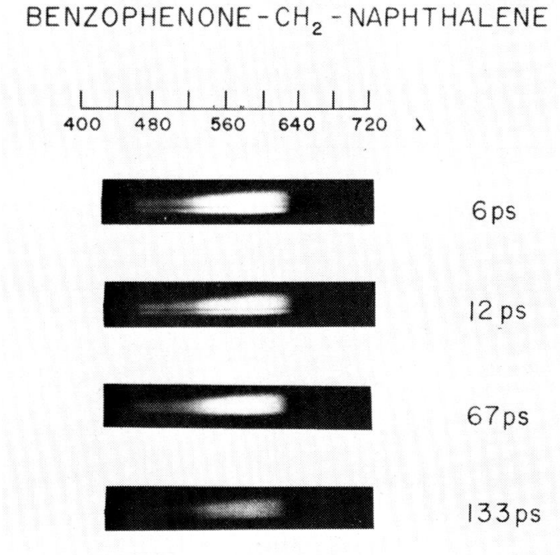

BENZOPHENONE - CH_2 - NAPHTHALENE

400 480 560 640 720 λ

6ps

12 ps

67ps

133ps

Fig. 5. Rise and fall of characteristic benzophenone triplet spectrum in the molecule 4-methylnaphthyl-benzophenone.

nitronaphthalene. The nitronaphthalene triplet is certainly formed in ca. 10 ps and it is formed with ca. 8000 cm^{-1} in excess of the zeropoint level of the triplet state. Notwithstanding the instantaneous energy excess the energy transfer still appears to require a thermal assistance*, indicating that the vibrational energy redistribution occurs much faster than the

---

* We have presented arguments [6] suggesting that the energy transfer is thermally assisted but this conclusion must remain speculative until temperature dependences have been studied.

triplet build-up time (i.e. much faster than ca. 10 ps): Our data probably would have indicated directly (and may have done so [6]) transients in the range of ca. 4 ps, and certainly energy transfer considerably faster than this would have been detected by a loss of our signal. We can therefore assume that the vibrational relaxation (energy redistribution) is at least an order of magnitude faster than energy transfer in this case. The pre-exponential quenching rate is 1.8×10^{11} sec^{-1} so vibrational relaxation takes less than ca. 0.5 ps. This is consistent with estimates of lifetime based on infrared linewidths in the overtone region.

5. Polarization Spectra:

We have devised a system to display simultaneously the two perpendicularly polarized spectra of transients produced by a single picosecond pulse. The apparatus works on the same principle as used in conventional polarization spectrography: A continuum beam was polarized at $\pi/4$ to the pump pulse polarization, passed through the sample and after passage through a Wollaston prism was focused onto the spectrograph slit (in future applications the photographic plate will be replaced by a diode array). The plate displays two continua, each with an absorption strip having orthogonal polarizations corresponding to the 0 and $\pi/2$ spectra of the transient. This technique may be used to study orientational relaxation, Förster energy transfer, polarization spectra of transients leading to state assignments, and the picosecond spectra of crystals. Figure 7 shows some of the spectra we have obtained with the apparatus shown in Figure 6 for xanthone in glycerol and benzene. In the more viscous solvent the polarization is about 2:1 in favor of perpendicular, as expected for a T-T absorption polarized perpendicular to the $S_o \rightarrow S_1$ transition. There was no measurable difference between the perpendicular and parallel absorption spectra in the benzene solvent, this effect probably being caused either by rotational relaxation or by saturation effects. A main point for the purpose of this presentation is to demonstrate a simple apparatus useful for studying polarized absorption spectra for very short-lived states as well as rotational relaxation.

6. Nitrogen Heterocyclics

N-Heteroaromatics usually have very low flourescence yields and

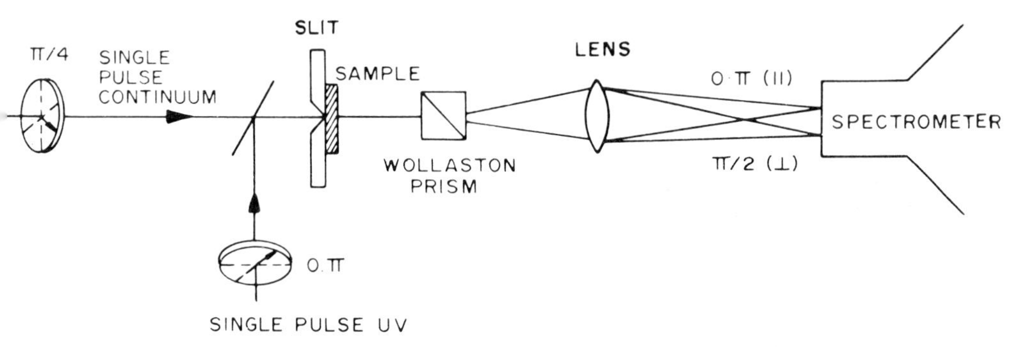

Fig. 6. Picosecond polarization spectrograph.

hence short fluorescence decay times. It is however relatively trivial to obtain the singlet excited state absorption spectra of such molecules using the apparatus described above. Figure 8 shows spectra for quinoxaline. For the case of quinoxaline we know that the fluorescence decay time is 41 ps [R. W. Anderson, Jr., unpublished results] so the 12 ps spectrum is that of the singlet ($^1n\pi^*$), and the 660 ps that of a naphthalene-like triplet state.

We have also studied 9,10-diazaphenthrene spectroscopically and with the polarization spectrometer. In this case the 353 nm excitation produces S_2 ($\pi\pi^*$) states but no evidence of an S_2 absorption was seen. However some indication of the transient existence of S_2 was obtained from polarization studies. The very rapid loss of polarization that was

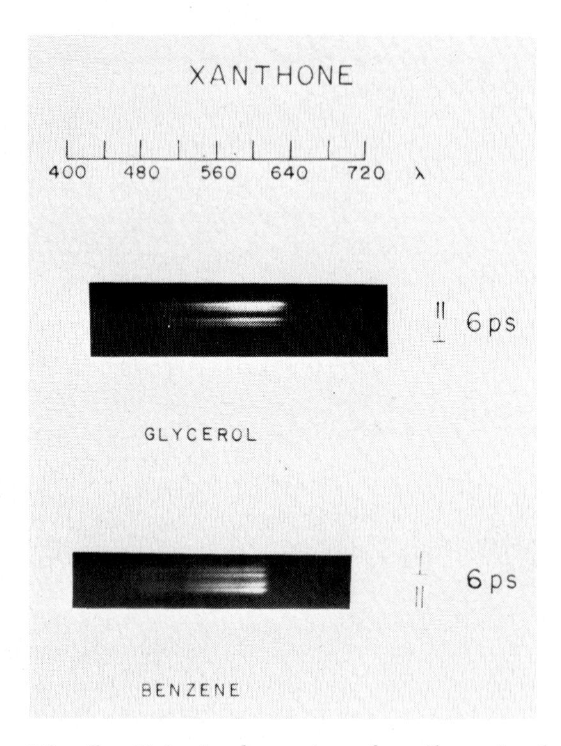

Fig. 7. Polarized spectra of xanthone in glycerol and in benzene.

observed could in this case be due to Forster dephasing - energy transfer amongst molecules in the sample. The energy transfer S_2 (on one molecule) to the nearby S_1 (on another molecule) calculates to be orders of magnitude larger than S_1 to S_1 transfer for molecules having weak but strong $S_0 \rightarrow S_1$ but strong $S_0 \rightarrow S_2$ transitions. These results and calculations will be published elsewhere.

7. Conclusions

By means of photographic spectroscopy with single picosecond pulses we have shown that the main assumptions of our previous work [5] are justified. In the case of benzophenone, the appearance of the spectrum produced with 353 nm excitation is not noticeably different at different

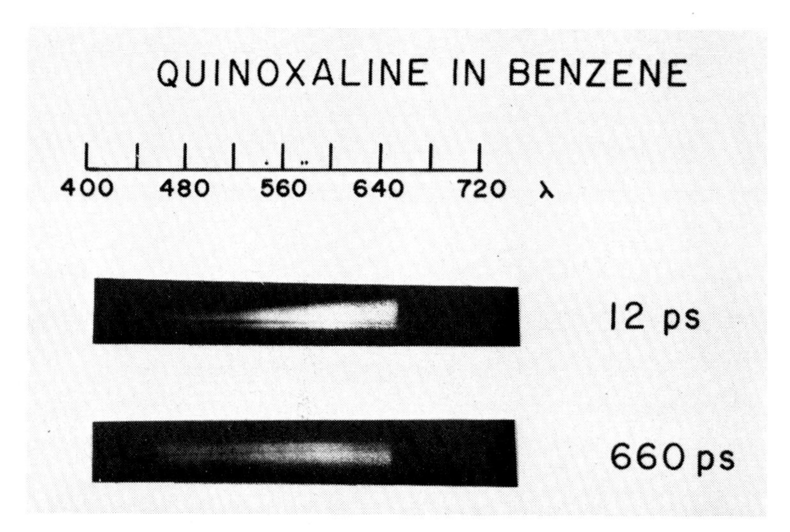

Fig. 8. Transient spectra of quinoxaline in benzene following 353 nm pumping.

times and in the limited spectral range studied we have seen no evidence of a singlet-singlet absorption. The spectra during energy transfer are also shown to be benzophenone as previously assumed [6] even in the case of the double molecule of benzophenone-naphthalene [15]. The study of nitronaphthalene [6] energy transferring to piperylene indicates that the energy excess available after intersystem crossing from the singlet state is not available for assisting energy transfer even though the energy required for thermal assistance of energy transfer is ca. ten times less than the energy excess. We take this as evidence that nitronaphthalene in solution has the energy excess distributed over a sufficient number of modes that within a time very short compared with our pulse width the molecule is relatively cool.

We have presented a polarization spectrograph that should be useful for studies of polarized spectra of short-lived states and for studies of rotational relaxation in liquids. Some preliminary results for xanthone demonstrate an unexpected rapid depolarization in benzene solutions.

REFERENCES

1. R. M. Hochstrasser and P. N. Prasad, in "Excited States" Ed. E.C. Lim (Academic Press, N.Y. 1974); p79.
2. S. Dym and R. M. Hochstrasser, J. Chem. Phys. 51, 2458 (1969).
3. R. M. Hochstrasser and T. Y. Li, J. Molec. Spectry., 41, 297 (1972).
4. J. Friedman and R. M. Hochstrasser, Chemical Physics, 6, 155 (1974).
5. R. M. Hochstrasser, H. Lutz and G. W. Scott, Chem. Phys. Letts. 24, 162 (1974).
6. R. A. Anderson, Jr., R. M. Hochstrasser, H. Lutz and G. W. Scott, J. Chem. Phys. 61, 2500 (1974).
7. R. A. Anderson, Jr., R. M. Hochstrasser, H. Lutz and G. W. Scott, Chem. Phys. Letts. 28, 153 (1976).
8. A. Nitzan, J. Jortner and P. M. Rentzepis, Chem. Phys. Letts. 8, 445 (1971).
9. R. M. Hochstrasser, G. W. Scott and A. H. Zewail, J. Chem. Phys., 58, 393 (1973).
10. D. Magde and M. W. Windsor, Chem. Phys. Letts. 27, 31 (1974).
11. P. M. Rentzepis, Science 169 239 (1970).
12. M. A. El Sayed and R. Leyerle, J. Chem. Phys. 62, 1579 (1975).
13. R. N. Nurmukhametov, L. A. Mileshina, and D. N. Shigorin, Opt. 22, 740 (1967).
14. V. L. Ermolaev and A. N. Terenin, J. Chim. Phys. 55, 698 (1958).
15. R. W. Anderson, Jr., R. M. Hochstrasser, H. Lutz and G. W. Scott, Chem. Phys. Letts. 32, 204 (1975).

DISCUSSION

R.M. HOCHSTRASSER, A.C. NELSON

C.R. Goldschmidt - The absorption spectrum $S_1^* - S_n^*$ you have shown in your ps measurement seems to be much broader than the spectrum published by Goldschmidt, Ottolenghi and Stein in Israel, J.of Chemistry, 8 (1970), 29. Could you comment?

R.M. Hochstrasser - We have not studied the coronene spectrum at times beyond a hundred or so picosecond, so we could not be certain whether the spectra should even be the same in the two situations. However our spectrum does not change significantly with time (other than intensity). The plate I showed has a rather large density of absorption so the contrast is not good enough for one to estimate the width by eye. The microdensitometer tracing does show a spectrum similar to that reported by Porter and Topp. The approximate half bandwidth of the green absorption band is mentioned in my text. After pumping with a 353 nm pulse it is easy to obtain coronene solutions that are totally absorbing across the whole visible spectrum.

O. de Witte - You assumed some decay channel for the population of the triplet state of benzophenone. First have you an idea of the path of energy transfer $S_1^* \rightarrow T^{**}$ or $S_1 \rightarrow T^*$. Secondly could the solvent effect on the rate constant K_{ST} give a partial answer to the first question?

R.M. Hochstrasser - The answer to your first question is that I do not know from which vibrational levels of S_1 the radiationless transition occurs. Our old linewidth measurements (Dyrn and Hochstrasser, J.Chem.Phys., 1968) indicated an increase in width with increasing excitation energy in the singlet manifold. More experiments with different wavelengths of UV exciting pulses are certainly needed in order to clear up this problem.

A. Laubereau - In your talk, you have discussed in short the dynamical information obtained from spectral linewidth data. In addition to the inhomogeneous broadening and energy relaxation mentioned, loss of phase correlation may also be important; In this case two line constants, T_1 and T_2 have to be distinguished. Can you please comment on this point ?

R.M. Hochstrasser - I do not fully understand how phase correlation effects will influence optical absorption lineshapes. However I can mention some experimental information that bears on the relation between linewidth and energy relaxation rates. Much of the available data for organic molecules in the condensed phase is brought out in a recent review ("Optical Spectra and Relaxation in Molecular Solids" by P.N.Prazach and R.M. Hochstrasser, in "Excited States", 1, 1974). In all known low temperature situations the observed relaxation times (e.g. fluorescence lifetimes) have come out to be roughly the same as the inverse lorenzian absorption widths. At higher temperatures there is significant thermal broadening (say above 5-10 K) and there are no experiments done yet (that I know of) to explore phase correlation contributions to the higher temperature situations. At the higher temperatures the T_2 process may significantly contribute to observed linewidths.

M. Stockburger - Some years ago we measured the 0-0 fluorescence band of the $S_1 \rightarrow S_0$ transition of benzaldehyde which has similar spectroscopic properties as benzophenone. This has been done using high spectroscopic resolution. At the same time we also recorded the 0-0 $T_1 \rightarrow S_0$ band. It turned out that the $S_1 - S_0$ band is considerably broader than the $T_1 - S_0$ band. From their broadening one deduces a lifetime of the S_1 vibrationless state of a few picoseconds. This homogeneous broadening should reflect intersystem crossing. It should be noted taht the broadening of a single peak of the rotational contour was considered.

R.M. Hochstrasser - One of the broadening effects that needs emphazing is the diffuseness introduced by congestion. For rotational states of molecules larger than say benzene this effect will nearly always obscure bandwidth measurements relating to picosecond channels.

R.A. Keller - In your experiments on endothermic energy transfer from nitronaphtalene are you able to determine the activitation energy as a function of the excess energy you put into the system? It would be interesting to determine if the excess energy affects the activation energy or the preexponential term.

R.M. Hochstrasser - I agree with your observation that it would be interesting to study these excess energy effects. Unfortunately in the nitronaphtalene case it is very likely that the singlet-triplet splitting is fixing the energy excess, so that, no matter where we excite in the singlet state, the excess energy in the triplet may be roughly constant. There does not appear to be much hope of selectively exciting the triplet state directly in this case.

ULTRASHORT PULSES. GENERATION OF SHORTER PULSES AND OF NEW FREQUENCIES VIA NONLINEAR OPTICAL PROCESSES

W. KAISER, A. PENZKOFER and A. LAUBEREAU
Technische Universität München, München, West Germany

In recent years solid state mode locked laser systems have received increasing attention by physical chemists. The reproducible generation of single bandwidth limited pulses of several picosecond duration has been achieved but the systems require still skilful experimentalists. It should be noted that pulse trains have widely varying parameters: The pulse duration, pulse shape, peak intensity and frequency bandwidth changes from the beginning to the end of the pulse train. Well-defined single pulses were obtained by cutting one pulse from the leading part of the pulse train using an electro-optic switch. As an example we list the various parameters of our ultrashort light pulses: Pulse duration $t_p \approx 6$ psec, pulse shape approximately Gaussian, peak power $\dot{P}_L = 10^9$W, frequency bandwidth $\Delta\nu = 3\ cm^{-1}$ (i.e. $t_p \times \Delta\nu = 0.6$; our pulses are essentially bandwidth limited), peak to background ratio 10^4, mode pattern TEM_{00}. With these pulses we have worked quite successfully in recent years. In this note three experiments should be discussed which deal with a reduction of the pulse duration and which provide new frequencies at a picosecond time scale.

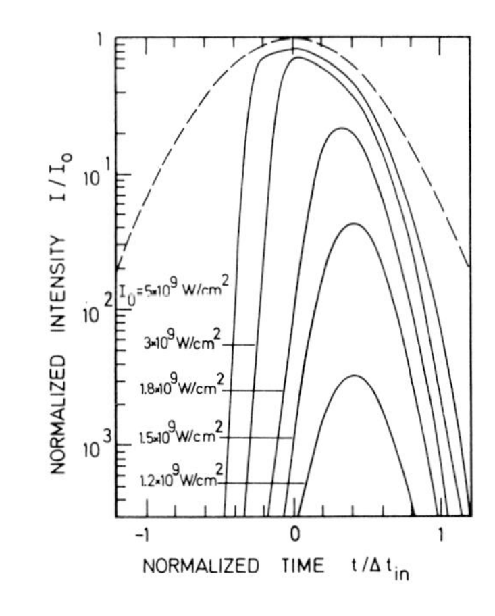

FIG. 1. Changes of the pulse shape after a single pass through a saturable absorber. Broken curve: normalized intensity of the input pulse (Δt_{in} = 8 psec, Gaussian pulse shape). Solid curves: normalized intensity of the transmitted pulse for several values of the input peak intensity I_0. Dye parameters: $T_0 = 10^{-7}$, $\sigma = 1.84 \times 10^{-16}\ cm^2$, $\tau = 9.1$ psec.

Picosecond light pulses experience considerable change in pulse shape when passing through a saturable absorber (e.g. a dye solution). In Fig. 1 calculated curves are presented. The normalized intensity of the input pulse (broken curve) and of various transmitted pulses is plotted versus time. The dye parameters chosen for this example are: initial transmission $T_0 = 10^{-7}$, lifetime of the excited state of the dye, τ = 9.1 psec and absorption cross-section of the dye molecule $\sigma = 1.8 \times 10^{-16}$ cm^2. It is readily seen from Fig. 1 that the shape of the transmitted pulse is substantially altered compared to the input pulse for a peak input intensity of I_0 = 5×10^9 W/cm^2. The leading part of the pulse is strongly absorbed by the bleachable dye while the peak and trailing part of the pulse are less effected. A pulse shortening of a factor of two is readily obtained by a single passage. It should be noted that pulses of low peak intensity ($I_0 < 1 \times 10^9$ W/cm^2) are unable to bleach the absorber sufficiently. As a result such a pulse is strongly attenuated over the whole pulse duration without significant change of pulse shape.

The pulse shortening discussed in Fig. 1 was confirmed experimentally for a variety of input intensities and different values of T_0. The idea has been extended to multiple transits through an amplifier-absorber system. Fig. 2 presents a schematic of our setup.

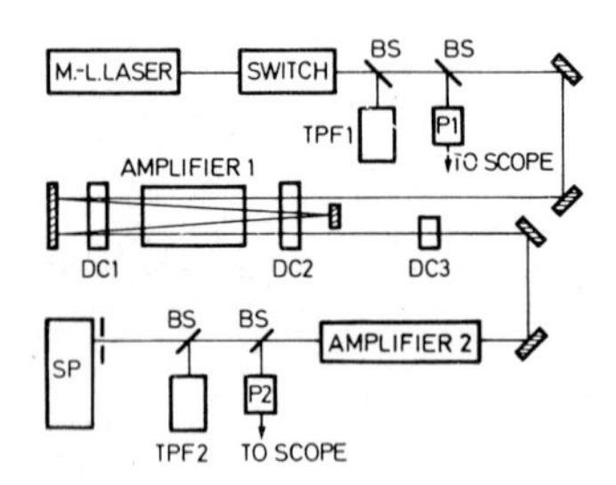

FIG. 2. Experimental setup for multiple transits through the absorber-amplifier system. Beam splitter BS; saturable absorber cells DC1 ($T_0 = 1.4 \times 10^{-2}$), DC2 ($T_0 = 1.4 \times 10^{-2}$), DC3 ($T_0 = 7 \times 10^{-3}$); photodetectors P1, P2; two-photon-fluorescence systems TPF1, TPF2; spectrometer SP.

After the pulse has passed one absorber, the peak intensity is amplified in order to bring the pulse to a power level for most effective pulse shortening in the subsequent absorber. With such a system we were able to shorten light pulses from 8 to 0.7 psec in five transits.

Solid state mode locked laser systems generate picosecond pulses at fixed frequencies. Nd-doped glass emitting at 1.06 μ has found most frequent application. For spectroscopic investigations tunable light sources are desired. As a first step in this direction, we have investigated three photon parametric fluorescence at high power levels in the infrared. This technique has subsequently been applied in the visible part of the spectrum.

In Fig. 3 a schematic of our experimental system is depicted.

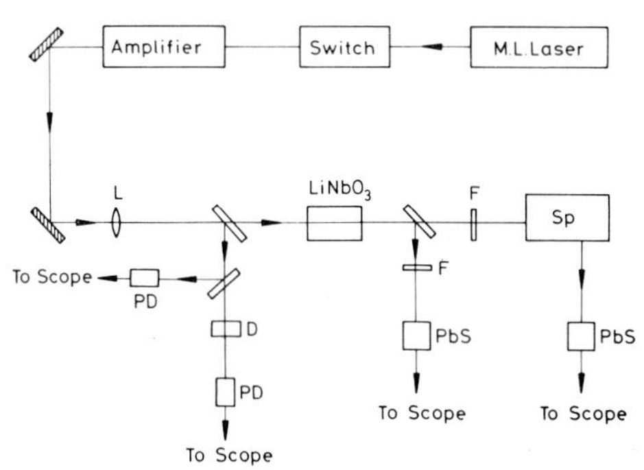

FIG. 3. Generation of tunable infrared pulses; generator crystal $LiNbO_3$, lens L, filter F, ir spectrometer Sp, ir detector PbS, nonlinear absorber D, fast photodiode PD.

A single picosecond pulse is switched out of the mode locked pulse train and amplified by double passage through an optically pumped Nd-glass rod. Signal and idler frequencies are generated when the intense pulse travels through a nonlinear $LiNbO_3$ crystal. The angle between the beam and the crystal axis determines the produced frequencies. Knowing the dielectric properties of $LiNbO_3$ it is possible to predict the generated frequencies. In Fig. 4 experimentally determined frequencies are compared with the calculated phase matching curve for $LiNbO_3$ at 293 K.

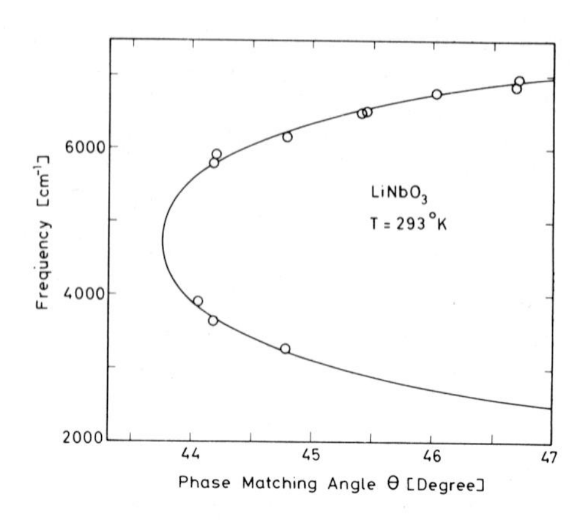

FIG. 4. Angular tuning curve of the picosecond parametric generator; experimental points and calculated curve.

The agreement between experimentally observed frequencies and the calculated curve is excellent. The frequency range extends from 2500 cm^{-1} to 7000 cm^{-1}. The lower frequency limit is determined by beginning infrared absorption of the $LiNbO_3$ crystal. In addition to the tuning range, the conversion efficiency of input power into signal power is of importance for practical applications. The signal intensity is plotted as a function of input peak intensity in Fig. 5.

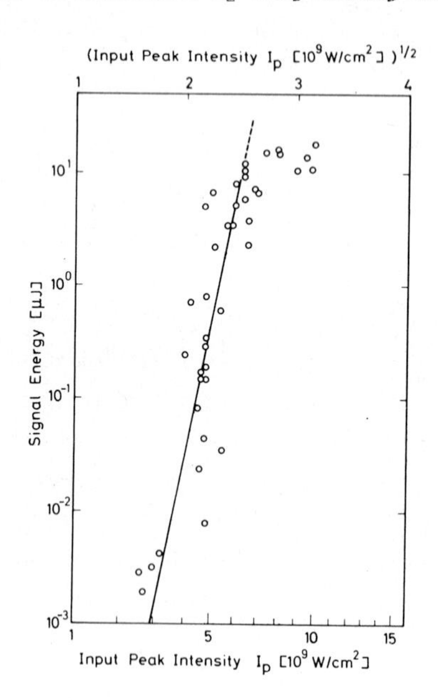

FIG. 5. Measured energy of signal pulses at 6500 cm^{-1} as a function of pump intensity I_p. Line calculated using experimental parameters.

It is readily seen from the Figure that the signal intensity rises first exponentially over several orders of ten and begins to saturate for very high input peak intensities. Power conversion efficiencies of several percent have been achieved. This high power conversion makes our travelling wave parametric amplifier system very attractive for the generation of picosecond pulses of tunable new frequencies. For an application see a subsequent paper by Laubereau et al.

We now turn to a third nonlinear process which allows the generation of very broad spectra extending from the infrared to the ultraviolet. In recent years broad frequency spectra of picosecond pulses have been observed by several authors and various explanations have been offered. For instance, self-phase modulation, stimulated Raman scattering and optical dielectric breakdown were considered to be responsible for the broad spectra. In most cases the experimental data are difficult to analyze because important pulse parameters (e.g. pulse duration and shape or input bandwidth) are not reported and most important, the question of self-focusing of the light beam was not investigated.

In the experiments discussed here we worked with well-defined input pulses (see above). We were certain that self-focusing and resulting beam distortion was not present in our experiments. A detailed analysis of our experimental data suggested that the broad frequency spectrum is generated by parametric four photon processes (e.g. in water).

The energy conversion per wave number is plotted versus input peak intensity for different frequencies in Fig. 6.

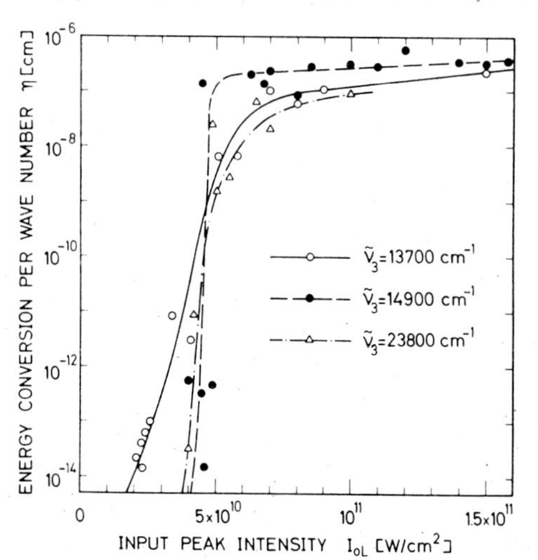

FIG. 6. Energy conversion η of laser emission ($\tilde{\nu}_L$ = 9455 cm^{-1}) into short-wavelength radiation at $\tilde{\nu}_3$. The curves below η = 10^{-9} are calculated.

The very rapid rise of the output signal over many orders of magnitude should be noted; it is characteristic of the nonlinear optical process discussed here. At higher input intensity a saturation of the output signal sets in. In Fig. 7 the energy conversion per wave number is plotted versus frequency for two input intensities.

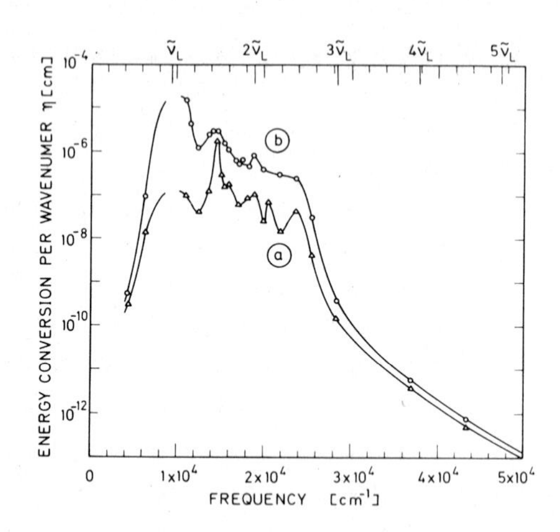

FIG. 7. Spectral distribution of parametric light in water (2 cm length) at a laser intensity of (a) 1×10^{11} W/cm^2 and (b) 5×10^{11} W/cm^2.

The spectrum extends from approximately 4000 cm^{-1} to 40,000 cm^{-1}. The energy conversion is strongest between ν_L and $2\nu_L$ where the primary four photon process occurs: $\nu_L + \nu_L \rightarrow \nu_3 + \nu_4$. Light at frequencies exceeding $2\nu_L$ is generated by frequency conversion $\nu_L + \nu_L + \nu'_3 \rightarrow \nu'_4$ and by higher order parametric four photon processes $\nu_L + \nu_3' \rightarrow \nu_3'' + \nu_4''$; here ν_3' represents the frequency of an electromagnetic wave which was generated in the primary or following parametric process. The frequency conversion process becomes less efficient at shorter wavelengths because of increased phase mismatch. Furthermore the efficiency of the parametric four photon process decreases with increasing order. The total energy conversion and the laser transmission are plotted as a function of input peak intensity in Fig. 8.

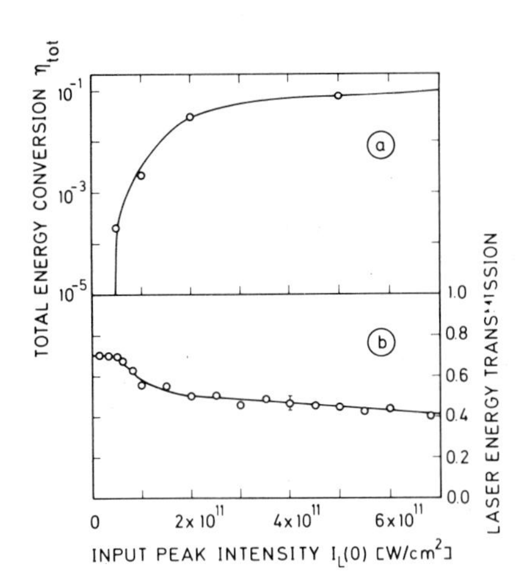

FIG. 8. a) Total energy conversion of laser light into parametric light. Single picosecond light pulse of $\Delta t_L \approx 6ps$, $\lambda L = 1.06\mu m$ in water (ℓ = 2cm).

A detailed analysis of the parametric process indicates that the absorption at the idler frequency in the infrared appears to be the major cause for the laser attenuation. The depletion of the laser intensity leads to the observed saturation of the conversion efficiency. The fairly smooth spectrum of Fig. 7 is of interest for dynamical absorption studies.

This report represents a brief summary of several publications. For details we refer the reader to the original papers:

A. Penzkofer, D. von der Linde, A. Laubereau and W. Kaiser
Appl. Phys. Letters 20, 351 (1972)

A. Laubereau, L. Greiter, and W. Kaiser
Appl. Phys. Letters 25, 87 (1974)

A. Penzkofer, A. Laubereau, and W. Kaiser
Phys. Rev. Letters 31, 867 (1973)

A. Penzkofer, A. Seilmeier, and W. Kaiser
Optics Communications (to be published)

DISCUSSION

W. KAISER, A. PENZKOFFER, A. LAUBEREAU

C. Varma - In general one expects that more than 90% of the total molecular polarizabilities is contributed by electrons. As a consequence the susceptibilities should increase towards the UV side. In connection with this, I wonder why the intensity of the four photon generated continuum drops in the blue part of the spectrum.

A. Laubereau - For our infrared laser pulse the electronic contribution to the non-linear susceptibility $\chi^{(3)}$ is quite small (for the substances I have discussed). The dominant contribution to the non-linear coupling term results from the infrared active vibrational transitions. $\chi^{(3)}$ is enhanced by single frequency resonances when the frequency of the idler component produced in the parametric interaction is close to the frequency of a vibrational mode of the molecules. On the other hand, the electronic transitions give rise to a small non-resonant part of the total non-linear susceptibility.

R.M. Hochstrasser - 1) What properties does the 1.06 pulse have after paramatric conversion process creating the infrared pulse in $LiNbO_3$? For example can the emergent 1.06 μ pulse be readily frequency doubled?
2) What is the angle between the emergent infrared and 1.06 pulses in your $LiNbO_3$ experiments?

A. Laubereau - 1) I think the 1.06 μm laser pulse could be used in the way you are suggesting. Your proposal does not appear to be advantageous however since the shape and peak intensity of the laser pulse may be affected by the parametric process in $LiNbO_3$, e.g. in the saturation region.
2) For normal incidence of laser pulse on the $LiNbO_3$ sample, the IR beam is emitted in a small cone (beam divergence $< 1°$ for our geometry) parallel to the axis of the laser beam.

M.W. Windsor - I note that your energy conversion efficiency is limited to about 10^{-6} per wave number. Does this imply an integrated conversion efficiency over the whole spectral range ($\geq$ 10 000cm^{-1}) of about 1%.
2) We observe lowest thresholds and best continuum intensities in the visible region for CCl_4; can you explain this?

A. Laubereau - 1) Indeed, the total energy conversion reachs 10% for the substances discussed and is limited by the loss connected with the infrared absorption of the idler components generated in the parametric interaction.
2) I am not aware of the absolute magnitude of the non-linear susceptibility $\chi^{(3)}$ of CCl_4. In this substance stimulated Raman scattering and possibly also self-focussing is important. The more complex situation for CCl_4 as compared with our data for H_2O should be noted; this makes an interpretation very difficult.

C. Reiss - In our experience, the continuum generated in water by a 1.06μ, 25 ps light pulse are not very reproductible, both in intensity and in the position of the various modes. Can you comment on this point?

A. Laubereau - In our experiments the observed spectral intensity distribution is reproducible. Obviously you are working under more complex experimental conditions, e.g. at considerably higher light intensities where dielectric breakdown may occur.

S. Leach - You eliminated the possible effects of dielectric breakdown by saying that your powers were an order of magnitude smaller than the power necessary for breakdown. Dielectric breakdown depend on whether the field is static or dynamic and also very much on the purity of the liquid. Under what conditions was the breakdown limit defined in your case? How does it depend on pulse length?

A. Laubereau - This point has been carefully checked by Dr. Penzhofer in our laboratory; he showed that dielectric breakdown occurs in the very same samples at a considerably higher intensity level of the incident laser pulse as compared with the parametric emission. The breakdown intensity level was measured under identical conditions, i.e. for the same laser pulses at 1.06 μm and for the identical samples we used in the parametric investigations. The dependence of the threshold intensity of dielectric breakdown on the pulse duration is expected to be fairly weak and has been discussed in the literature for solids by Bloembergen et al.

CHRONOGRAPHY IN MOLECULAR PHYSICS ON THE PICOSECOND TIME SCALE : TIME RESOLVED FLUORESCENCE, MOLECULAR ORIENTATIONAL RELAXATION TIMES IN LIQUIDS

by G. MOUROU and M.M. MALLEY
Ecole Polytechnique, Paris, France
San Diego University, Calif. U.S.A.

SUMMARY

We report a straightforward technique to resolve the kinetics of spontaneous fluorescence of dye solutions and to determine the molecular orientational decay of liquids.

This technique uses an ultra fast light shutter driven by 1.06μ picopulses and an optical multichannel analyser (O M A). Both the rise and the decay of the fluorescence emission are photographed on the OMA in one laser shot.

From the "grow in time" analysis of the selected wavelenght fluorescence vibrational or solvent orientational relaxation are readily deduced with 1 ps accuracy, while the direct observation of the fluorescence decay provides direct excited state lifetime measurements.

Molecular orientational decay of liquids are readily deduced by substituting the liquid under study for the CS2.

INTRODUCTION

Since the discovery of picosecond light pulses, a number of experiments have been carried out to analyse physical phenomena ranging in the picosecond time scale. Unfortunately in this time domaine few chronographic techniquesare available. As a matter of fact real time oscilloscope and crossed field photomultipliers are limited to several hundred picoseconds whereas streak camera have time resolutions which can reach 1 ps |1| but are very expensive and so far have a limited dynamic range. Most of the experiments were designed to deal with ultra fast phenomena by the optical sampling technique first introduced by Duguay and Hansen |2| by means of their optical Kerr shutter. However this time consuming technique requires a large number of laser shots and because of the poor laser reliability the results are often inaccurate.

In this paper, we will introduce a chronographic technique based on a optical multichannel analyser coupled with an optical Kerr cell. This method requires only one laser shot instead of several hundred as in the prior sampling experiment. Moreover this technique offers good time resolution ∿ 1 ps, wide dynamic 10^2, and high sensitivity. We applied it in molecular physics, especially to resolve the fluorescence of dye solutions, to determine vibrational decay times and finally to measure orientational decay times of pure liquids.

I MODE-LOCKED ACTINIC PULSE ANALYSIS. |3|

For simplicity we will first describe the technique when it is used to determine the time duration of mode-locked pulses. The set up shown in the Fig. 1 looks the same as that used by Duguay and Hansen to photograph pulses in flight |4|, except that instead of film, we make use of a linear array of 500 diodes, each 25μ wide, in an optical multichannel analyser. This device provides, sensitivity, linearity and eases considerably the photometric processing.

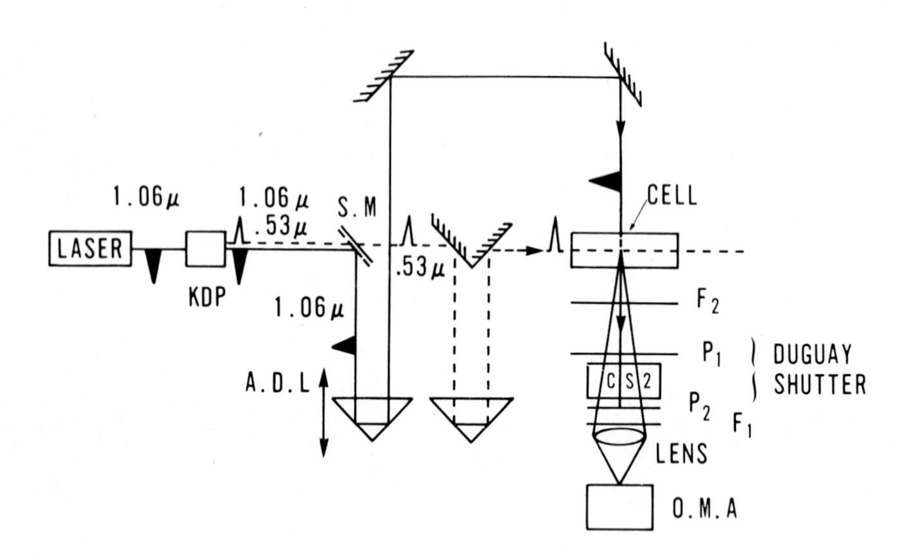

Fig. 1. Experimental apparatus : A D L adjustable delay line, S M semi dielectric mirror O M A optical multichannel analyser , P_1 P_2 crossed polarizers.

A mode-locked Nd : glass laser generates a train of about 30 pico-pulses of 10 - 15 ps duration, and each contains approximately 10^{-2} joule. The 530 nm pulses are obtained by harmonic generation in a potassium dihydrogen phosphate crystal. A dielectric mirror reflects the 1.06µ into an adjustable delay line and passes the 530 nm light through a fixed delay line and the diffusing cell filled with a weakly scattering medium (milk in water). The 1.06µ picopulses travel through a sample filled with CS2 located between two crossed polarizers. With the polarizers crossed the device has a spatial average transmission of 10^{-5}. When the linearly polarized 1.06µ picopulses travel through the CS2 cell, as is well known, they orient the anisotropic CS2 molecules. The partial orientation of the molecules in the liquids givesrise to a transitory birefringence which raises the transmission of the optical Kerr cell from 10^{-5} to 10^{-1}. The framing time of the optical gate involves the 1.06µ pulse duration and the orientational response of the CS2. It is roughly of 10 ps

Now, if we adjust the optical legs of the 1.06µ and the 530 nm in order to open the gate when the pulse is in the proper position in the cell, we briefly image by means of a lens located against the second polarizer the track of the 530 nm travelling through the diffusing cell along the 500 diodes of our detector. Then the electric signal of each diode is digitized and stored in a console to be displayed either on a monitor display oscilloscope or an X Y recorder.

A typical result appears on the Fig. 2. As it has been mentionned above , owing to the framing time of the shutter the signal is blurred due to the linear motion of the green pulse in water during the framing time. As a matter of fact the signal collected by the diodesis given by the convolution integral

$$I\ (t) = \int_{-\infty}^{+\infty} I_{530}\ (t')\ T\ (t-t')\ dt'$$

where I_{530} (t) is the intensity of the green signal and T(t) the transmission response of the optical Kerr shutter. In the case of CS2 where the orientational decay of the molecule is much shorter than the pulse duration we have

$$T\ (t)\ \alpha\ I^2_{1.06\mu}(t)\ \alpha\ I_{530nm}\ (t)$$

Therefore, it turns out that the experimental curve I(t) corresponds merely to the second order autocorrelation of the 530 nm intensity or the fourth order autocorrelation of the 1.06µ intensity. Despite of the relatively long framing time this technique can give valuable information overall extend of the 1.06μ pulse.

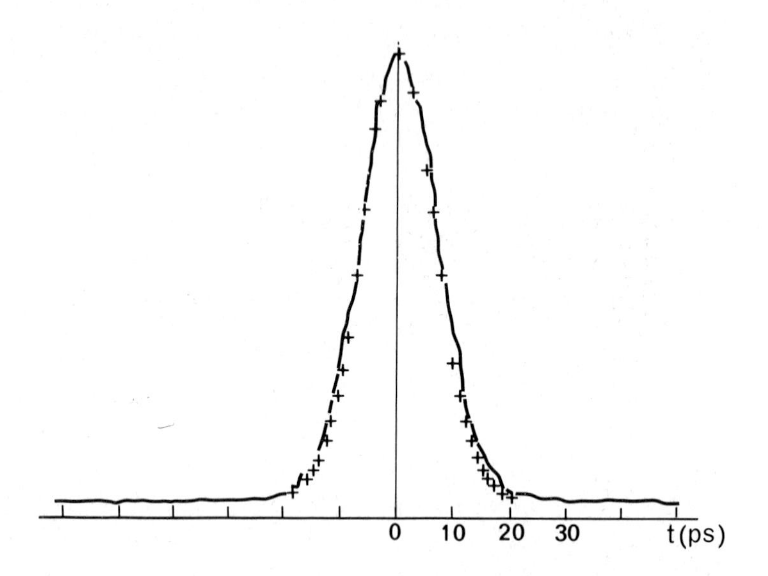

Fig. 2. Time resolved mode locked pulse.

The second order autocorrelation of the 530 nm intensity has been calculated numerically for an assumed gaussian-shaped 1.06μ pulse. We merely note the excellent agreement between the experimental and calculated curves, which confirms that the assumption of a gaussian-shaped pulse holds very well. By means of this technique the pulse width of the 1.06μ has been deduced. We found that those pulses lie between 12-16 ps.

II TIME RESOLVED FLUORESCENCE. |5|

We made use of the same basic method to resolve on a picosecond time scale subnanosecond fluorescence of solutions (Fig. 3). This was readily accomplished by substituting the dye solutions for the diffusing medium. The 530 nm pulses, then leave behind excited molecules. By means of an adjustable delay line we open the optical shutter at the appropriate time,

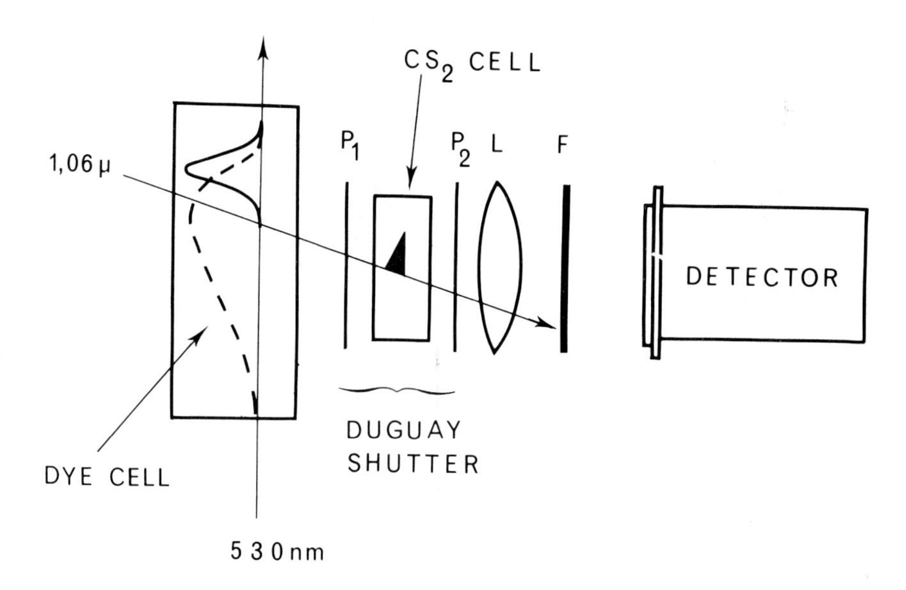

Fig. 3. Experimental set up used to photograph in one stroke the fluorescence track induced by the 530 nm pulses travelling in the solution cell.

when the exciting pulse is in the cell and briefly photograph the rising and the falling part of the excited molecule distribution. From this fluorescence profile we can readily deduce the fluorescence decay of the molecules studied. We have to point out that the observed intensity are corrected for the small (.5 db/cm) attenuation of the pulse since the high sensitivity of our detector enables us to work with very dilute solutions.

Furthermore, it is worth noting that the method is very convenient for comparing the exciting pulse and the fluorescence emission since no jitter affects the technique. The figure 4 shows simultaneously the exciting pulse and the corresponding excited molecule distribution for erythrosin B in water . This has been obtained by means of two shots, one in the diffusing cell and the second in the dye solution under study.As expected the fluorescence rise looks like the integral of the exciting pulse the half maximum intensity corresponding to the maximum of the 530 nm intensity. Lifetimes of 57 ps ± 6 ps and 140 ps ± 10 ps for erythrosin in water and in methanol have been readily deduced by this technique.

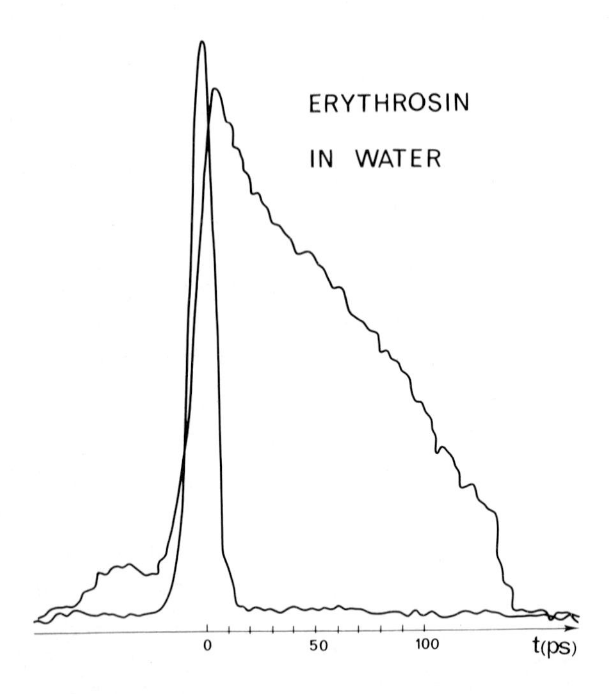

Fig. 4. Time resolved fluorescence of the erythrosin in water

III VIBRATIONAL DECAY IN THE FIRST EXCITED VIBRONIC MANIFOLD. |3|

We have mentionned above that the technique enables us to determine accurately the time delay between the excitation and the emission. This time corresponds to a direct measurement of the radiationless relaxation in the vibronic manifold of the first excited state. It may be useful to emphasize this point. The spectral shift between absorption and fluorescence spectra schematically indicated in the figure 5 arises from the change in the configuration of the excited molecules, for example rearrangement of nuclear coordinates, rotation of substituant groups, reordering of solvent shell about the excited molecules. These inter and intra-molecular relaxation processes are driven by the change in the π electron distribution caused by the electronic transition. If τ_1 is the time for the molecule to reach the stable excited state after the electronic transition it can be shown that τ_1 may be represented by the delay between the maximum of the

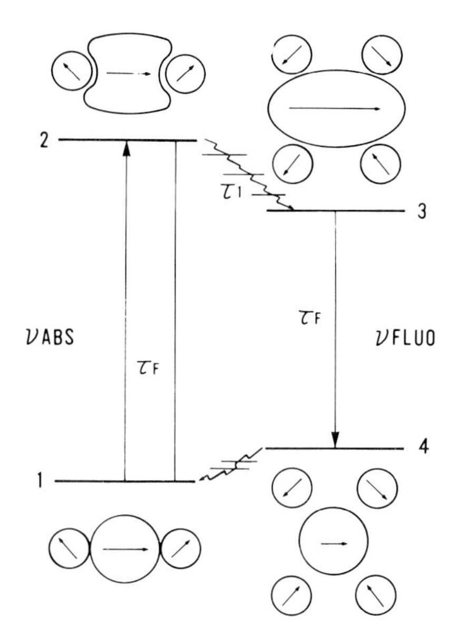

Fig. 5. Diagram of the four-level of a dye molecule in solution. The equilibrated ground state, 1, is surrounded by a solvent cage. Upon transition to Franck-Condon excited state, 2, the nuclear configuration of the dye and the solvent cage remains stationary. Resonance fluorescence from the Franck-Condon state is shown. Equilibration in the excited state of the dye molecule and the surrounding solvent is obtained in level3. The major portion of the fluorescence takes place between level 3 and 4, although emission also occurs continuously from the intermediate level.

actinic pulse and the half intensity position of the fluorescence rise. To determine τ_1, the set up of the figure 1 is modified slightly to magnify just the rising part of the fluorescence track. Furthermore an interference filtered centered on the maximum of the fluorescence spectrum is set on the front of the O M A.

The figure 6 shows the appearance of the stokes-shifted emission from rhodamine 6 G in water and in ethanol and for the 530 nm actinic pulse. The dotted line and crossed are the calculated curves using proper convolution integrals on account of the light propagation pulse width and the optical Kerr shutter response. The break in the fluorescence corresponds to the front edge of the cell. The best fit to the data was obtained for a decay time τ_1 equal to zero. A calculation for a decay time τ_1 of 5 ps is shown for comparison. Those results imply that the delay between excitation and the stokes-shifted emission is less than one ps which give support to a vibrational decay in the first excited manifold shorter than one ps for those dye solutions.

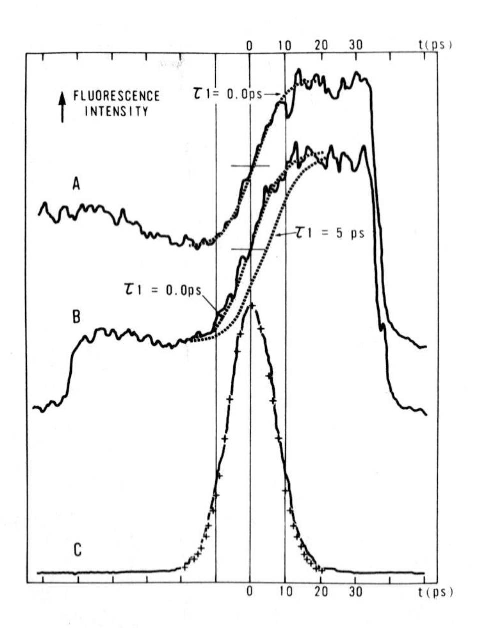

Fig. 6. The appearance of the Stokes-shifted emission from rhodamin 6 G in water A, and in ethanol B, and the .53μ actinic pulse C, are shown above. A calculation for a delay time of five picoseconds in shown for comparison.

IV ORIENTATIONAL DECAY IN PURE LIQUIDS |6|

In the first part of this paper we stated that orientational effects are largely responsible for the birefringence induced by the strong polarized 1.06μ light and the aperture time of the gate was controlled by the 1.06μ pulse width but also by the orientational response making up the liquid considered. So the time resolved transmission of the Duguay cell provides direct measurements of the molecular orientational decay. On a practical level the cell located between the two crossed ploarizers is filled with the liquid under study. The scanning in time is performed needless to say at the speed of light by the green pulses in the diffusing cell. Because the light is collected along the 500 diodes of our optical multichannel analyser we have in one shot the transmission versus time of the Duguay cell, which enables us by means of the convolution integralsof the chapter I to come up to the orientational decay of the liquid considered.

In the figure 7, the transmission of the Duguay shutter in time is scaled up for nitrobenzene and CS2. We can see the striking difference of the two experimental curves. If we consider the CS2 curve as the prompt

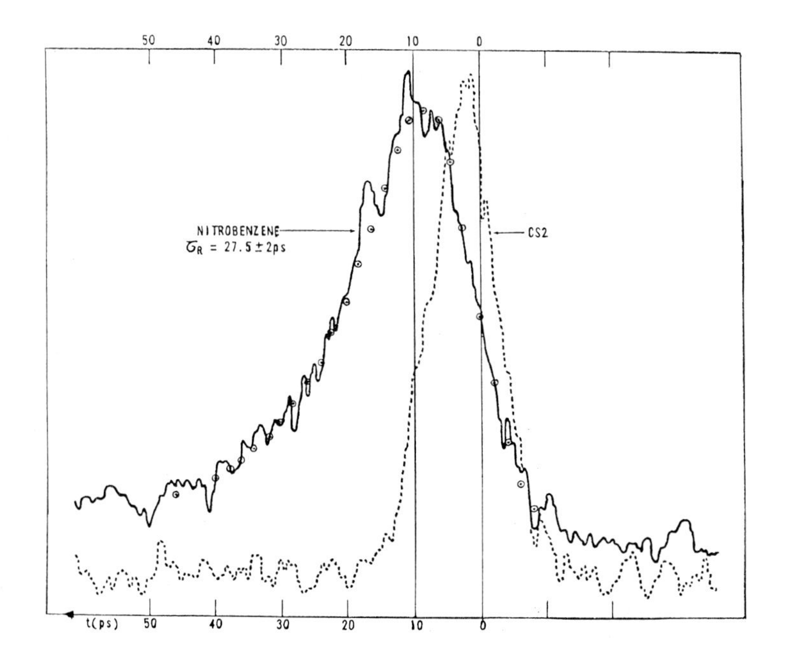

Fig. 7. Transmission of the Duguay shutter in time for nitrobenzene and CS2

response of our system which is not completely true owing to the orientational decay of the CS2 (∿ 2ps) |7|, the transmission peak of the nitrobenzene is shifted toward larger delay and an exponential tail clearly displays the effect of orientational relaxation. We would like to point out that because τ_R for nitrobenzene is larger than the pulse width of the 1.06μ represented by the CS2 response, the Duguay cell trasmission looks like the integral of the 1.06μ pulse. Using the appropriate convolution integral, we have numerically fitted the experimental curve to obtain the orientational decay of the nitrobenzene $\tau_R = 27.5 \pm 2$ ps.

Orientational decay has been determined also for time smaller than the 1.06μ pulse width. Fo instance, the figure 8 displays time-resolved transmission curves obtained for bromobenzene, toluene, benzene. It is readily seen that the maximum shifts as the orientational decay increases. The time value have been numerically computed for each liquid and appears beside each curve.

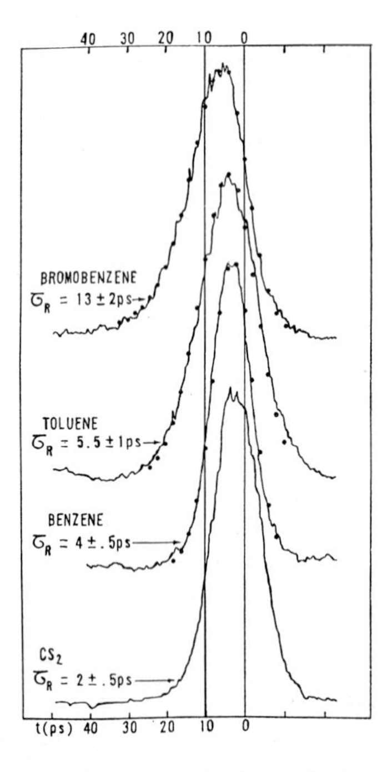

Fig.8. Transmission of the Duguay shutter in time for bromobenzene, toluene, benzene.

REFERENCES

1. D.J. Bradley, Laser Focus , (1975) 28.
2. M.A. Duguay and J.W. Hansen, Opt. Com. 1, (1969), 254.
3. G. Mourou and M.M. Malley, Chem. Phys. Lett. 16, (1972), 617.
4. M.A. Duguay and J.W. Hansen, IEEE. Quant. Electron. QE-7,(1971), 37.
5. G. Mourou and M.M. Malley, Opt. Com. 11, (1974), 282.
6. G. Mourou and M.M. Malley, Opt. Com. to be published.
7. S.H. Shapiro and H.P. Broida, Phys. Rev. 154, (1967), 129.

DISCUSSION

G. MOUROU, M.M. MALLEY

E.P. Ippen - In a previous publication you observed the temporal change of the fluorescence spectrum of Rh 6G following picosecond excitation. Why does that change not manifest itself in these experiments?

G. Mourou - This is our jigsaw puzzle. By means of time resolved spectroscopy we were able to see a shift and a broadening of the fluorescence spectrum. Unfortunately these results do not seem in good agreement with the result obtained by the study of the Stokes shifted fluorescence rise.

B. Nickel - The fluorescence spectrum of large molecules does not sensitively depend on the vibrational excess energy. The major changes with increasing excess energy are 1) a relative increase of hot-band emission, 2) a broadening and a small red-shift of the vibrational bands. It is expected that the intensity of the fluorescence in the maximum of the fluorescence spectrum is rather insensitive to vibrational relaxation. Therefore the fact that no delay of the fluorescence was found, does not necessarily imply that the vibrational relaxation is an extremely fast process.

G. Mourou - The measurement has been performed also, not just in the center of the fluorescence maximum but too on the 0-1 fluorescence band and we found $\tau_1 \approx 0$. However in this case the signal to noise ratio was small.

D. Lavalette - In your time delay measurements you are observing the 530 nm pulse and the fluorescence at a different wavelength. Is there any need for a group velocity correction or is this correction negligible?

G. Mourou - The group velocity difference between 530 nm and fluorescence intensity is very small and may be neglected.

J. Langelaar - Do you expect that non-linear effects, due to the high power excitation pulse, will affect the spontaneous fluorescence time behaviour?

G. Mourou - We work with weak 530 nm intensity, in order to stay always in the domain where the fluorescence intensity is proportional to the excitation intensity.

PICOSECOND FLASH PHOTOLYSIS : 3,3' - DIETHYLOXADICARBOCYANINE IODIDE (DODCI)

J.C. MIALOCQ, J. JARAUDIAS, A.W. BOYD*, J. SUTTON
Département Recherche et Analyse, CEN Saclay, GIF-sur-Yvette 91190 (France)

ABSTRACT

The excitation of DODCI in ethanol solution by a train of picosecond pulses emitted by a mode locked dye laser at 605nm leads to the formation of a photoisomer. Absorption spectroscopy measurements made over the entire visible region enable the steady state concentration of each species to be determined and fluorescence measurements made at 630nm show the existence of two decay processes, the lifetime of the faster, τ = (420 ± 60)ps, being assigned to the relaxation of the excited state of the DODCI photoisomer.

INTRODUCTION

3,3' - diethyloxadicarbocyanine iodide, DODCI, is employed [1,2] for mode locking rhodamine 6G or rhodamine B dye lasers emitting between 585 and 645nm. Since DODCI does not absorb above 625nm the mode locking action between 625 and 645nm is attributed [3,4,5] to the presence of a photoisomer of which the absorption maximum lies at 620nm [5]. New [6] has also drawn attention to the fact that these lasers furnish pulses as short as 2.5 ps even though no relaxation time of DODCI lower than 250 ps has ever been observed. In the meantime Bush, Jones and Rentzepis [7] have studied the time dependence of the photobleaching and recovery of DODCI using 530nm single pulse excitation and have given a recovery time for the absorption of about 10 ps; they did not find photoisomer formation. Arthurs, Bradley and Roddie [3,4] who excited DODCI with the complete train of laser pulses (each pulse having 50-100μJ of energy), observed the formation of the photoisomer, measured

* On leave of absence from Physical Chemistry Branch, Atomic Energy of Canada Limited, Chalk River, Ontario, Canada, 1974-5

the relaxation time of bleaching of the fundamental state and the fluorescence lifetimes of the excited states. According to these authors the two excited isomers have the same lifetime, τ = (330 ± 40) ps. Using single pulse excitation (2-3 mJ at 530nm) Magde and Windsor [8] have measured the fluorescence lifetime of the excited state, its absorption relaxation time ($S_n \leftarrow S_1$) and also the recovery relaxation time of the fundamental state and found τ = (1.15 ± 0.15)ns. More recently Shank and Ippen [9], using weak pulses (a few nJ), found τ = (1.2 ± 0.1) ns.

These authors having worked under very different conditions it seemed necessary to us to seek an explanation of the disparity amongst the results by attempting to measure the individual fluorescence lifetimes of the two photoisomers. Two approaches were opened to us, namely, to study the fluorescence at different wavelengths in order to separate the contributions of the two isomers, using excitation by the total pulse train from the laser (which leads to a partial transformation to the photoisomer and hence the possibility of exciting both forms simultaneously and measuring their fluorescence decay) or to study the emission provoked by a single pulse which enables the fluorescence of the normal form of DODCI to be measured.

Experimental method

The DODCI supplied by Eastman Kodak was used without further purification. The solvent, ethanol, UVASOL 95% for spectroscopy, was also used as received from the supplier, Merck. The concentration of DODCI in ethanolic solution was between 1 and 4 x 10^{-4}M in the 1mm thick cell C; this value was chosen in order that the greater part of the exciting light would be absorbed. The solution was excited by a beam of 1mm diameter from an Electrophotonics model SUA-10, rhodamine 6G dye laser (DL). Mode locking was achieved using a 2mm thick cell containing a 3 x10^{-5}M ethanolic solution of DODCI. The emission wavelength was 605nm, the pulse train duration was about 2μs and the cavity transit time, 3.6 ns. The total energy of the pulse train measured near the laser exit window with two different microcalorimeters (Cilas BT 16 and Hadron 108) was 60 mJ.

Figure 1 shows the apparatus used for the fluorescence studies.

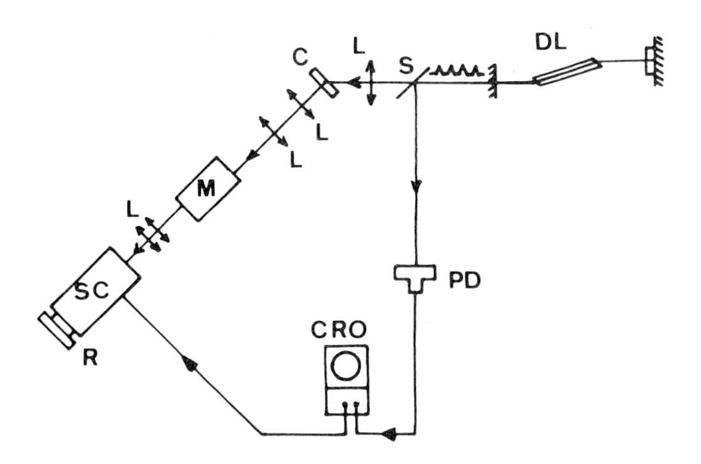

Fig. 1 : Diagram of the apparatus used for the fluorescence study.
DL : dye laser, S : splitter, PD : photodiode, CRO : cathode ray oscilloscope, L : lens, C : cell, M : monochromator, SC : streak camera, R : 35mm film camera.

The fluorescence decays were examined after passing the emitted light through a Bausch and Lomb 25cm grating monochromator. The streak camera (Electrophotonics model ICC 512) (SC) was triggered using a fast photodiode (PD) in series with a Ferisol oscilloscope (Model OZ 100C) (CRO) which supplied the necessary voltage pulse and also enabled the simultaneous observation of the train of pulses from the laser. The sweep speed of the camera was selected between 2 and 10ns per cm, the light signal on the output screen being photographed on Kodak 2475 recording film (1000 ASA). The optical densities, d, of the recorded signals were measured by microdensitometry. For values of d, between 0.4 and 1.5 the relationship (I) was obeyed :

$$d = \gamma \log_{10} E \qquad (I)$$

where E is the exposure.

Outside these limits the optical densities were only qualitative. The contrast γ is a function only of the film and the development conditions. It was determined to better than 10% accuracy using neutral filters of known optical density. The exposure E is proportional to the fluorescence intensity, that is to $d[S_1]/dt$ where $[S_1]$ is the concentration of the excited state of the compound being observed, so that :

$$-\frac{d[S_1]}{dt} = k[S_1] = k[S_1]_o e^{-kt} \qquad \text{(II)}$$

where $k = \tau^{-1}$, τ being the lifetime of the excited state. Combining (I) + (II) one obtains (III) :

$$d = -\frac{\gamma \cdot t}{2.3 \cdot \tau_F} + \text{constant} \qquad \text{(III)}$$

From the slope, p, of the microdensitometric recordings (Figure 2) the value of the fluorescence lifetime is obtained using

$$\tau_F = \frac{\gamma}{2.3 \cdot P} \qquad \text{(IV)}$$

Under our experimental conditions $\gamma = (0.85 \pm 0.15)$.

Figure 3 shows the schema of the apparatus devised by Clerc et al [10], which was used with little modification for absorption measurements. The analysing light beam from a xenon arc (XBO 450 W) is focused down to 1mm within the volume of liquid irradiated by the laser train; after leaving the cell the analysing beam is focused onto the entrance slit of the monochromator. The monochromatic beam is then focused onto the entrance slit of the streak camera and the streak images are photographed on polaroid film (speed 3000 ASA). The time between the start of the laser and the triggering of the camera could be varied by means of a delay unit.

Experimental results

Fluorescence

When excited by the laser pulse train the DODCI solution

fluoresced at wavelengths above 585nm. Figures 4.1 and 4.2, obtained with a sweep speed of 10 ns. cm^{-1}, show the scattered laser pulses at 605nm and the fluorescence signal at 632nm. The microdensitogram of figure 2 obtained at 630nm shows two linear sections corresponding to two separate decay kinetics. While it is possible to determine a mean relaxation

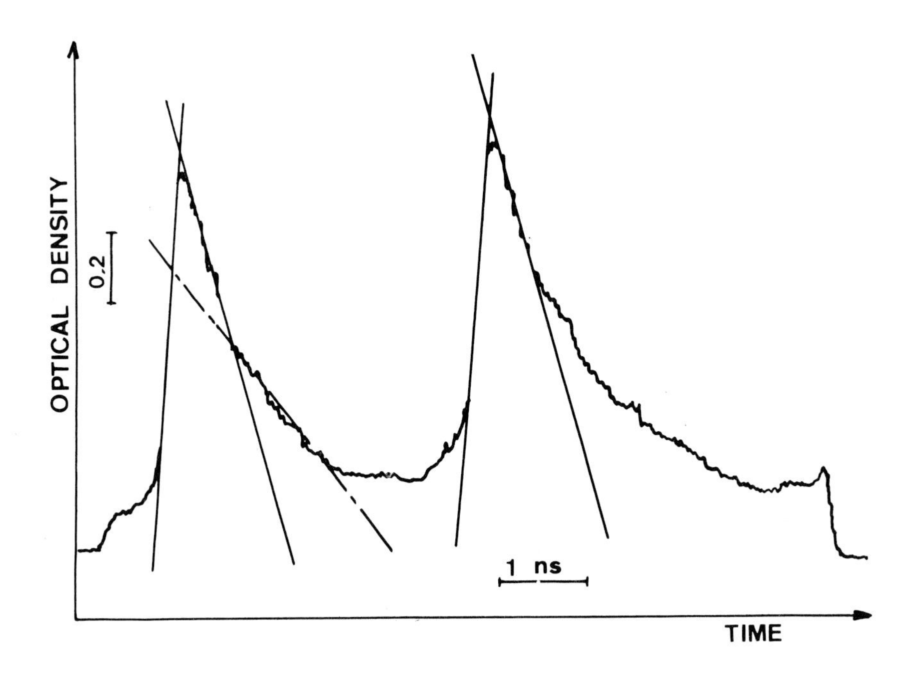

Fig. 2 : Microdensitometer recording of the fluorescence of DODCI in ethanol.

time for the fast fluorescence decay from the results of several sweeps, a kinetic treatment of the second decay is illusory. The duration of the laser pulses is less than 20 ps whereas the risetime of the fluorescence is about 100 ps. This value results from a combination of at least three factors, the pulse duration, the camera slit width, and the streak camera image granularity. The intervention of this last factor has been previously indicated by Arthurs et al [4]. It is necessary to eliminate the instrumental response time τ_I by

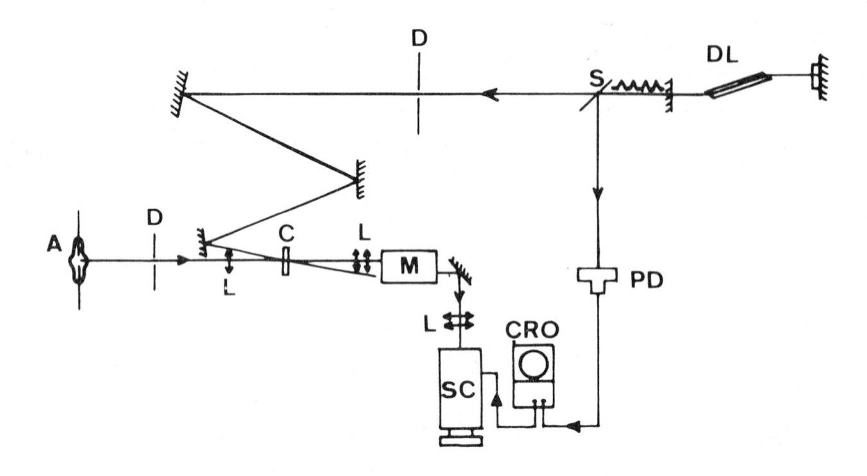

Fig. 3 : Diagram of the apparatus used for the absorption study : DL : dye laser; S : splitter; PD : photodiode; CRO : cathode ray oscilloscope; SC : streak camera; D : diaphragm; A : Xenon arc; L : lens; C : absorption cell; M : monochromator.

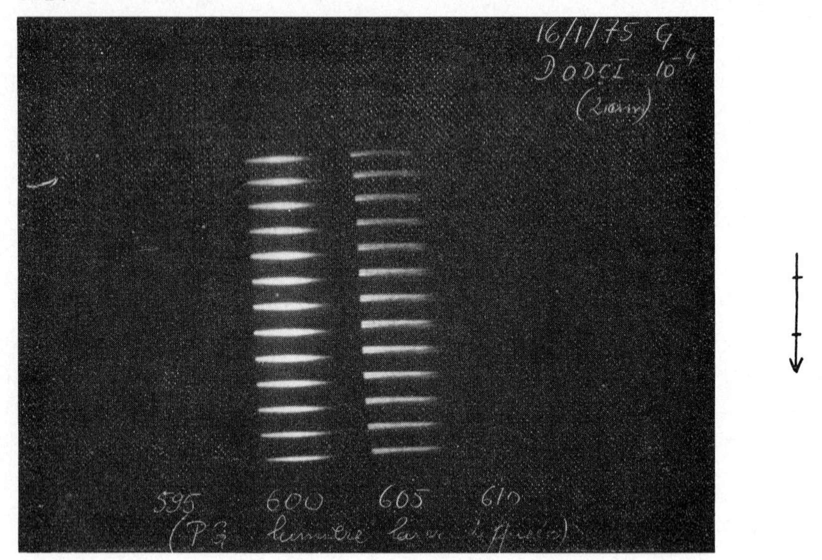

Fig. 4.1 : Laser pulses (speed : 10 ns. cm^{-1})

deconvolution using the relationship (V) :

$$\tau_F = \tau_f + \tau_I \qquad (V)$$

where τ_F is the relaxation time given by (IV), and τ_f the true relaxation time. The value of τ_f obtained from several readings is $\tau_f = (420 \pm 60)$ ps, the absolute error on the measurement being the sum of the graphical error and that associated with the determination of γ.

At 590 nm and 650 nm the fluorescence intensities are too low to enable the decay kinetics to be determined.

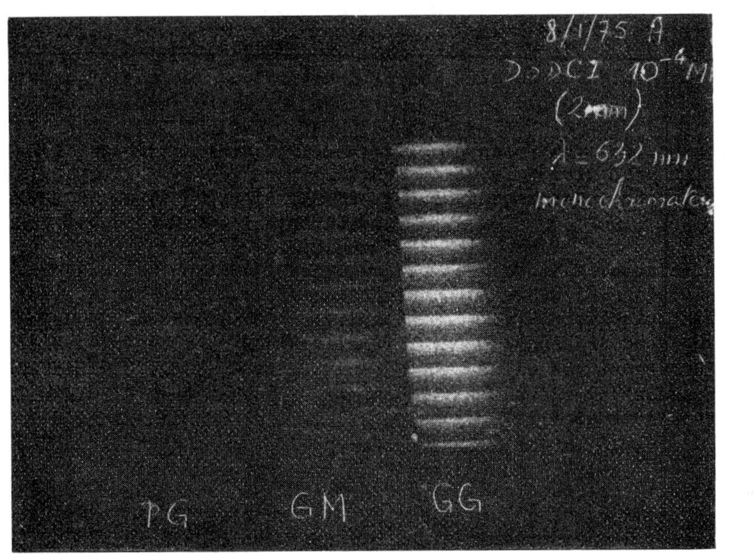

Fig. 4.2 : Fluorescence traces from 10^{-4} M DODCI recorded at 632 nm (speed : 10 ns. cm^{-1})

Absorption

The absorption spectrum of DODCI extends from 480 to 620nm [5] with a peak at 580 nm where the extinction coefficient has the value $\varepsilon_{580nm} = 2.2 \times 10^5\ M^{-1}\ cm^{-1}$. In these experiments a diminution of the absorption of the normal form of DODCI and the appearance of a new absorption band near 620nm were remarked. This latter result was expected : it corresponds to the already established absorption spectrum [5] of the photoisomer. Figure 5 shows two sweeps (speed 3.8ns.cm^{-1})

recorded with the camera at 565nm using a 2×10^{-4}M ethanolic solution of DODCI. The left hand image a, was produced by the analysing light traversing the solution without laser beam excitation and the image b was obtained in the same way but about 250ns after the start of laser excitation. The transmission of the solution is increased by the reduction in concentration of the normal isomer.

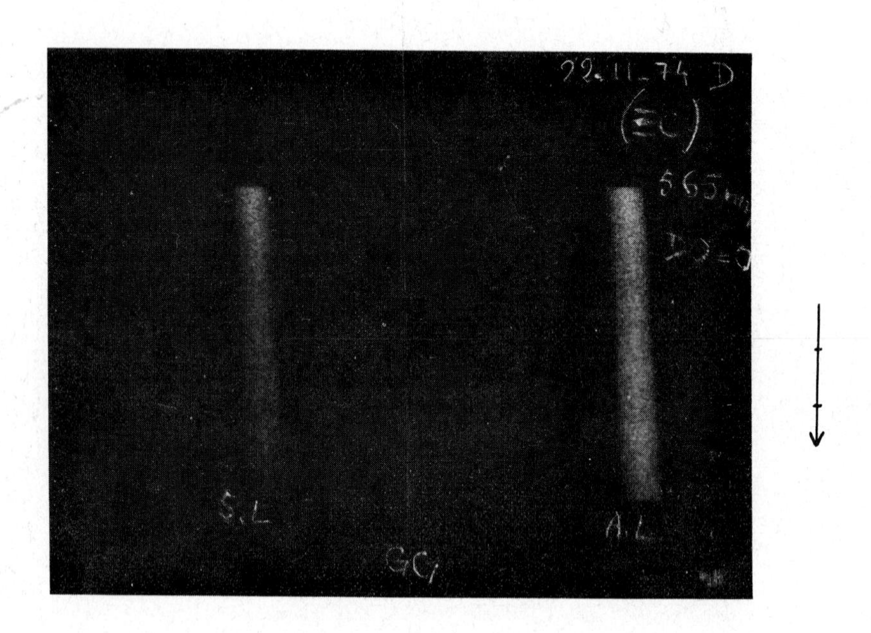

a b

Fig. 5 : Absorption study with analysing light of 565nm : a. without laser excitation; b. 250 ns after start of laser excitation (speed : 3.8 ns. cm^{-1})

Data obtained in the same way at 620nm are shown in figure 6.1 but here the transmission of the solution decreases because of the absorption of the photoisomer. The change in optical density measured at 250ns and 1.5 μs after the start of the pulse train is constant : the photoisomer concentration thus attains a stationary value during the excitation period. The optical density of its absorption, obtained by comparison with a neutral attenuator, is less than 0.5 which enables its upper concentration limit to be calculated equal to 3 x 10^{-5}M taking a value of the molar extinction coefficient

ε_{620nm} = 1.84 x 10^5 M^{-1} cm^{-1} [5].

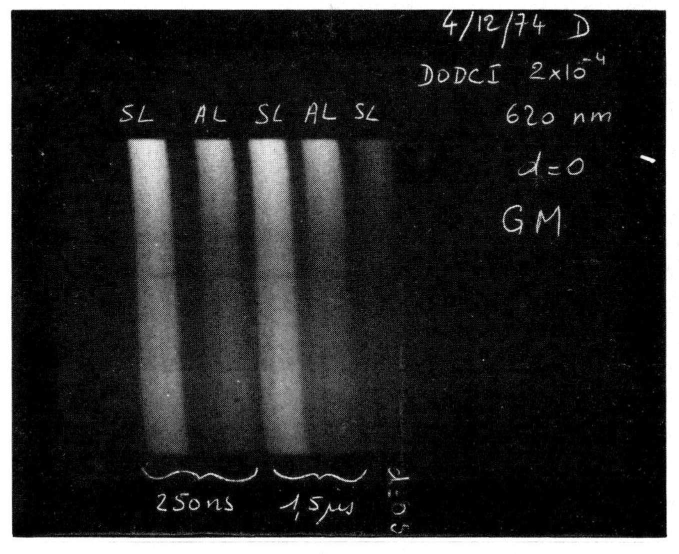

a b c d e

Fig. 6.1 : Absorption study with analysing light of 620nm :

a, c and e . without laser excitation

b. 250 ns after start of laser excitation

d. 1.5 μs after start of laser excitation

e. same as a with filter of density 0.5 interposed (speed : 3.8 ns. cm^{-1})

Figure 6.2 shows, in order, recordings of the analysing light, the scattered light from the laser pulses and the analysing light in presence of laser excitation 250ns after the start of the pulse train, at 605nm. The last image shows the attenuation of the transmission after the laser pulses resulting from the superposition of a light gain due to the partial disappearance of the normal DODCI isomer (N) and a slightly greater loss due to the absorption of the photoisomer (P).

We were not able to measure the variations in optical density after each pulse : they are probably less than 0.1.The following table summarises the data concerning the concentrations of the two isomers. d_N and d_P represent their individual optical densities at 605nm and d_{TOT}, their sum. As figure 6.2

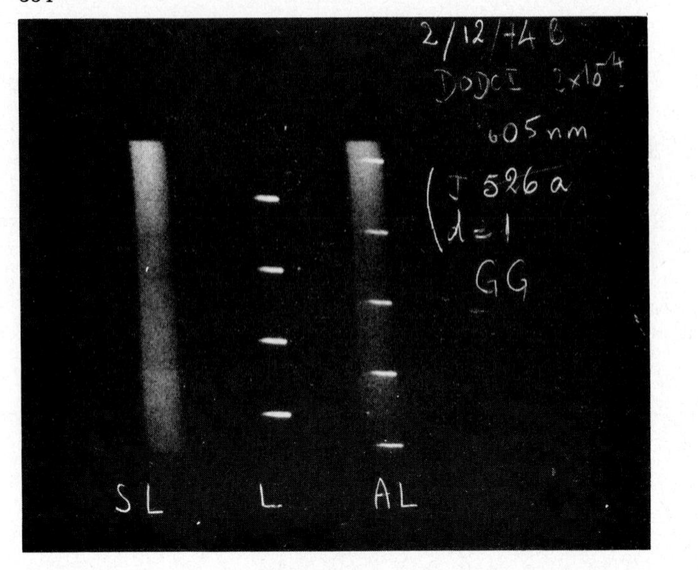

a b c

Fig. 6.2 : Absorption study with analysing light at 605nm :
a. without laser excitation; b. laser pulses only;
c. 250ns after start of laser excitation
(speed : 3.8 ns. cm^{-1})

shows, the transmission observed 250ns after the beginning of the pulse train is less than the initial value. One also sees from the tabulated results that the normal isomer absorbs twice as much light as the photoisomer under the stationary conditions. The fluorescence resulting from the 605nm excitation thus comes from both isomers.

Table

Time (ns)	[DODCI(N)] (M)	[DODCI (P)] (M)	ε_{605}(N) (M^{-1} cm^{-1})	ε_{605}(P) (M^{-1} cm^{-1})	d_N	d_P	d_{TOT}
0	$2x10^{-4}$	0	$4x10^4$	$1,2x10^5$	0,8	0	0,8
250	$\sim 1,7x10^{-4}$	$\sim 3x10^{-5}$			$\sim$0,68	$\sim$0,36	$\sim$1,04

Discussion

We have observed the variation of transmission of a solution of DODCI produced during excitation at 605nm by a laser pulse train and shown the production of the photoisomer absorbing at 620nm, the lifetime of this latter being 1.3 ms [5]. Estimates of the concentrations of the two isomers during the pulse train show that both contribute to the absorption and that the fluorescence must derive from the relaxation of excited states of both isomers.

The relaxation time measured at 630 nm, τ = (420 ± 60) ps, represents an upper limit since the contribution of the slower process is not negligible. This value is close to that found by Arthurs et al [4], τ = (330 ± 40) ps. Though these authors consider that the excited states of both isomers have the same lifetimes, we think that they have measured the lifetime of the excited photoisomer. Our data show the existence of two kinetics, the more rapid corresponding to the photoisomer and the slower, we believe, to the normal species.

Using a single 10ps pulse at 530nm Magde and Windsor [8] were able to produce a certain concentration of photoisomer but were unable to excite it : they have then measured the relaxation time of the normal DODCI, finding τ = (1.15 ± 0.15) ns. Shank and Ippen [11] have confirmed this value, τ = (1.2 ± 0.1) ns. The relative positions of the absorption spectra of the two isomers enable us to conclude that at 590nm the fluorescence is due to the normal form of DODCI : it is an anti-Stokes emission since excitation is at 605nm. This may be explained in terms of excitation from vibrational states other than v = 0. A variation of 20nm around 600nm corresponds to an energy of 1.6 $Kcal.mole^{-1}$ and the calculation of the Boltzman distribution indicates that 7% of the DODCI molecules will be in a suitable initially excited vibrational state. Even though the anti-stokes fluorescence of the normal form thus appears possible at 585nm, that of the photoisomer appear much less probable. On the long wavelength side of the exciting frequency the emissions of the two isomers are more difficult to differentiate.

If our conclusion is exact, taking into account the hypothesis of photoisomer formation and the time constant determined by Arthurs et al [4], all the data in the literature lie within the framework of the model predicted by Dempster et al [5] with the single exception of the relaxation time of 10 ps found by Busch et al [7]. Dempster et al [5] found, in effect that using the Forster expression the radiative lifetime of the excited state of the normal isomer was τ_r = (2.54 ± 0.25) ns and hence taking a value of ϕ = (0.49 ± 0.02) for the fluorescence quantum yield, $\tau = (1.25 \pm 0.15)$ ns. They also found a value of 300 ps for the relaxation time of the excited state of the photoisomer.

We are at present carrying out a single pulse experiment to determine the lifetime of the excited state of the normal DODCI isomer.

REFERENCES

1 W. Schmidt, F.P. Schäfer, Phys. Lett. V26 A, N 11, (April 1968), 558

2 E.G. Arthurs, D.J. Bradley, A.G. Roddie, Appl. Phys. Lett V 20, N 3, 1972, 125

3 E.G. Arthurs, D.J. Bradley, A.G. Roddie, Opt. Comm V8, N 2, 1973, 118

4 E.G. Arthurs, D.J. Bradley, A.G. Roddie, Chem. Phys. Lett V 22, N 2, 1973, 230

5 D.N. Dempster, T. Morrow, R. Rankin, G.F. Thompson, J. Chem. Soc, Faraday II, 68, 1972, 1479

6 G.H.C. New, Opt. Comm V6, N 2, 1972, 188

7 G.E. Busch, R.P. Jones, P.M. Rentzepis, Chem. Phys. Lett V 18, N 2, 1973, 178

8 Douglas Magde, Maurice W. Windsor, Chem. Phys. Lett V27, N 1, 1974, 31

9 C.V. Shank, E.P. Ippen, Appl. Phys. Lett V 26, N 2, 1975,62

10 M. Clerc, P. Goujon, J. Sutton, Rapport DRA/SRIRMa (CEN-Saclay) 74/627/JS/JD (12/3/1974)

11 I.H. Munro, I.A. Ramsay, J of Scient. Instr (Journal of Physics E) 1968, Series 2, volume 1, p 147

12 J.G. Calvert, J.N. Pitts, Jr, Photochemistry, John Wiley and Sons, N-Y, 1966.

DISCUSSION

J.C. MIALOCQ, J. JARAUDIAS, A.W. BOYD, J. SUTTON

C. Shank - Have you considered the effects of stimulated emission on your measurement? Could stimulated emission be responsible for the initial rapid decay of 0.4 ns followed by a 1-2 ns decay when the population inversion disappears.

J.C. Mialocq - The initial decay $\tau = (420 \pm 60)$ ps followed by a slower decay cannot be considered as an effect of stimulated emission. We tell that the slower decay is due to the normal DODCI fluorescence. As we have shown, the lifetime obtained when exciting the solution with a single pulse (this result will be published later by J.C.Mialocq, A.W. Boyd, J. Jaraudias and J. Sutton) is (1.2 ± 0.3) ns. As we measured the fluorescence lifetimes by an identical method (using a streak camera and negative film) with both a single pulse (where we obtained only the normal DODCI fluorescence lifetime) and a train of pulses, it would have been possible to see stimulated emission on the decay of the single pulse experiment.

M.W. Windsor - Your abstract leaves the impression that your lifetime measurement for DODCI and ours disagree. However, from your talk I gather that we are now in agreement. All three groups (Mialocq et al. Ippen and Shank, Magde and Windsor (1) agree that the lifetime of the excited state of the parent DODCI molecule is about 1.2 ns. Your value of 420 ± 60 ps refers to the excited state of the photo-isomer rather than that of parent DODCI and shows fair agreement with the 300 ps value reported by Arthurs et al. Chem.Phys.Letter 22(1973)230 and Opt.Commun. 8(1973), 118.

(1) Magde and Windsor, Chem.Phys.Lett.27(1974) 31.

J.C. Mialocq - The abstract was sent before having the single pulse experiment results, so it was not precise enough and did not attribute the two decays. Now, we are able to tell that the fluorescence lifetime $\tau = (1.2 \pm 0.3)$ ns obtained with the single pulse experiment refers to the normal DODCI, in fair agreement with your measurement and the one of Shank and Ippen.
The lifetime $\tau = (420 \pm 60)$ ps obtained with the train of pulses experiment is only the upper limit for the fluorescence lifetime of the photo-isomer since we have interference of the normal DODCI fluorescence. This value seems to be in good agreement with the (330 ± 40) ps value reported by Arthurs et al. ; we do believe that this value refers to the photo-isomer fluorescence lifetime.

O. de Witte - I think it would be very interesting to look at the photo-isomer formation in highly viscous solvents because its rate of formation should be much longer.

J.C. Mialocq - The data we have on the photoisomerisation quantum yield ($\varphi = 0.08$ - Dempster et al.; $\varphi = 0.01$ - Arthurs et al) refer only to one solvent : ethanol. We have no evidence for the photoisomerisation process. We agree that it is still interesting to look at fluorescence in other solvents and try to understand the isomerisation process. We know from the work of Dempster et al. that the excited singlet state is involved and not the triplet state.

A. Laubereau - The question has been raised in the discussion that stimulated emission may be important for the faster relaxation time of DODCI you have reported; in such a situation, two non-exponential decays would be expected. Do your data clearly indicate two distinct relaxation times independent of the degree of excitation?

J.C. Mialocq - We did not try to measure the fluorescence lifetimes at different degrees of excitation, but our data clearly indicate two distinct relaxation when using a train of pulses and only a single relaxation time when exciting with a single pulse.

E.P. Ippen - From your measurements of the relative fluorescence intensities do you have a measure of the rate (or quantum efficiency) of photo-isomer generation? Before one can deduce a photo-isomer fluorescence lifetime one must take into account that the photo-isomer excited state is being fed by a normal species excited state that is decaying in time. Have you included this consideration in your determination of photo-isomer lifetime? In addition to the fluorescence lifetime measurement that you referenced Shank and Ippen (Appl. Phys. Lett. 1975) have also reported an absorption recovery measurement with the same result.

J.C. Mialocq - We do not believe that the photo-isomer fluorescence we observe, comes from the photo-isomer excited state fed by a normal species excited state, but is the result of excitation of the photo-isomer from its ground state. We showed that we had a steady state concentration of the DODCI photo-isomer, 250 ns after the beginning of the pulse train, so the two isomers contribute to the absorption of the exciting light. Moreover, the photo-isomerisation quantum yield is too low to see fluorescence of the photo-isomer directly after excitation of the normal form of DODCI.

SPECTRAL AND KINETIC INVESTIGATIONS IN EXCITED MOLECULES USING A TRAIN OF PICOSECOND PULSES

J. FAURE, J.P. FOUASSIER and D.J. LOUGNOT

Laboratoire de Photochimie Générale, Equipe de Recherche Associée au CNRS N° 386, 3 rue Alfred Werner 68093 MULHOUSE Cedex (FRANCE)

SUMMARY

In photochemistry, processes occuring in the subnanosecond range until several nanoseconds cannot be investigated by flash techniques.

The case of a mode-locked ruby laser, with a train of six to seven pulses of 300 ps width is described in this paper. It is emphazise that a convenient picosecond flash spectroscopy may be applied to the problem of transients determination and intramolecular rate constants evaluation.

Two kinds of investigations have been choosen as typical examples : the transients of a laser dye and the triplet lifetime of the CO group yielding to a NORRISH-II type elimination.

INTRODUCTION

In photochemistry, several intramolecular processes occur in the time scale lying from few hundred picoseconds to about ten nanoseconds : for instance, intersystem crossing, photoisomerisation, ...

We have adapted a mode locked ruby laser, which can be mounted with a frequency doubler, in order to investigate these non radiative processes. Among the possibilities offered by picosecond spectroscopy which has been developped during last years, we have excluded an excitation by ultrashort pulses. It is possible to obtain a fairly good mode locking, in the case of a ruby laser, using a cyanine dye as saturable absorber, and adapting the thickness of the exit resonant reflector.

The scheme of the apparatus is indicated on figure 1. A single pulse selector can be adapted between the oscillator (OSC) and the amplifier (A) ; nevertheless the aim of this paper is to emphazise the advantage to use the overall train of pulses. This train is constituted by six to seven pulses, the width of which roughly may be choosen among four possibilities, by rotating the device supporting the interferometer flat acting as an exit mirror, up to some hundred picoseconds. The most part of components have been obtained from QUANTEL-FRANCE.

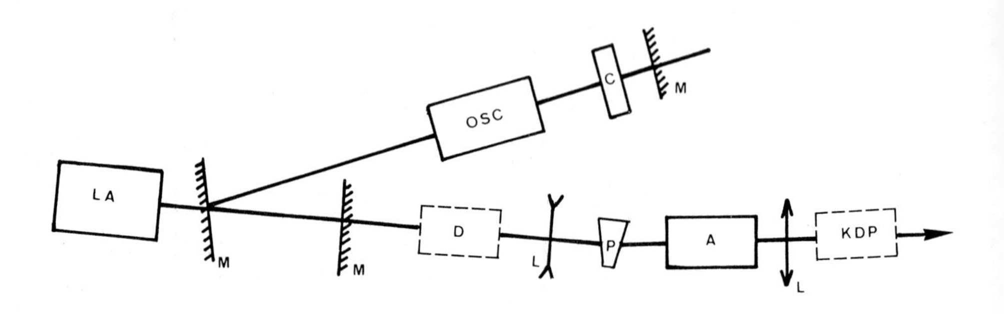

Fig.1. Schematic representation of the mode locked ruby laser : M : mirrors ; OSC : ruby oscillator ; C : saturable absorber ; D : pulse selector ; P : prism ; L : lens ; A : ruby amplifier; KDP : second harmonic generation crystal ; LA : He-Ne laser.

We report two kinds of experiments :

- Direct measurement of a very short triplet state produced in the U.V. range by excitation of a carbonyl group yielding to a NORRISH-II type photoelimination.
- The study of a laser dye excited by the fundamental 694 nm, in which several intramolecular rate constants, molar extinction coefficients of transients states, spectra of photoisomer and triplet state have been either directly measured or deduced from a computation based on the kinetic equations of population and evolution of the first excited singlet state.

I. INVESTIGATION OF A LASER CYANINE DYE EMITTING IN THE NEAR INFRARED

1.3.3.-1'.3'.3'-hexamethyl indotricarbocyanine (HITC) is known as a very convenient laser dye emitting at 860-880 nm [1] [2][3] and has been extensively studied.

Both kinetic and spectral properties of excited states are

involved in the laser emission process. In order to explain why the emission occurs at these wavelengths in the overall fluorescence band (about 500 nm with a maximum at 780 nm in DMSO) we tried to determine the absorption spectra of all transients which can be detected in the time scale of the flash pumping operation of the laser dye.

The type of measurement involves an analysis beam produced by the laser emission of several fluorescent dyes choosen in order to investigate the spectral range from 700 to 930 nm [4].

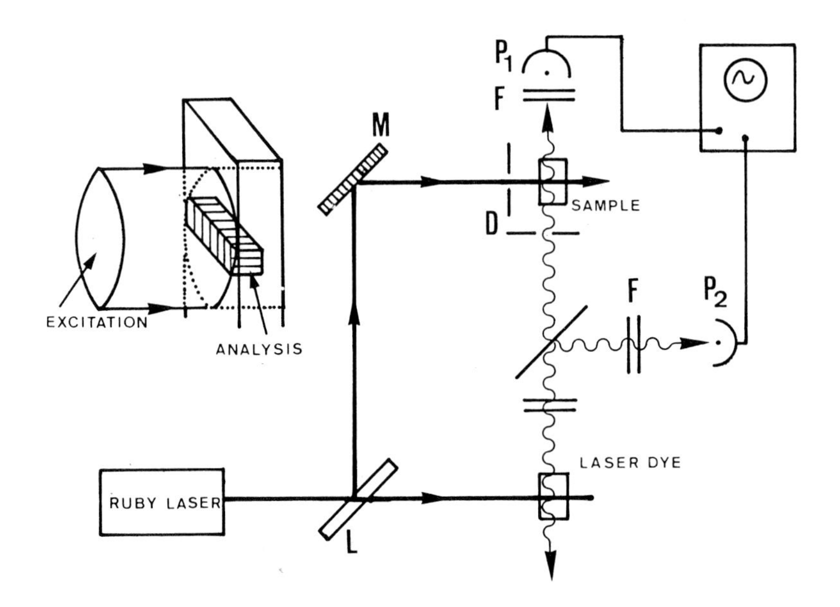

Fig.2. Experimental arrangement for absorption picosecond spectroscopy. M : mirrors ; F : filters ; L : beam splitter ; D : diaphragms ; P_1, P_2 : photodiodes

The scheme of the device is depicted in figure 2 and is analogous to a spectroscopic flash apparatus in which the problem of the risetime of the detection system would be turned out. The experimental procedure involves the transmission measurements for each pulse of the train, once with pumping train blocked before the sample (T_0), then under excitation (T). The difference between the transmitted probe pulses in these two cases contains the information on the excited state relaxation, transients absorption and intermediary states generation from

S_1. Then, we compare the experimental results with the transmission value calculated from an excited states model of the molecule. We take into account three excited states : singlet (S_1), triplet (T) and photoisomer (P). We can write :

$$\frac{dN_1}{dt} = P_0\sigma_{01}^{\ell}N_0 - \frac{N_1}{\tau}$$

$$\frac{dN_T}{dt} = K_{ST}N_1 - \frac{N_T}{\tau_T}$$

$$\frac{dN_p}{dt} = K_pN_1 - \frac{N_p}{p}$$

$$N_0 + N_1 + N_T + N = N$$

$$\text{Log } T = \ell[(\sigma_e - \sigma_{1n})N_1 - \sigma_{TT}N_T - \sigma_pN_p - \sigma_{01}N_0]$$

Here σ_{01}^{ℓ} is the absorption cross section at ruby laser wavelength ; N_0, N_1, N_T, N_p the population density (cm^{-3}) on S_0, S_1, T, P. The symbol "σ" represents the cross section at the probing light wavelength. P_0 is the pumping power in phot. $cm^{-2}s^{-1}$, if the excitation pulse is assumed to behave as a δ function.

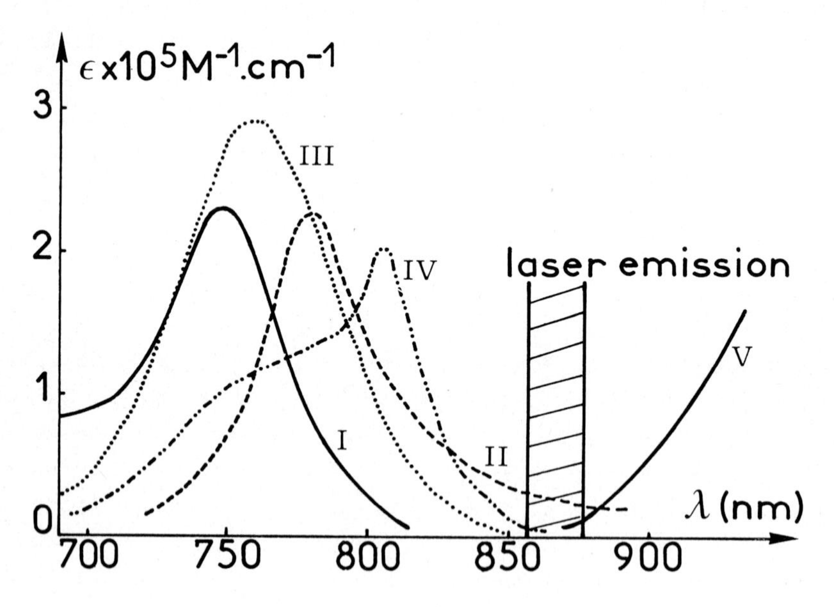

Fig.3. HITC absorption and emission spectra : S_0-S_1(I), S_1-S_0 (II), S_1-S_n(III), photoisomer(IV), triplet(V)

The method of resolution is described in ref.4. The spectra are reported in figure 3 ; we obtained the values of 10^8 s^{-1} for the photoisomerisation process and $2-3.10^7$ s^{-1} for the intersystem crossing rate constant.

We may explain the following points :

- The laser emission range of HITC is lying in a narrow window (860-880 nm [3]) in which the transient absorptions are the less important.
- The oxygen dependence of the laser efficiency [3] is related to the presence of the strong triplet-triplet absorption.
- The presence of a photoisomer generated via the first excited singlet state and absorbing in the 800 nm range allows to describe the DOTC mode locking by HITC [3] in the same manner it was pointed out by ARTHURS [5] in the case of the R6G mode locking by DODCI.

II. INVESTIGATION OF THE CARBONYL TRIPLET STATE [6]

The second type of experiment we would like to report is the direct measurement of a very short triplet lifetime : the triplet lifetime of carbonyl group involved in the NORRISH-II photoelimination process. This reaction which occurs in the monoarylketones is one of the most extensively studied. The final products are known to arise from the triplet state.

The photodegradation of such material in solution has been studied previously almost exclusively through steady-state experiments that is to say by STERN-VOLMER experiments ; this type of measurement gave only values for the term $\tau_T K_Q$ (τ_T : triplet lifetime - K_Q : quenching constant), and from several assumption concerning K_Q, it was possible to get an indirect estimation of τ_T. In such a way, WAGNER [7] estimated a triplet lifetime shorter than 10 ns. Recently, a laser spectroscopic determination of τ_T has been measured and is in desagreement with WAGNER's [8]. It appears important to get a direct measurement of τ_T, in order to complete the estimation of the quenching rate constant [6].

The experimental set up is schematically depicted in figure 4. The U.V. beam obtained from the ruby laser after frequency conversion, used as pumping source to induce S_0-S_1 transition in the sample A, serves simultaneously as probing beam and interrogating beam.

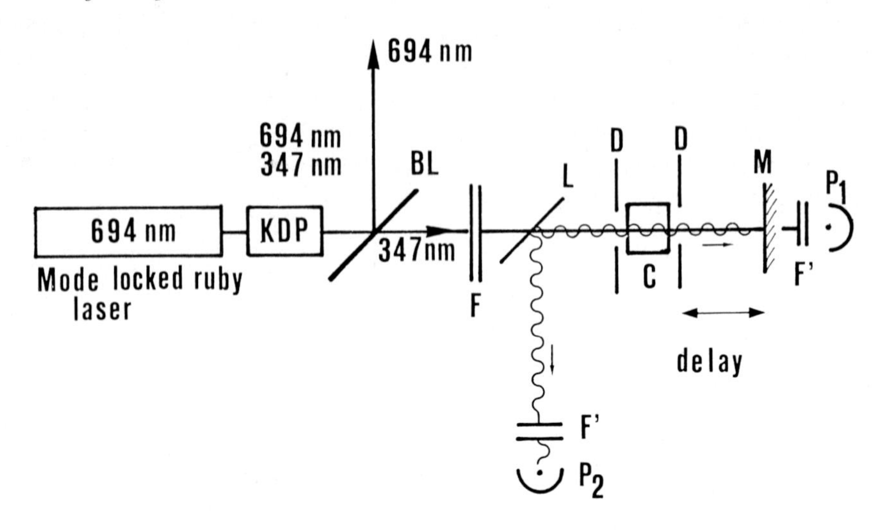

Fig.4. Experimental set up for triplet lifetime measurement in the nanosecond range. BL : beam splitter ; F : 694 cut off filter ; L : slab ; D : diaphragms ; C : sample cell ; M : mirror; F' : filters ; P_1, P_2 : photocells

This ligth, consisting in a train of 6 to 7 pulses, 10 ns separated, 300 ps half-duration at λ=347 nm, is first directed to the beam splitter BL to remove the red part not converted in the KDP crystal, then crosses over a sample and is received by the mirror M ; the weak part transmitted by this mirror is collected by the photocell P_1 ; the other part after attenuation returns through the sample and reaches a slab which directs a part of this light to the photocell P_2. From the output ratio of these two cells, first with the sample containing the solvant only and then with the solution present, it is possible to record the optical transmission. Thus, if the distance L separating the sample and the mirror M is varied, the transmission may be recorded as a function of the optical delay $\frac{2L}{c}$ (c : light velocity), that is to say as a function of time. Working in such a way we could get the change of transmission in the case of valerophenone after laser excitation. This curve is reported on the figure 5(a) in which one see in a first step the triplet population corresponding to very fast decrease of the transmission and then in a second step the triplet state corresponding to a decrease of the transmission.

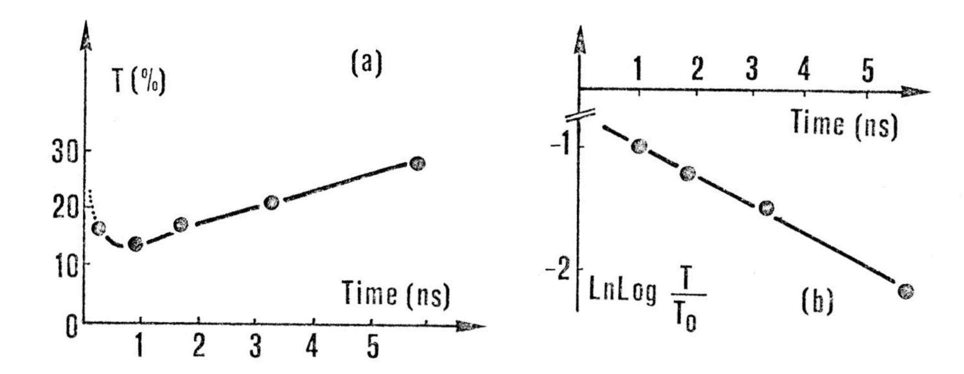

Fig.5. (a) Transmission of a valerophenone solution in benzene versus the optical delay - (b) Plot of $LnLog(T/T_0)$ versus time t (the slope of the straight line gives the value $1/\tau$)

The kinetic analysis of the transient state according to a first order decay leads to the values τ_T=4 ns±0,2 (Fig.5(b)) and σ_{TT}~7000 $M^{-1}cm^{-1}$.

Many other experiments have been performed upon polymeric material exhibiting the same photochemical behaviour ; in these cases the kinetic analysis are more difficult in relation to the accumulation process which may take place since the triplet lifetime becomes longer than 5 ns. Nevertheless a mathematical analysis of a three level model give the opportunity to use a trial and error computation which leads to the values of τ_T, K_{ST} and σ_{TT} [9].

For instance, in the case of polyvinylphenylketone in solution in aromatic solvents we found τ_T=9±3 ns and σ_{TT}~5000 $M^{-1}.cm^{-1}$; the intersystem crossing rate constant K_{ST} was noted to be solvent dependent when going from benzenic to trichloromethylbenzenic solution, the experimental values lying between 3.10^9 to 10^8 s^{-1}. This change seems to be connected to the viscosity of the solvent.

CONCLUSION

We have pointed out the interest of the use of the whole train of pulses for the determination of spectral and kinetic properties, in the subnanosecond range in the case of a dye laser and of a photosensitive carbonyl compound. The rather long duration of the picosecond pulses permits us to avoid most of the inconvenient due to an ultrashort excitation. More details for each measurement have been published elsewhere [4][6] [9].

ACKNOLEDGEMENTS

We feel undebted to Dr R. SALVIN for helpfull discussions on the carbonyl triplet state investigations.

REFERENCES

[1] H. FURUMOTO, H. CECCON
IEEE, 6(1970)262

[2] A. ERANIAN, P. DEZAUZIER, O. DE WITTE
Opt. Comm., 7(1973)150

[3] A. HIRTH, J. FAURE, D.J. LOUGNOT
Opt. Comm., 7(1973)339

[4] J.P. FOUASSIER, D.J. LOUGNOT, J. FAURE
Chem. Phys. Lett., in press

[5] E.G. ARTHURS, D.J. BRADLEY, A.G. RODDIE
Appl. Phys. Lett., 20(1972)125

[6] D.J. LOUGNOT, J.P. FOUASSIER, J. FAURE
J. Chim. Phys., 72(1975)125

[7] P.J. WAGNER
Acc. Chem. Res., 4(1971)168

[8] G. BECK, G. DOBROWOLSKI, J. KIWI, W. SCHNABEL
Macromol., 8(1975)9

[9] J. FAURE, J.P. FOUASSIER, D.J. LOUGNOT
To be published

DISCUSSION

J. FAURE, J.P. FOUASSIER, D.J. LOUGNOT

W.S. Schnabel - We also measured lifetimes of intermediates of aromatic ketones produced by 347 nm light (from a ruby laser). The pulse width was 25 ns. We found with benzene solutions of butyrophenone and polyphenylvinylketone an intermediate absorbing between 300 and 600 nm, which we consider corresponding to the excited triplet state. The lifetime was about 100 ns in both cases. When we started our experiments the absorption decayed within the 15 ns flash. However, after having purified carefully the solvent we found the lifetime mentioned above. Therefore, I would like to ask how the solvents were purified in your case and whether the shorter lifetime you find is not due to impurity quenching.

J. Faure - Yes, the solvents are purified as far as possible, but I don't think the question of the discrepancy between your results and ours can be explained by such a problem of solvent purity.

S. Schneider - Don't you think that the variation in ISC-rates may be due to external heavy atom effect induced by the solvent molecules rather than to viscosity effects?

J. Faure - It is a very important question because at this point, we are not able to separate the effects of the two parameters on the ISC rates. I think we shall have to study the viscosity dependence of this process by changing the solvent out of this external heavy atom problem.

PICOSECOND FLASH PHOTOLYSIS AND SPECTROSCOPY AND KINETICS OF INTERMEDIATES IN BACTERIAL PHOTOSYNTHESIS

Maurice W. Windsor and Mark G. Rockley
Department of Chemistry, Washington State University, Pullman, WA 99163

and

Richard J. Cogdell and William W. Parson
Department of Biochemistry, University of Washington, Seattle, WA 98195

SUMMARY

We have used the techniques of picosecond flash photolysis and kinetic spectroscopy to examine the primary photochemical reaction of bacterial photosynthesis. Immediately following excitation of reaction centers (RC's) of Rhodospeudomonas Sphaeroides with 530 nm wavelength laser pulses of 8 ps duration, a transient state is observed that decays with an exponential lifetime of 246 ± 16 ps. As the transient state decays, the radical cation of the RC bacteriochlorophyll complex appears. The absorption spectrum of the transient state shows that it is identical with a state (P^F) that has been detected previously in RC's that are prevented from completing the photo-oxidation, because of chemical reduction of the electron acceptor. Thus the results demonstrate conclusively that P^F is a direct intermediate in the photo-oxidation of the bacteriochlorophyll.

INTRODUCTION

We have developed a picosecond flash photolysis apparatus, capable of photographic recording of the spectra of transient intermediates for exploratory studies, coupled with provisions for subsequent detailed kinetic studies at any wavelength in the range 400 to 950 nm using a vidicon detector coupled to an optical multichannel analyzer [1]. The sample is excited with a single optical pulse of 5-8 ps duration at a wavelength of 530, 694 or 1060 nm and monitored by a time-delayed continuum pulse, also of picosecond duration. Time resolution of about ± 4 ps is obtainable over the range from a few ps to a few ns.

We have used this apparatus to study the short-time photophysics and photochemistry of a variety of dyes and other large molecules, some of which have applications as laser media and/or saturable absorbers for laser pulse-shaping. Measurements of both ground-state repopulation (GSR) and excited-state absorption (ESA) are used to obtain kinetic data. In the molecule, crystal violet, we observe a pronounced effect of viscosity on the excited state relaxation that appears to be connected with solvent-hindered rotation of the phenyl rings [2]. The molecule bis-(4-dimethylaminodithio-benzil)-Ni(II), (BDN) exhibits a large and interesting dependence of internal conversion rate on solvent. The extreme values of the excited state lifetime are 9 ns in benzene and 220 ps in iodoethane [3]. We have discussed the significance of these data for laser mode-locking [4].

In the area of photobiology we have determined the absorption spectra of both the first excited singlet S_1 and the lowest tripiet T_1 of several porphyrins, together with extinction coefficients and kinetic and quantum yield data [5]. Picosecond flash photolysis is especially useful for studying systems such as these that have very low fluorescence yeilds (<1% for octa-ethyl porphyrin $SnCl_2$) and would therefore be very difficult to study by other techniques.

Most recently, we have made a picosecond study of the spectroscopy and kinetics of intermediates in bacterial photosynthesis [6]. This is the main subject of this report.

Photosynthesis by Plants and by Bacteria [7,8]

In both plants and bacteria the photosynthetic apparatus consists of units in which organized arrays of up to several hundred chlorophyll molecules (the antenna or light-harvesting system) collect solar energy and funnel the excitation non-radiactively to a reaction centre (RC) containing a special chlorophyll complex. Here the energy is converted to and trapped as high-energy chemical intermediates. These comprise a primary oxidant and a primary reductant. Each primary reactant initiates an enzymatic electron-transport-chain. In its barest essence the outcome is that electrons are removed from an oxidizable substrate and, with the aid of light energy, a reductant is made that is capable of reducing CO_2 to carbohydrate.

This is the basic metabolic process by which both plants and bacteria grow. But the details differ. In green plants, photosynthesis is more complex than in bacteria. Two photochemical pigment systems, PSI and PSII are involved; light energy is used twice to boost the reduction potential. A downhill electron-transport-chain couples the weak reductant produced by PSII with the oxidizing side of PSI. The strong primary oxidant of PSII removes electrons from H_2O via a manganese-containing enzyme, thus liberating molecular O_2. Light excitation in PSI produces a strong reductant capable of reducing NADP. Electron flow through PSI and PSII also drives the phosphorylation ADP to produce ATP. Both the stable reductant NADPH and ATP are required for the conversion of CO_2 to sugar.

Photosynthetic bacteria, in contrast to plants, have only a single photosystem [7]. In some ways it resembles PSI in green plants. In both, the primary oxidant produced an illumination has a midpoint potential of about +450mV. This is not strong enough to remove electrons from water and, lacking PSII, bacteria must rely on other oxidizable substrates, e.g., sulfide, thiosulfate, or succinate as a source of electrons. On the reducing side the potentials generated appear to differ considerably. In plants PSI has a reducing potential of about -600 mV. In general, the primary reductant in bacterial chromatophores appears to have a midpoint potential near -100 mV.

Despite these differences, photosynthetic bacteria perform much of the same chemistry and photochemistry engaged in by green plants. As experimental subjects for photochemical studies, they have the following advantage over green plants:

1) They are easier to grow and harvest and more reproducible in their properties.
2) Reaction center preparations, free of antenna chlorophyll, can be made. These constitute the smallest entities that are capable of carrying out the characteristic photochemistry in more or less complete fashion.
3) They have only a single photosystem, thus simplifying the interpretation of photochemical behavior.
4) Different species of bacteria and readily available mutant forms provide a useful range of variation in composition and properties.

By suitable detergent treatment Clayton and others have shown that it is possible to remove the bulk amount of antenna chlorophyll to produce reaction center (RC) preparations [9]. The RC's are still capable of carrying out the characteristics photochemistry of the cell. The availability of RC's is an enormous advantage for photochemical and spectroscopic studies aimed at searching for transient intermediates; otherwise the strong absorption of the much greater number of antenna chlorophylls tends to obscure any transient optical changes taking place in the pigment molecules of the RC.

Recent Nanosecond and Picosecond Studies of RC's

Although a great deal is now known, especially about the sequence of enzymatic intermediates, much is still a mystery. For example, virtually nothing is known concerning the nature of the primary electron acceptors. The details of the charge separation process by which the energy of an excited singlet state of chlorophyll is converted into chemical potential are still far from understood.

The primary photochemical reaction of bacterial photosynthesis is the transfer of an electron from an excited bacteriochlorophyll complex, P, to an acceptor, X, whose identity is still uncertain [9,10,11]. The electron transfer reaction occurs with a quantum yield of essentially 100% [12] and it occurs with great speed, even at temperatures below 4°K. Parson, et al., [13] studied RC preparations in which the electron transfer reaction was blocked by the chemical reduction of X with sodium dithionite. When such preparations were excited by a 20 ns laser flash, two short-lived intermediates could be detected via their optical absorbance changes. One, called "P^F," appeared instantaneously and decayed with a half-time of about 6 ns. More recent measurements with a 6 ns laser flash give a value of 10 ± 2 ns [14]. The high quantum yield of P^F suggested it might be an intermediate in the electron transfer reaction. The other transient state, P^R, was of less interest because of its slower formation and lower yield (∿10%). Subsequent studies showed that P^R is formed as P^F decays in the chemically blocked system, but that it is not formed under conditions that permit electron transfer to occur. P^R has been tentatively identified as the lowest triplet state of the BChl complex. Thus formation of P^R appears to be an alternative chennel for the excitation energy that functions only when the primary electron transfer channel is blocked.

There remains the question of the identity and role of state P^F. It is unlikely to be the lowest excited singlet-state, P^*, of the BChl complex, because fluorescence yield measurements show that the lifetime of P^* is only 20 to 40 ps when X is in the reduced form and even less when X is not reduced [15,16]. It is appealing to imagine that the primary electron transfer reaction goes via P^F, rather than proceeding directly from P^* to P^+, because this would then explain a longstanding anomaly between the quantum yield of photochemistry and fluorescence [11-13].

There was, however, still the possibility that P^F also was just a side-product that appears only when the normal reaction is blacked. To settle this question, we needed to know whether or not P^F was formed _under conditions that permit electron transfer to occur._ Nanosecond time resolution was inadequate for this purpose. We, therefore, decided to embark on a study using our recently developed techniques of picosecond flash photolysis and kinetic spectroscopy [1].

Experimental Details

Reaction centers were obtained from cells of Rhodopseudomonas sphaeroides strain R-26 by the method of Clayton and Wang [11]. The detergent lauryl-dimethylamine oxide was replaced by Triton X-100 by dialysis [6,12].

The apparatus used in the psec experiments was essentially that of Magde and Windsor [1]. We shall summarize the approach only briefly here for clarity; ref [1] provides additional details. A single pulse from a mode-locked Nd^{+3}/glass laser was frequency-doubled to provide a 530 nm excitation flash lasting about 8 psec. This illuminated only a small band across the center of the sample. A measuring flash of white light, which was generated by self phase modulation of residual 1060 nm light from the original pulse, traversed an optical delay path and interrogated the sample at an adjustable time after the actinic flash. The relatively weak measuring pulse passed through both the center of the sample and the unexcited regions above and below the center, allowing a measurement of the absorbance differences caused by the excitation. After passing through the sample, the measuring pulse proceeded to a spectrograph which was equipped with a camera, or with a silicon vidicon detector coupled to an optical multichannel analyzer (SSR Instruments, Inc., Model 1205 A). To correct for spatial inhomogeneities in the measuring beam, we first measured the apparent absorbance of the central region of the sample in the absence of the excitation flash. This was then compared with a measurement that included the excitation flash.

RESULTS AND CONCLUSIONS

A complete account of this work including detailed spectral and kinetic data will appear shortly [6]. The main new points are as follows:

1) Immediately following excitation of unblocked RC preparations of Rps. Sphaeroides with 530 nm laser pulses of 8 ps duration, there appears a transient state characterized by new absorption bands near 500 and 650 nm, by a bleaching of bands near 540, 600, 760

and 870 nm and by a blue shift of a band near 800 nm.

2) The absorption spectrum of the transient state shows it to be identical with a state P^F, detected previously only in reaction centers that are prevented from completing photo-oxidation, because of chemical reduction of the electron acceptor.
3) The state P^F decays with an exponential decay time, τ, of 246 $\pm$ 16 ps.
4) As P^F decays, the absorption spectrum of the radical cation, P^+, of the BChl complex appears.
5) The rate of disappearance of P^F and the rate of formation of P^+ are identical within experimental error. There is no reason to believe that the process requires more than a single step.
6) The P^F spectrum agrees tolerably well with the calculated spectrum for an anion-cation biradical, $BChl^+$-$BChl^-$.
7) Finally, bleaching of an absorption band at 800 nm that decays very rapidly ($\tau \sim 30$ ps) is observed. This short-lived transient may reflect the excited singlet state, P^*, of the BChl complex or possibly the relaxation of interactions among the four BChl molecules of the RC during the process of formation of P^F.

While the present work was in progress, Kaufman, *et al.*, [17] (KDNLR), using somewhat different techniques of picosecond spectroscopy, also observed an intermediate, the spectrum of which agreed well with that of P^F. They reported a decay time of about 120 ps for their transient, but failed to follow the appearance of P^+. For technical reasons, we believe they also underestimated the decay time [6]. Nevertheless, after taking account of this, we find essential agreement between their results and ours.

In summary, then, it appears established that Parson's intermediate P^F is the immediate precursor of the cation P^+ of the BChl complex in bacterial RC's of *Rps. Sphaeroides*. Further, there is some suggestion from the spectra that P^F is to be identified with the cation-anion biradical, $BChl^+$-$BChl^-$. Although more data are needed before this assignment can be considered firmly established (or rejected), this charge-transfer species does make an attractive half-way house, energetically, (between the initial excited singlet, $BChl^*$, and the final state of complete photo-oxidation, ($BChl^+$ +X^-), in which the electron has been completely transferred to the acceptor.

ACKNOWLEDGEMENT

This research was supported in part by the Office of Naval Research and by National Science Foundation Grant 93-30732X.

REFERENCES

1. M. D. Magde and M. W. Windsor, (Picosecond Flash Photolysis and Spectroscopy: 3,3'-Diethyloxadicarbocyanide Iodide (DODCI)), Chem. Phys. Lett., 27(1974)31.
2. M. D. Magde and M. W. Windsor, (Picosecond Internal. Conversion in Crystal Violet), Chem. Phys. Lett., 24(1974)144.
3. M. D. Magde, B. A. Bushaw, and M. W. Windsor, (Picosecond Flash Photolysis and Spectroscopy: Bis-(4-dimethlaminodithio-benzil)-Ni(II), (BDN)), Chem. Phys. Lett., 28(1974)263.
4. M. D. Magde, B. A. Bushaw, and M. W. Windsor, (Q-Switching and Mode-Locking the Nd^{3+}-Glass Laser with the Nickel Dithienes), IEEE J. Quant. Electronics, QE10(1974)394.
5. M. D. Magde, M. W. Windsor, D. Holten, and M. Gouterman, (Picosecond Flash Photolysis: Transient Absorptions in SN(IV), Pd(II), and Cu(II) Porphyrins), Chem. Phys. Lett., 29(1974)183-188.
6. M. G. Rockley, M. W. Windsor, R. J. Cogdell, and W. W. Parson, (Picosecond Detection of an Intermediate in the Photochemical Reaction of Bacterial Photosynthesis), Proc. Nat. Acad. Sciences, (1975) in press.
7. R. K. Clayton, Light and Living Matter, Vol. 2, McGraw-Hill, New York, 1971.
8. R. K. Clayton, Proc. Nat. Acad. Sci. U.S., 69(1972)44-49.
9. R. K. Clayton, Ann. Rev. Biophys. Bioeng., 2(1973)131-156.
10. W. W. Parson, Ann. Rev. Microbiol., 28(1974)41-59.
11. W. W. Parson, and R. J. Cogdell, Biochim. Biophys. Acta. (Revs. in Bioenergetics), 416(1975)105-149.
12. C. A. Wraight, and R. K. Clayton, Biochim. Biophys. Acta., 33(1974)246-260.
13. W. W. Parson, R. K. Clayton, and R. J. Cogdell, Biochim. Biophys. Acta., 387(1975)265-278.
14. R. J. Cogdell, J. G. Monger and W. W. Parson, Biochim. Biophys. Acta., (1975) submitted for publication.

15. K. L. Zankel, D. W. Reed, and R. K. Clayton, Proc. Nat. Acad. Sci. U.S., 61(1968)1243-1249.
16. L. Slooten, Biochim. Biophys. Acta., 256(1972)452-466.
17. K. J. Kaufmann, P. L. Dutton, T. L. Netzel, J. S. Leigh, and P. M. Rentzepis, Science, (1975) in press.

DISCUSSION

M.G. ROCKLEY, M.W. WINDSOR, R.J. COGDEL, W.W. PARSON

J. Joussot-Dubien - Since radical cation and radical anion absorption spectra are usually quite similar, how well known is the spectrum of chlorophyll radical anion to be taken with confidence to account for the "manufactured" $Chl^{+}...Chl^{-}$ absorption?

M.W. Windsor - The cation and anion radicals of bacteriochlorophyll (BChl) can be produced chemically by oxidation and reduction respectively of BChl. We used the reported spectra of Fajer et al. (Proc.Nat. Acad.Sci.,U.S., 71 (1974) 994-998) and Fajer et al. (J. Am.Chem.Soc. 95 (1973) 2739-2741).

E. Roux - 1) Do you have any idea about the parts of bacteriochlorophyll molecule which carry the positive and the negative charge?
2) What are your hypotheses about the electron migration after charge separation? 3) Have you studied the evolution of the dimer difference spectrum according to pH and redox potential?

M.W. Windsor - 1)No. But your question reminds me of recent studies of iron porphyrins by Dolphin (U. of British Columbia). He believes that the photoexcited state initially undergoes oxidation in which a π electron is removed from the peripheral π-orbitals of the ligand. Subsequently an electron jumps from the Fe^{II} central atom to fill the hole, thus oxidizing the Fe^{II} to Fe^{III}. I would suspect that in BChl the charge would be delocalized over the ligand.
2) We believe that the initial charge separation step is the formation of the charge-transfer dimer ($BChl^{+}$ - $BChl^{-}$). Subsequently with a time constant of about 240 ps the cation $BChl^{+}$ plus the reduced acceptor X^{-} are formed by electron transfer from $BChl^{-}$ to X. The identity of X, the acceptor, is not, however, known.
3)No, but I think this would be valuable to do. We plan in future to make such measurements.

J.K. Eweg - Do you have experimental evidence for the dimer of two chlorophylls you showed in one of your slides?

M.W. Windsor - As to ourselves, only that which I showed in the slide- namely the considerable measure of agreement between our experimental spectrum and the synthetic spectrum ($BChl^{+}$ + $BChl^{-}$). Katz and Norris, from EPR linewidth measurements, also have evidence for dimer formation (I.J. Katz and J. Norris, Curv.Top.Bioenergetics, 5 (1973) 41-75), in particular a dimer in which one H_2O molecule is sandwiched between two chlorophylls.

P. Mathis - The idea of a dimer of bacteriochlorophyll originates in a comparison of the width of the ESR line of $BChl^{+}$ in solution and in reaction centers. It is smaller by about $\sqrt{2}$ in reaction centers, indicating delocalization of the spin over two molecules (ref.Norris and Katz).

C. Varma - Could you say a few words on the method you use to find the synchronization between the probe light and the excitation pulse?

M.W. Windsor - We replace the sample by a solution of molecules that exhibit significant ground-state depletion, e.g. crystal violet or BDN. The removal of molecules from the ground-state by the excitation pulse must be a prompt (no delay) process. Therefore the rate of bleaching of the ground-state must be coincident with the rise of the excitation pulse.

P. Mathis - I understand that there is a transient species which last about 30 psec and which precedes the formation of P_F. Is it possible to correlate this species with the singlet excited state of bacteriochlorophyll?

M.W. Windsor - We are not yet sure. Most of the spectral features we attribute to P^F appear immediately ($\leqslant$ 10 ps). But then the bleaching of the 800 nm absorption band only decays with $\tau \sim 30$ ps after the flash. This relatively minor change may correspond to relaxation of the BChl reaction center complex. We must remember that the RL contains 4 BChl + 2 bacteriopheophytin molecules. We plan further experiments that may answer this question - that is whether or not we are observing $BChl^*$ the excited singlet state.

C. Reiss - What is the flatness of your continuum? How do you correct for the spatial inhomogeneity of the continuum?

M.W. Windsor - Our continuum appears flat because it is somewhat over-exposed. Variations in continuum intensity with wavelength do occur. These are not big enough to cause us trouble. Laubereau will talk in detail about the generation of picosecond continua. We monitor the unexcited spectral region on both sides of the excitation region using the vidicon and optical multichannel analyzer. These measurements would show up any spatial inhomogeneities. If severe spatial inhomogeneities occur, we reject that shot. Otherwise we average the "above" and "below" intensity to obtain a value for I unexcited.

G.R. Goldschmidt - In view of the high accuracy $\Delta OD = 0.003$ of your measurements at ps time resolution and even at low absorption signals; does it not require a tremendously high monitoring light intensity in order to overcome statistical fluctuations?

M.W. Windsor - I cannot answer this question in detail without giving further thought to the matter. However, from a pragmatic standpoint, the reproducibility of our experimental data show that we do have acceptable signal/noise ratios and that statistical fluctuations are correspondingly not a problem.

EXEMPLES D'UTILISATION DES LASERS DANS LES RECHERCHES SUR LA PHOTOSYNTHESE

J. BRETON, E. ROUX
Département de Biologie, CEN Saclay, BP N°2, 91 Gif-sur-Yvette
France.

RESUME

Dans cette revue les auteurs font apparaître l'intérêt que présente l'utilisation des lasers dans les recherches en photosynthèse portant

1) sur les mécanismes présidant au transfert de l'énergie de la molécule de chlorophylle primitivement excitée vers le centre réactionnel siège d'une ionisation,

2) sur la structure du milieu au sein duquel ces évènements se déroulent et en particulier sur l'orientation et l'environnement des molécules captant l'énergie lumineuse et de celles jouant le rôle de centre réactionnel.

INTRODUCTION

Le passage de la biologie du stade descriptif au stade explicatif est lié à une intervention chaque jour plus importante des connaissances de la chimie et de la physique dans la recherche Biologique.

En conséquence tout progrès technique réalisé en Physico-chimie entraine des possibilités nouvelles pour le biophysicien et c'est ainsi que l'avènement des lasers a été le point de départ d'une ère nouvelle pour la photobiologie aussi bien dans le domaine de l'étude des structures que dans celui du fonctionnement des Photorécepteurs.

La cellule chlorophyllienne est limitée par une membrane cellulosique, contient en son centre une vacuole et entre la vacuole et la membrane se trouve un milieu protéique plus ou moins fluide : le protoplasme.

Au sein du protoplasme on distingue :

- un noyau porteur de l'information génétique,
- des mitochondries responsables de la respiration,
- des chloroplastes en nombre variable de 1 à 10 au niveau desquels se déroule la photosynthèse, c'est-à-dire l'ensemble des réactions induites par la lumière et qui ont pour conséquence

l'absorption de CO_2 et le dégagement d'oxygène.

Ces chloroplastes sont des corpuscules ovoïdes de 3μ à 5μ non homogènes ; on distingue à leur intérieur :

- Une partie non pigmentée, protéique, le stroma peu structuré à l'intérieur de laquelle se différencient des lamelles qui en fait sont des sacs limités par une double couche lipidique.
- Au sein de cette double couche essentiellement formée de galactolipides se trouvent des pigments chlorophylliens et

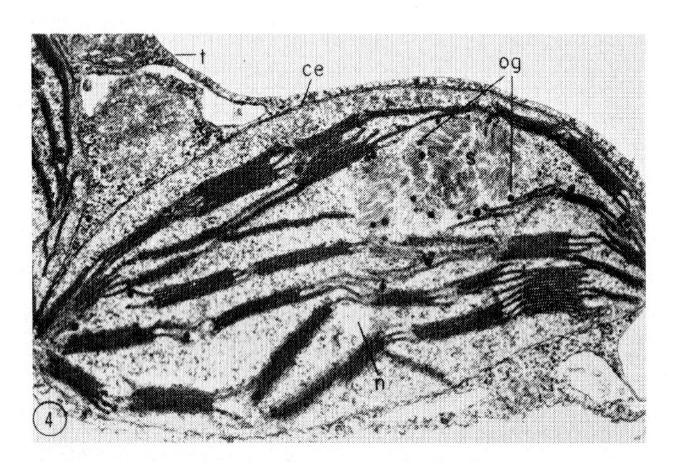

caroténoïdes ainsi que des protéines jouant pour la plupart un rôle dans les migrations électroniques.

Le modèle de Singer [1] correspond à l'hypothèse la plus récente émise sur l'organisation globale de cette double couche lipidique.

C'est à l'intérieur de cette double couche lipidique que se déroule l'acte photochimique primaire de la photosynthèse, c'est à dire l'ensemble des évènements qui sont la conséquence directe et immédiate de l'absorption du photon.

Le chloroplaste nous apparaît en fait comme étant un double transformateur d'énergie :

- il transforme tout d'abord l'énergie électromagnétique de la lumière en énergie électronique
- puis il est capable de transformer cette énergie d'électrons mobiles en l'énergie chimique de l'ATP et du $NADPH_2$.

La première transformation seule nous intéresse ici et pour en aborder le mécanisme il importe de faire le point de nos connaissances relativement à la structure de la lamelle chloroplastique.

Les possibilités concernant l'arrangement des mono ou digalactosyl dilinoleyl glycerides sont limitées et l'hypothèse de Singer semble la plus vraisemblable.

Il en est de même en ce qui concerne le positionnement des protéines au sein de cette double couche lipidique.

Par contre pour ce qui est des pigments la question se posait de savoir comment ils étaient disposés les uns par rapport aux autres et aussi par rapport au plan des lamelles.

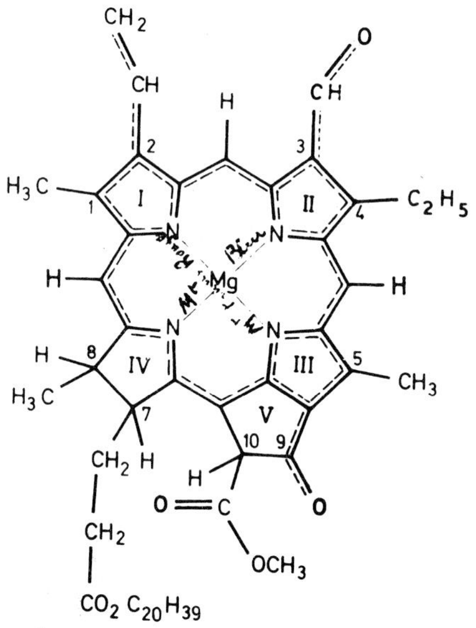

Lorsqu'une molécule possède un ou plusieurs moments de transition dont la direction intramoléculaire est connue, les mesures de dichroïsme linéaire permettent de positionner cette molécule par rapport à un plan de référence.

C'est précisément le cas pour la molécule de chlorophylle et des mesures de dichroïsme linéaire [2] effectuées sur des chloroplastes orientés mécaniquement ou en champ magnétique (les lamelles s'orientent alors perpendiculairement au champ magnétique) ont permis de montrer que la majorité des molécules de chlorophylle reposent au sein de la double couche lipidique avec leur moment de transition responsable de l'absorption dans le rouge parallèle au plan de la membrane et celui responsable de l'absorption dans le bleu faisant un angle d'environ 45° avec ce même plan.

Pour ces mesures nul besoin des lasers et c'est précisément pour faire apparaître la nécessité de leur utilisation que nous devons aller plus avant dans l'exposé de nos connaissances relatives à l'acte photochimique primaire.

L'étude en lumière discontinue de la variation du rendement quantique de la photosynthèse (nombre de molécules de CO_2 assimilées par photon absorbé) en fonction de la fréquence des éclairs en même temps que des recherches portant sur la dépola-

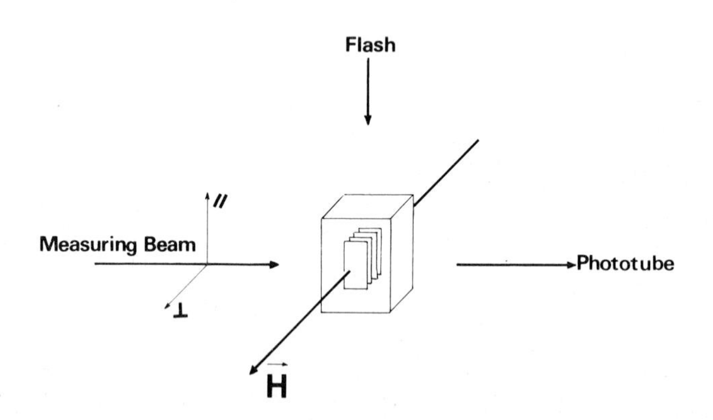

risation de la fluorescence émise par des suspensions de chloroplastes éclairés en lumière linéairement polarisée ont conduit à admettre que l'énergie absorbée par l'une quelconque des 500 molécules de chlorophylle appartenant à un ensemble appelé unité photosynthétique (UPS) est susceptible de provoquer l'ionisation d'une molécule privilégiée appelée trappe appartenant à cet ensemble.

En d'autres termes, ce n'est pas la molécule initialement excitée qui est le siège d'une ionisation ou d'une réaction photochimique mais au contraire l'énergie d'excitation de cette molécule migre à travers quelques centaines de molécules de chlorophylle vers celle jouant le rôle de trappe, cette dernière étant le siège d'une séparation de charges.

Ces trappes sont de deux types et chacun des deux types de trappes possède sous la forme réduite un spectre d'absorption différent de celui qu'elles possèdent sous forme ionisée.

Pour l'une d'elle cette variation d'absorption entre les deux formes est maximum à 7000 Å (d'où son nom de P_{700}) mais elle demeure relativement faible. C'est pourquoi les mesures du dichroïsme caractéristique des changements d'absorption photo-induits au niveau de cette trappe nécessitent un peuplement maximum en trappes ionisées, peuplement maximum qui ne peut être réalisé qu'à la suite de l'absorption d'un éclair saturant de 0,5 µs. L'expérience réalisée [3] est donc la suivante :

Les chloroplastes sont en suspension dans un tampon saccharose 0,4M tris 20 mM KCl 20 mM en présence de valinomycine

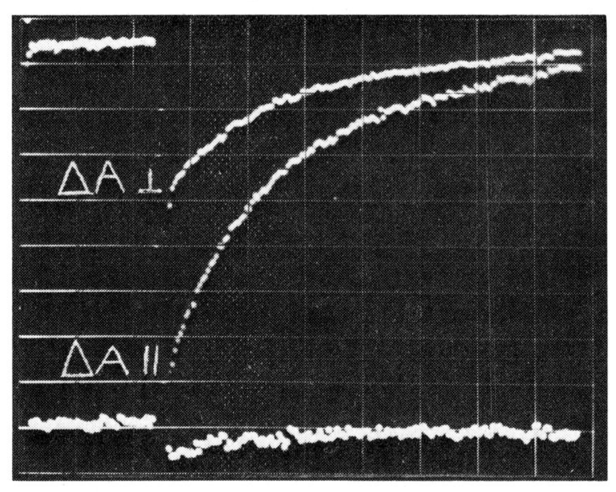

10^{-6}M et de ferricyanure de K 10^{-3}M.

La suspension est placée dans un champ magnétique de 10000 Gauss qui fait s'orienter les chloroplastes de façon telle que le plan des lamelles soit perpendiculaire au champ magnétique.

La lumière excitatrice saturante qui est absorbée par les "molécules antennes" est produite par un laser à Rhodamine 6G. L'éclair dure 500 ns, son énergie 20 mJ et la longueur d'onde 6000 Å . La fréquence des éclairs est de 4 Hz.

L'intensité du faisceau de mesure à 7000 Å est telle qu'elle ne produit aucun effet notable par elle même sur la variation d'absorption. Ce faisceau d'analyse est polarisé par un Rochon puis focalisé sur une fente de 2 mm située 10 cm derrière la suspension de façon à rendre minimum la lumière de fluorescence reçue par le photomultiplicateur situé lui même 30 cm derrière la fente et protégé par un filtre interférentiel ne laissant passer que λ = 7000 Å. Le signal du photomultiplicateur est amplifié puis envoyé à un échantillonneur Didac dont la résolution est 1ms/canal.

En absence de champ magnétique les changements d'absorbance sont les mêmes quelle que soit la polarisation du faisceau d'analyse : $\Delta A_{\parallel} = \Delta A_{\perp}$

Par contre, en présence de champ magnétique un rapport $\frac{\Delta A_{\parallel}}{\Delta A_{\perp}} = 2,5$ rend compte du dichroisme. Les cinétiques étant identiques pour les deux polarisations on peut raisonnablement admettre qu'elles correspondent à un changement d'absorbance unique.

Le dichroïsme élevé observé à 7000Å indique que le moment de transition responsable de ce changement d'absorbance est orienté presque parallèlement au plan des lamelles chloroplastiques, un rapport dichroïque de 2,5 correspond bien à une inclinaison d'environ 20° par rapport au plan des lamelles mais cette valeur est certainement supérieure à la valeur réelle compte tenu de l'alignement non parfait de toutes les lamelles dans le champ.

Cette orientation du moment de transition rouge du P_{700} confirme les résultats précédemment obtenus relatifs au dichroïsme des molécules de chlorophylle jouant le rôle d'antennes, ce dichroïsme augmentant au fur et à mesure que les molécules de chlorophylle antennes absorbent davantage vers le rouge lointain

Ces résultats confirment ceux de Junge qui utilisant une technique élégante de photosélection détermina le premier sur des chloroplastes non orientés le dichroïsme des changements d'absorption du P_{700} [9].

Nous savons donc désormais comment sont orientées par rapport au plan des lamelles chloroplastiques à la fois les molécules de chlorophylle antennes collectrices de photons et la trappe correspondante.

Voyons donc maintenant quel peut être l'apport des lasers à l'étude des mécanismes des phénomènes se déroulant dans ces structures au cours des quelques centaines de picosecondes qui suivent l'absorption du photon.

Mr Lutz vous exposera comment la spectroscopie Raman de résonance permet de caractériser les substituants du noyau tétrapyrrolique impliqués in vivo dans les interactions chlorophylle-chloroplaste.

Il nous suffit pour le moment de savoir qu'in situ dans le chloroplaste ces molécules de chlorophylle ne sont soumises qu'à des interactions de Van der Waals faibles certes mais cependant essentielles.

Elles ont en effet pour conséquence de lever la dégénérescence des niveaux excités de la chlorophylle faisant ainsi apparaître des fréquences distinctes de résonance électronique pour les différentes molécules de chlorophylle de l'unité photosynthétique.

Ces différentes fréquences d'excitation jouent le rôle d'un potentiel pour la migration de l'énergie d'excitation électronique et permettent précisément de canaliser cette dernière vers le centre réactionnel siège de l'ionisation [5].

Par analogie avec les processus de migration d'énergie se déroulant au sein des cristaux moléculaires des recherches théoriques qui auraient grand besoin d'être etayées par une expérimentation de qualité ont conduit à penser qu'au sein de l'unité photosynthétique les migrations d'énergie s'effectuent suivant un mécanisme excitonique, les excitons étant localisés.

La description de l'exciton en terme de quasi particule[4] lui fait attribuer une"masse effective" m et une vitesse V qui sont reliées à la longueur d'onde de l'onde que la mécanique ondulatoire associe au mouvement de toute particule par la relation $\lambda = \frac{h}{mV} = \frac{2\pi}{K}$ d'où $V = \frac{hK}{2m\pi}$ h = Constante de Planck
K = vecteur d'onde.

Par ailleurs le produit de l'incertitude Δr sur la localisation de la particule par l'incertitude ΔV sur sa vitesse ne peut être inférieur à $\frac{h}{2\pi}$ donc $\Delta r \cdot \Delta V \geqslant \frac{h}{2\pi}$ ou encore

$$\Delta r \cdot \Delta K \cdot \frac{h}{2m} \geqslant \frac{h}{2\pi}$$

d'où $\Delta r \cdot \Delta K \geqslant$ nombre fini.

En conséquence les excitons caractérisés par un vecteur d'onde K parfaitement défini sont des excitons non localisés puisque $\Delta K = 0$ entraine Δr infini.

Mais dans l'unité photosynthétique telle qu'elle a été précédemment définie l'interaction de l'onde d'excitation avec les vibrations correspondant aux mouvements nucléaires provoque un mélange des états K créant ainsi une incertitude sur la valeur de K et en conséquence induit une localisation de l'exciton - il devient dès lors important de mesurer:

a) le temps pendant lequel l'excitation demeure sur un site moléculaire défini,

b) le temps nécessaire à l'excitation pour passer d'une molécule à une molécule voisine.

a) Evaluation de la durée pendant laquelle l'excitation demeure sur un site moléculaire donné :

Au sein de l'unité photosynthétique les molécules de chlorophylle sont distantes d'une dizaine d'Angstroms et l'énergie de liaison U entre molécules voisines est de l'ordre de 0,015 eV [6].

Il résulte du principe d'incertitude que U est relié au temps t nécessaire pour que l'excitation localisée sur une molécule bien définie se déplace vers une molécule voisine par la relation : $Ut \geq \frac{h}{2\pi}$ pour U = 0,015 eV $t \geq 10^{-13}$ s.

Ainsi, même dans le cas des interactions moléculaires faibles caractéristiques de ces structures l'excitation se propage très rapidement et le temps pendant lequel l'excitation demeure sur une molécule donnée bien que très court est notablement plus long que celui correspondant à une période des mouvements vibrationnels des atomes du réseau. C'est là une autre façon d'exprimer le fait que l'exciton est localisé.

b) Evaluation du temps que nécessite le saut de l'énergie d'une molécule à une molécule voisine.

Le calcul conduisant à cette détermination n'est pas immédiat et nécessite la connaissance de :

- l'énergie d'interaction de deux molécules voisines (U = 0,015 eV)
- la demi-largeur de la bande rouge de la chlorophylle in vivo $\delta\Omega = 450\ cm^{-1}$
- déplacement de Stokes entre maximum d'absorption et maximum d'émission in vivo $\Delta\Omega = 200\ cm^{-1}$.

Cette faible valeur du déplacement de Stockes signifie que l'énergie d'activation impliquée dans les transferts intermoléculaires d'énergie est très faible $E_a = \frac{1}{4}\Delta\Omega = 50\ cm^{-1}$

et tout calcul fait on trouve pour le temps nécessaire au saut de l'énergie d'une molécule à une autre $t_{f} = \frac{12}{4\Omega} = 0{,}3\ 10^{-12}$ s.

Des mesures précises de l'évolution de l'intensité et de la polarisation de la fluorescence émise par une suspension de chloroplastes à la suite de l'absorption d'un éclair lumineux visible et de très courte durée (1 à $5 \cdot 10^{-12}$ s) devraient permettre de vérifier si ces prévisions théoriques rendent fidèlement compte du mécanisme présidant au transfert intra chloroplastique de l'énergie.

Un premier essai dans ce sens a consisté à utiliser un laser à mode bloqué délivrant un train d'impulsions de 3 à 5.10^{-12}s à 6100 Å et une caméra à balayage ICC 512 ayant un temps de résolution de 5.10^{-12}s.

Avec cet appareillage le temps de vie de fluorescence de la chlorophylle in vivo (chloroplastes d'épinards en suspension) a été trouvé égal à 55 ± 10 ps. Cette valeur bien qu'en bon accord avec les résultats récents obtenus par Kollman, Shapiro et Campillo [8] sur des algues unicellulaires est très faible comparée aux valeurs obtenues sur le même matériel par fluorimétrie de phase ($0,5.10^{-9}$s).

Cette différence peut être due au fait que ces mesures dans le domaine de la picoseconde sont faites, jusqu'à présent, en utilisant la totalité des impulsions émises par le laser, c'est à dire sur un milieu très perturbé au sein duquel la création d'états triplets peut inhiber la fluorescence.

Ces mesures doivent être répétées en utilisant un sélecteur d'impulsion unique.

REFERENCES

1 S.J. Singer, Membrane Structure and its biological applications. D. Green editor. Annals of the N.Y. Academy of Science Vol 195 (1972) 16-25

2 J. Breton et E. Roux, Biochem. Biophys. Res. Comm. 45 (1971) 557-563

3 J. Breton et E. Roux, J. Whitmarsh,Biochem. Biophys. Research Communication sous presse.

4 G. Paillotin, IInd International Congress on Photosynthesis (1971) 331-336

5 G. Paillotin, Thèse Université d'Orsay (1974)

6. G.W. Robinson, Broohaven Symposium Biology (1967) 19-16

7 J. Breton, M. Clerc, P. Goujon, J. Whitmarsh, Resultats non publiés.

8 Kolman, Shapiro, Campillo, Biochem. Biophysics Research Communication 63 (1975) 917-923

9 W. Junge and A. Eckoff, Biochimica Biophysica Acta 357 (1974) 103-117.

DISCUSSION

E. ROUX

J. Joussot-Dubien - The model of membrane that is becoming to be known as Singer's model was in fact proposed independently by P. Bothorel and C. Lussan, but unfortunately escaped the attention of biologists (Comp.Rendus.Acad.Sci.)

DISSECTING THE CELL WITH A LASER MICROBEAM

M. W. BERNS
Department of Developmental and Cell Biology, University of California, Irvine, California 92664, U. S. A.

INTRODUCTION

The use of a ruby laser in microbeam irradiation was first described by Bessis [1] in 1962. Since that time, there have been numerous publications and several reviews on the use of lasers in biological microirradiation [2-4]. Laser microbeam studies fall into two categories: (a) studies that describe laser effects on the cell, (b) studies that use the laser microbeam as a tool for answering biological questions. Both of the above are not mutually exclusive. There are studies in which a structural effect is described and a biological question is answered. However, for the large part, investigations have fallen into one or the other category.

The general approach that we have been taking is to study cell function by laser microirradiation of various subcellular structures [4]. However, to fully understand how laser microirradiation alters cell function, the ultrastructural effects of the irradiation must be characterized as well. As a result, considerable effort has been devoted to electron microscope analysis of microirradiated cells. In this manuscript, we will briefly review past investigations and present electron microscope data from some more recent experiments.

MATERIALS AND METHODS

Laser Instrumentation

Three laser microbeams have been employed. The argon laser microbeam employs a Hughes 3030H pulsed argon ion laser (primary wavelength 488 and 514 nm). The pulse duration is 50 μsec and the peak power is 34 watts. In all experiments, the laser was focused to a spot of 2-0.25 μm. Energy density in the focused spot varied from 25 $\mu J/\mu m^2$ to 1,000 $\mu J/\mu m^2$. The size of the spot was varied by using different microscope objectives, and the energy density was controlled by attenuating the raw laser beam with calibrated neutral density filters. The entire argon laser system with associated microscope and videotape have been described in detail elsewhere [5].

The second laser microbeam employs a flashlamp pumped tunable organic dye laser [6]. The available wavelengths span the visible region of the spectrum from 450 to 640 nanometers. Output of the visible wavelengths can be as high as 1 megawatt. However, due to problems of dye stability and flashlamp variability, the output varied as much as 20% between shots [4]. This laser was also mounted above a Zeiss phase contrast photo-microscope (Fig. 1).

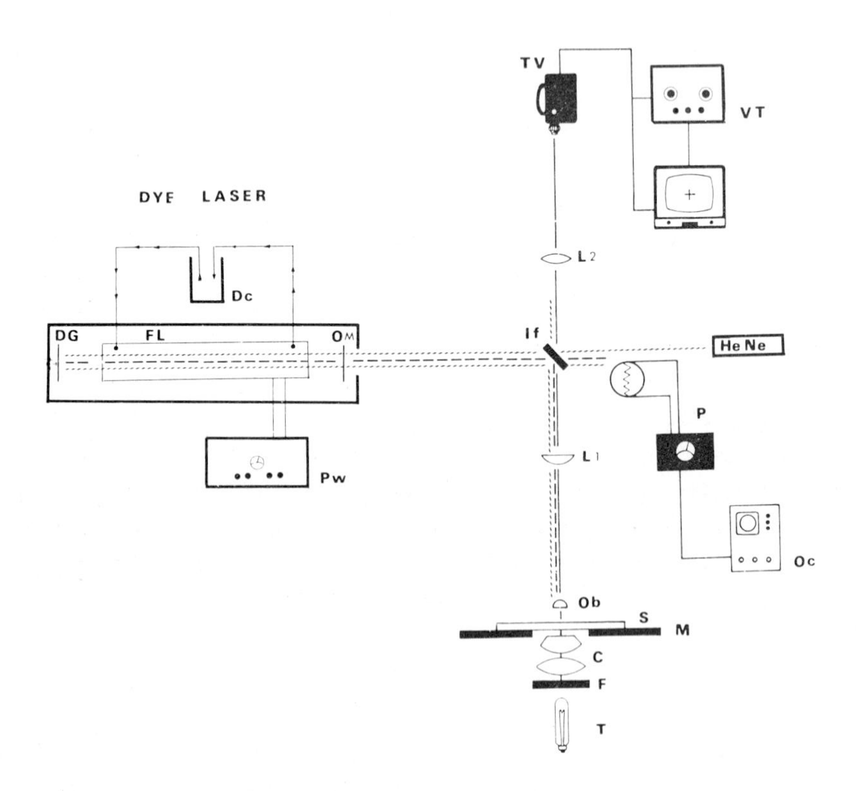

Fig. 1. Dye laser microbeam system. DC, Dye container; DG, diffraction grating; FL, flash lamp; OM, output mirror; Pw, laser power supply; TV, television camera; VT, videotape recorder; L_2, focal length image correction lens; If, dichroic filter; L_1, X1 ocular; Ob, microscope objective (either Zeiss X40 or X100 Neofluar); S, specimen chamber; M, microscope stage; C, microscope condenser; F, substage filter; T, tungsten light source; HeNe, helium neon laser alignment beam; P, power meter; Oc, oscilloscope; ---, dye laser beam; ——, substage illumination containing specimen image; ∿∿, alignment laser beam; dye laser purchased from Synergetics, Inc., Princeton, New Jersey.

The third laser microbeam system employed a neodymium YAG laser with intracavity doubling optics. The frequency shifted output of this laser was used to pump either an organic dye laser or was further frequency shifted by passing it through a fixed doubling apparatus using an ADP crystal. The organic dye laser was pumped with any of the frequency shifted YAG wavelengths, and the dye wavelengths were further frequency doubled with ADP crystals. The entire system is diagrammed in Fig. 2. This system provided laser wavelengths from 265 to 679 nanometers. Output varied from 3 watts to 3 KW depending upon the wavelength and its method of generation.

Cell Culture Procedures

In all studies, vertebrate cells in culture were irradiated. Chromosome irradiation studies were performed in primary cultures of salamander (*Taricha*) lung [7] and established cultures of various rat kangaroo (*Potorous tridactylis*) cell lines [8]. Early nucleolar irradiation studies were performed on a malignant human cell line, CMP [9], and later studies have been performed on the rat kangaroo lines [10]. Mitochondrial irradiation studies have been performed on primary cultures of newborn rat heart ventricles as initially described by Mark and Strasser [11].

In all the organelle-irradiation studies, the cells were grown in Rose culture chambers on number one thickness coverglasses in supplemented Eagle's medium. The procedures for irradiation were identical to those already described.

Analytical Procedures

The primary procedures employed to study the irradiated cells were time-lapse microscopy, cytochemical staining, radioautography, and electron microscopy. The time-lapse studies employed either a 16mm Sage Model #500 system in a 37°C hot room or an especially modified Sony half-inch videotape system with time-lapse capabilities. This latter system was combined with the laser microbeam system (Fig. 1) and allowed for continual taping prior, during, and after microirradiation.

Standard Feulgen [12] for DNA and alkaline fast green [13] for histone proteins staining procedures were used to analyze chromosome alterations. Radioautography employing H^3-uridine was used to analyze nucleolar irradiated cells [10].

Cells that were to be studied with the electron microscope were grown in Rose chambers on coverslips precoated with 0.1% of Siliclad. This was necessary to facilitate subsequent separation of the cells from the glass.

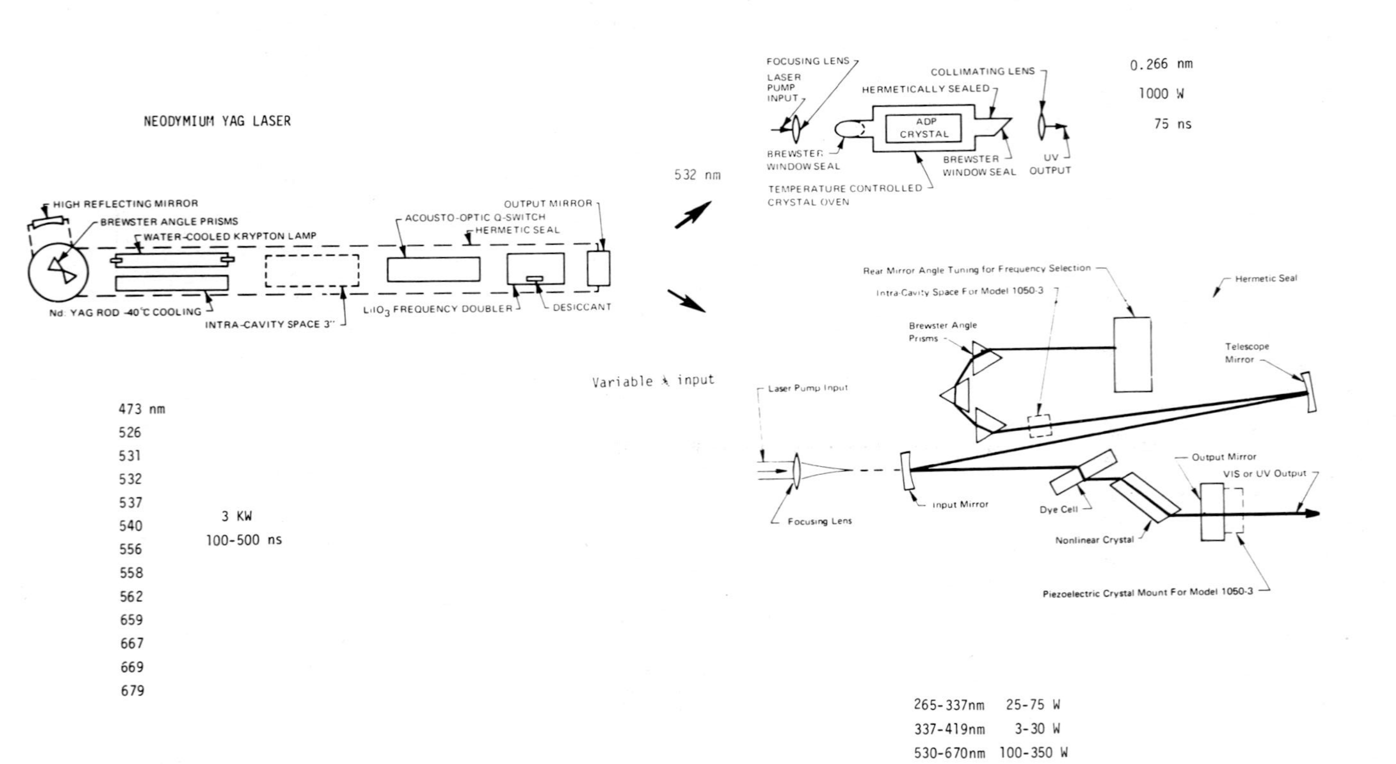
NEODYMIUM YAG LASER
HIGH REFLECTING MIRROR
BREWSTER ANGLE PRISMS
WATER-COOLED KRYPTON LAMP
Nd: YAG ROD -40°C COOLING
INTRA-CAVITY SPACE 3"
ACOUSTO-OPTIC Q-SWITCH
HERMETIC SEAL
OUTPUT MIRROR
LiIO3 FREQUENCY DOUBLER
DESICCANT
473 nm
526
531
532
537
540
556
558
562
659
667
669
679
3 KW
100-500 ns
532 nm
Variable λ input
FIXED FREQUENCY DOUBLER
FOCUSING LENS
LASER PUMP INPUT
BREWSTER WINDOW SEAL
HERMETICALLY SEALED
ADP CRYSTAL
COLLIMATING LENS
BREWSTER WINDOW SEAL
UV OUTPUT
TEMPERATURE CONTROLLED CRYSTAL OVEN
0.266 nm
1000 W
75 ns
Rear Mirror Angle Tuning for Frequency Selection
Intra-Cavity Space For Model 1050-3
Hermetic Seal
Brewster Angle Prisms
Telescope Mirror
Laser Pump Input
Focusing Lens
Input Mirror
Dye Cell
Nonlinear Crystal
Output Mirror
VIS or UV Output
Piezoelectric Crystal Mount For Model 1050-3
265-337nm 25-75 W
337-419nm 3-30 W
530-670nm 100-350 W
30-35 ns
DYE LASER

Fig. 2

Cells were fixed, flat embedded and serially thin sectioned as described originally by Brinkley et al. [14] and modified by Rattner and Berns [15]. Thin sections were cut in the silver range and examined on either a Siemens Elmiskop 1A or a Zeiss 9S electron microscope.

RESULTS AND DISCUSSION

Chromosomes

Chromosome irradiation studies have either involved structural dissection or functional inactivation. The structural studies have been aimed at trying to selectively damage the molecular components of the chromosome. By combining this capability with subsequent analysis over varying time periods, a better understanding of chromosome organization and replication could be attained.

Employing blue-green wavelengths, it is possible to produce lesions as small as 0.25 μm in preselected chromosomes. When viewed through the phase microscope, all these lesions appear as a "lightened" or "paled" spot (Fig. 3). The paled lesions have been produced by either irradiation with laser densities of 500-1,000 $\mu J/\mu m^2$ to untreated cells or by irradiation with lower energy densities (50 $\mu J/\mu m^2$) of cells that have been pretreated with light-absorbing compounds that either bind to the chromosome or are incorporated as part of the chromosomal DNA.

The agent most extensively employed is the amino acridine dye, acridine orange (AO). This compound has been shown to bind directly to the DNA molecule by intercalation between the nitrogenous base-pairs [16]. When chromosomes that have been exposed to this dye for five minutes are irradiated with focused blue-green laser light, lesions of varying degrees of severity can be produced depending upon the output of the laser and the concentration of the dye [7]. Viewed with the electron microscope, the lesions appeared to be confined to the chromosome region that appeared pale when viewed with the phase contrast microscope. The lesion area appeared to contain small spherical electron dense bodies 0.05-0.15 μm in diameter [15].

In contrast to the above finding, chromosome paling produced with high laser energies (500 $\mu J/\mu m^2$) without acridine orange treatment contained larger and interconnected electron dense aggregates (Fig. 4). When lesions produced under both irradiation conditions were cytochemically analyzed for DNA and histone-protein, it was found that the AO-treated chromosomes irradiated with low and moderate laser energies were negative for DNA and positive for histone in the lesion area. Chromosomes irradiated with high energies (500 $\mu J/\mu m^2$) without AO treatment were positive for DNA and negative for histone (Table 1). Based upon the ultrastructural, cytochemical, and subsequent functional studies (see later part of this paper), it is apparent that by varying appropriate parameters, the laser microbeam can be used to make selective molecular alterations.

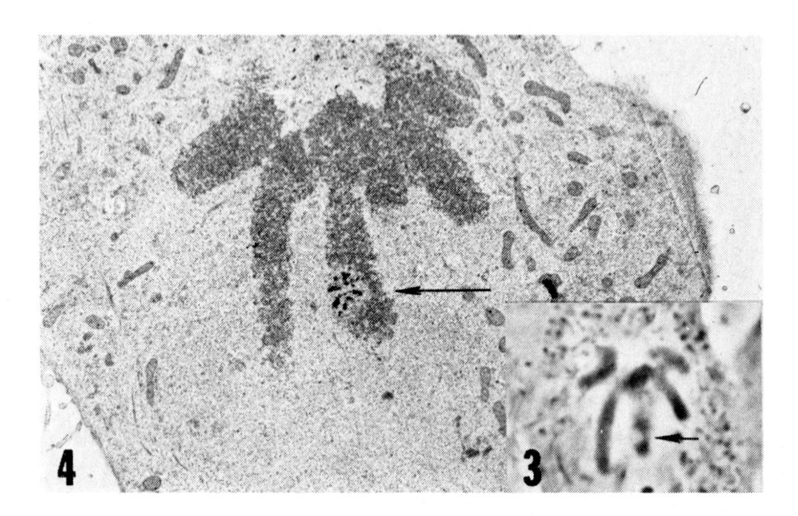

Fig. 3. Typical chromosome paling following laser microirradiation with 500-1,000 $\mu J/\mu m^2$.
Fig. 4. High power electron micrograph of lesion area from chromosome irradiated with high laser energy without acridine orange treatment. Note large electron dense aggregate (X40,000).

In addition to the above studies employing acridine orange, other preliminary investigations using ethidium bromide (also an intercalary agent) have shown selective laser sensitivity. Ethidium bromide has been used in numerous studies as a metabolic inhibitor [17,18], and it has been suggested that its effects are reversible upon removal [19]. Because of its reversible properties and selective binding activity, it could be potentially useful in laser microbeam studies. Preliminary studies have shown that ethidium bromide does selectively sensitize the chromosomes to the laser microbeam. A ten minute exposure of the cells to 10 $\mu g/ml$ of ethidium bromide resulted in sensitization to the laser and the production of the typical "paled" lesion. In addition, within two hours of the ethidium bromide treatment, the chromosomes had lost their laser sensitivity. The results were further corroborated by fluorescent microscopy which demonstrated a complete loss of ethidium bromide fluorescence by 8 hours following treatment.

Table 1

	Pre-treatment Acridine Orange 0.1 mg/ml 5 min.	Wavelength nanometers	Energy $\mu J/\mu^2$	Light microscope morphology	Electron microscope morphology	Alkaline fast green	Feulgen reaction
ARGON LASER	−	488 514	1,000	Large pale region	Central electron dense mass + peripheral aggregates	−	−
	−	488 514	500	Small pale region	Electron dense aggregates .08 -.19μ	−	+
	+	488 514	500	Large pale region	Numerous electron dense aggregates 0.5 -.15μ	+	−
	+	488 514	50	Small pale region	Numerous electron dense aggregates .05 -.15μ	+	−
DYE LASER	−	460	not available	Large pale region	Central electron dense mass + peripheral aggregates	not available	not available
	−	460	not available	Small pale region	Electron dense aggregates .08 -.19μ	not available	not available
	+	460	not available	Large pale region	Numerous electron dense aggregates .05 -.15μ	not available	not available
	+	460	not available	Small pale region	Numerous electron dense aggregates .05 -.15μ	not available	not available

Another compound that results in increased sensitivity of chromosomes to argon laser microirradiation is BrdU (bromodeoxyuridine). When cells were exposed to BrdU for 48 hours prior to irradiation, the BrdU was incorporated into the new DNA during S phase of the cell cycle in place of thymine. Numerous studies have indicated that BrdU-treated cells are very susceptible to chromosome breakage when exposed to visible or near visible light [20,21]. Apparently this finding can be carried down to the subchromosomal level. By focusing the laser microbeam onto the BrdU-substituted chromosomes, it is possible to produce the same kind of chromosome paling observed with AO and ethidium bromide. Ultrastructural studies have yet to be carried out to ascertain if the fine structure of the lesions are similar to the AO-induced paling.

Functional studies on chromosomes have been aimed at selective inactivation of the nucleolar organizer genes and selective removal of whole chromosomes followed by cell cloning [22,23]. Irradiation of the secondary constriction regions of mitotic salamander and *Potorous* cells resulted in concomitant reduction of nucleoli formed after mitosis. When all the major secondary constrictions were irradiated, the post-mitotic cells did not produce any major nucleoli but instead produced numerous small micronucleoli [8]. Current studies are being conducted to determine the metabolic activity and fine structure of these micronucleoli by incubation in H^3-uridine and electron microscopy.

More precise studies have been undertaken to determine whether the nucleolar ribosomal DNA is located directly in the secondary constriction or immediately adjacent to it. Irradiation of the adjacent region consistently resulted in a reduction of post-mitotic nucleoli whereas direct secondary constriction irradiation resulted in nucleolar reduction in 50% of the cases [24]. These studies suggested that the adjacent region is important to nucleolar formation. Current studies employing *in situ* cytological molecular hybridization [25] will help answer the question of the precise location of the ribosomal (nucleolar) DNA.

In addition to the above studies, a series of investigations have been undertaken to selectively damage the nucleolar DNA followed by cell cloning. The purpose of these experiments is to establish a series of cell lines deficient in varying amounts of nucleolar DNA. Preliminary results have demonstrated that post-mitotic cells are consistently lacking in nucleoli. In subsequent divisions of the irradiated cells, some have regained the normal number of nucleoli and others maintained the deficiency (Witter and Berns, unpublished). Whether or not the regained nucleoli represent a repair of damaged ribosomal DNA, the expression of previously repressed DNA, or the incomplete destruction of a block of rDNA, must await quantitative rDNA determination of cell lines derived from the cloned cells.

Studies on the selective removal of whole chromosomes have been reported elsewhere [22]. Chromosomes irradiated in their centromere regions lost their attachment to the mitotic spindle and were frequently lost from one or both daughter cells. A surprising finding was that the clonal descendants of these cells had the normal number of the particular chromosome removed. Several mechanisms were proposed for replacement of the lost chromosome. Current experiments have been aimed at elucidating this mechanism and further repeating the original results (Table 2).

Table 2
Summary of cloning results - PTK_2 cells

	Total Attempts	At Least One Additional Mitosis	More Than One Additional Mitosis	Successful Clones	Cloning %*
Lasered Chromosomes**	102	37	21	9	9%
Lasered Non-Chromosome Region (Control)	9	9	9	6	66%
No Laser (Control)	26	25	24	21	80%

* Cloning % = $\frac{\text{Successful Clones}}{\text{Total Attempts}}$

** This class represents both partial chromosome deletion of DNA and entire chromosome removal by kinetochore irradiation.

Nucleoli

Nucleolar irradiation studies have employed the vital dye quinacrine hydrochloride as a selective sensitizing agent [9,26]. Quinacrine treatment (0.1 μg/ml) followed by laser microirradiation is effective in producing lesions as small as 0.25 μm in single nucleoli. Cells incubated in H^3-uridine after the laser microirradiation have a reduced ability to synthesize RNA.

More recent studies [10] have been aimed at elucidating the functional role of the two major nucleolar fine structural elements, the fibrillar and granular components. Cells were initially treated with the drug actinomycin D for two hours (0.5 μg/ml). This resulted in the temporary separation (segregation) of the fibrillar and granular elements [27]. The cells were next sensitized with quinacrine hydrochloride and then laser microirradiated either in the granular or the fibrillar regions [10]. The cells were then incubated in H^3uridine for thirty minutes in order to determine the effect on RNA synthesis. The results of these experiments indicated only a temporary reduction in RNA synthesis when the granular region was irradiated and a more severe and permanent reduction in RNA synthesis when the fibrillar region was irradiated. These data supported the idea of initial RNA synthesis in the fibrillar region and subsequent movement to the granular region [28].

Preliminary electron microscope studies reveal the lesion material as spherical electron dense bodies. The lesion material appears to extend considerably away from the irradiated nucleolus suggesting that the damaged nucleolar material moves away from the nucleolus within a few minutes of irradiation. Fixation always occurred within three minutes of the irradiation.

Heart Cell Mitochondria

The large mitochondria of contracting ventricular myocardial cells absorb the visible blue-green laser energy readily because of the presence of the cytochrome respiratory pigments. Several types of lesions have been produced, and these have been correlated with specific changes in cellular contractility [29,30].

Ultrastructurally, the lesions fall into two categories - weak and moderate. The weak lesions involve damage to rather localized regions within the mitochondrion. The moderate lesions involve several concentric zones of damage and result in a more generalized effect to the mitochondrion. However, cells with both lesion types do not appear to contain any obvious damage outside the irradiated mitochondrion. Only cells with the moderate lesion undergo contractility changes (Table 3).

Table 3
Correlation of lesion type, laser energy density, and contractility

	Laser energy**			Change	Remains
Lesion type	6.7μJ	10.0μJ	12.6μJ	in contraction	rhythmic
Weak	12*	5	1	0	18
Moderate	0	2	15	14	3

* Number of cells

** Energy density in 0.5 μ focused spot

More recently preliminary studies have involved irradiation with the longer wavelengths (red) of the dye laser. The lesions produced appeared similar to the weak lesions produced with the argon laser. However, much higher energy densities had to be used, and contractility changes were not produced. This contrasts with an earlier study [31] where a red ruby laser did not produce mitochondrial alterations unless a vital dye, Janus green B, was first bound to the mitochondria.

Centrioles

A series of investigations has been undertaken to elucidate the molecular organization of centrioles as well as their role in the mitotic process. Irradiation of the centriolar duplexes with high (1,000 $\mu J/\mu m^2$) argon laser energies without employing a light-sensitizing agent did not result in any apparent structural alteration. This result suggests that the centriolar complex does not contain substantial basic histone protein. Additional studies have employed acridine orange (1-0.1 μg/ml for five minutes) as a sensitizing agent. In these investigations, it appears possible to damage the centriolar region. However, it is possible that the acridine orange is binding to the cytoplasm surrounding the centrioles and not to the centrioles directly. Whether or not this approach will be useful in the study of centriole structure and function must await further ultrastructural studies and studies where irradiated cells have been cloned.

CONCLUSION

In summary, several laser microbeams are being used to selectively alter various cell organelles. In combination with accessory techniques, such as, electron microscopy, autoradiography, time-lapse microscopy, cytochemistry, and tissue culture cloning, numerous problems in cell function and organization are being investigated.

ACKNOWLEDGMENTS

The research reported here has been supported by grants from the National Science Foundation, 36646, 43527; U. S. PHS 15740; and the University of California Cancer Coordinating Committee. I am greatly indebted to Dr. J. B. Rattner and Stephanie Meredith for the electron microscopy reported here.

REFERENCES

1 M. Bessis, F. Gires, G. Mayer and G. Nomarski, Compt. Rend. Acad. Sci.,

225(1962)1010.
2 M. W. Berns and D. E. Rounds, Ann. N. Y. Acad. Sci., 168(1970)550.
3 M. W. Berns and C. Salet, Int. Rev. Cytol., 33(1972)131.
4 M. W. Berns, Int. Rev. Cytol., 39(1974)383.
5 M. W. Berns, Exp. Cell Res., 65(1971)470.
6 M. W. Berns, Nature, 240(1972)483.
7 M. W. Berns, D. E. Rounds and R. S. Olson, Exp. Cell Res., 56(1969)292.
8 M. W. Berns, A. D. Floyd, K. Adkisson, W. K. Cheng, L. Moore, G. Hoover, K. Ustick, S. Burgott and T. Osial, Exp. Cell Res., 75(1972)424.
9 M. W. Berns, R. S. Olson and D. E. Rounds, J. Cell Biol., 43(1969)621.
10 L. Brinkley and M. W. Berns, Exp. Cell Res., 87(1974)417.
11 G. E. Mark and F. O. Strasser, Exp. Cell Res., 44(1966)217.
12 C. Leuchtenberger, in J. F. Danielli (Editor), General Cytochemical Methods, Vol. 1, Academic Press, New York, 1958, p. 219.
13 A. D. Deitch, in G. L. Wied (Editor), Introduction to Cytochemistry, Academic Press, New York, 1966, p. 327.
14 B. R. Brinkley, P. Murphy and L. C. Richardson, J. Cell Biol., 35(1967) 279.
15 J. B. Rattner and M. W. Berns, J. Cell Biol., 62(1974)526.
16 L. S. Lerman, Proc. Nat. Acad. Sci. (U.S.), 49(1963)94.
17 A. L. Snyder, H. E. Kann, Jr. and K. W. Kohn, J. Mol. Biol., 58(1971) 555.
18 E. Zylber, C. Vesco and S. Penman, J. Mol. Biol., 44(1969)195.
19 G. Soslau and M. M. K. Nass, J. Cell Biol., 51(1971)514.
20 T. C. Jones and W. F. Dove, J. Mol. Biol., 64(1972)409.
21 E. H. Y. Chu, N. C. Sun and C. C. Chang, Proc. Nat. Acad. Sci. (U.S.), 69(1972)3459.
22 M. W. Berns, Science, 186(1974)700.
23 M. W. Berns, Cold Spring Harbor Symp., 38(1974)165.
24 M. W. Berns and W. K. Cheng, Exp. Cell Res., 69(1971)185.
25 J. G. Gall and M. L. Pardue, Proc. Nat. Acad. Sci. (Wash.), 63(1969)378.
26 M. W. Berns, S. El-Kadi, R. S. Olson and D. E. Rounds, Life Sci., 9(1970)1061.
27 H. Busch and K. Smetana, The Nucleolus, Academic Press, New York, 1970.
28 N. Granboulan and P. Granboulan, Exp. Cell Res., 38(1965)604.
29 M. W. Berns, D. C. L. Gross, W. K. Cheng and D. Woodring, J. Mol. Cell. Cardiol., 4(1972)71.
30 M. W. Berns, D. C. L. Gross and W. K. Cheng, J. Mol. Cell. Cardiol.,

4(1972)427.

31 R. Storb, R. K. Wertz and R. L. Amy, Exp. Cell Res., 45(1967)374.

DISCUSSION

M. BERNS

R. Mohan - Our group from Düsseldorf and Munich (prof. R. Kaufmann and his group) has developed a compact laser mico dissection instrument which has been recently put in the market for the purpose of the various kinds of interesting work reported by you. We are also developing a laser microsonde with a mass spectrum analyser for estimating the concentrations of Ca, Na and K within each small organelles, e.g. within the muscle fibre system.

M. Berns - That is very interesting because we think onepossibility to explain the results we observe in heart cell mitochondria, is the release of Ca^{++} from the irradiated mitochondrion. We would be greatly interested in an instrument that could detect localized differences in Ca^{++} distribution.

O. Svelto - What is the mechanism by which the DNA is destroyed by the laser beam?

M. Berns - We do not know. However, it could be a photochemical process, a two-photon effect, heat, dielectric breakdown or plasma formation. It is a very difficult problem to solve because of the extremely small area that we are irradiating.

M.W. Windsor - You are focussing your laser into a quite small volume. Perhaps local heating is the effective damage mechanism.

M. Berns - Yes, that is a definite possibility. In fact the electron microscope pictures of the mitochondrial damage look very much like a thermal burn, with the greatest damage in the center.

S. Ameen - The mechanism of damage to DNA of the chromosomes can also be explained on the basis of the factthat UV and visible electromagnetic radiations cause ionization of some of the aromatic moieties of the nucleic acids (like of tyrosine, tryptophane, histidine, etc.). In case longer wavelengths (as IR radiation) can be employed I would not expect similar damage; however damage due to other cause may or may not occur. It may so happen that the high intensity of the laser beam may cause serious conformational changes in the molecular structure and moieties attached which sterically forbids the reaction with dye acridine orange. I will suggest an experiment to irradiate first and then react with dye to test this clearly.

S. Leach - With focal spots of the order of those used (e.g. 0.25 μ) the irradiation power density is well sufficient to create a plasma at the point of impact. One should therefore consider the possibility of plasma physics and plasma chemistry processes in explaining
1) the mechanism of production of lesions and also, perhaps,
2) subsequent structure effects with their implications on the biological level. Purely photochemical processes may be relatively unimportant.

M.W. Windsor - If a plasma were formed, I would expect formation of a gas bubble that would cause destruction more widespread than that which you observe.

M. Berns - I agree. The electron microscope pictures show very little secondary damage outside the lesion proper.

J.K. Eweg - In connection with Dr. Windsor's question about the mechanism of damaging the chromosomes or mitochondria, I would like to ask you if it is possible, after separation of mitochondria from the cell, to apply the same technique. In this case you know the absorbing specied and you will be able, in principle at least, to analyse photoproducts.

M. Berns - This is certainly a possibility and we have thought of it. However we will need a laser powerful enough to approximate the same energy density that we have in a 0.25 μm spot.

C.B. Harris - Just a comment. This work opens up a great many new experiments, particularly when you consider the possibilities of selective photoaffinity labeling.

LASER STUDY OF OXYGEN DIFFUSION IN HEMOPROTEINS

B. ALPERT, Institut de Biologie Physico-chimique
13, rue Pierre et Marie Curie, 75005 - Paris , and
L. LINDQVIST, Laboratoire de Photophysique Moléculaire du
C.N.R.S., Université de Paris-Sud, 91405 - Orsay .

SUMMARY

A method of studying structural changes in hemoproteins in solution is described. The protein structure determines the rate of oxygen diffusion through the protein towards the porphyrin; this diffusion rate was obtained by measuring the rate of quenching of the triplet state of the Fe-free hemoprotein by oxygen. The second harmonic (529 nm) of a Q-switched Nd glass laser was used to excite the Fe-free hemoprotein; the triplet formation and decay were measured by kinetic spectrophotometry. The method was used to study structural variations in hemoglobin due to pH variations and phosphate fixation.

INTRODUCTION

In previous papers [1, 2] we presented a method for the study of oxygen diffusion in hemoproteins, using the triplet state as a probe; a similar approach has also been used by Geacintov et al. in a study of the oxygen diffusion in DNA [3] . It is well-known that molecules in the triplet state are quenched efficiently by oxygen; measurements of the decay of the triplet state of the porphyrin group in the hemoprotein thus yield the rate of oxygen diffusion through the protein towards the active site of the heme group. To obtain a triplet state lifetime sufficiently long to allow the oxygen diffusion study, it was necessary to remove the Fe(II) ion from the heme group. The Fe-free hemoproteins were exposed to a short pulse of laser light populating the porphyrin triplet state, and the triplet decay was followed by kinetic spectrophotometry.

The results showed that the rate of oxygen diffusion to the heme pocket is an order of magnitude faster than the rate of oxygen fixation in hemoglobin and myoglobin, which indicates that the diffusion is not the rate-limiting step in the fixation process.

In the present study, we demonstrate that the method described above can be used to reveal structural variations in the proteins. Such variations were induced by varying the pH and by adding regulators (organic and inorganic phosphates).

EXPERIMENTAL

A Q-switched Nd^{3+}-doped glass laser with second harmonic generation produced light pulses of wavelength 529 nm and FWHM 30 ns; the pulse energy was in the 10-100 mJ range in the experiments described. The sample solution, contained in 1 x 10 or 5 x 10 mm spectrophotometry cells, was maintained at 2°C.

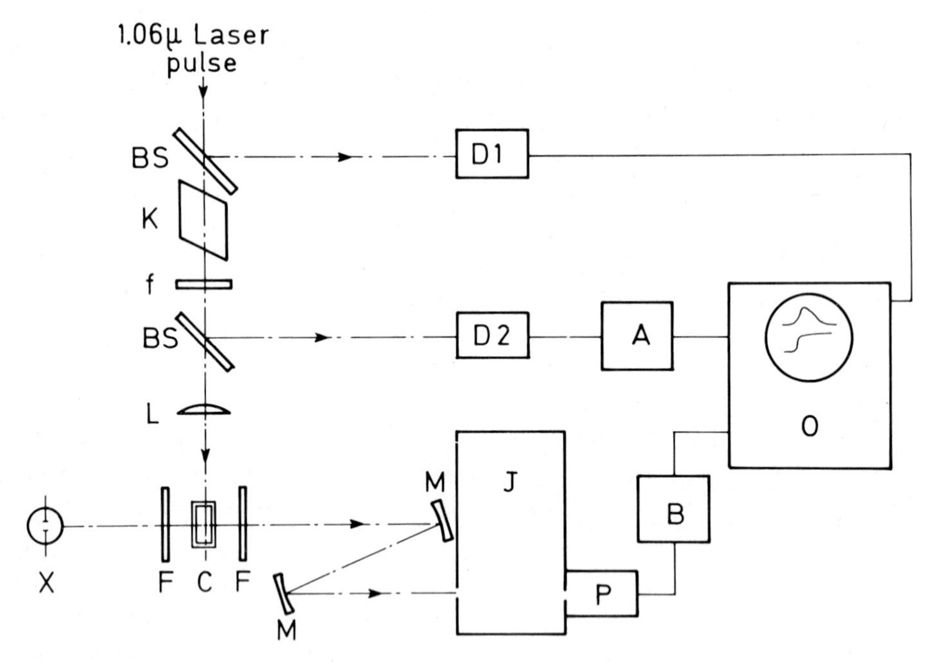

Fig.1. Laser apparatus. A, integrating circuit; B, dc compensating circuit; BS, beam splitter; C, sample cell in temperature-controlled cell compartment; D1, photodiode triggering oscilloscope sweep; D2, photodiode monitoring laser energy; F, filters; f, 4mm BG 38 Schott filter; J, 25 cm Jarrell-Ash monochromator (bandwidth 1.6 nm); K, KDP crystal; O, 100 MHz double-beam oscilloscope; P, photomultiplier tube; X, Xenon flash lamp (FWHM 1 ms).

The laser light, incident through one of the shorter cell sides, produced transient optical transmission changes which were measured as a function of time at various wavelengths using the set-up shown in Fig. 1. The laser and the monitoring flash lamp were synchronized such that the laser pulse was obtained during the peak of the flash lamp emission.

Human hemoglobin A and the main component of horse myoglobin were isolated by the usual methods [4, 5]. The globins of these proteins were associated with protoporphyrin IX in solutions which were rigorously free of phosphates [6]. The excess porphyrin and the denatured proteins were eliminated by chromatography. The eluted peak gave an homogeneous component of Fe-free hemoprotein on electrophoresis on starch gel.

The solutions used were 0.5 - 1 x 10^{-4} M with respect to porphyrin. Ultracentrifugation showed that the Fe-free myoglobin was present as a monomer and the Fe-free hemoglobin as a tetramer at these concentrations. The solutions were buffered using alternatively 0.005 - 0.01 M tris or bis-tris, 0.1 M ammonium sulfate, 0.1 M K/Na phosphate, or 0.1 M sodium borate. Sodium 2,3-diphosphoglycerate (DPG) solution [7] was added in some of the runs. The ionic strength was kept at 0.2 by adding NaCl. The pH was adjusted at 2°C. All solutions were used within 24 hours of preparation. The absorption spectra of the Fe-free hemoproteins were independent of pH and added salts in the conditions of the present study.

RESULTS AND DISCUSSION

It was shown in our previous studies [1, 2] that laser excitation at 529 nm of Fe-free hemoglobin ($Hb_{des\ Fe}$) or myoglobin ($Mb_{des\ Fe}$) produces transient changes in light absorption due to the porphyrin triplet state (populated by intersystem crossing from the excited singlet state), and absorption spectra of the proteins in the triplet state in phosphate-buffered solution were reported. In the present study, the absorption spectrum of the triplet state of Hb_{desFe} was measured from pH 6.2 to 9.5, in the presence and absence of phosphate (DPG or inorganic phosphate). It was found that these different conditions do not influence the triplet absorption spectrum (Fig. 2).

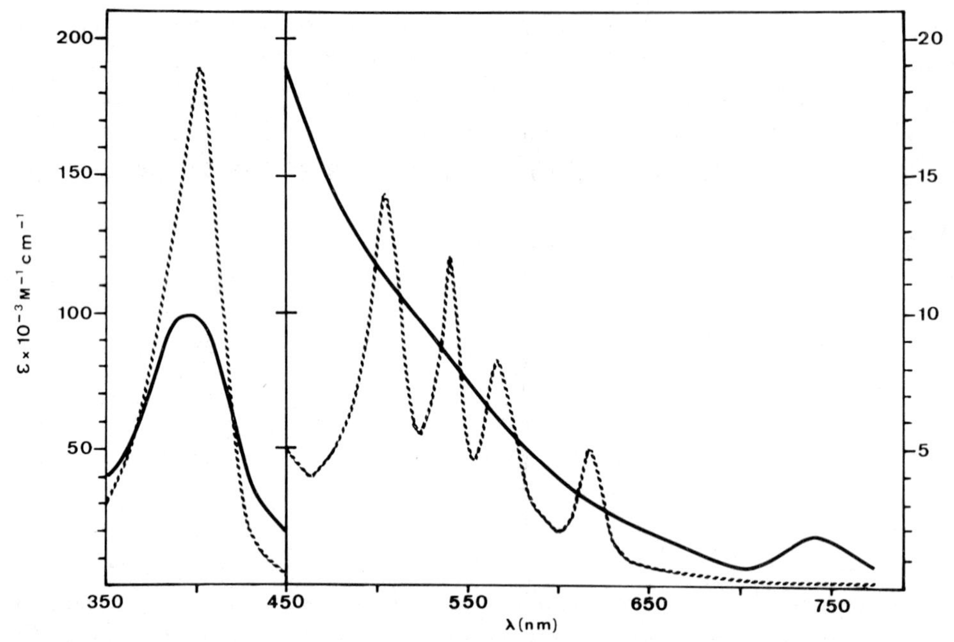

Fig. 2. Absorption spectrum of $Hb_{des\ Fe}$ in the ground state (-----) and in the triplet state (———). These spectra were obtained in the pH range 6.2 - 9.5, in the presence and absence of phosphate.

The triplet (^3P) is rapidly deactivated in the presence of air at atmospheric pressure, by the process

$$^3P + {}^3O_2 \xrightarrow{k_q} P + {}^1O_2 \quad (\text{or } {}^3O_2) .$$

The decay due to other processes may be neglected compared to the rate of this quenching reaction [1] . Then, one may express the triplet decay after laser excitation by the relation

$$-d\left[^3P\right] /dt = k_q \left[^3O_2\right] \left[^3P\right] . \qquad (1)$$

The rate constant k_q can be written as the product $k_q = p \times k_d$ where k_d is the rate of diffusion of oxygen from the solution and through the protein towards the porphyrin group, and p is the probability of quenching during the porphyrin-oxygen encounter. This probability may vary with the nature of the triplet molecule; however, in the present investigation, the porphyrin was the same throughout the study, so p should be a constant. The oxygen diffusion rate k_d will depend upon the structure of the protein particularly in the environment of the porphyrin group; variations in k_q (which is proportional to k_d) thus give an indication of structural changes in this region.

The triplet quenching study was made in the pH range 6 - 9.5 on air-saturated solutions of $Hb_{des\ Fe}$ containing

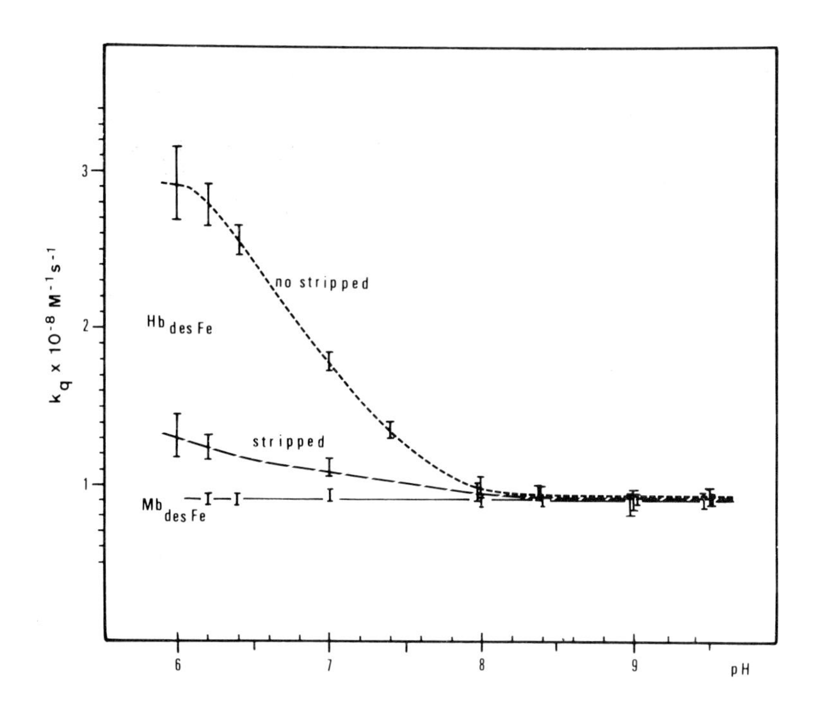

Fig. 3. pH dependence of the rate constant of porphyrin triplet quenching by oxygen (k_q) for $Mb_{des\ Fe}$ (———) and for $Hb_{des\ Fe}$ stripped (— — —) or no stripped (- - - -). The bars indicate the dispersion of the results.

phosphate (DPG or inorganic), on the stripped (phosphate-free) protein and on $Mb_{des\ Fe}$ in phosphate-free solutions. Porphyrin triplet decays measured after laser excitation were found to be first-order, in agreement with Eq.(1). This equation was used to calculate k_q, using the value of 4.3×10^{-4} M for the oxygen concentration at 2°C [8] . The results are presented in Fig. 3.

It is seen that k_q for $Mb_{des\ Fe}$ is independent of pH whereas that of stripped $Hb_{des\ Fe}$ increases by approx. 40% when the pH is lowered from 9.5 to 6. The k_q values for the stripped protein are the same in tris, bis-tris, ammonium sulfate or borate buffers. They were also found to be independent of the ionic strength (varied from 0.01 to 0.2 in a set of runs).

The presence of phosphate (DPG or inorganic) does not influence k_q of $Hb_{des\ Fe}$ noticeably at pH 8 - 9.5; however, a strong enhancement of k_q is observed at lower pH's, reaching a ratio of 2.2 at pH 6.0.

As mentioned above, the quenching rate constant is proportional to the constant of diffusion of oxygen to the porphyrin. The observed variations in k_q with pH for $Hb_{des\ Fe}$ may be related to the variations of the oxygen fixation in normal hemoglobin with pH (the Bohr effect). The ionization of some of the peptide chain aminoacids in the hemoglobin tetramer due to pH modulates the formation of the $Fe(II)-O_2$ complex; this ionization is accompanied by a slight conformational change at the termini of the α and β chains [9]. A corresponding effect does not exist in myoglobin. The strong pH dependence of k_q for $Hb_{des\ Fe}$ (which is in the T structure [10, 11]) apparently reflects the small protein readjustments associated with the ionization reactions. The readjustments influence the rate of diffusion of the oxygen towards the porphyrin in the active site. For $Mb_{des\ Fe}$, on the other hand, there is no variation in k_q.

Fig. 2 reveals a striking difference in k_q for $Hb_{des\ Fe}$ in the presence and absence of phosphate, at pH 6 - 7.5. It has been shown previously that the fixation of DPG (or inorganic phosphates) between the β chains in the Hb T state slightly perturbs the N-terminal groups [12] . This perturbation is transmitted through the whole chain and modifies the magnetic properties of the Fe ions of the four chains [13] . Our results indicate that this perturbation also affects the accessibility of oxygen to the porphyrin.

In conclusion, the present study shows that oxygen diffusion rate measurements by triplet quenching provide a means of revealing even subtle changes in hemoprotein properties.

REFERENCES

1 B. Alpert and L. Lindqvist, International Conference on the Excited States of Biological Molecules, Lisbon (April 18th-24th, 1974).
2 B. Alpert and L. Lindqvist, Science 187 (1975) 836.
3 N.E. Geacintov, W. Moller and E. Zager, International Conference on the Excited States of Biological Molecules, Lisbon (April 18th-24th, 1974).
4 D.L. Drabkin, J. Biol. Chem. 164 (1946) 103.
5 P. George and D.H. Irvine, Biochem. J. 52 (1952) 511.
6 M. Coppey and B. Alpert, C.R. Acad. Sci. Paris (sous-presse).
7 R. Benesch, R.E. Benesch and C.I. Yu, Proc. Natl. Acad. Sci. U.S. 59 (1968) 526.
8 Landolt-Börnstein, Zahlenwerte und Funktionen (1962) Springer-Verlag, Berlin.
9 M.F. Perutz, H. Muirhead, L. Mazzarella, R.A. Growther, J. Greer, J.V. Kilmartin, Nature 222 (1969) 1240.
10 R. Noble, G.L. Rossi, R. Berni, J. Mol. Biol. 70 (1972) 689.
11 S. Ainsworth and A. Treffry, Biochem. J. 137 (1974) 331.
12 A. Arnone, Nature 237 (1972) 146.
13 Y. Alpert, R. Banerjee and J. Denis, Nature New Biol. 243 (1973) 80.

DISCUSSION

B. ALPERT, L. LINDQVIST

C. Salet - Est-il difficile d'ôter Fe? Quel est le rapport entre la structure de l'hémoglobine avec et sans Fe?

B. Alpert - Oui, il est difficile d'ôter Fe. En fait, on enlève l'hème et on le remplace par la protoporphyrine. L'hémoglobine sans Fe doit être dans la structure de la desoxyhémoglobine. Nous n'avons pas de preuve certaine pour cela, mais de fortes indications. La cinétique de certain mercuriel (PMB) sur les groupes SH est identique sur Hb désoxy et Hb desFe. L'angle de rotation à 233 nm de la lumière est le même pour ces deux composés. Enfin la constante d'équilibre tétramère-dimère de Hb_{desFe} est pratiquement celle de Hb désoxy. Ce dernier fait est important puisque l'oxyhémoglobine est beaucoup plus dissociée que la désoxyhémoglobine.

J.A. McCray - From the DPG titration curve you have given, can one conclude that oxygen diffuses more easily into the heme pocket when the tetramer Hb_{desFe} is in the state corresponding to the Hb_4deoxy-state (T-state) as compared with the tetramer oxy-state (R-state).

B. Alpert - The oxygen diffuses more easily into the heme pocket when the DPG is present. I don't konw which is the state of Hb_{desFe}, probably in the deoxy-state (T-state). I cannot conclude that oxygen diffuses into the protein more easily in the T-state compared with R-state, because T and R-states are not observed in $Hb_{des\ Fe}$.

PULSED TUNABLE LASER IN CYTOFLUOROMETRY: A STUDY OF THE FLUORESCENCE PATTERN OF CHROMOSOMES

A.ANDREONI and C.A.SACCHI
Laboratorio di Fisica del Plasma ed Elettronica Quantistica del C.N.R., Istituto di Fisica del Politecnico, Milan, Italy.
S.COVA
Istituto di Fisica del Politecnico, Milan, Italy.
G.BOTTIROLI and G.PRENNA
Centro di Studio per l'Istochimica del C.N.R., Istituto di Anatomia Comparata dell'Università di Pavia, Italy.

SUMMARY

An apparatus, based on a nanosecond-pulsed tunable dye laser and pulse electronic technique, is described to allow measurements of the main parameters of the fluorescent transition in the single bands of the chromosome fluorescence pattern. The apparatus can measure the fluorescence decay time, the fluorescence saturation intensity and the quantum yield. The aim of the present investigation has been essentially to check this laser technique of measuring fluorescence parameters in subcellular organelles to explain the variation in the QM fluorescence intensity along metaphase chromosomes. Some experimental results are reported concerning Vicia Faba M-chromosomes stained with Quinacrine Mustard (QM) and DNA binding agent.

INTRODUCTION

One of the current problems in molecular biology is to explain the banding patterns of the metaphase chromosomes to provide information about their composition. Up to the present time, several authors have proposed specific dyes (Proflavine, Acriflavine, Quinacrine, etc. and derivatives) to label DNA components, owing to their different binding mechanisms [1-5]. As far as our initial experiments are concerned, we chose Quinacrine Mustard (QM) (Fig. 1) to stain Vicia Faba M-chromosomes.

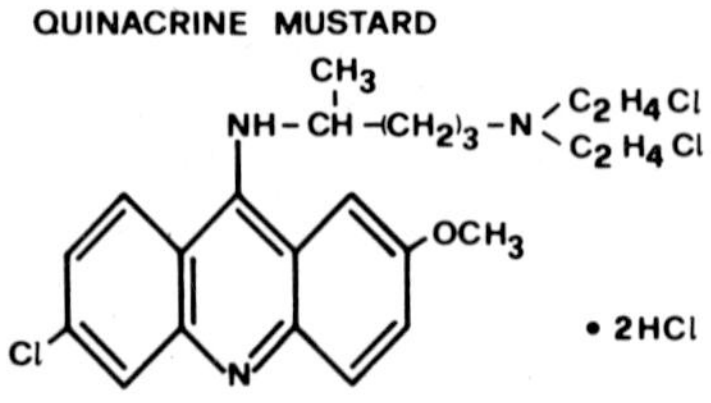

QUINACRINE MUSTARD

CH_3
$NH-CH-(CH_2)_3-N<^{C_2H_4Cl}_{C_2H_4Cl}$
OCH_3
Cl N • 2HCl

Fig. 1 - QM structure

This dye binds to the bihelical DNA of the chromosome in the following way [1,3]: (i) It intercalates two DNA adjacent base pairs thus being held by dipolar forces of the bases; (ii) The intercalation process in further strengthned by an io-nic bond between the basic heterocyclic nitrogen atom of QM and the DNA phosphate; (iii) When at least one of the two base pairs is GC, a covalent bond between the alkylating group in the QM side chain and the 7-amino group in guanine is also ef-fective for the intercalation. Detailed experimental investi-gations of the fluorescent emission of QM bound to synthetic mononucleotides and polymucleotides [3-7] and to natural DNA [2,3] have shown that the fluorescence quantum yield is stron-gly dependent on the type of binding. When QM intercalates two AT base pairs, in fact, the emitted fluorescence power is high while it is about 30 times smaller when QM is bound to a gua-nine residue [7] .

So far, despite the large amount of experimental works on this subject, the fundamental question has been left unanswe-red as to whether the banding fluorescence patterns of chromo-somes are due to either one of the following circumstances: (i) different dye intake of the DNA; (ii) different base pairs com-position of the DNA in each band. After some feasibility expe-riments involving the measuring of fluorescence parameters in single cells, we have set up an apparatus which should allows us to contribute in finding an answer to this question.

EXPERIMENTAL APPARATUS

Our experimental apparatus is shown in the block diagram in Fig. 2. In our first feasibility experiments we used a dye-

-laser solution of Rhodamine 6G in methanol (10^{-3}M). The dye--laser is transversally pumped by a Nitrogen laser which gives

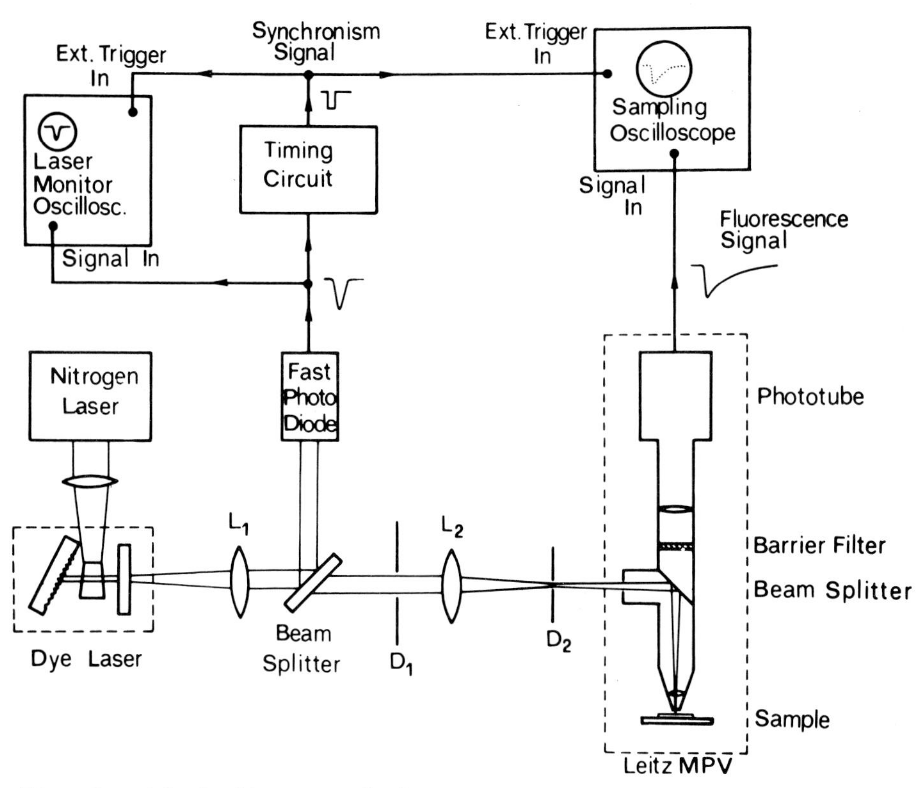

Fig. 2 - Block diagram of the experimental apparatus.

pulses of up to 2MW peak power and about 2.5 nsec of duration, with a repetition rate of up to 100 p.p.s. at λ=337.1 nm. The dye laser resonator consists of a diffraction grating (1200 grooves/nm with a first-order blaze wavelength of 750 nm) in Littrow mounting and of a flat mirror. The laser pulses have a duration of about 2 nsec, a peak power up to 100 KW and a band-width of about 1 nm; their wavelength can be tuned from 560 to 630 nm simply by rotating the grating. The lens L_1 is positioned in such a way as to give a plane beam, while L_2 focuses the beam on the diaphragm D_2. The beam then enters a Leitz MPV cy-tofluorometer whose objective forms an image of D_2 on the plane of the biological sample. The laser spot size on this plane can be changed by changing the lenses L_1 and L_2 and, as the beam is practically diffraction-limited, it can be concentrated down to the resolving spot of the microscope (about 0.5μm). A beam splitter placed after L_1 sends a small fraction of the laser light on a fast photodiode (Hadron TRG-105C; risetime $<$0.3nsec); its electrical signal serves the dual purpose of monitoring the laser power on a Tektronix 519 travelling-wave oscilloscope and of providing a suitable signal for a precision timing circuit. In order to monitor the fluorescence light, a fast photomultiplier tube (RCA-70045 D; risetime 0.7 nsec) is mounted at the top of the Leitz MPV cytofluorometer. The high PMT gain enables us to detect also single-photons; on the other hand, when we have to measure multi-photons fluorescence pulses, we must limit the peak intensity of the pulse within the linear range of our PMT; however, we can choose the optimum operating conditions for both cases. When single-photon pulses have to be observed, the electrical PMT output is sent to a conventional fast oscilloscope but, in most measurements it is sent to a sampling oscilloscope with a storage tube (Tektronix 564), thus making use of its internal facilities. However, in both cases, the oscilloscope is externally triggered by the output pulse from the timing circuit, thus obtaining synchronous measurements down to single-photon levels.

It is worth pointing out that the characteristics of our excitation laser play an important role both choosing the electronic apparatus and in determining the quality and, in some cases, even the feasibility of the measurements. In fact, our dye laser emits pulses of short duration (~2 nsec) with a very reproducible waveform and a low intensity jitter ($<$5%). These properties allow, for instance, the PMT current waveform to be measured simply by using a sampling oscilloscope, where each sample corresponds to a different laser pulse. So our system is well suited for measurements of fluorescent lifetimes even in subnanosecond range. Moreover, the repetition rate of 100 p.p.s. is sufficient to perform eventually averaged measurements in relatively short times. Actually it may be recalled that, if the number of detected photons is sufficiently high, as in our first feasibility experiments, the statistical fluctuations in the PMT currents are sufficiently low compared with the average value and even a measurement on a single waveform ca give significant data. Otherwise, as in the present case of the chromosomes fluorescence, averaging must be performed over many waveforms. Finally the high intensity available in the laser beam makes it possible to perform synchronous measurements of the excitation and fluorescence waveform and other measurements otherwise not possible as the study of the fluorescence saturation behaviour and the measurement of signals given by samples containing very little fluorescent material. Moreover, there are some characteristics peculiar to the laser beam which we have been able to use to our advantage: 1) because of its spatial coherence, the laser light can be focused in subcellular dimensions as, for example, in a single band of the chromosome fluorescence pattern; 2) the light is monocromatic and tunable so that we can excite the dye with which the biological sample has been stained at the peak of its fluorescence excitation spectrum.

EXPERIMENTAL INVESTIGATIONS

I Feasibility experiments

The biological sample used in our first feasibility experiments was made of frogs' erythrocytes (Rana Esculenta) stained by the conventional Feulgen reaction [8] (Pararosaniline-SO_2) for DNA demonstration (sample A) or by a 0.3% water solution of Rhodamine 3GO (pH 1.2) (sample B) [9] . "In vitro" solutions were also used for comparison with histochemical samples.

We carried out our investigations in single cells in three main directions: a) fluorescence waveform and related physical parameters (lifetimes,etc.); b) fluorescence saturation behaviour and related parameters (saturation intensity , quantum yield, etc.); c) high sensitivity fluorometry to see if our experimental apparatus is sensitive enough to measure fluorescence signals given by cellular samples in which the fluorescent chromophores are of low concentration and/or low quantum yield, as in the present case of the weakly fluorescing chromosome bands.

In our previous paper [9], measurements as at points a),b)

and c) of the above are widely reported. Here let us mention that in the feasibility experiments we measured the PMT current waveform simply with a sampling oscilloscope in single-sweep mode, thus taking alternate measurements of the fluorescence pulse and of the laser pulse reflected towards the PMT by a glass slide placed in the object plane after removing the barrier filter shown in Fig. 2. Typical waveforms of both fluorescence and laser pulses are shown in Fig. 3. We evaluated the upper level lifetime τ, by measuring the relative shift of the barycenter of the two waveforms.

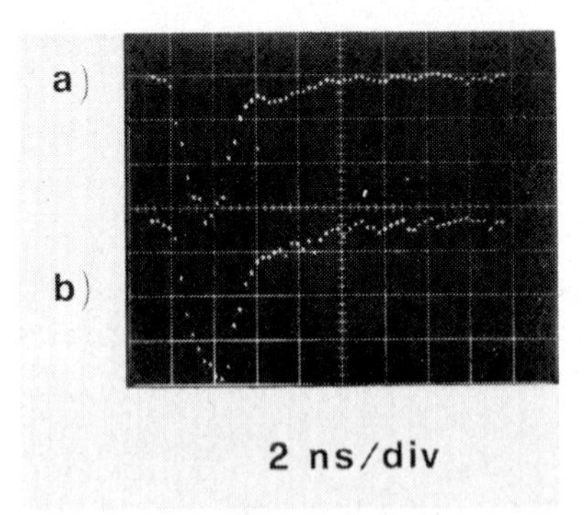

Fig. 3 - Typical PMT current waveforms as displayed by the sampling oscilloscope in single-sweep mode. a) Rhodamine 6G laser pulse; b) fluorescence pulse given by a single cell nucleus of sample B (see text).

In this experiment τ turned out to be very short thus proving the accuracy (0.2 nsec) of our apparatus in measuring short light pulses. Furthermore, the high intensity available in the laser beam makes it possible to saturate the fluorescence transition. A plot of the fluorescence power P_F collected by the PMT vs the peak laser intensity I_{Lmax} is drawn in Fig. 4. The independent evaluation of τ and of the saturation intensity I_S (Fig. 4) allows the absorption cross-section in single cell or subcellular organelles as well as the quantum yield of the fluorescent transition under examination to be measured. In ref. [9] we report all these measurements on the biological samples A and B and on the corresponding "in vitro" solutions.

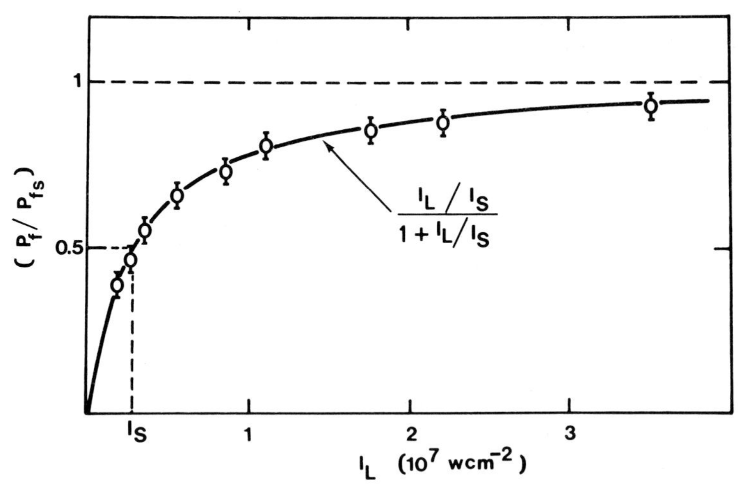

Fig. 4 - Fluorescence power P_F vs the laser peak intensity $I_{L\ max}$ showing the saturation of the fluorescent transition in sample A (see text). The theoretical curve and the experimental points are shown.

II Investigations on chromosome fluorescence patterns

As previously outlined, we have been able to use our experimental apparatus, without making any important changes in it, to study the fluorescent emission from Vicia Faba M-chromosomes stained with QM. The excitation has been accomplished by a suitable dye-laser which differs from the dye-laser used in our feasibility experiments only in the dye cell design. With such a design, light pulses as short as 0.4 nsec can be generated [11] . In particular, to excite the chromosomes, we used as an active medium a 1.5×10^{-3}M POPOP solution in toluene thus obtaining pulses of 0.8 nsec duration, of up to 100 KW peak power and with an excellent stability level, as shown in Fig.5, where about 100 pulses are superimposed.

The pulse wavelength ($\lambda_{exc}\simeq$419 nm) coincides with the peak of the QM absorption spectrum as drawn in Fig. 6.

We usually focus the excitation beam in our microscope's object plane on a spot of about 0.7 μm of diameter. Actually (consider also that the chromosomes are about 0.6 μm thick) we have to collect and to monitor fluorescence signals coming from 0.5μm^3.

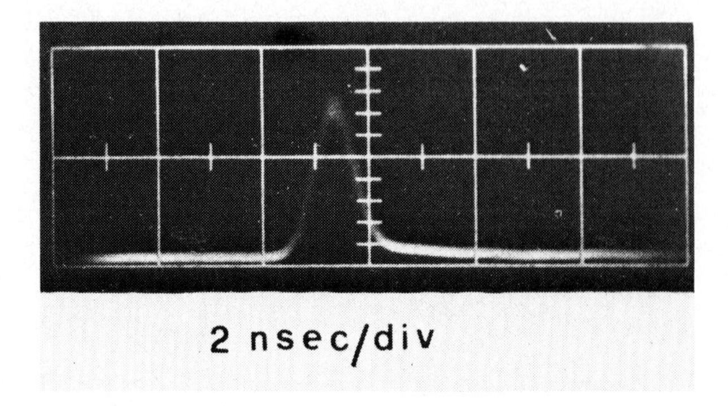

Fig. 5 - Excitation pulse generated by our POPOP dye laser [11] as detected by a fast photodiode (TRG, 105 C) and displayed by a 519 Tektronix travelling wave oscilloscope (overall risetime 0.41 nsec).

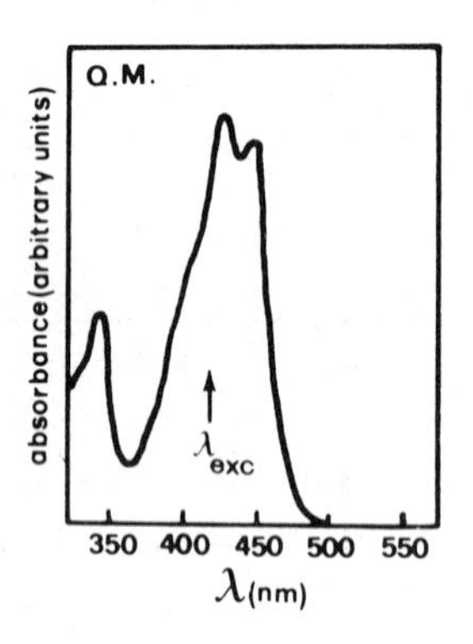

Fig. 6 - QM absorption spectrum and excitation wavelength (λ_{exc}=419 nm).

That is why very low light levels have to be measured and an averaging technique is therefore necessary. In particular, we have to average samples corresponding to the same instant (sample time) of many repetitions. Digital averaging was preferred to analog averaging due to the relatively low repetition rate of the laser pulsing and to the dependence of the average on the stability of this rate [12] . Commercial systems are currently available for digital averaging but some of them are not well suited to low repetition rates, as in our case, and all of them are very expensive. An automated system is therefore being developed in our laboratory and a very simple preliminary system has been used until now (Fig. 7) with point by point averaging obtained by manually scanning the sampling oscilloscope. At a given time position of the sample, the d.c. voltage of the

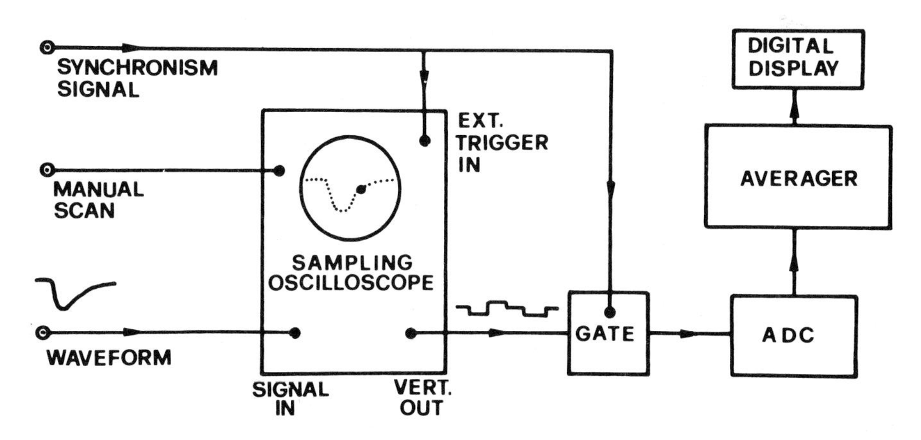

Fig. 7 - Block diagram of the averaging system.

vertical output is sampled by a slow gate, open for a given interval after the laser pulse. This analog pulse is converted to a digital measure, which is stored in a digital memory.The process is repeated for a predetermined number of laser pulses and the average of these repetitions is obtained. The sample time is then shifted to the next position and the operation repeated. The fluorescence waveforms obtained by exciting a chromosome "in band" or "out of band" are plotted in Fig. 8. Both waveforms show two different decay components: a fast decay time of 1.0 ± 0.2 nsec and a slower decay time the value of which we can estimate to be about 10 nsec. These results suggest that the QM bound to G residues is responsible for the fast decay while the QM intercalating two AT base pairs is responsible for the slowerdecay. From the strong fluorescence waveform, by extrapolating the slow exponential tail to the origin [13] , we can find that the AT percentage in the strong bands is 60% of the total DNA, which is in complete agreement with the accepted biological data. At the present time, a similar measurement

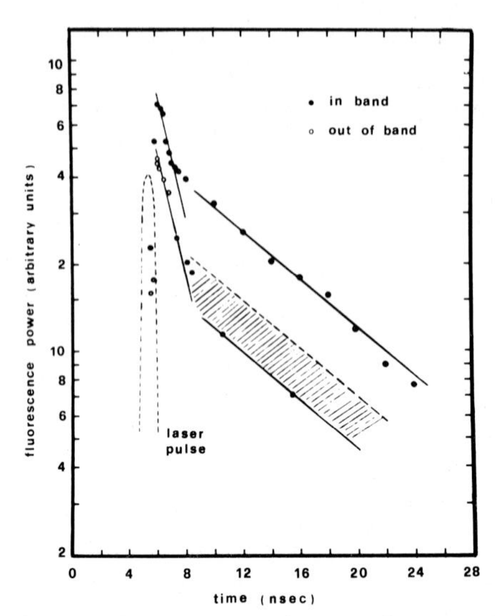

Fig. 8 - Plot of the fluorescence waveforms obtained by "in band" or "out of band" excitation of a Vicia Faba M-chromosome stained with QM. The laser pulse is also shown.

is not possible out of band since the low light level in the tail of the exponential decay results in an accuracy of 10% which is insufficient for our purpose. From the peak values, however, we have been able to find a relative measurement of the total concentrations of the dye bound by either mechanisms (see Introduction) to the chromosome DNA. We have found that the "out of band" dye concentration is 1.7 times smaller than the "in band." This ratio, i.e. the different dye intake possibility of the DNA, is not sufficient to explain the intensity ratio between "in band" and "out of band" of 2÷2.3 times, as found by CW excitation. In order to completely agree with this intensity ratio, a further contribution from a different base pairs composition would be necessary; in particular, the "out of band" percentage of AT should be [13] about 52% of the total DNA. Therefore, the aim of our work will be to improve the accuracy of our apparatus to make possible a measurement even "out of band" to find this result.

Finally, it is worth pointing out that our short laser pulses and our fast electronic system allow measurements of fluorescence signals down to a fraction of a nanosecond (0.2 nsec after deconvolution). Measurements of this kind are rich in information, even more so than simple spectral measurements, and should prove to be very useful in other fields of the molecular biology and even in photochemistry.

REFERENCES

1. T.Caspersson, L.Zech, E.J.Modest, G.E.Foley, V.Wagh and E.Simonsson, Exptl.Cell Res., 58 (1969) 128
2. B.Weisblum and P.L.de Haseth, Proc.Nat.Acad.Sci.-USA, 69 (1972) 629
3. Ritva-Kajsa Selander, Biochem.J., 131 (1973) 749
4. U.Pachmann and R.Rigler, Exptl.Cell Res., 72 (1972) 602
5. E.J.Modest and S.K.Sengupta, Fluorescence Techniques in Cell Biology, ed.by A.A.Thaer and M.Sernetz, Springer Verlag, Berlin, 1973
6. J.Barbet, B.P.Roques and J.B.Le Pecq (in press) Proc.Nat. Acad.Sci. (august 1975)
7. A.M.Michelson, C.Monny and A.Kovoor, Biochemie,54 (1972) 1129
8. C.A.Sacchi, O.Svelto and G.Prenna, Histochem.J.,6 (1974) 251
9. S.Cova,G.Prenna,C.A.Sacchi and O.Svelto, presented at the "Internationl Conference on Excited States of Biological Molecules",Lisbon,Portugal, April 18-24th, 1974 and to be published in the Wiley-Interscience Sèries of Monographs in Chemical Physics (J.B.Birks ed.)
10. O.Svelto, Principi dei Laser, Tamburini,Milano,1972,pp. 47-54
11. A.Andreoni, P.Benetti and C.A.Sacchi, Appl.Phys.,7 (1975) 61
12. S.Cova,M.Bertolaccini and C.Bussolati, Phys.Stat.Sol.(a), 18 (1973) 11
13. A.Andreoni, C.A.Sacchi,O.Svelto (to be published).

DISCUSSION

A. ANDREONI, C.A. SACCHI, O. SVELTO, S. COVA, G. BOTTIROLI, G. PRENNA

R. Mohan - With such a small laser probe volume and if you enable yourself to fluorescent label only a few hundreds of molecules, you should be able to follow the number fluctuations correlation spectroscopy, which is very informative on the localized moelcular motions of conformational changes, translational and rotational diffusion. This beautiful technique has been developed by Berne, Elson, Magde and Webb, and been reviewed by Webb et al. in Ann. Rev. Biophys. and Bioengng.Vol.4, 1974. We intend to apply this technique for the study of the cross bridge motions in the muscle fibre.

Mme Andreoni - I think that the measurements you are suggesting could be in principle performed by means of our apparatus, perhaps even without any improvement of its sensitivity.

C.B. Harris - Do you foresee a combination of saturation spectroscopy and the measurement of the fluorescence lifetime being useful in a fast flow cell for rapid differential chromosomes or cell analysis?

Mme Andreoni - Though I appreciate your suggestion, I think that our technique of measuring fluorescence parameters cannot be applied to a fast flowing biological sample because of the difficulty of centering the laser spot on the same part of the sample for the time we need to perform averaging of the fluorescence waveform.

M. BernsWhat is your conclusion concerning the mechanism of quinacrine mustard binding?

Mme Andreoni - The reasons for the banding fluorescence patterns of chromosomes come out from our experiments to be 1) a different QM concentration in each band. In fact, the ratio between the QM concentrations in band or out of band is 1.7; this reason is not sufficient to explain the corresponding fluorescence power ratio of 2 $\div$ 2.3 as found by CW excitation. This fact suggest a further reason. 2) The AT base pairs concentration is different "in band" or "out of band". This concentration was measured in band and it turned out to be 60%. In order to obtain the right ratio of fluorescence power, this concentration is expected to be 52% out of band. We are trying now to perform this measurement.

EFFET DE RESONANCE RAMAN LASER SUR LES BASES DE SCHIFF PROTONEES DU RETINAL

par P. V. HUONG*, R. CAVAGNAT et F. CRUEGE
Laboratoire de Spectroscopie Infrarouge et Raman, associé au C. N. R. S., Université de Bordeaux I,
351, Cours de la Libération - 33405 - TALENCE, France

Abstract :

The resonance Raman Scattering Effect of retinal, retinylidene 1, 5 diamino pentane and protonated retinylidene 1, 5 diaminopentane, chosen as a model of visual pigments, has been analyzed in connexion with the chemical structure of the chromophore, using ten Laser Exciting lines covering the visible region.

An important resonance effect has been recorded resulting from the vibronic coupling between the C=C vibrationand the singulet π, π^* transition of the polyene chain. New results for the non protonated and protonated Schiff base show that the excitation profile of the Raman line does not follow simply the electronic absorption profile, but presents additional peaks which can be connected with the triplet π, π^* transition of the conjugated chain of the chromophore.

Finally, it is shown that the specific interaction $\overset{+}{N} - H\ldots A^-$ between the protonated chromophore and anions, such as COO^-, may play a role is the red-shift of the visible absorption maximum of the chromophore approaching the absorption range of visual pigments.

INTRODUCTION

Il est hautement probable, selon de nombreux auteurs (1,2), que la vision est associée à l'absorption de la lumière par des pigments comme rhodopsin dans lequel le rétinal, un polyène-aldéhyde, est lié à la protéine opsin. Il est aussi quasi-certain que cette liaison se fait par l'intermédiaire d'une base de Schiff avec un résidu lysine (3-5). Toutefois les rétinylidène-amines modèles n'absorbent, comme d'ailleurs le rétinal

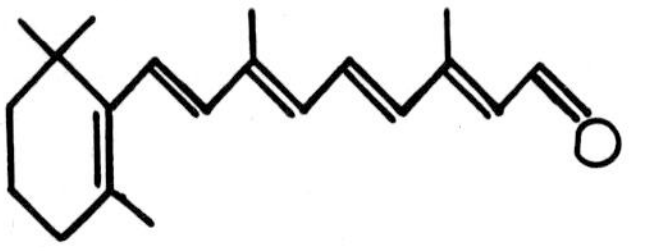

RETINAL

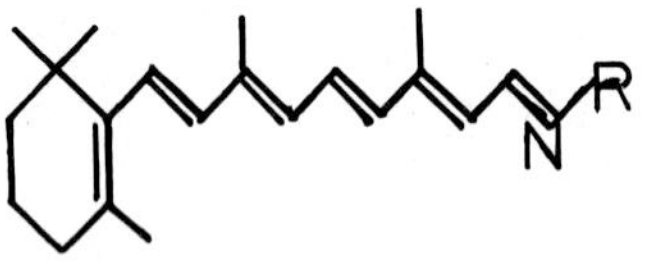

RETINYLIDENE-AMINE

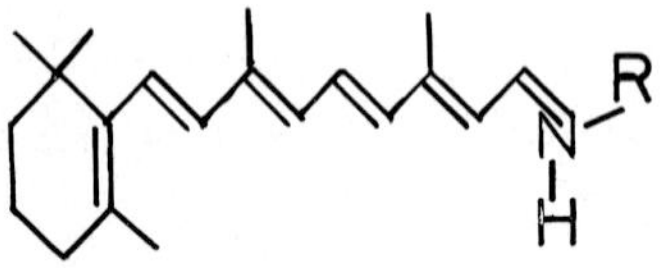

RETINYLIDENE-AMINE PROTONEE

lui-même, qu'au bord du visible (6-9), à la limite de l'ultraviolet. Dès lors, le problème de la structure chimique du chromophore, protoné ou non protoné, liée aux propriétés d'absorption et de diffusion lumineuses, se pose (8-10).

A ce premier problème s'ajoute une autre question concernant la photoisomérisation du chromophore (6, 11-13). Certains auteurs pensent que l'isomérisation est uniquement cis-trans (1) ; d'autres statuent qu'elle se fait par l'intermédiaire des états triplets (11-14). D'autres suggèrent tout récemment (15, 16) que des conformations intermédiaires du chromophore rétinal doivent jouer aussi un rôle dans le mécanisme de la vision.

Des études plus fines montrent enfin qu'un troisième facteur doit être tenu en compte, ce sont les interactions entre le chromophore et son environnement. Toutefois, si des idées ont été émises que ces interactions peuvent être soit un transfert de charge (17), soit des forces électrostatiques (18, 19), aucune localisation précise n'a été montrée.

La spectroscopie Raman de résonance est une bonne technique pour étudier ce problème, en particulier pour répondre aux première et dernière questions posées.

Rappelons que la diffusion Raman de résonance dans une molécule apparaît lorsque la fréquence de la lumière excitatrice tombe à l'intérieur d'une bande d'absorption électronique de cette molécule (20, 21). Elle se manifeste avec une exaltation souvent très importante de l'intensité Raman des modes vibrationnels qui sont couplés avec la transition électronique. Cette exaltation est telle qu'on arrive à obtenir le spectre Raman des chromophores à des concentrations voisines des conditions biologiques (10^{-5} - 10^{-6} mole/litre). Elle permet en outre de voir seulement le spectre vibrationnel du chromophore, au moment où, dans les mêmes conditions expérimentales, le spectre du reste de l'échantillon n'a qu'une intensité quasi-nulle.

Cette technique récente a été déjà appliquée à l'étude des rétinals (16, 22-26). Des résultats importants ont été déjà obtenus ; nous en discuterons dans la suite de notre mémoire.

Dans ce travail, nous nous intéresserons d'une part aux propriétés de diffusion lumineuse liées à la structure des chromophores modèles et d'autre part aux effets spectroscopiques résultant des interactions entre le chromophore et son environnement.

PARTIE EXPERIMENTALE

Les rétinals du commerce (Fluka) ont été utilisés sans purification. Les solvants ont été généralement distillés avant usage. Parmi les carbures saturés, l'isopentane se révèle comme bon solvant "inerte" pour l'étude de l'effet Raman de résonance. Il en est de même pour le solvant autoassocié et polaire méthanol. Le tétrachlorure de carbone est le solvant dans lequel la stabilité est la moins grande.

Les rétinylidène-amines ont été préparées par longue agitation, à l'abri de la lumière, de la solution de rétinal additionnée d'amine choisie. La petite quantité d'eau résultant de la réaction de Schiff ne gêne pas le spectre Raman de résonance.

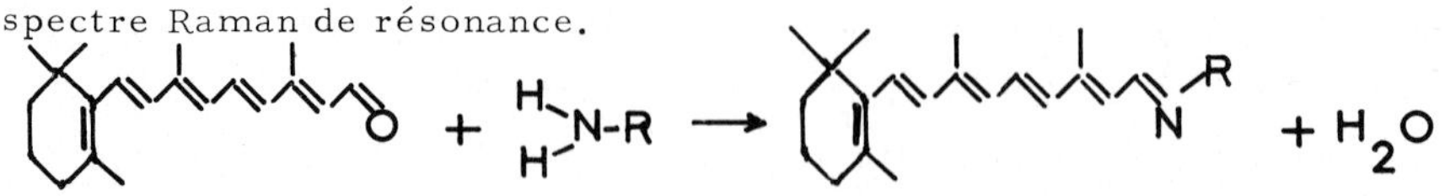

$$+ \; H_2N\text{-}R \longrightarrow \quad N\text{-}R \; + \; H_2O$$

Pour simuler les fonctions d'amines biologiques, la 1,5-diaminopentane et la n-hexylamine ont été utilisées.

Les rétinylidène-amines protonées ont été préparées à partir des rétinylidène-amines correspondantes, débarrassées de leur solvant par évaporation sous vide, lorsque ce solvant est non polaire. L'acide protonant peut être soit ClH, Cl_3CCOOH soit $CH_3(CH_2)_4COOH$ ou acide caproique. Ce dernier a été choisi à cause de sa ressemblance avec la chaîne lysine et parce qu'il possède le groupement COO^-, couramment rencontré en milieu biologique. Lorsque le solvant est polaire, la protonation peut se faire par addition directe de l'acide dans la solution de rétinylidène-amine.

Les spectres d'absorption dans le visible ont été enregistrés sur un spectromètre Beckman DK 2A. Les spectres Raman sont obtenus à l'aide d'un appareil Coderg T 800, triple-monochromateur. L'excitation lumineuse a été assurée par dix raies laser couvrant le visible : 454,5 ; 457,9 ; 465,8 ; 472,7 ; 476,5 ; 488,0 ; 496,5 ; 501,7 et 514,5 nm avec la source argon ionisé Spectra Physics modèle 164 et 632,8 nm avec la source hélium-néon Spectra Physics modèle 125 A.

Pour réduire la dégradation des substances chimiques par échauffement local provoqué par l'impact du laser, nous avons utilisé une cuve Raman tournante à 2000 tours/minute dont le principe a été décrit antérieurement (27) (fig. 1). Le niveau de l'échantillon liquide est toujours maintenu plus bas que l'incidence lumineuse lorsque la cuve est au repos, pour éviter des dégâts éventuellement causés par l'évaporation indésirée du solvant surchauffé, en cas de fausse manœuvre.

Les mesures d'intensité des bandes de diffusion Raman, avec des raies du solvant non résonnant comme référence, sont corrigées de la fraction de produit dégradé. Bien que la dégradation soit minime, on peut noter de petites modifications spectrales lorsque la source laser est trop intense ou lorsque l'échantillon est longuement exposé à l'irradiation.

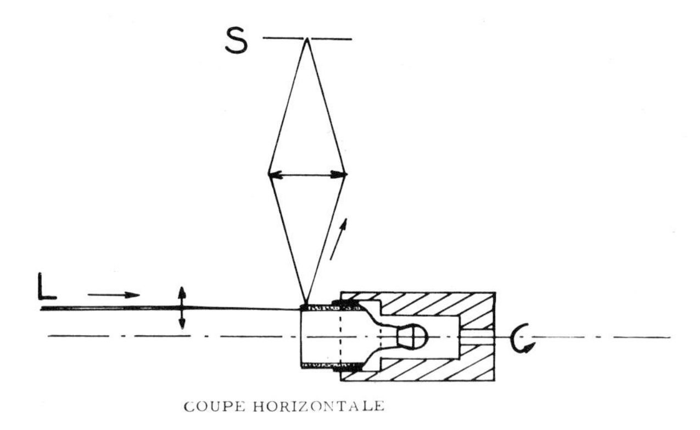

Fig. 1 - Schéma de la cuve Raman tournante 2000 tours/mn.

RESULTATS ET DISCUSSION

Nous avons donc étudié la diffusion Raman de résonance, sous irradiation laser, du rétinal dans différents solvants organiques; isopentane, cyclohexane, chloroforme, méthanol, éthanol. Nous nous sommes ensuite intéressés aux variations de ces effets de diffusion de résonance du chromophore rétinal lorsqu'il est en interaction par base de Schiff avec la 1,5-diaminopentane à cause de sa ressemblance avec les fonctions aminées de la lysine. Ensuite, l'étude de la rétinylidène-1,5-diaminopentane protonée va nous montrer une résonance encore plus accrue. Pour analyser l'effet d'environnement sur les propriétés optiques du chromophore protoné, nous avons utilisé d'une part des solvants de natures différentes, d'autre part des anions différents : Cl^-, CCl_3COO^- et $CH_3(CH_2)_4COO^-$ provenant des acides qui ont servi à la protonation de la rétinylidène-amine.

Des mesures, pour comparaison, avec l'hexylamine comme base ont été aussi faites.

EFFET DE DIFFUSION RAMAN DE RESONANCE ET STRUCTURE CHIMIQUE DU CHROMOPHORE

Avec les solutions de rétinal dans divers solvants, le spectre Raman obtenu présente des bandes dont l'intensité relative, par rapport à celle des bandes du solvant, varie beaucoup selon la longueur d'onde de la lumière excitatrice. La figure 2 reproduit, à titre d'exemple, les spectres Raman enregistrés avec les deux raies laser 454,5nm et 632,8 nm, d'une

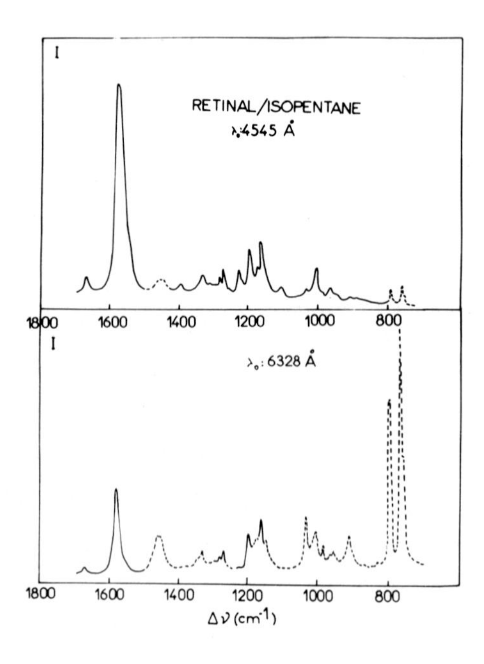

Fig. 2 - Spectres Raman du rétinal à 10^{-3}mole/ℓ dans l'isopentane, avec deux excitatrices de longueur d'onde différente. (Les pointillés représentent le spectre du solvant).

même solution de rétinal dans l'isopentane, à une concentration de l'ordre de 10^{-3} mole/litre. Si avec l'excitatrice 632,8 nm on voit, et d'une intensité prédominante, le spectre du solvant (en pointillé), en particulier les raies vers 800 cm^{-1}, avec l'excitatrice de 454,5 nm le spectre du solvant a pratiquement disparu et l'on ne voit que celui du rétinal. Pour ce dernier, l'intensité comparée à celle du solvant varie selon les vibrations, celle qui s'intensifie le plus est située à 1578 cm^{-1} ; elle représente essentiellement la vibration ν(C=C) de la chaîne polyène du rétinal. Son intensité, avec l'excitatrice de 454,5 nm, est de l'ordre de

10^4 fois plus élevée que celle d'une double liaison d'un composé asymétrique incolore. Notons que la vibration ν(C=O), conformément aux données de Rimai et al. (22-24), se situe à 1660 cm^{-1}. Notons toutefois que les auteurs antérieurs ont opéré avec une cuve fixe et souvent avec le tétrachlorure de carbone ; les deux conditions peuvent favoriser la dégradation du rétinal. Dans un solvant donneur de proton comme le chloroforme, la fréquence ν(C=O) devient 1651 cm^{-1}, l'abaissement par rapport à la fréquence observée dans l'isopentane est attribuable à la liaison hydrogène C-H...O=C. Bien que l'intensité Raman de la bande ν(C=O) soit faible, son attribution est certaine. Nous l'avons systématiquement vérifiée à l'aide du spectre infrarouge sur lequel la bande ν(C=O) est la plus intense.

En travaillant avec la polarisation Raman, nous avons obtenu un rapport de dépolarisation "anormale", pour la bande ν(C=C) par exemple, ρ est de 0,30.

L'évolution de l'exaltation d'intensité en fonction de la longueur d'onde de l'excitatrice et la valeur du rapport de dépolarisation obtenu montrent que l'effet Raman enregistré est fortement en résonance (20).

Le même phénomène se retrouve avec les solutions de rétinylidène-1, 5-diaminopentane. Il est à noter que seul le spectre de la moitié rétinylidène est apparent. La disparition totale de la bande à 1663 atteste que la fonction aldéhyde a complètement disparu ; par contre, il apparaît à 1624 cm^{-1}, dans l'isopentane, une bande attribuée à la vibration ν(C=N). Cette dernière varie légèrement selon le solvant : 1623 dans le cyclohexane et 1612 dans le méthanol. Il est remarquable que la vibration ν(C=C) reste pratiquement au même endroit que pour le rétinal lui-même (fig. 3). Ce fait prouve que la formation de la base de Schiff ne perturbe pas l'arrangement électronique de la chaîne polyène du chromophore.

Avec la rétinylidène-1, 5-diaminopentane protonée, nous assistons à un plus grand changement spectral. D'une part, une bande nouvelle, responsable du groupement $-C=\overset{+}{N}(H)-$, apparaît à 1650 cm^{-1} au moment où la vibration ν(C=C) se déplace vers 1550 cm^{-1} (fig. 3). Ce grand déplacement traduit une délocalisation plus importante des électrons π ; l'abaissement de sa fréquence indique une perte plus grande de son caractère de double liaison, par rapport au rétinal et à la base de Schiff non protonée.

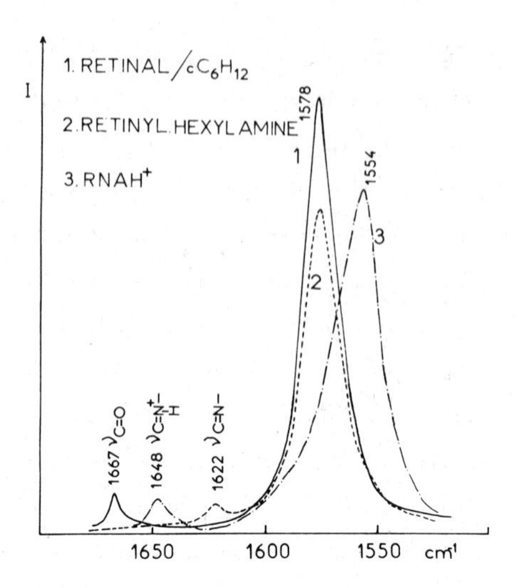

Fig. 3 - Détails des spectres Raman des dérivés du rétinal.

D'autre part, l'intensité de l'ensemble du spectre, en particulier de la bande ν(C=C) est encore fortement exaltée. Son facteur de dépolarisation étant égal à 0,30, son intensité est de l'ordre de cent fois plus élevée qu'avec le rétinal. C'est ainsi que nous avons pu opérer avec des concentrations très faibles, de l'ordre de 10^{-5} à 10^{-6} mole/litre.

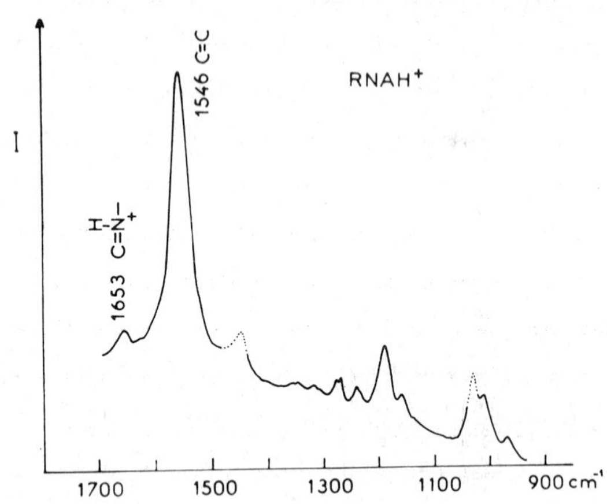

Fig. 4 - Spectre Raman de la rétinylidène-1,5 diaminopentane protonée. λ_o = 4545 Å.

Naturellement, la protonation entraîne des modifications dans le reste du spectre du chromophore (Fig. 4).

PROFILS D'EXCITATION - COUPLAGE VIBRONIQUE

Pour le rétinal et ses dérivés, lorsque la fréquence de l'excitation s'approche du maximum de la bande d'absorption singulet π-π^* dans le visible (fig. 5), l'exaltation d'intensité de la bande de diffusion Raman prouve que le couplage entre le niveau vibrationnel, essentiellement ν(C=C), et cet état singulet S_0-S_1 est responsable de la résonance de l'effet Raman enregistré.

Toutefois, les profils d'excitation n'épousent pas simplement le profil d'absorption comme le laisse prévoir la théorie simple. En effet, avec toutes les solutions étudiées, il apparaît sur chaque profil d'excitation un pic supplémentaire.

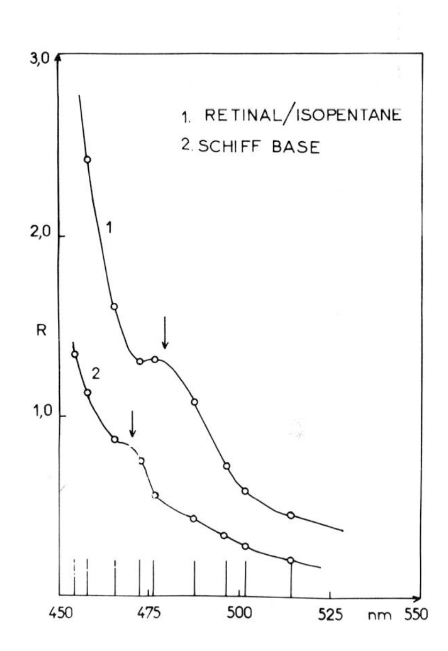

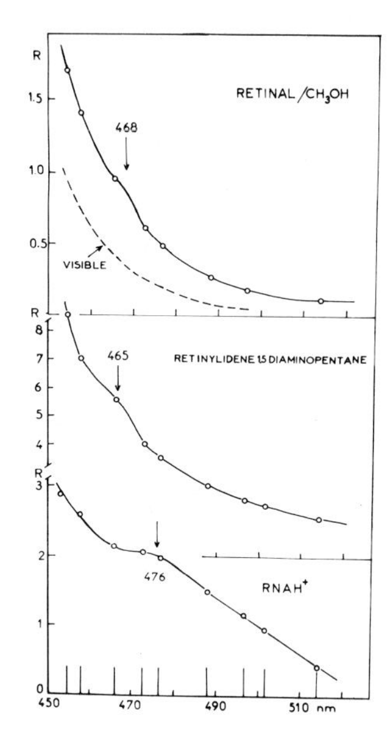

Fig. 5 - Profils d'excitation de la raie Raman ν(C=C) des dérivés du rétinal en fonction des longueurs d'onde de l'excitatrice laser.

Ce phénomène a été remarqué pour la première fois par Rimai et al. (29) pour la solution de rétinal dans le benzène. Nous l'avons retrouvé avec les solutions de rétinal dans le méthanol et aussi dans un solvant inerte comme l'isopentane. Sur la figure 5, nous avons aussi reproduit des résultats nouveaux sur le profil d'excitation de la rétinylidène-amine et de son homologue protoné. Signalons que l'intensité apparente de ces pics semble diminuer par exposition prolongée de l'échantillon au Laser ; c'est le cas de nos solutions dans le méthanol. Nous observons aussi une variation de la fréquence de ce pic lorsque le pH de la solution varie. Pour le rétinal, le pic se situe à 480 nm ou 20.833 cm^{-1} dans l'isopentane et à 468 nm ou 21.367 cm^{-1} dans le méthanol. Pour la rétinylidène-1, 5 diaminopentane, le pic se trouve à 471 nm ou 21.231cm^{-1} dans l'isopentane et à 465 nm ou 21.505 dans le méthanol. Pour le chromophore protoné dans le méthanol, le pic se situe vers 476 nm ou 21.000 cm^{-1}. La précision est de l'ordre de $\pm$100 cm^{-1} pour les deux premiers et de $\pm$ 200 cm^{-1} pour l'espèce protonée.

Si l'on peut relier, comme le suggèrent Gill et al. (29), ces pics à des transitions triplet $\pi\pi^*$, nous signalons que la fréquence de ces pics "extra" sur les profils d'excitation ne sont pas exactement celles des transitions triplet récemment mises en évidence par Fisher et Weiss (8) et par Ottolenghi et al (30), à l'aide de la technique de flash-photolyse à Laser rubis pulsé.

C'est un exemple montrant que la spectroscopie Raman de Résonance, qui s'intéresse d'abord aux états vibrationnels fondamentaux permet d'apporter, indirectement, certaines informations concernant les états excités.

EFFETS D'ENVIRONNEMENT

Rappelons que l'effet bathochrome de la protonation sur la base de Schiff du rétinal n'est pas suffisant pour amener le maximum d'absorption vers la zone d'absorption du rhodopsin naturel qui se situe vers 500-550 nm. Divers auteurs ont alors pensé à un effet supplémentaire qui s'exercerait sur le chromophore protoné. En effet, sans pouvoir préciser, Mendelsohn a envisagé une interaction par transfert de charge entre le chromophore et l'autre moitié de la protéine porteuse (17). T. Suzuki et Kito orientent plutôt les recherches vers les effets des solvants (18), alors que H. Suzuki et al. proposent de tenir compte des forces électrostatiques avec divers sites (19) et cela à l'aide de calculs théoriques.

Nos résultats vont montrer que l'effet d'interaction spécifique entre le chromophore protoné et les anions doit jouer un rôle important.

En effet, nous remarquons, à la fois sur le spectre de diffusion Raman et sur le spectre d'absorption visible (Fig. 6) qu'il y a un effet de solvant ; de plus, dans chaque solvant on enregistre des changements notables lorsque la nature de l'anion en présence varie.

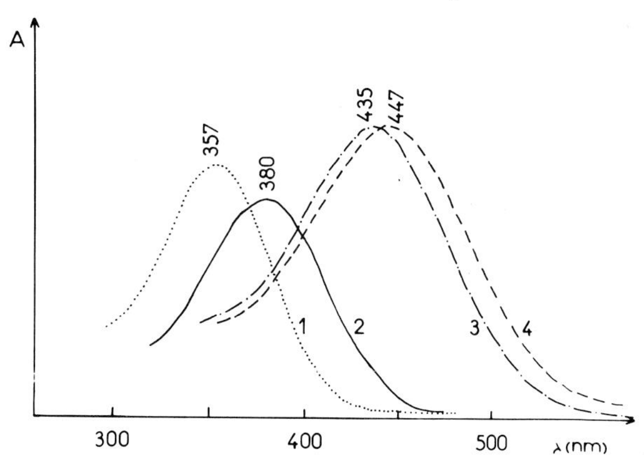

Fig. 6 - Courbes d'absorption visible des dérivés du rétinal :
1 : rétinyldidène-1,5 diamine ; 2 : rétinal ;
3 : rétinylidène-amine + HCl ; 4 : rétinylidène-amine + acide caproique.

Par exemple, avec les anions provenant des acides qui ont servi à la protonation de la rétinylidène-amine, les fréquences des vibrations ν (C=$\overset{+}{\underset{H}{N}}$ -) et ν(C=C) dans le spectre de diffusion Raman sont abaissées de 5cm^{-1} lorsque l'on passe de Cl^- à $CH_3(CH_2)_4COO^-$. De même sur le spectre d'absorption visible, l'écart est d'une quinzaine de nanomètres (Fig. 6). Il est à noter aussi qu'avec l'ion COO^-, l'effet bathochrome est plus important qu'avec Cl^-, ce qui ramène l'absorption du chromophore étudié vers la zone d'absorption du rhodopsin naturel. Ceci n'est pas surprenant si l'on pense que l'anion COO^- est très couramment rencontré en milieu biologique.

CONCLUSION

Nous avons tout d'abord analysé l'effet de diffusion Raman de résonance en relation avec la structure chimique d'un modèle de chromophore responsable de la vision : le rétinylidène.1, 5diaminopentane non protonée et protonée.

Outre l'effet de résonance important enregistré, qui est dû au couplage entre la vibration caractérisant essentiellement la liaison C=C de la chaîne polyène et la transition électronique singulet π, π^* du chromophore, nous avons montré, en particulier avec les résultats nouveaux sur la base de Schiff et son homologue protoné, que le profil d'excitation de la raie Raman n'épouse pas simplement le profil d'absorption électronique, mais présente des pics supplémentaires reliables à une transition triplet $\pi\pi^*$ de la chaîne conjuguée du chromophore.

Enfin, nous avons montré expérimentalement que l'interaction localisée $\overset{+}{N}$ - H . . . A^- entre les chromophores et les anions, en particulier l'ion COO^-, doit contribuer à l'explication de son effet bathochrome ramenant son absorption visible vers la zone d'absorption des pigments visuels.

BIBLIOGRAPHIE

1 R. HUBBARD et A. KROPF, Proc. Nat. Acad. Sci. U. S. 44 (1958) 130

2 G. WALD, Science, 162 (1967) 230

3 D. BOWNDS, Nature, 226 (1969) 1178

4 M. AKHTAR, P. T. BLOSSE et P. B. de WHURST, Chem. Commun. 1967, 631

5 R. P. POINCELOT, P. G. MILLER, R. L. KIMBEL et E. W. ABRAHAMSON, Nature, 221 (1969), 256

6 A. KROPF et R. HUBBARD, Photochem, Photobiol, 12 (1970) 249

7 R. S. BECKER, K. INUZUKA et D. E. BALKE. J. Am . Chem. Soc., 93 (1971), 38

8 M. M. FISHER et K. WEISS, Photochem. Photobiol., 26, (1974)423

9 T. A. MOORE et P. S. SONG, J. Mol. Spectr., 52 (1974) 224

10 E. W. ABRAHAMSON et J. M. JAPAR in Handbook of Sensory Physiology, Vol. VII, H. J. Dartnall Ed. Springer, Heidenberg(1972)

11 R. A. RAUBACH et A. V. GUZZO, J. Phys. Chem. 77(1973)889

12 T. ROSELFELD, A. ALCHALEL et M. OTTOLENGHI, Photochim. Photobiol. 26 (1974) 121

13 R. BENSASSON, E. D. LAND et T. G. TRUSCOTT, Photochim. Photobiol. 17 (1973) 53

14 D. LANGLET, B. PULLMAN et H. BERTHOD, J. Chim. Phys. 66(1969) 1616

15 B. HONIG et M. KARPLUS, Nature 229 (1971)558

16 A. R. OSEROFF et R. H. CALLENDER, Biochem., 13 (1974) 4243

17 R. MENDELSOHN, Nature 243 (1973) 22

18 T. SUZUKI et Y. KITO Photochem. Photobiol. 15 (1972) 275

19 H. SUZUKI, T. KOMATSU et H. KITAJIMA, J. Phys. Soc. Japan 37 (1974) 177

20 J. BEHRINGER in Raman Spectroscopy, H. A. SZYMANSKI Ed. Plenum. New-York 1967, Chapiter 6

21 J. TANG et A. C. ALBRECHT in Raman Spectroscopy, Vol. 2 H. A. SZYMANSKI Ed. Plenum, New York, 1970.

22 L. RIMAI, GILL et J. L. PARSONS, J. Am . Chem. Soc. 93-(1970)353

23 D. GILL, M. E. HEYDE et L. RIMAI, J. Am. Chem. Soc. 93(1971)6288

24 M. E. HEYDE, D. GILL, R. G. KILPONEN et L. RIMAI, J. Am. Chem. Soc. 93 (1971) 6776

25 A. LEWIS, R. S. FAGER et E. W. ABRAHAMSON, J. Raman Spectr. 1 (1973) 465

26 R. MENDELSOHN, A. L. VERMA, H. J. BERNSTEIN et M. KATES Canad. J. Bioch. 52 (1974) 774

27 W. KIEFER et H. J. BERNSTEIN, Appl. Spectr. 25 (1971) 500

28 P. V. HUONG, Résultats non publiés

29 L. RIMAI, M. E. HEYDE, H. C. HELLER et D. GILL, Chem. phys. Letters 10 (1971) 207

30 ROSENFELD, A. ALCHALEL et M. OTTOLENGHI, J. Phys. Chem. 78 (1974)336.

SYMMETRY, ORIENTATION AND ROTATIONAL MOBILITY OF HEME A_3 OF CYTOCHROME-C-OXIDASE IN THE INNER MEMBRANE OF MITOCHONDRIA

WOLFGANG JUNGE + AND DON DEVAULT

Johnson Research Foundation, University of Pennsylvania, Philadelphia (USA)

SUMMARY

The complex between heme-a_3 of the cytochrome-c-oxidase and CO is photolysed on excitation with a flash from a rhodamin 6G - laser. Excitation with the linearly polarized laser flash of nonsaturating energy photoselects an anisotropic subset of hemes from an originally isotropic ensemble. The photoinduced linear dichroism of the absorption changes of heme-a_3 is studied. The extent of the dichroism under conditions where rotational diffusion of the hemes is inhibited approximates the theoretical expectation for a circularly degenerate chromophore. The same extent is observed with the cytochrome-c-oxidase in situ, when photoselection experiments are carried out with suspensions of mitochondria. The only detectable relaxation of the linear dichroism can be attributed to the rotational diffusion of whole vesicles in the suspension.

Therefrom it is concluded that the cytochrome-c-oxidase is either totally immobilized in the inner membrane of mitochondria or it rotates anisotropically around a single axis coinciding with the symmetry axis of heme a_3.

These alternatives are discussed in the light of the literature. The most probable interpretation is rotational diffusion of the enzyme which is limited to the axis perpendicular to the plane of the inner membrane of mitochondria, with heme a_3 coplanar to the membrane. This opens the way for similar studies on the orientation and rotational mobility of the other hemes in the respiratory chain.

+ permanent adress: Max-Volmer-Institut, Technische Universität, Berlin (Germany)

INTRODUCTION

In contrast to some very specific applications of lasers to biophysical problems this is one where the requirements are not that stringent, just 100mJoule within 0.3μsec linearly polarized output at 585nm.

The biological question is directed towards the orientation and the rotational mobility of an enzyme in a submicroscopic biological membrane. This enzyme, the cytochrome-c-oxidase, mediates the terminal electron transfer step in the respiratory chain from cytochrome-c to dioxygen. There is evidence that the oxidase conserves part of the redox-energy for the formation of ATP which is supplied to the cell to perform biological work. At least three independent lines of experiments have revealed that the enzyme is transmembrane in the inner membrane of mitochondria [1-5]. Moreover there is some evidence that electron transfer across the oxidase charges the membrane electrically [6,7], which possibly represents an intermediate step for ATP-synthesis [8]. So one may visualize the oxidase (MW 140 000 [9]) as spanning the membrane with one of its two hemes, namely a_3, located nearer to the m-side, where it reacts with dioxygen, while the other one, heme-a, reacts with cytochrome-c at the other side. This configuration is illustrated in Fig.1.

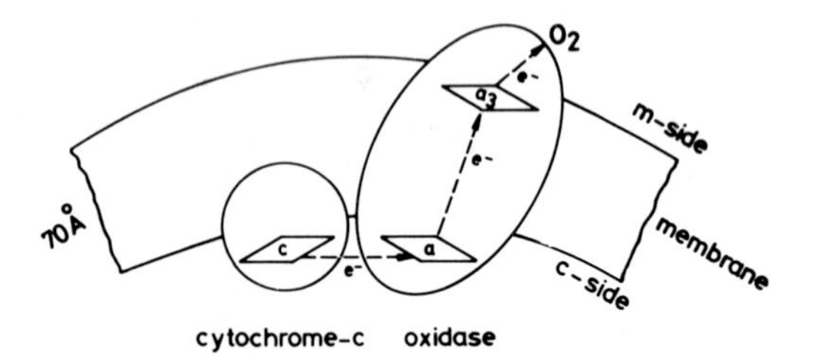

Fig.1. A possible configuration of the cytochrome-c-oxidase in the inner membrane of mitochondria. The relative orientations of hemes c and a are arbitrary, for heme-a_3 see text.

There are two interesting questions relating to the orientation and the rotational mobility of the oxidase in the membrane:

1. Does cytochrome-c react with the oxidase by diffusion and collision just as they would in solution or are they locked to each other, or even locked within the membrane?
2. Which is the mutual orientation of hemes a and a_3, which mediate vectorial electron transport across the membrane?

The experimental technique for tackling these questions is linear dichroism by photoselection. We studied suspensions of pigeon heart mitochondria in a buffer medium. The mitochondria were run anaerobic and carbon monoxyde was supplied as ligand for heme-a_3 of the cytochrome-oxidase [10,11]. The complex between heme-a_3 and CO can be photolysed by a flash from a rhodamin 6G-laser (585nm). At the usual CO-concentrations in these experiments the recombination of the photolysed complex takes several 100 msec at room temperature. Photolysis produces a shift of the Soret-band of the CO-complexed heme from 445 nm towards shorter wavelength. The rapidly induced but slowly relaxing absorption changes at 445 nm are examined for photoinduced linear dichroism. The principle of photoselection experiments is illustrated in Fig.2. An isotropic solution of chromophores (represented by 3 orthogonal transition moments) is excited with a linearly polarized flash of light. This excites and photolyses by preference those chromophores with their transition moments in parallel to the E-vector of the exciting light. In consequence absorption changes will result from the same photoselected anisotropic ensemble. The absorption changes will differ depending on whether they are monitored with light polarized parallel or perpendicular to the exciting flash.

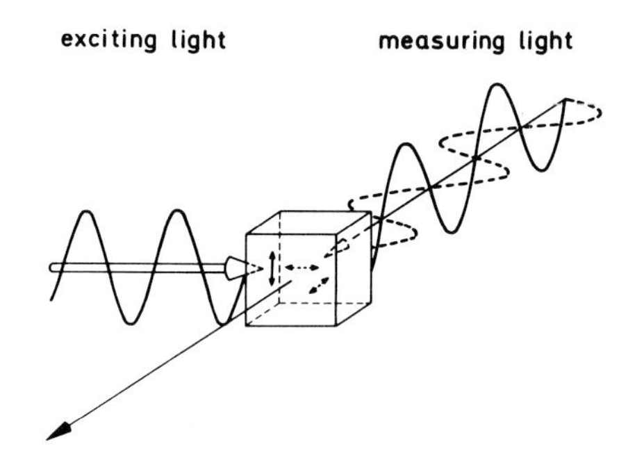

Fig.2. Principle of photoselection

The extent of the linear dichroism reflects the symmetry properties of the chromophore and the mutual orientation of the excited and the interrogated oscillator, the relaxation of the dichroism yields information on the rotation of the chromophore with respect to the lab system [12,13].
A necessary condition for photoselection to occur is that the sample is excited with a flash of non-saturating energy. The dependence of the apparent dichroic ratio $\Delta A_{\parallel}/\Delta A_{\perp}$ of the absorption changes on the relative saturation $\Delta A_{\parallel}^{obs}/\Delta A_{\parallel}^{max}$ is illustrated in Fig.3 for the special case of a circularly degenerate oscillator. It is obvious that up to 20% saturation the apparent

dichroic ratio does not dèviate too much from the ideal one at zero excitation energy.
For further details on the experimental procedure, especially on the check-up for possible artifacts, please, see our extended paper [11].

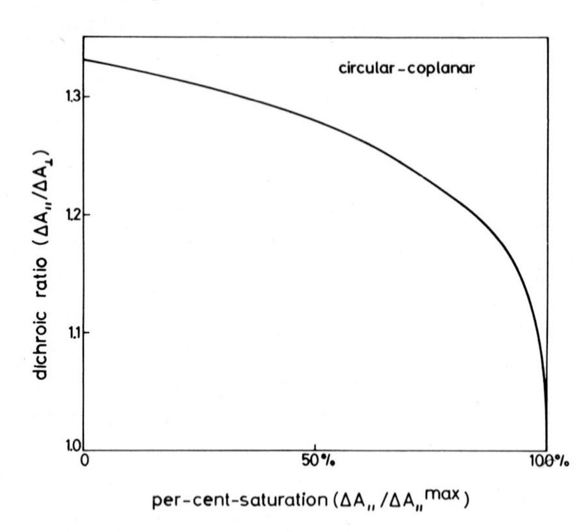

Fig.3. Dependence of the apparent dichroic ratio on the relative saturation by the exciting flash of light in photoselection. Excited and interrogated oscillators circularly degenerate and coplanar.

RESULTS AND DISCUSSION

As a check-up for possible artifacts we first studied solutions of isolated hemoproteins where the expectation was clearer. Fig.4 shows the absorption changes at 445 nm for parallel (above) and perpendicular orientation (below) of the two light beams. Linear dichroism with a ratio of 1.3 is obvious from the two traces in the left, while no dichroism is observed under the other conditions. This can be interpreted as follows:
The left traces were obtained with the isolated cytochrome-c-oxidase suspended just in phosphate buffer. This is known to produce polymerization of the enzyme into large units [14]. In contrast to this,solubilization in the presence of a detergent (Tween-80) keeps the enzyme active and supposedly monomeric. According to Taoś rule of thumb for the rotational diffusion time of a spherical globular protein in solution [15],the monomeric oxidase should rotate at about 20 nsec, while hemoglobin at about 5 nsec in aqueous solution. Such a high rotational diffusion dissipates the information on photoselection by three orders of magnitude more rapid than resolvable in our set-up.

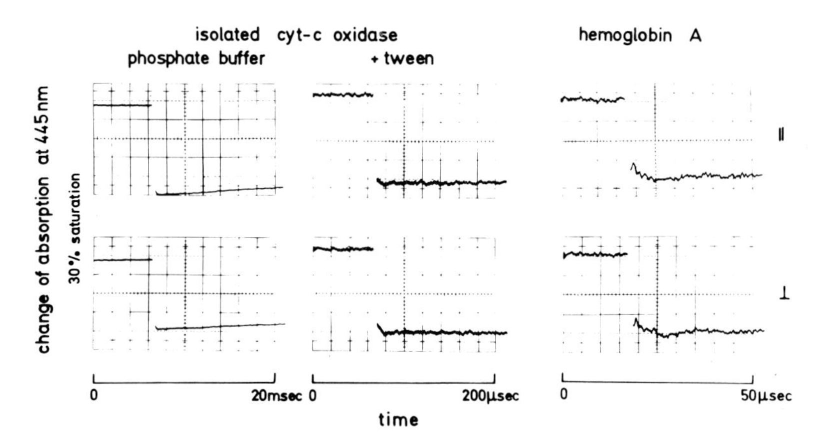

Fig.4. Linear dichroism by photoselection with isolated hemoproteins in aqueous solution (for details, see text).

The question arises whether the observed dichroic ratio of 1.3 already results from the totally immobilized cytochrome-c-oxidase or whether the true dichroic ratio is greater but has partially relaxed due to limited rotational mobility of the enzyme within the polymer. To check this we froze submitochondrial particles down to liquid nitrogen temperature in a glycerol glass. The average dichroic ratio obtained for this system was 1.35 ± 0.1 [11]. Comparison of these dichroic ratios of heme-a_3 with the theoretically expected ones for chromophores of various symmetries [12] suggests that heme-a_3 behaves like a circularly degenerate chromophore for which the expected dichroic ratio is 4/3. Circular degeneracy is probable from the fourfold symmetry of hemes of the a-type. Comparative EPR-studies have shown that the rhombicity of the "EPR-visible" other heme (a) in the oxidase is lower than the already low one of cytochrome-c [16] (see also [17]).

Next we studied the photoinduced linear dichroism with isotropic suspensions of mitochondria. As obvious from inspection of Fig.5 a dichroic ratio of 1.27 appears with no evidence for any relaxation in the range from 20 μsec to 5 msec. In the light of the above results, this can be interpreted in two alternative ways:

(a) either the enzyme is totally immobilized in the membrane,

(b) or it carries out anisotropic rotational diffusion limited to a single axis which coincides with the symmetry axis of heme-a_3.

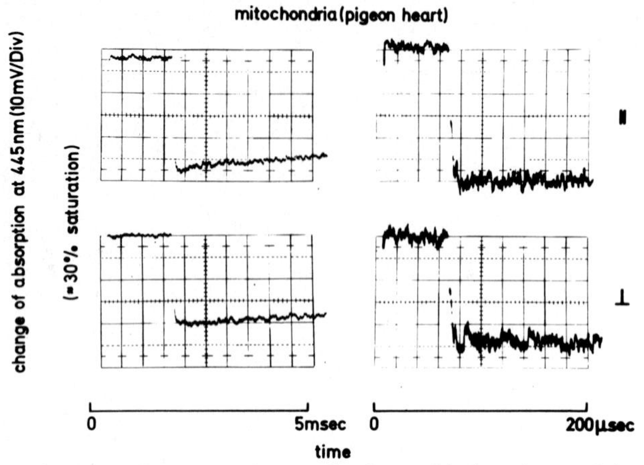

Fig.5. Linear dichroism by photoselection of heme-a_3 with isolated mitochondria in aqueous suspension.

It has to be mentionned that dichroic ratios that high as in Fig,5 were observed only at rather low protein concentrations in mitochondrial suspensions. At higher concentrations depolarizarion by light scattering in the turbid suspension decreased the apparent dichroic ratio as documented in Table 1. This is the reason why no dichroism was detected in a prior study [18].

Protein concentration (mg/ml)	Dichroic ratio ($\Delta A_{\parallel}/\Delta A_{\perp}$)
0.3	1.25
0.65	1.08
1.2	1.00

Table 1. The effect of light scattering on the apparent dichroic ratio in photoselection experiments with mitochondria.

In Fig.5 one feature is striking, the absence of dichroic relaxation even at a low time resolution. One would expect to detect the effect of the rotational diffusion of the whole organelle, at least. In order to demonstrate this effect, we repeated the experiment with submitochondrial particles (SMP) from beef heart. These are smaller in size and thus rotate more rapid than the photolysis signal at 445 nm relaxes. SMPs obtained by sonication of beef heart mitochondria revesiculate to yield particles with a size distribution ranging from 500 to 2000 Å in diameter [19]. This according to Perrinś formula for the rotational diffusion of spheres [20] implies tumbling times between 0.8 and 50 msec. The result of a photoselection experiment with SMPs is depicted in Fig.6. In the left they were suspended in aqueous buffer with a viscosity around 1 cPoise, in the right in a medium with increased viscosity (8 cPoise). It is obvious from Fig.6 that the partial relaxation of the dichroism in the low viscosity medium (halftime 1 msec) is slowed down in the high viscosity medium. The relaxation time of 1 msec and the fact that there is a wider spectrum of relaxation times are compatible with the above data on the size distribution of SMPs.

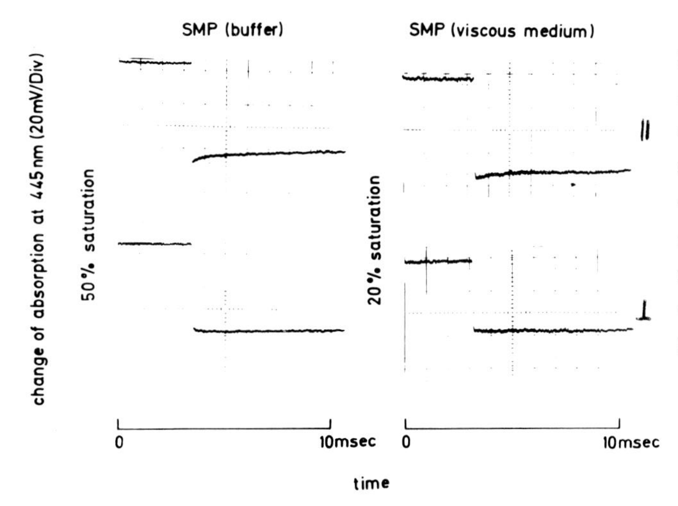

Fig.6. Linear dichroism by photoselection of heme-a_3 with submitochondrial particles in aqueous suspension (left) and in a medium with increased viscosity (8 cPoise by dextran), resp.

CONCLUSIONS

The above results give evidence for a circular degeneracy of heme-a_3 in the cytochrome-c-oxidase. Although the present experiments are limited to one pair of wavelength (585 nm for excitation and 445 nm for interrogation), results from the literature cited above lead to the expectation that circular degeneracy will be closely approximated for other wavelength, too.
From the photoinduced linear dichroism of heme-a_3 in situ, it can be concluded that the cytochrome-c-oxidase is either locked in the inner membrane of mitochondria - or that it is free to carry out uniaxial rotational diffusion, the axis being parallel to the symmetry axis of heme-a_3.
Considering the so far spurious evidence for a high fluidity of the inner membrane of mitochondria, which resulted from fluorescence depolarization studies of a Ca^{++}-ionophore [21], we consider the uniaxial rotational diffusion as the more probable interpretation of our results.

How to understand uniaxial rotational diffusion in terms of molecular structure? The evidence for a transmembrane configuration of the oxidase has been cited above, already. Singer and Nicolson in their membrane model [22] suggested that transmembrane proteins should have a polarity distribution (di)polar - hydrophobic - (di)polar in the direction perpendicular to the membrane. This distribution for energetic reasons would in fact limit their rotational mobility to a single axis which is perpendicular to the membrane.
If this holds true for the cytochrome-c-oxidase our results imply that the plane of heme-a_3 is coplanar to the membrane. Anyhow, the absence of dichrioc relaxation in photolysis experiments with isotropic suspensions of mitochondria reveals that photoselection of certain orientations of heme-a_3 selects certain orientations of the mitochondrial inner membrane as well. It will be interesting to interrogate the orientations and the rotational mobilities of other hemes (a,c,c_1) in the respiratory chain by a similar technique extended by a stopped flow system.

The applicability of photoselection to mitochondria is most important, as their odd shape resists to any other method to create an anisotropic ensemble with a reasonable degree of macroscopic orientation.

AKNOWLEDGEMENT

Thanks are due to Britton Chance, C.P.Lee, R.Colonna, T.Yonetany, D.Wilson, K.Fairs, K.Petty and others in the Johnson Foundation for encouragement and help with these experiments.
Financially support was provided in part by grants from the National Science Foundation, GB 27600 and from the National Institute of Health, GM 12202. W.J. was the recipient of a travel grant from the Technische Universität Berlin.

REFERENCES

1 R.A.Capaldi Biochim.Biophys.Acta 303(1973)237
2 E.Racker,C.Burstein,A.Loyter and O.Christiansen, in J.M.Tager, S.Papa, E.Quagliarello and E.C.Slater (Editors), Adriatrica Editrice, Bari,1970, p. 235
3 D.L.Schneider,Y.Kagawa and E.Racker J.Biol.Chem. 247(1972)4074

4 P.Mitchell and J.Moyle in J.M.Tager, S.Papa, E.Quagliarello and E.C.Slater (Editors), Electron Transport and Energy Conservation, Adriatica Editrice, Bari, 1970, p.575

5 S.Papa, F.Guerrieri and M.Lorusso Biochim.Biophys.Acta 357(1974)181

6 L.L.Grinius,A.A.Jasaitis,Yu.P.Kadzianuskas,E.A.Liberman,V.P.Skulachev, V.P.Topali,L.M.Tsofina and M.A.Vladimirova, Biochim.Biophys.Acta 216(1970)1

7 L.E.Bakeeva,L.L.Grinius,A.A.Jasaitis,U.V.Kuliene,D.O.Levitzky,E.A.Liberman,I.I.Severina and V.P.Skulachev, Biochim.Biophys.Acta 216(1970)13

8 P.Mitchell Biol.Rev. 41(1966)445

9 W.W.Wainio,T.Laskowska-Klita,J.Rosman and D.Grebner, J.Bioenergetics 4(1973)455

10 B.Chance and M.Erescinska Arch.Biochem.Biophys.143(1971)675

11 W.Junge and D.DeVault Biochim.Biophys.Acta - in press -

12 A.C.Albrecht J.Mol.Spectrosc. 6(1961)84

13 G.G.Belford,R.L.Belford and G.Weber Proc.Natl.Acad.Sci.-US 69(1972)1392

14 T.Yonetani J.Biol.Chem.236(1961)1680

15 T.Tao Biopolymers 8(1969)609

16 J.Peisack,W.E.Blumberg,S.Ogawa,E.A.Rackmidewitz and R.Oltzik, J.Biol. Chem.246(1971)3342

17 W.A.Eaton and R.M.Hochstrasser J.Chem.Phys. 46(1967)2533

18 W.Junge FEBS Letters 25(1972)109

19 C.H.Huang,E.Keyhani and C.P.Lee Biochim.Biophys.Acta 305(1973)455

20 F.Perrin J.Phys.Radium 7(1926)390

21 G.D.Case,J.M.Vanderkooi and A.Scarpa Arch.Biochem.Biophys. 162(1974)174

22 S.J.Singer and G.L.Nicolson Science 175(1972)720

DISCUSSION

W. JUNGE, D. DEVAULT

C. Varma - Analyzing the induced dichroism, you implicitely assumed that all sample molecules undergo the same electronic transition. Is there experimental evidence that all absorbing species are identical and do not have bands of different polarization overlapping at the excitation wavelength?

W. Junge - In a suspension of pigeon heart mitochondria heme-a_3 of the cytochrome-oxidase is the only species which complexes CO. P 450 and hemoglobin were negligible in the preparations used. Thus it is obvious that the absorption changes induced by flash excitation in the presence of CO resulted from heme-a_3.
A minor contribution from heme-a of the oxidase is conceivable, due to the coupling between the two hemes in the enzyme. However, there is no experimental evidence for such a contribution, yet.

R. Mohan - With your detailed experiments on cytochrome oxidase under different environmental conditions, could you comment on the controversy as to 1) the membrane is rigid, or 2) the transmembrane rotation is restricted by the thermodynamic situation across the membrane? Could you suggest as to how this controversy could be experimentally resolved?

W. Junge - Conservatively speaking our experiments by themselves do not allow us to discriminate between a rigid situation of the cytochrome-oxidase and an uniaxial rotational mobility.
However, probability is in favour of the latter. First there is the above cited evidence for a high fluidity of the mitochondrial inner membrane from studies on the rotational mobility of a Ca^{++}-ionophore. This, however, does not exclude the possibility that the oxidase is located in rigid patches within the membrane. These eventual patches must be very large, indeed, say with diameters in the order of 400 Å units, since we found no dichroic relaxation up to the range of 50 msec in experiments with mitochondria.
As yet, I cannot see any straight forward way for an experimental discrimination between the two above possibilities.

D. Lavalette - 1) Regarding Dr. Varma's questions, I think that overlapping transitions do not really matter in the case of circular degeneracy. In fact, the most convincing evidence for the existence of circular degeneracy would be to check that the polarization ratio remains constant (p=1/7) all over the absorption spectrum, by tuning the excitation wavelength. I wonder whether this could be done in the present work?
2) I suppose one of the difficulties you just mentioned was the need for keeping the ground state depletion to a low value and therefore to deal with small absorbance changes. It is well known that a large ground state depletion can decrease the measured dichroism which eventually tends to zero. Can you give an estimate of the amount of the ground state depletion in your measurements?

3) Did you have any troubles with scattering depolarization in the case of the mitochondria suspensions and did you apply any corrections for it?

W. Junge - 2) In the figures the ground state depletion after the exciting flash is given at the ordinate in terms of say "20% saturation". This stands for : the observed extent of the absorption change under parallel polarization amounts to 20% of the maximum value, which is obtained under excitation with a saturating flash of light. At this saturation the dichroic ratio of circularly degenerate systems still comes close to the value under zero flash energy (see Fig. 2, right).
3) The table gives the influence of depolarization by light scattering in the observed dichroic ratio with suspensions of mitochondria. This influence is rather drastic. We did not apply any theoretical correction but instead worked at protein concentrations so low that the effect of scattering became negligible. This, of course, was paid for by a loss in signal-to-noise-ratio. The fact that the dichroic ratios observed with turbid mitochondria were close to those observed with much less turbid submitochondrial particles and isolated enzymes proves that the protein concentrations chosen were low enough to overcome scattering depolarization effects.

R.M. Hochstrasser - While it is true that cytochrome-c is an "essentially circular absorber" the elegant experiments of W.A. Eaton on this and on many other heme-protein crystals have generally indicated only that the frequency averaged polarization of the π-π^* transitions is essentially due to splittings of the degeneracies in the underlying vibronic states. It occurs to me therefore that you may distinguish between a fixed and axially rotating heme by investigating whether the dichroism you observe is the same at a number of probing frequencies, especially in the region of the 0 - 0 band of the Q-system where, as I recall, the deviations from circularity are often significant.

W. Junge - In contrast to detailed information on cytochrome-c the rhombicity of heme-a_3 is unknown, yet. We are at present working on this. Preliminary experiments indicated that it is rather low. Moreover the prognosis is rather against a strong rhombicity. As cited above EPR-studies revealed that the other heme in the oxidase(heme-a) has an even lower rhombicity than the already low rhombicity of cytochrome-c.
If the rhombicity is small it will be very difficult to resolve the small fraction of the linear dichroism which relaxes due to uniaxial rotation from the noise background in our experiments.

M. Poliakoff - We have studied the phtoselection of the unstable molecule $Cr(CO)_5$ in methane matrices at 20°K. This carbonyl is an excellent model compound for the heme-a_3, since it has a square pyramidal structure with complete circular degeneracy (C_{4v} symmetry). Photolysis with polarized light in the E transition of this molecule produces linear dichroism almost identical in magnitude to that observed by you (Chem. Commun., 1975, 157).

RESONANCE RAMAN SPECTROSCOPY OF THE CHLOROPHYLLS IN PHOTOSYNTHETIC STRUCTURES AT LOW TEMPERATURE

Marc LUTZ
Département de Biologie, C.E.N. Saclay, Gif sur Yvette, France.

INTRODUCTION

Little is presently known about the interactions assumed by chlorophyll with surrounding molecules in photosynthetic structures.

Electronic absorption spectra of chloroplasts of green plants show that the 630-730 nm region, corresponding to the first excited singlet of chlorophyll (Q_y band), is of greater complexity than expected from the contribution of two molecular species only, chlorophyll a (Chl a) and Chl b. This has generally been taken as an indication that Chl a *in vivo* is divided into distinct populations, differenciated by the nature of their interactions, or binding with their environment [1]. However, the mere enumeration of these different populations remains controversial [1-4]. This structural heterogeneity is also evidenced by other techniques [5] and is believed to be related to the differences in functional properties assumed by chlorophyll *in vivo* [6].

Studying the chlorophylls among the complex medium of the chloroplast necessitates some kind of selectivity.

Owing to the fact that a very few molecular species of the chloroplast have electronic transitions in the visible, we recently showed that resonance effect at the level of the Soret bands of chlorophylls could be used to selectively enhance Raman scattering from vibrational modes of either Chl a or Chl b in spinach chloroplasts [7,8]. Furthermore, these resonating modes include stretching motions of the 9-ketone (Chl a and b) and 3-aldehyde (Chl b) carbonyl groups as well as vibrations involving the central Magnesium atom [9,10],(Fig.I). These atoms are known to play a predominant role in intermolecular associations of chlorophyll *in vitro* [11]. Room temperature Resonance Raman (RR) spectra of spinach chloroplasts indicated that the Mg atoms of many Chl a and b molecules are in coordination states analogous to those encountered in dry or wet chlorophyll oligomers *in vitro*. The carbonyl stretching regions (1600-1750 cm^{-1}) appeared as complex superpositions of bands, most of them downshifted from the free C=O frequencies. This suggested simultaneous existence *in vivo* of different types of intermolecular binding of the aldehyde and/or ketone carbonyls of Chl b and Chl a respectively [8].

In the present study, we attempted to resolve the 1600-1750cm^{-1} clusters of bands into individual components by cooling the samples. We extended our previous observations to other organisms, including species containing only a few or no Chl b molecules, whose aldehyde carbonyl vibrations were suspected to interfere with the Chl *a* contribution in spectra obtained with 441,6 nm light [8].

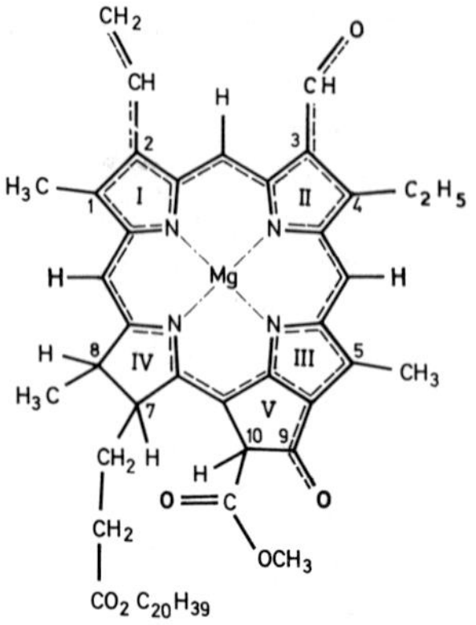

Fig.1- Molecular structure of chlorophyll *b*.
Broken lines indicate the extension of the conjugated π electron system. Chlorophyll *a* is 3-desformyl 3-methyl chlorophyll *b*.

EXPERIMENTAL

Lasers. The red fluorescence of chlorophyll did not allow observation of Raman spectra at wavelengths longer than 600 nm, and resonance effect was thus sought for by illumination of the samples in or near the Soret bands, using Argon and Helium-Cadmium continuous wave lasers. Coherent beams, in addition to their well known advantages in producing Raman spectra, allow to build RR experiments in which absorption of both incident and scattered light is kept to a minimum for a given Raman signal, and thus also unwanted secondary effects, as heating and photochemical effects. In the present work, luminous power incident on the sample was less than 5 mW.

Samples. The monocellular algae *Chlorella vulgaris* and *Botrydiopsis alpina* were studied as intact whole cells simply centrifugated from their culture medium. Whole chloroplasts of maize mesophyll, with Chl *a*/ Chl *b* molar ratios higher than 6 were prepared from plants grown under intermittent light [12]. Whole chloroplasts of spinach were obtained following usual methods [13].

Centrifugation pellets were deposited on a coverslip and directly immersed in a flow of gazeous Helium at ca 35°K which insured efficient evacuation of the heat produced at the irradiation site. Spectroscopic methods were described elsewhere [9]. In the study of the 1550-1750 cm^{-1} carbonyl stretching region, summation of 5 to 20 spectra was achieved using a multichannel analyser.

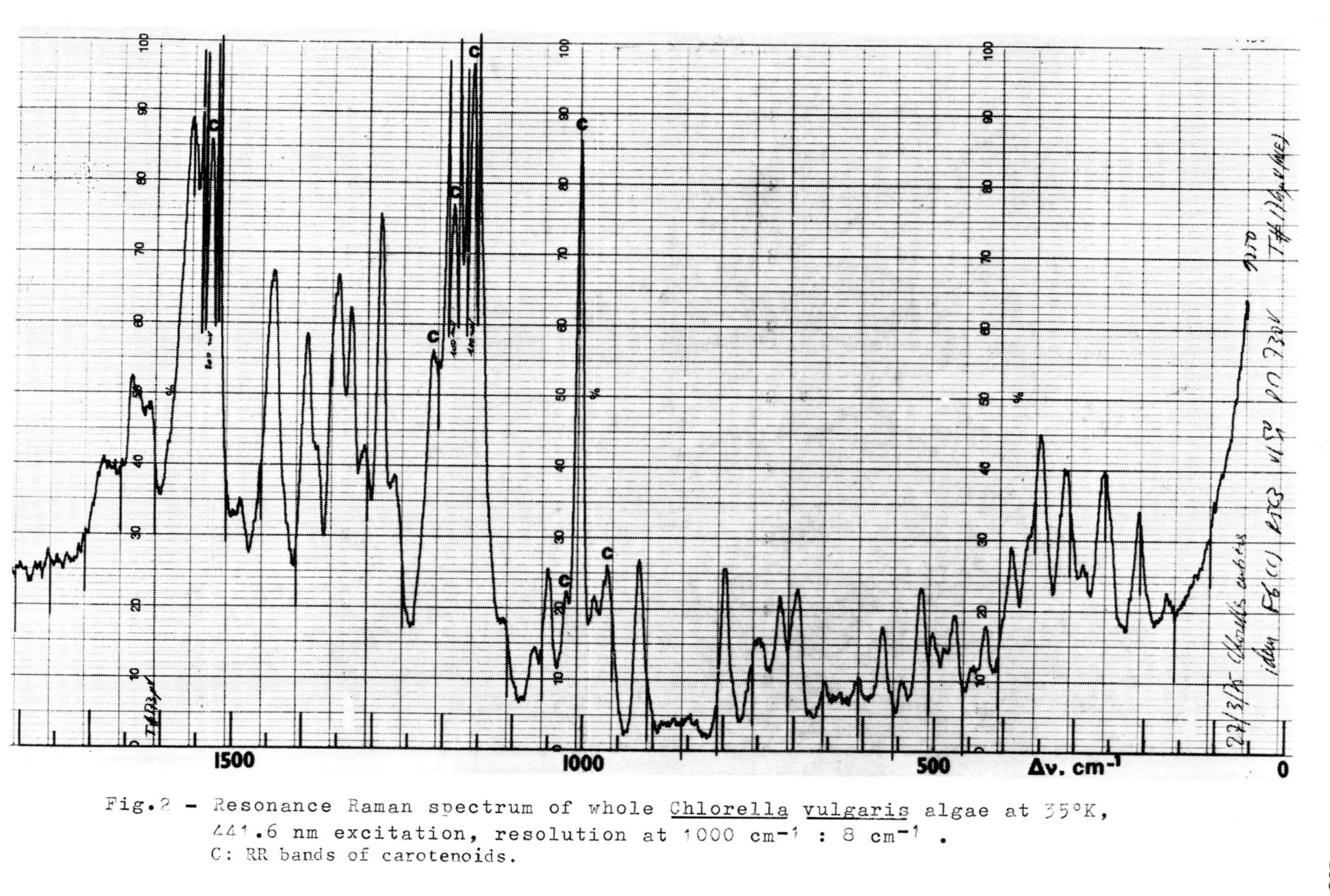

Fig.2 - Resonance Raman spectrum of whole <u>Chlorella vulgaris</u> algae at 35°K, 441.6 nm excitation, resolution at 1000 cm^{-1} : 8 cm^{-1} .
C: RR bands of carotenoids.

RESULTS AND DISCUSSION

At 35°K, the highest relative enhancements of RR spectra of Chl a and of Chl b with respect to those of carotenoids were obtained at 441.6 nm and at 465,8 nm respectively. A typical spectrum is reproduced in Fig.2.

While the frequencies and relative intensities of chlorophyll bands are almost unaffected by cooling, the relative intensities of a number of carotenoid bands, which were unobserved or extremely weak at room temperature, notably increased in spectra obtained with 450-480 nm excitations. Although most of these bands closely correlate with lattice and group modes previously observed in vitamin A type molecules [14], some of them depend on the nature of the carotenoids present, particularly in the low frequency region, and may eventually yield information on carotenoid states *in vivo*. Fortunately, chlorophyll bands in the 0-900 and 1620-1750 cm^{-1} regions remain unaffected by this contribution.

Carbonyl region.

Carbonyl stretching vibrations of Chl a and Chl b in their monomeric states give rise to RR bands of about 12 to 15 cm^{-1} halfwidth, and showing no structure under 5 cm^{-1} resolution.

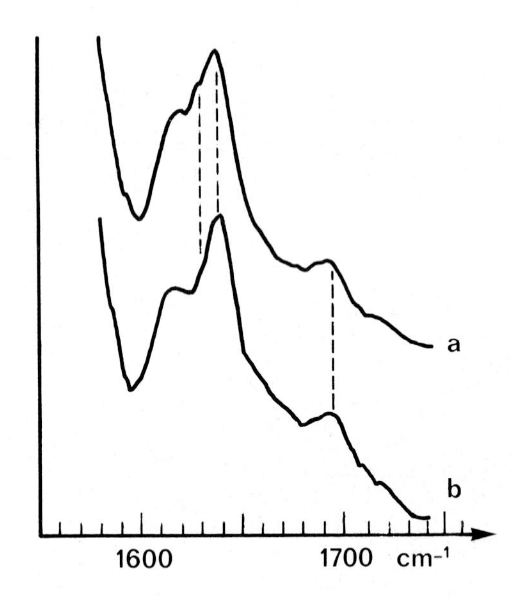

Fig.3 - Partial RR spectra of a) spinach chloroplasts, and b) Chlorella, excitation 465.8 nm, resolution 5 cm^{-1}, T = 35°K.
Multichannel analyser traces after a), 5 runs, b), 8 runs. Broken lines indicate stretching frequencies of Chl b carbonyls common to the two species.

Table 1.
Resonance Raman frequencies observed in the carbonyl stretching region for chlorophyll *a* and chlorophyll *b* in intact chloroplasts at 35°K.

	SPINACH		CHLORELLA		MAIZE	BOTRYDIOPSIS	
Exc.	465.8nm	441.6	465.8	441.6	441.6	441.6	Attribution
	1618 m	1618 m	1616 m	1615 m	1616 m	1614 m	phorbin
		1625vwsh		1625vwsh			?
	1629 sh	1631 sh	1630wsh	1630 sh	1629 wsh	1638?ew	Chl *b* 3C=0
	1639 m	1640 m	1640 m	1641 m	1641 sh		Chl *b* 3C=0
	1655bsh		1655bsh				Chl *b* 9C=0?
				1654?ew	1655vwsh	1651vwsh	Chl *a* 9C=0
		1660 sh		1662 vw		1660 m	Chl *a* 9C=0
		1671 w		1670 wsh	1669 sh	1670 sh	Chl *a* 9C=0
		1680 w		1680 w	1682 w	1683 m	Chl *a* 9C=0
	1686?ew		1685?ew				Chl *b* 9C=0?
		1689 w		1688 sh	1691 sh	1691 sh	Chl *a* 9C=0
	1694 w		1695 w				Chl *b* 9C=0
		1700 sh		1700 sh	1703 sh	1704 sh	Chl *a* 9C=0

Frequencies in cm^{-1}. m : medium relative intensity. w : weak. v : very. e : extremely. sh : shoulder. b : broad.

Chlorophyll *b*. The 1550-1750 cm^{-1} regions of RR spectra of spinach and Chlorella under 465.8 nm illumination are reproduced in Fig.3, and the frequencies observed are listed in Table 1. We previously attributed, in RR spectra of spinach at room temperature, the strongest component of this region, at *ca* 1635 cm^{-1}, chiefly to 3-aldehyde C=O stretching, as far as this grouping gives rise, in RR spectra of monomeric Chl *b*, to a band about 4 times more intense that the 9-ketone carbonyl band [9]. On the same basis, we attribute the 1630 and 1640 cm^{-1} components of the 35°K spectra to aldehyde carbonyl stretching modes of Chl *b*, downshifted from the 1668 cm^{-1} frequency of their free state by extramolecular binding. It is worth noting that the relative intensities of these two bands are in inverse order with respect to their room temperature values. The 1695 cm^{-1} bands present in both spectra are unambiguously attributed to stretching vibrations of free 9-ketone carbonyls of Chl *b*, while the weak features at *ca* 1655 and 1685 cm^{-1} could arise from such groupings interacting with undetermined partners.

The remarkable correlations existing between the carbonyl frequencies present in the RR spectra of spinach and in those of Chlorella strongly suggest that the at least two pools of Chl *b* existing in both organisms are differentiated by the same sets of interactions with surrounding molecules in the chloroplasts of both species.

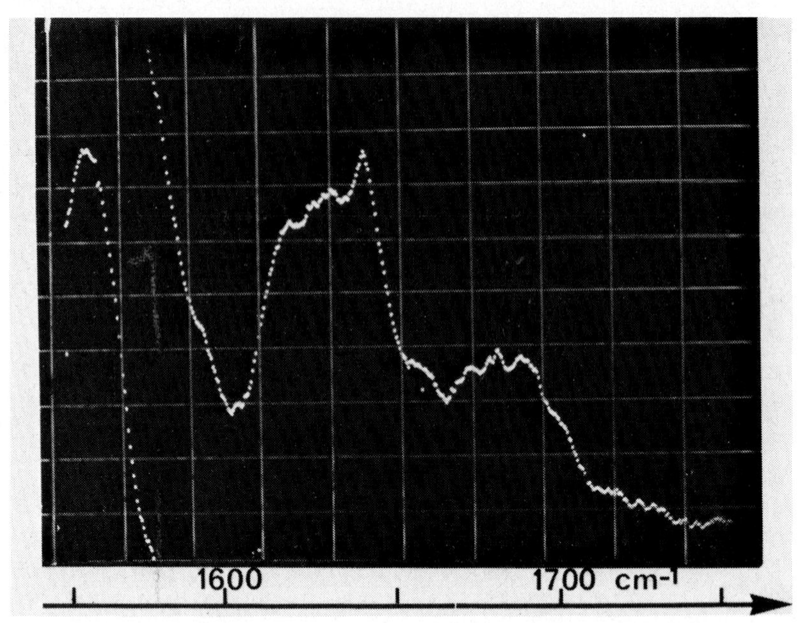

Fig.4- RR spectrum of spinach chloroplasts at 35°K, 1550-1750 cm^{-1} region, excitation 441.6 nm, resolution 5 cm^{-1}. Multichannel analyser trace after 5 runs.

Chlorophyll *a* . The carbonyl stretching region of spinach chloroplasts under 441.6 nm light is reproduced on Fig.4, as observed on the screen of the multichannel analyser. Fig.5 correlates the spectra obtained at 35°K for the four organisms studied.

The spectrum of Chl *b* free Botrydiopsis contains no band of significant intensity either at 1630 or at 1640 cm^{-1}, while weak bands only occur at these positions for maize chloroplasts with Chl *a*/Chl *b* molar ratios higher than 6. These two bands are then unlikely to involve any significant Chl *a* contribution in the Chl *b* containing species. This point confirms our previous remark that no important pool of antenna Chl *a* may consist in $(Chl\text{-}nH_2O)_m$ aggregates of the *in vitro* type, which yield characteristic *ca* 1640 cm^{-1} RR and IR bands [9,15].

The RR spectrum of Botrydiopsis does not either include any strong feature around 1625 cm^{-1}. It may thus be concluded that, as in solution, the 2-vinyl group of Chl *a in vivo* is not strongly conjugated with the phorbin skeleton, most probably because its C=C bond does not lay in the phorbin plane [16].

The *ca* 1587 and 1617 cm^{-1} bands of the four species are undoubtedly arising from planar vibrations of the phorbin skeleton [9].

Thus, the 9C=O stretching modes of most of the Chl *a* molecules of the chloroplasts here studied are confined in the 1650-1705 cm^{-1} cluster of bands. Cooling brought partial resolution of this cluster for spinach and Botrydiopsis, and the observed

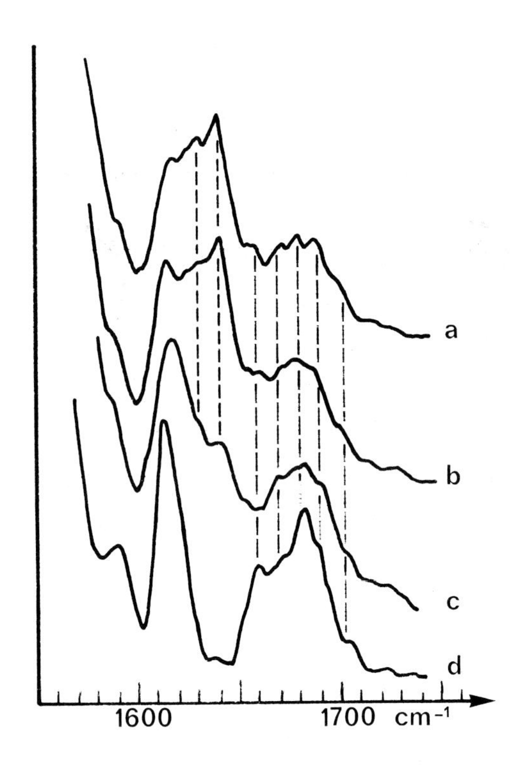

Fig.5 - RR spectra of,
a) spinach chloroplasts, 5 runs,
b) Chlorella, 5 runs,
c) chloroplasts of greening maize mesophyll Chl a/Chl b = 6.6, 9runs,
d) Botrydiopsis alpina, 13 runs.
Other conditions identical to those of Fig.4. Short dashed lines : 3C=O stretching frequencies of Chl b . Long dashed lines :9C=O stretching frequencies of Chl a .

frequencies correlate one by one within experimental uncertainty (Fig.5 and Table 1). In the less resolved spectra of Chlorella and of greening maize, shoulders on the main 1680-1682 cm^{-1} component appear to match closely with the above features.

Thus, at least five discrete pools of Chl a appear to exist in vivo, particularized by different extramolecular interactions of their 9-ketone carbonyls. Moreover, as for Chl b, each pool of Chl a is likely to be located in the same environment in each of the four species here studied.

Comparison with data from electronic absorption. The present results bring qualitative confirmation to the interpretations of low temperature studies of the red band of chloroplasts given by different authors [1,2], who presented arguments in favour of several universal forms of Chl a in green plants and algae. Our study indicates that this proposal could be equally valid for Chl b [17]. No correlation, however, may be attempted at this stage between classes of the two sets, as far as the latter were defined on the basis of criteria which do not necessarily bear any simple relation.

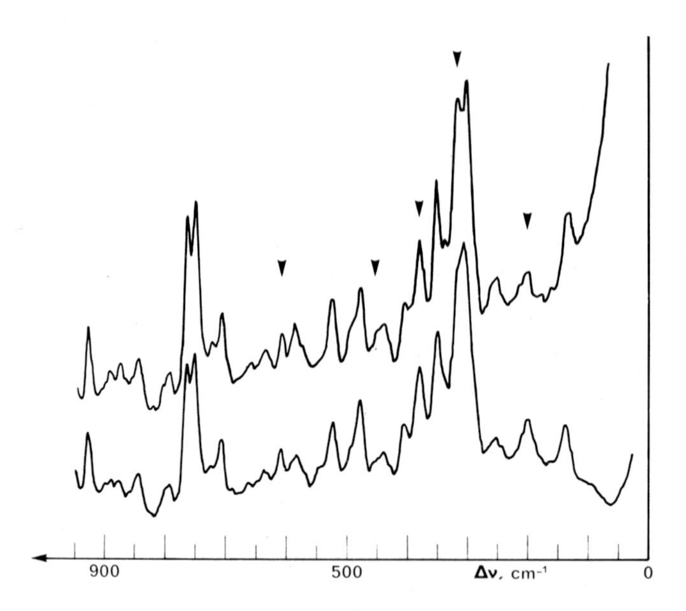

Fig.6 - RR spectra of Chlorella (upper trace) and of spinach (lower trace) at 35°K, 0-950 cm^{-1} region. Excitation 465.8 nm, resolution 7 cm^{-1}.

Lower frequencies region.

In the 0-950 cm^{-1} region of RR spectra of the four species, all the bands of significant intensity observed at 35°K under 441.6 nm excitation arise from Chl a modes, while all those observed for Chlorella and spinach under 465.8 nm excitation arise from Chl b modes (Fig.6 and 7).

Some of these bands, indicated by arrows in Fig.6 and 7, are absent from RR spectra of chlorophyll in polar solvents and in chloroplasts disorganized by heating [8,18], but correspond to modes active in RR spectra of chlorophyll aggregates in vitro [9]. We previously showed that the presence of these bands could be related to extramolecular interactions on the central Mg atom rather than to interactions on the peripheral groups [9].

It is thus questionable whether the simultaneous presence of these frequencies and of ca 1655 cm^{-1} carbonyl components are demonstrative of the existence of Chl-Chl interactions in vivo, through Mg··O=C(9) binding, and so much more so as one of them, at 310 cm^{-1} for both Chl a and b, closely corresponds to an aggregation dependent IR band which has been attributed to Mg···O stretching [19].

However, we already indicated that out of the phorbin plane, unconjugated Mg-Ligand bonds are unlikely to give rise to resonance enhanced Raman bands in chlorophyll spectra [9]. Furthermore, no 310 cm^{-1} band is apparent in RR spectra of acetone solutions of monomeric chlorophyll, whose Mg atom is then most probably bound to the carbonyls of two acetone molecules, one one each side of the phorbin plane [20].

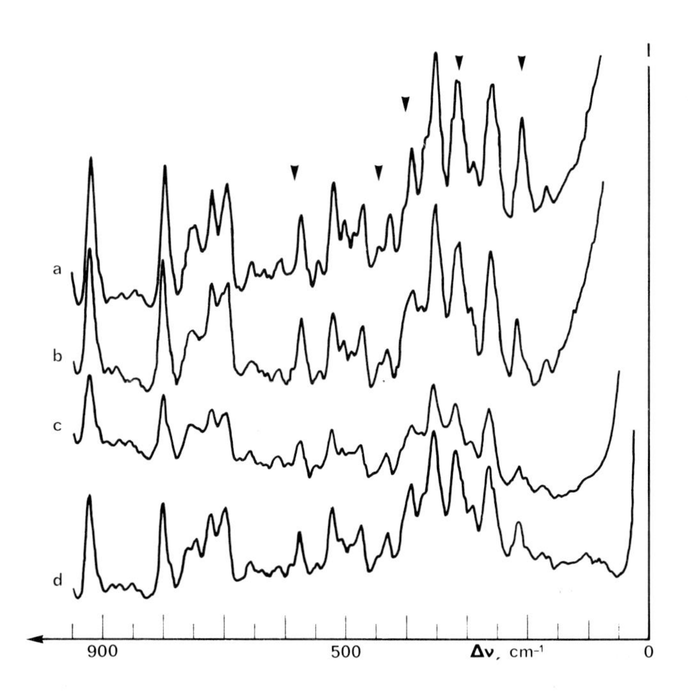

Fig.7 - RR spectra of a) Chlorella, b) Botrydiopsis, c) maize, d) spinach at 35°K, 0-950 cm^{-1} region. Excitation 441.6 nm, resolution 7.5 cm^{-1}.

Thus, a <u>necessary</u> condition for the 310 cm^{-1} band to appear is that one extramolecular ligand only is bound to Mg. In such a case the Mg atom is known to be drawn out of the phorbin plane [21]. Among the low frequency, interaction dependent set of RR bands, the 310 cm^{-1} band is found close to a <u>ca</u> 300 cm^{-1} band which involves a vibrational mode of the $Mg-N_4$ group [9,10], and the 210 cm^{-1} band is also close to a <u>ca</u> 200 cm^{-1} band which is likely to involve such a contribution [9]. We thus suggest that at least these two bands arise from vibrations of the $Mg-N_4$ group distorted from its nearly planar conformation by binding of only one extramolecular ligand, whatever the nature of this ligand may be.

The simultaneous presence of the low frequency, interaction dependent set, and of _ca_ 1655 cm^{-1} components in RR spectra of chlorophyll _in vivo_ cannot be taken as conclusive evidence for one pool to consist in Chl-Chl aggregates. Merely, the low frequency sets indicate that many of the Chl _a_ and Chl _b_ molecules have their Mg atoms bound to only one extramolecular ligand. This ligand, finally, may well have a different nature, depending on the chlorophyll pool considered, as differences in relative intensities and frequencies occur for the bands of the low frequency set between the different species studied (Fig. 6 and 7).

CONCLUSION

Resonance Raman spectroscopy appears to constitute an efficient new method of investigation of chlorophyll in intact photosynthetic structures. The present low temperature study showed that discrete pools of Chl _a_ and of Chl _b_ exist _in vivo_, particularized by different interaction states of their carbonyl groups, and, possibly, of a fifth ligand bound to their central Mg atoms. Conformational information has also been drawn out from the present spectra.

More complete exploitation of the information content of RR spectra of chlorophyll _in vivo_ is in part dependent on further technological progress on blue emitting, continuous wave tunable lasers, which are needed for detailed exploration, and refined use in the resonance process, of the Soret band of chlorophyll.

ACKNOWLEDGEMENTS

The author is indebted to Mr J. Farineau for chloroplasts of greening maize, to Mr G. Girault for spinach chloroplasts preparation, and to Mrs Guyon, Laboratoire de Photosynthèse, CNRS GIF, for cultures of Botrydiopsis and Chlorella algae.

REFERENCES

1 - C.S. French, J.S. Brown and M.C. Lawrence, Plant Physiol., 49 (1972) 421.

2 - F.F. Litvin, B.A. Gulyaev and V.A. Sineshchekov, Dokl. Akad. Nauk SSSR, 199 (1971) 1428, Eng. transl. p. 95.

3 - J.J. Katz, in C.O. Chichester (Editor), The Chemistry of Plant Pigments, Academic press, New York, 1972, p.103.

4 - L. Szalay, J. Hevesi and E. Lehoczki, Acta Phys. Chem., 19 (1973) 403.

5 - M.E. Deroche and C. Costes, Ann. Physiol. Vég., 8 (1966) 223.

6 - R.K. Clayton, in L.P. Vernon and G.R. Seely (Editors), The Chlorophylls, Academic press, New York, 1966, p 609.

7 - M. Lutz, C.R. Acad. Sci. Paris, sér. B, 275 (1972) 497.

8 - M. Lutz and J. Breton, Biochem. Biophys. Res. Comm., 53 (1973) 413.

9 - M. Lutz, J. Raman Spectrosc., 2 (1974) 497.

10 - M. Lutz, 4th Int. Conf. on Raman Spectroscopy, Brunswick, Maine, 1974.

11 - J.J. Katz, R.C. Dougherty and L.J. Boucher, in L.P. Vernon and G.R. Seely (Editors), The Chlorophylls, Academic press, New York 1966, p. 185.

12 - C. Tuquet, T. Guillot-Salomon, J. Farineau and M. Signol, Physiol. Vég., (1975) in press.

13 - M. Avron, Anal. Biochem., 2 (1961) 535.

14 - L. Rimai, D. Gill and J.L. Parsons, J. Am. Chem. Soc., 93 (1971) 1353.

15 - K. Ballschmiter and J.J. Katz, J. Am. Chem. Soc., 91 (1969) 2661.

16 - M. Lutz and J. Kléo, C.R. Acad. Sci. Paris, sér. D, 279 (1974) 1413.

17 - J.B. Thomas and F. Bretschneider, Biochim. Biophys. Acta, 205 (1970) 390.

18 - B. Ke, in L.P. Vernon and G.R. Seely (Editors), The Chlorophylls, Academic press, New York, 1966, p. 427.

19 - L.J. Boucher, H.H. Strain and J.J. Katz, J. Am. Chem. Soc., 88 (1966) 1341.

20 - C.B. Storm, A.H. Corwin, R.R. Arellano, M. Martz and R. Weintraub, J. Am. Chem. Soc., 88 (1966) 2525.

21 - M.S. Fischer, D.H. Templeton, A. Zalkin and M. Calvin, J. Am. Chem. Soc., 94 (1972) 3613.

DISCUSSION

M. LUTZ

M.W. Windsor - Why would Nature want 5 discrete pools of chlorophyll *a* in the antenna system? I can understand that chlorophylls situated *within* the pool would appear differently by resonance Raman from those on the periphery of the pool, but why as many as (or perhaps as few as) 5 different subsets?

M. Lutz - The word "pool" used in this work may be somewhat misleading. RR scattering differentiates (at least) five discrete categories of Chl *a* by the different natures of the extramolecular intereactions to which their 9-keto carbonyl groups are submitted. But RRS does not tell anything on the relative spatial arrangments of these categories of Chl *a*. The existence of only one pool of antenna Chl *a* consisting of repeated juxtapositions of Chl *a* molecules of the five RR defined types appears nevertheless less likely than that of separate assemblies of Chl *a* molecules all belonging to one RR type only, since the relative proportions of molecules of the different RR types were seen to be different depending on the organism studied. Finally, the two partners of a O=CChl Mg...O=Cchl Mg dimer, if existing *in vivo* , would enter into two different RR defined pools.

W. Junge - I wonder why you apply this powerful method to whole chloroplasts, where you have a hotchpotch of chlorophylls. This certainly makes your interpretation less specific. Why did not you try bacterial reaction centers, having only four chlorophylls plus two pheophytins?

M. Lutz - We have undertaken a RR study of bacteriochlorophyll (B Chl) in solution and in chromatophores and reaction centers of different photosynthetic bacteria (M. Lutz, F. Reiss-Husson, unpublished work, 1974). However, the very high photooxydability of B Chl in solution constitues a major problem for *in vitro* work, and the quality of RR spectra of the biological samples suffers from the presence of overlapping fluorescence bands of different origins in the 400-600 nm range.

M.W. Windsor - I just want to add that nobody has yet succeeded in making reaction centers from plant chloroplasts. The best that has been done (by I. Ikegami and S. Satoh, Biochim. Biophys. Acta, 367 (1975), 588-92) is to make a subchloroplast preparation with a ratio of antenna chlorophyll to reaction center chlorophyll of 6:1. Even so, I think it would be valuable to look at this system using your techniques.

M. Lutz - In the working hypothesis of the presence of a Chl-H_2O-Chl aggregate in P 700, the possible presence of C=O bands from remaining Chl *b* molecules in the 1640 cm^{-1} region has proven to constitute an obstacle to the study of system I enriched preparations (M. Lutz, J.S. Brown, unpublished work, 1974). Informations are expected from the study of chloroplastic fractions of Chl *b* per organisms.

M. Stockburger - Did you irradiate with the laser lines into the Soret band? If so, the Raman lines you found are expected to be polarized lines. Did you prove this?

M. Lutz - Polarization measurements were made on RR spectra of Chl *a* and *b* in solution, in both preresonance and full resonance conditions relatively to the Soret systems of electronic bands (cf. ref. 9) All measurable bands were effectively found to be polarized throughout the spectral range explored, with the only exception of a weak Chl *a* band (1265 cm^{-1}), which may gain intensity, in part, from resonance on an upper vibrational sublevel of one of the Q bands.

PRIMARY REACTIONS OF PHOTOSYNTHESIS IN GREEN PLANTS. A STUDY OF PHOTOSYSTEM-2 AT LOW TEMPERATURE

P. MATHIS, A. VERMEGLIO and J. HAVEMAN
Service de Biophysique, Département de Biologie
Centre d'Etudes Nucléaires de Saclay, BP 2, 91190 Gif-Sur-Yvette, France

SUMMARY

The measurement of laser flash-induced absorption changes allows to identify the following components involved in the primary photochemical reactions of photosystem-2, at -170° :

- the primary electron donor is a chlorophyll molecule whose cation presents a peak at 825 nm.
- the primary acceptor is a plastoquinone molecule whose radical anion is detected at 320 nm and whose reduction is accompagnied by a shift of a pigment absorbing around 546 nm.
- several secondary donors can function, including cytochrome b_{559} and probably chlorophyll molecules.

We describe the kinetic relationships between these constituents.

INTRODUCTION

The primary reactions of photosynthesis have recently been the object of a greatly increasing interest [1-4]. A major reason is that photosynthetic structures constitute a machinery which converts the energy of visible light into stable chemicals, with a rather good yield : this ability is of a great potential importance for a better use of solar energy. Photosynthesis can be considered both as a model for the realization of artifical energy converters and as an actual energy storage process susceptible of improvement. The recent studies on primary reactions were made possible thanks to new powerful tools , and especially thanks to the lasers and to fast electronic devices (see e.g. ref. 5).

For a brief outline of the photosynthetic reactions we can start with the absorption of a photon, at any place in an array of light-harvesting pigments. The electronic excitation energy is rapidly transferred to a special chlorophyll molecule (that is named a trap), in less than 0.1 ns. The excited trap chlorophyll transfers an electron to a primary electron

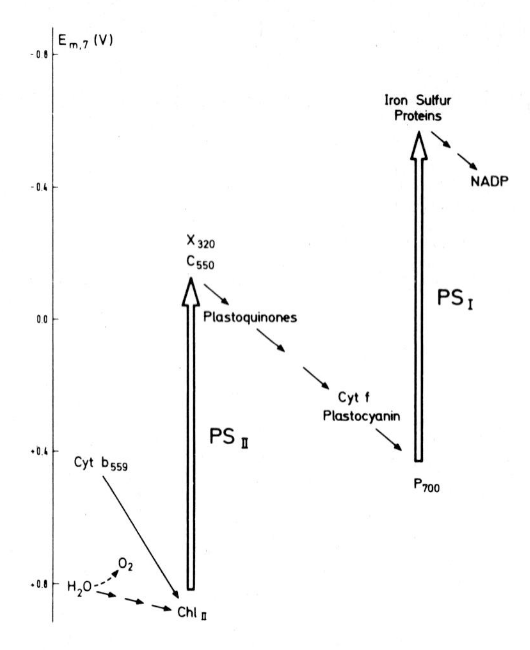

Fig. 1. Schematic representation of the photosynthetic electron transfer chain in green plants. The large arrows pointing upward indicate the photoreactions. The vertical scale stands for the mid-point potential of the components of the chain.

acceptor ; it is then reduced by an electron donor. The ensemble constituted by the trap chlorophyll, its direct partners of the electron transfer chain, together with suitable proteins and lipids involved in maintaining a favorable structure, makes up a "reaction center". In green plants there are two types of reaction centers which, with their respective antenna pigments, are named photosystem-1 and photosystem-2, and which are operating in series for the transfer of electrons from water (ultimate electron donor) to $NADP^+$, the electron acceptor whose reduced form is necessary for the reduction of CO_2 (the ultimate electron acceptor).

In this paper we will consider as "primary" the reactions concerning the trap chlorophyll. A convenient tool for their study is a flash of light which is both very short (compared to secondary dark reaction) and intense (in order to saturate all the reaction centers present in the sample under study). In that respect pulsed lasers are the considerable interest. The use of low temperatures revealed to be an additional interesting tool, as they allow nearly normal primary reactions to occur while secondary ones are largely blocked. In this paper we are mainly concerned with the reactions occurring at the reaction center of photosystem-2, around -170°. The reaction partners are identified and some of their kinetic relationships are studied on the basis of characteristic light-induced (mostly laser flash-induced) absorption changes.

MATERIALS AND METHODS

As biological material we used spinach chloroplasts, suspended in a mixture of buffer (40 %) and glycerol (60 %). The suspension is contained in a special cuvette (1 mm thick) fitting in a partly unsilvered Dewar flask half filled with liquid N_2. Differential absorption spectra were recorded with a double-beam spectrophotometer. Kinetics of flash-induced absorption changes were recorded as basically described in fig. 2. The laser was a Q-switch ruby laser constructed by Quantel. A more detailed description of the materials, methods and techniques is given in other publications [6-11].

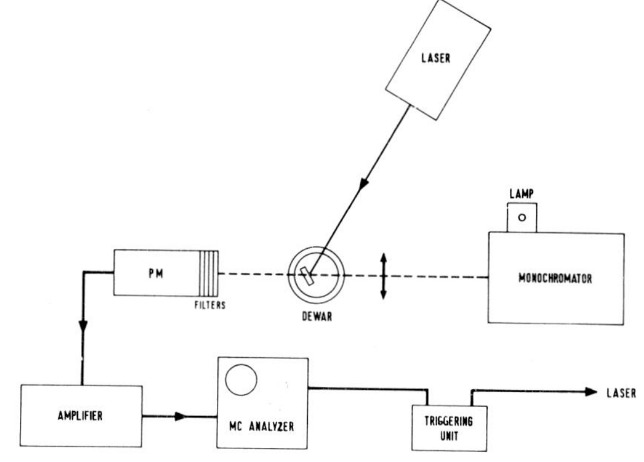

Fig. 2. Block diagram of the basic elements of the equipment which is used for the measurement of flash-induced absorption changes in chloroplasts suspensions.

RESULTS AND DISCUSSION

The primary electron donor of photosystem-2

In photosystem-1 it has been known for some time that a special chlorophyll (named P_{700}) is the primary donor. The major argument is the occurrence of a light-induced EPR spectrum which is nearly identical to the EPR spectrum of the radical cation of chlorophyll (Chl^+) which can be formed in vitro. In optical absorption spectroscopy this radical cation presents a characteristic band at 825 nm.

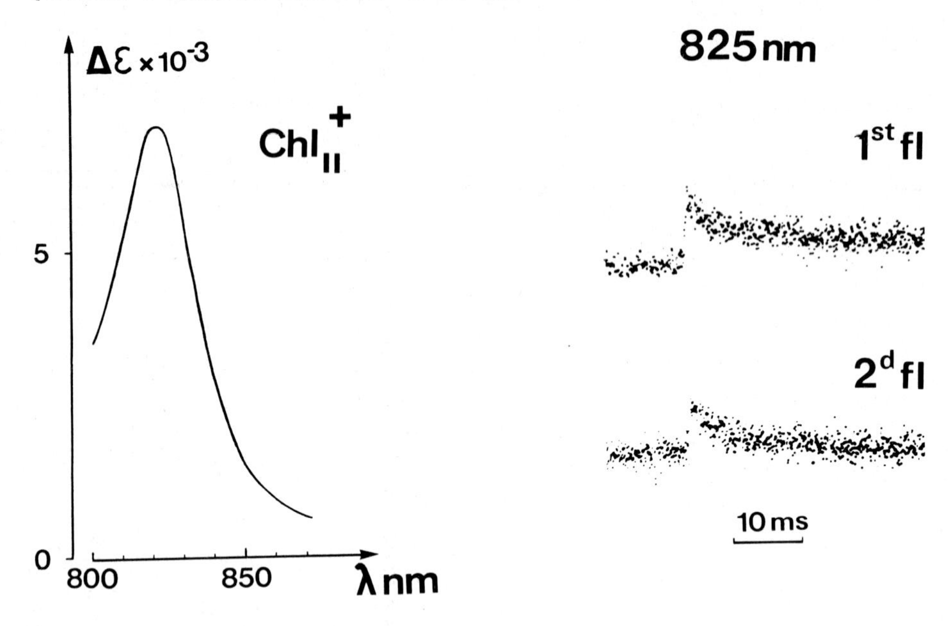

Fig. 3. Kinetics of laser flash-induced absorption changes, at 825 nm, with a chloroplasts suspension, at -170° (on the right). Effect of the first and of the second flash (separated by 30 s). The maximum absorbance change on the first flash is : ΔA_{max} 1st fl = + 1.1 x 10^{-3}. Chlorophyll concentration : 0.9 $mg.ml^{-1}$.
On the left : spectrum of the rapidly decaying phase, attributed to Chl_{II}^{+}. The molar extinction coeficient was calculated assuming one Chl_{II}^{+} per molecule of cytochrome b_{559} oxidized by continuous light at -170° (extinction coefficient at 556 nm : 20,000).

We made use of this last property for studying the primary electron donor of photosystem-2 (Chl_{II}) (see also ref. 11). In fig. 3 we present the long-wave-length absorption peak of Chl_{II}^{+} and the kinetics of its formation and decay after a laser flash, at -170°. Chl_{II}^{+} is formed in less than 80 μs and decays by reduction with a half-time of 3.0 ms. In fig. 3 the non-reversible absorption increase is due to P_{700}^{+} as shown by control experiments in which photosystem-2 is inactivated by a treatment with DCMU and hydroxylamine. Within a series of successive laser flashes (separated by 30 s) the amount of Chl_{II}^{+} which is formed by a flash decreases progressively as if a pool were progressively exhausted.

Analogous properties were observed independently by Malkin and

Bearden, by flash EPR [12]. Other studies have been performed by absorption spectroscopy : an absorption decrease in the red absorption band of chlorophyll _a_ has been interpreted as an oxidation of Chl_{II} (which is named also Chl a_2 or P_{680}) [13, 14]. It has never been possible to prevent Chl_{II}^+ from dark reduction, even at low temperature and under very oxidizing conditions. This is a major difference between the trap of photosystem-2 and the traps of photosystem-1 or of photosynthetic bacteria whose oxidized state can be fairly stable. This difference may result from the very high potential (greater than + 0.80 V) of the couple Chl_{II}/Chl_{II}^+ ; this is much higher than the potential of the Chl/Chl^+ couple in vitro (+ 0.52 V).

The primary electron acceptor of photosystem-2

Early information on the acceptor was obtained indirectly by monitoring the fluorescence yield of chlorophyll : it is low when the acceptor is oxidized (efficient photochemistry) and increases when it is reduced (no photochemistry) [15]. Later on, the acceptor was studied by absorption spectroscopy and shown to be a plastoquinone which is photoreduced into its radical-anion form [16, 17]. The radical-anion presents a characteristic absorption peak around 320 nm [18].

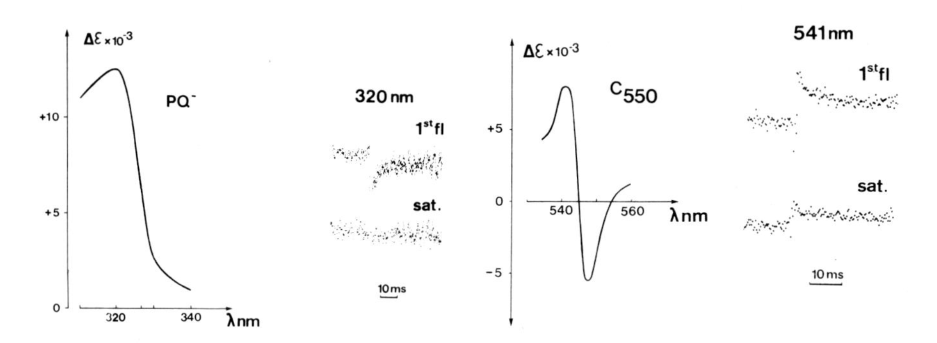

Fig. 4. Kinetics of laser flash-induced absorption changes, at -170°, due to the reduction of the primary acceptor, and corresponding difference spectra. For the evaluation of Δε, we refer to cytochrome b_{559}, as in fig. 3. For the kinetics we present the effect of the first flash and the effect of one flash after a saturating illumination of the cuvette by continuous light.
320 nm : ΔA_{max} 1st fl = + 0.9 x 10^{-3} (for technical reasons the direction of ΔA is inverted). Chlorophyll concentration : 0.25 $mg.ml^{-1}$.
542 nm : ΔA_{max} 1st fl = + 1.7 x 10^{-3}. Chlorophyll concentration : 0.79$mg.ml^{-1}$.

At low temperature the reduction of the primary acceptor has been characterized by an absorption shift around 540-550 nm, a shift attributed to a species named C-550 [19,1]. In agreement with previous observations made at room temperature, we have recently shown that C-550 and the plastoquinone molecule behave identically at -170°. Difference spectra for both species are presented in fig. 4, together with the kinetics of corresponding laser flash-induced absorption changes. An identical absorption change can be induced by one flash or by continuous light. After illumination by continuous light the acceptor remains reduced for at least 15 minutes ; after one flash, however, it is largely reoxidized by a back-reaction with Chl_{II}^{+}, with a half-time of 3.0 ms. About 25 % of the acceptor remains reduced, and the stable reduced state progressively accumulates within a series of successive flashes (see also refs 9, 10, 20).

The absorption change peaking at 320 nm fits fairly well with in vitro properties of the plastoquinone anion-radical [18]. However, the origin of C-550 is not understood, as neither plastoquinone nor its anion-radical have an absorption band in the 500-550 nm range. It has been suggested that pheophytin or some carotenoïd were involved in C-550 [1, 17].

Secondary electron donors of photosystem-2

At low temperature the best characterized secondary donor is cytochrome b_{559} [21]. This cytochrome is largely reduced in fresh chloroplasts ; around 2 molecules per reaction centerare present in photosystem-2, and about half of it is oxidized by continuous light at -170° 22 . Under illumination with continuous light, the rate of oxidation of cytochrome b_{559} is about twice slower than the rate of reduction of the primary acceptor [7, 23]. After one flash, a small fraction of the photooxidizable cytochrome b_{559} is oxidized with a half-time of 3.0 ms [9,11]. These results are compatible with cytochrome b_{559} functioning as a secondary electron donor.

In oxidizing conditions (e. g. upon addition of ferricyanide to the chloroplasts before cooling) cytochrome b_{559} is oxidized. However the behaviour of Chl_{II}^{+} as well as of the primary acceptor are practically unaffected. This indicates that another efficient donor replaces cytochrome b_{559}. The stable aborption increase reported in fig. 5 is attributed to this alternative donor (D in fig. 5). It might be a chlorophyll molecule (from the bulk of light-harvesting pigments) as a typical EPR spectrum has been observed in the same conditions in several laboratories, which is identical to the spectrum of the radical-cation Chl^{+} produced in vitro [24-26].

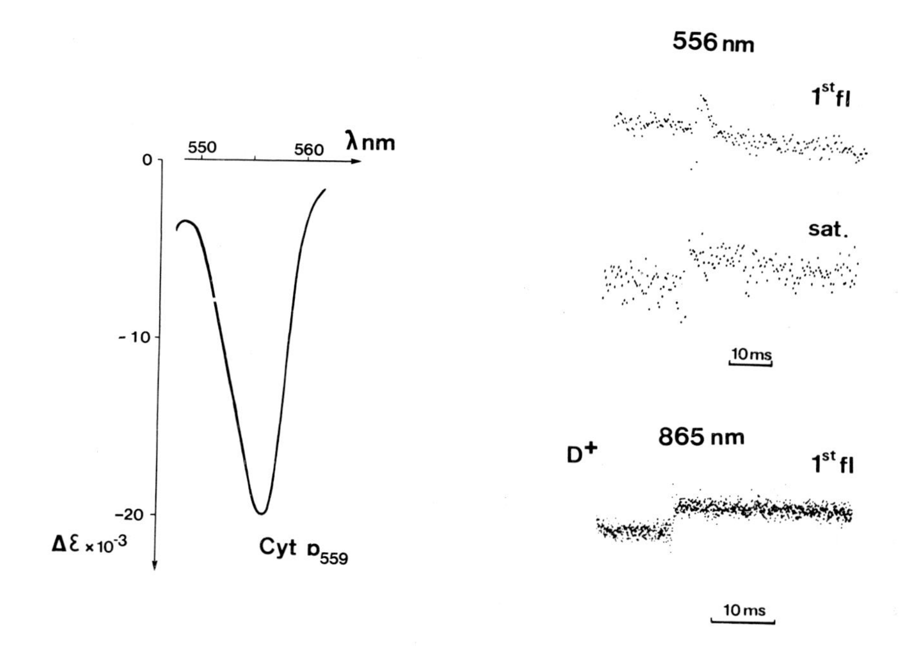

Fig. 5. Kinetics of laser flash-induced absorption changes, at -170°, due to the oxidation of secondary electron donors in photosystem-2.
556 nm : cytochrome b_{559}. First flash (initial ΔA_{max} = + 3.6 x 10^{-4}) and one flash after a saturating illumination by continuous light. Chlorophyll: 0.79 $mg.ml^{-1}$. No addition.
865 nm : alternative donor D. ΔA_{max} 1^{st}fl = + 3.2 x 10^{-4}. Chlorophyll : 0.97 $mg.ml^{-1}$. Addition of 10 mM of potassium ferricyanide. Difference spectrum due to the oxidation of cytochrome b_{559} (α band) at -170° (on the left).

It must be pointed out that there is no evidence for an identity between the donors operating at liquid N_2 temperature and the physiological donors. At -50° we observed that the chemical nature of the donor was a function of the number of positive charges accumulated on the oxygen-evolving side of Chl_{II} [6]. This is an indication that photosystem-2 functions similarly at -50° and at physiological temperatures. In the last conditions the donors are not chemically identified although a newly discovered EPR signal has been attributed to the normal secondary donor [27].

CONCLUSIONS

The reactions occurring in photosystem-2 at liquid nitrogen temperature are best described by the following model [9, 11, 28, 29] :

$$\left(\begin{matrix}PQ\\C\text{-}550\end{matrix}\right) - Chl_{II} - Cyt\ b_{559} \underset{}{\overset{h\nu}{\rightleftharpoons}} \left(\begin{matrix}PQ\\C\text{-}550\end{matrix}\right)^{-} - Chl_{II}^{+} - Cyt\ b_{559}$$

$$\swarrow$$

$$\left(\begin{matrix}PQ\\C\text{-}550\end{matrix}\right)^{-} - Chl_{II} - Cyt\ b_{559}^{+}$$

The light-driven step probably involves directly the singlet excited state of chlorophyll, as there is a competition between fluorescence and photochemical processes. The intermediate state in the above model is detected by the simultaneous presence of PQ^-, $C\text{-}550^-$ and Chl_{II}^+. This state is unstable and it evolves to the final state (stable) or, with a greater probability, to the initial state, by a back-reaction. This model is satisfactory in accounting for flash-induced absorbance changes, but it fails to explain the kinetics of reactions induced by continuous light. This discrepancy led us to propose a model with two light reactions at the same reaction center (see refs. 10, 22 for a more complete discussion). The partners of the primary reactions are inserted in the chloroplast membrane. There are now many evidences for a sidedness (internal versus external) of the membrane. At room temperature, a consequence of the oriented arrangement is the coupling of proton movement with electron transport and the occurrence of a light-induced transmembrane electric field. This one is detected by an electrochromism of some pigments (especially the carotenoïds). We have observed and studied this light-induced electrochromism in low-temperature photoreactions [8, 9].

A great deal of informations have been obtained recently on the primary reactions in photosynthetic bacteria, by a conjonction of biochemical methods (isolation of purified reaction centers) and of spectroscopic techniques, especially with pulsed lasers [3]. In the case of photosystem-2 of green plants, the approach is more difficult because reaction centers have not yet been isolated so that we have to extract small effects at the level of the center from a large amount of constituents (mainly the light-harvesting pigments). So it is necessary to ally very fast techniques with high sensitivity, a difficult alliance which however constitutes the condition to further progress.

REFERENCES

1 W. Butler, Acc. Chem. Res., 6 (1973) 177.
2 A.J. Bearden and R. Malkin, Quart. Rev. Biophys., 7 (1975) 131.
3 W.W. Parson and R.J. Cogdell, Biochim. Biophys. Acta, 416 (1975) 105.
4 K. Sauer, in Govindjee (Editor), Bioenergetics of Photosynthesis, Academic Press, New-York, 1975, p. 115.
5 J.S. Leigh, T.L. Netzel, P.L. Dutton and P.M. Rentzepis, FEBS Lett., 48 (1974) 136.
6 A. Vermeglio and P. Mathis, Biochim. Biophys. Acta, 314 (1973) 57.
7 P. Mathis, M. Michel-Villaz and A. Vermeglio, Biochem. Biophys. Res. Comm., 56 (1974) 682.
8 A. Vermeglio and P. Mathis, Biochim. Biophys. Acta, 368 (1974) 9.
9 P. Mathis and A. Vermeglio, Biochim. Biophys. Acta, 368 (1974) 130.
10 A. Vermeglio, Thesis, Orsay 1974,
11 P. Mathis and A. Vermeglio, Biochim. Biophys. Acta, in press.
12 R. Malkin and A.J. Bearden, Biochim. Biophys. Acta, in press.
13 G. Döring, G. Renger, J. Vater and H.T. Witt, Z. Naturforsch., 24 b (1969) 1139.
14 R.A. Floyd, B. Chance and D. de Vault, Biochim. Biophys. Acta, 226 (1971) 103.
15 L.N.M. Duysens and H.E. Sweers, in Studies on Microalgae and Photosynthetic Bacteria, University of Tokyo Press, Tokyo, 1963, p. 353.
16 H.H. Stiehl and H.T. Witt, Z. Naturforsch., 24 b (1969) 1588.
17 H. van Gorkom, Biochim. Biophys. Acta, 347 (1974) 439.
18 R. Bensasson and E.J. Land, Biochim. Biophys. Acta, 325 (1973) 175.
19 D.B. Knaff and D.I. Arnon, Proc. Nat. Acad. Sci. U.S., 63 (1969) 963.
20 A. Vermeglio and P. Mathis, Biochim. Biophys. Acta, 292 (1973) 763.
21 D.B. Knaff and D.I. Arnon, Proc. Nat. Acad. Sci. U.S., 63 (1969) 956.
22 A. Vermeglio and P. Mathis, in M. Avron (Editor), Proc. Third Intern. Congress on Photosynthesis, Elsevier, Amsterdam, 1974, p. 323.
23 W.L. Butler, J.W.M. Visser and H.L. Simons, Biochim. Biophys. Acta, 292 (1973) 140.
24 R. Malkin and A.J. Bearden, Proc. Nat. Acad. Sci. U.S., 70 (1973) 294.
25 B.Ke, S. Sahu, E. Shaw and H. Beinert, Biochim. Biophys. Acta, 347 (1974) 36.
26 J.W.M. Visser and C.P. Rijgersberg, in M. Avron (Editor), Proc.

Third Intern. Congress on Photosynthesis, Elsevier, Amsterdam, p. 399.
27 R.E.Blankenship, G.T. Babcock, J.T. Warden and K. Sauer, FEBS Lett., 51 (1975) 287.
28 W.L. Butler, Proc. Nat. Acad. Sci. U.S., 69 (1972) 3420.
29 N. Murata, S. Itoh and M. Okada, Biochim. Biophys. Acta, 325 (1973) 463.
30 A. Trebst, Ann. Rev. Plant Physiol., 25 (1974) 423.

DISCUSSION

P. MATHIS, A. VERMEGLIO, J. HAVEMAN

M.W. Windsor - Do you have any information on the rate of formation of the radical cation (Chl^{+}_{II})?

P. Mathis - Our time resolution is 50 μs. With this resolution we cannot resolve the formation of Chl^{+}_{II}, which is obviously expectable.

A PHOTOOXYGENATION ACTINOMETER FOR LASER INTENSITY MEASUREMENTS*

J. N. DEMAS, E. W. HARRIS and R. P. McBRIDE
Chemistry Department, University of Virginia, Charlottesville, Virginia 22901 (U.S.A.)

SUMMARY

The ways in which chemical actinometry supplement and surpass photoelectric and thermal laser power measurements are discussed. A new almost ideal chemical actinometer for laser power measurement, the tris(2,2'-bipyridine)ruthenium(II) sensitized photooxygenation of tetramethylethylene, is described.

INTRODUCTION

This paper is presented somewhat in the spirit of a missionary before cannibals. Photochemists and spectroscopists using high power lasers almost exclusively measure powers with calibrated photodiodes or thermal detectors [2,3]. We will present and defend a minority opinion that not only are electronic meters not the only way to measure powers, they are not always the best. The alternative method is chemical actinometry, or light intensity measurements from the rate of a light induced chemical reaction. We shall explain why actinometry complements and sometimes supercedes electronic detectors, and we will describe a new class of actinometers specifically designed for use with high power lasers.

Besides their high cost, there are two major disadvantages to electronic power meters. First, the sensitive areas tend

---

*This article in part summarizes a full paper on the actinometer [1]. A detailed discussion of the calibration and use of the actinometer is available from J. N. Demas.

to be small, 1 cm^2 or 1 in in diameter being typical values; this can prevent their use in many photochemical and spectroscopic experiments. Second, most workers use commercial factory-calibrated meters and are completely at the mercy of the manufacturer and of sensitivity shifts caused by aging electronics or sensor degradation. Even self-calibrating bolometers such as the ones made by Scientech can only correct for electronic drift but not for target degradation. Customary solutions include ignoring the problem, refinishing the bolometer surface regularly, sending the meter back to the manufacturer for frequent recalibration--assuming he maintains a good radiometry laboratory, or keeping one or more spare power meters just for monitoring long term stability of the working one(s). Clearly, none of these solutions is completely satisfactory because of labor, time or cost.

Chemical actinometers, on the other hand, are usually solution and can thus be placed into any size or shape actinometer cell. For example an actinometer of 100 cm^2 area in almost any shape could easily be fabricated. In fact the cell could almost completely surround a luminescent or scattering sample and thus intercept nearly 4 π steradians of radiation. Further, because the actinometer solution is homogeneous, the sensitivity can be made virtually <u>absolutely</u> uniform over the entire surface, regardless of size. These possibilities are impossible or prohibitively expensive by thermal or photoelectric methods.

Further, because of the inherent absolute reproducibility of a chemical reaction, actinometers offer inexpensive long term monitors for power meter stability and inexpensive calibration of homemade meters. Finally, data referenced either directly or indirectly to an actinometer, then can be corrected if the actinometer's yield is ever updated. Data taken from an improperly calibrated electronic instrument not traceable to an actinometer, however, would usually be lost.

DISCUSSION

Our actinometer is the tris(2,2'-bipyridine)ruthenium(II), $Ru(bipy)_3^{2+}$, sensitized photooxidation of tetramethylethylene, TME, in methanol. It is the first successful one specifically designed for use with high power lasers. It is especially suitable for the Ar^+ laser lines, but preliminary tests suggest that it will be equally useful with high power pulsed N_2 laser and most pulsed or cw lasers with $\lambda < 550$ nm. The key processes are

$$Ru(bipy)_3^{2+} \xrightarrow[\phi'_{isc}]{h\nu} {}^*Ru(bipy)_3^{2+} \qquad [1]$$

$$^*Ru(bipy)_3^{2+} + O_2 \xrightarrow{k_{et}} Ru(bipy)_3^{2+} + {}^1O_2 \qquad [2]$$

$$^1O_2 + TME \xrightarrow{k_{rx}} (CH_3)_2COOHC(CH_3) = CH_2 \qquad [3]$$

where $^*Ru(bipy)_3^{2+}$ is the complex in its long-lived charge transfer excited state, ϕ'_{isc} is the efficiency of population of its emitting state following optical excitation, and $^1O_2$ is metastable singlet oxygen.

The complete actinometer system consists of a photolysis cell containing an O_2-saturated methanol solution of TME and $Ru(bipy)_3^{2+}$ connected to a gas buret filled with pure O_2. Radiation is absorbed by the intensely orange-colored $Ru(bipy)_3^{2+}$ which is quenched by dissolved oxygen to form $^1O_2$ which in turn is consumed by the TME. Oxygen uptake is read directly from the buret. With O_2 pressures near 1 atm and $[TME] \sim 0.11$ M, the efficiency of O_2 consumed is ~ 0.76 moles of O_2/Einstein which gives ~ 5 ml/min uptake for 1 W at 500 nm.

The observed or effective quantum yield during an irradiation, ϕ'_{obs}, is given by

$$\phi'_{obs} = (0.855) \left\{ \frac{K_{SV}\overline{[O_2]}}{1 + K_{SV}\overline{[O_2]}} \right\} \left\{ \frac{\overline{[TME]}}{\beta + \overline{[TME]}} \right\} \qquad [4]$$

where $K_{SV} = 1400\ M^{-1}$, $\beta = 0.0027$ M, and $\overline{[O_2]}$ and $\overline{[TME]}$ are the average oxygen and TME concentrations in the solution during the irradiation. Average rather than initial concentrations are used because some TME is consumed and O_2 does not redissolve as rapidly as it reacts; the computational uncertainties are small, however, and little error results. From ϕ'_{obs} the laser intensity is then easily calculated [1]. Equation 4 which is based on accurate spectrofluorimetry and absolute bolometry is certainly accurate to better than 5% over the 280-560 nm range [4].

This system represents a nearly ideal chemical actinometer for high power sources [3]. There is no evidence for reciprocity failure over the 0.1-1.5 W range with cw Ar^+ lasers, and with a suitable cell cw powers to beyond 100 W are feasible. The $Ru(bipy)_3^{2+}$ absorbs strongly ($\epsilon \sim 10^3$-10^5) from the vacuum uv to >530 nm which minimizes or eliminates transmission corrections which frequently plague actinometry. Because ϕ'_{isc} is constant to better than 5% over at least the 280-560 nm region and is probably absolutely constant over this region [4], the actinometer response will be quantum flat and independent of the spectral distribution of the light source within this range. Thus, the actinometer directly yields quantum intensities from polychromatic sources unlike thermal detectors from which this quantity is frequently difficult to obtain.

The experimental apparatus is inexpensive and simple. A circulating constant temperature regulator, a magnetic stirrer, a thermostated photolysis cell and a gas buret are all that is needed (Fig. 1). The photolysis cell is easily fabricated by a competent glass blower and has no thermostating liquid in the optical paths [3]. Alternatively an unjacketed cell in a thermostated block should work equally well and cost less.

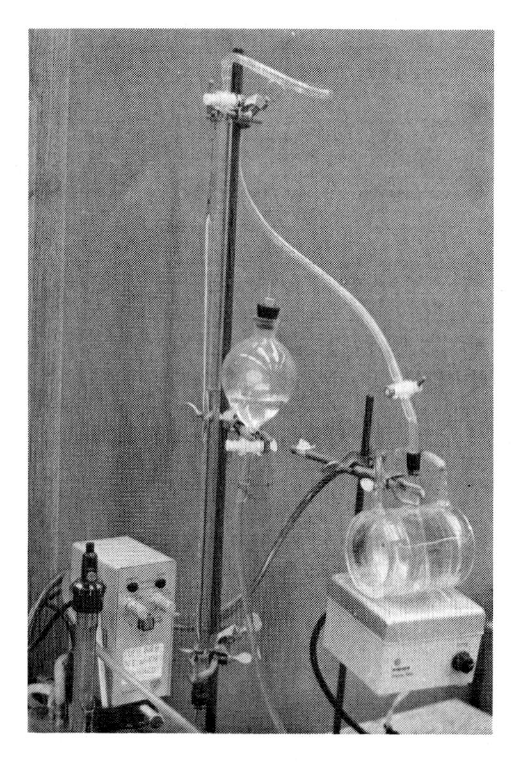

Fig. 1. Photolysis Apparatus. The constant temperature regulator is at the left, the gas buret made from an inverted 100 ml liquid buret and a 240 ml separatory funnel leveling bulb is in the center, and the photolysis cell with associated magnetic stirrer is on the right. Gas connections are made with Tygon tubing. The defocusing lens is not shown.

To summarize, this actinometer system is easy to use, is very reproducible, uses commercial chemicals and requires no analytical instrument such as a spectrophotometer. Also, unlike most actinometers, repetitive runs are easily made on the same solutions. For example, seven runs on a single solution over a two-day period showed no variation in yield.

The equipment shown here is probably the simplest but not the most useful configuration. The equilibration times before and after a photolysis are rather long ($\sim$ 30 min) although little attention is required. The need for using $\overline{[O_2]}$ rather than the saturation value adds some experimental and computational complexity. Also, the laser beam must be defocussed to prevent hole burning in the O_2 concentration and the resulting unpredictable variations in $\overline{[O_2]}$. Flow cells would eliminate all of these problems. A gas recirculation pump for bubbling O_2 through the cell would accelerate replacement of the consumed O_2 and equilibration. A fluid flow pump for circulating the solutions into a spray vessel would act similarly. Also if the cell were made like a laser dye cell with the actinometer solution pumped at high velocity down the tube, the O_2 hole burning problem would be eliminated and collimated laser beams of 10's of watts could be measured directly without defocussing.

This system's wavelength range is somewhat limited ($\lambda < \sim 550$ nm). The system's long-wavelength cut-off can easily be tailored, however, by replacing $Ru(bipy)_3^{2+}$ with other metal complexes or organic dyes. For example, iridium(III) complexes should yield only uv sensitivity while osmium(II) complexes should be quantum flat responses to beyond 750 nm [5]. Very carefully purified dyes like Rose Bengal and methylene blue may yield operation to beyond 600 and 700 nm respectively if the purification and dimerization problems can be overcome. Finally, cyanine dyes may yield quantum flat responses to $\lambda > 1{,}000$ nm [6]. Work in both cell and sensitizer design is in progress.

ACKNOWLEDGEMENTS

We gratefully acknowledge the donors of the Petroleum Research Fund, administered by the American Chemical Society, the Research Corporation (Cottrell Research Grant), and the National Science Foundation. We also thank F. S. Richardson, H. Dewey and R. A. Keller for the use of their lasers, A. W. Norvelle, Jr. for constructing the power meter and E. W. West for a preprint and his kind gift of Manganin wire used in the power meter.

REFERENCES

1 J. N. Demas, R. P. McBride and E. W. Harris, submitted for publication.

2 E. D. West, in W. R. Ware (Editor), Creation and Detection of the Excited State, Vol. III, Marcel Dekker, New York, in press.

3 J. N. Demas, ibid, in press.

4 J. N. Demas and G. A. Crosby, J. Am. Chem. Soc., 93(1971) 2841.

5 J. N. Demas, E. W. Harris and C. M. Flynn, Jr., J. Am. Chem. Soc., in press.

6 R. A. Nathau and A. H. Adelman, Chem. Comm., (1974)674.

DISCUSSION

J.N. DEMAS, E.W. HARRIS, P.R. McBRIDE

J. Joussot-Dubien - How does the actinometer that you propose compare with the classical chemical actinometer?

J.N. Demas - There are two commonly used actinometers which appear potentially useful for laser power measurements : ferrioxalate and Reineckate. Ferrioxalate has a wavelength dependent yield and is useful for $\lambda \leqslant 500$nm; it is the most sensitive common actinometer. Reineckate is about an order of magnitude less sensitive than ferrioxalate, but it does have a nearly wavelength independent yield and is useful over the range $340 \leqslant \lambda \leqslant 750$ nm. Reineckate does have a fairly rapid thermal dark reaction, is awkward to use especially for the longer wavelengths and calculations indicate that reciprocity failure can arise at relatively low powers. At this time, we definitively do not recommend the use of ferrioxalate. The measured intensity can be strongly influenced by the order of addition of reagents, the time between additions and the developing times. These problems are not given in any of the current literatures. Finally, using an absolute bolometer, we have preliminary results which indicate the accepted yields, at least for the 0.15 M system with $\lambda > 450$ nm, may be in error by 10-15 %. We are currently examining the questions of reproductibility and absolute yields for this system. Our system is nowhere near as sensitive as either ferrioxalate or Reineckate. Its major advantages are very flat response with wavelength (i.e. probably $280 \leqslant \lambda \leqslant 530$ nm), freedom from transmission corrections, high reproducibility, and in a properly designed cell the almost unlimited power handling capability.

J. Langelaar - What is the accuracy of an absolute intensity measurements?

J.N. Demas - We would be surprized if the actinometer calibration were off by much more than 5%. This limit is determined mainly by the absolute bolometer calibration and the uncertainty between the equivalence of electrical and optical heating.

J.K. Eweg - Your actinometer is, as far as reproducibility of the instrument is concerned, strongly dependent on the efficiency of $^1O_2$ formation being constant. Can this efficiency be maintained during a long time, even when the system needs to be refilled? Is there no need for recalibration after refillment?

J.N. Demas - The efficiency of $^1O_2$ production appears to be independent of the degree of photolysis, standing between irradiations or the operator. We have made up to seven intensity determinations using the same solution, including several buret refillings and permitting the solution to stand overnight between some of the runs; there was no variation in $^1O_2$ production efficiency within the accuracy of our measurements. In normal use, one would not have to calibrate the actinometer at all. Our measured quantum yield is usable over a wide range of sensitizer, substrate and oxygen concentrations.

SHORT TIME BEHAVIOUR OF DIFFUSION-CONTROLLED REACTIONS

M.J. PILLING
Physical Chemistry Laboratory, South Parks Road, Oxford, England.

SUMMARY

The behaviour of diffusion-controlled reactions at short times, before the distribution of reactant pairs has relaxed to its steady state form, is reviewed. Generally the kinetics are accelerated during this period, and their time dependence allows both the relative diffusion coefficient and the encounter distance to be evaluated. Interactions arising from equilibrium processes (e.g. Coulomb forces between ions) or from processes intrinsic to the reaction (e.g. dipole-dipole interactions in singlet-singlet energy transfer) modify this transient behaviour. Experiments on a wide range of time scales have permitted investigation of all these effects, and their results and implications are discussed.

THE SMOLUCHOWSKI TREATMENT

Most theories of diffusion controlled reactions have been based on Smoluchowski's [1] treatment of the coagulation of colloid particles. For the reaction

$$A + B \rightarrow \text{products}$$

Fick's laws are applied to the diffusion of B towards a central sink, A, and, in the ensemble average, this leads to a gradient in the concentration of B molecules around A, which drives the reaction, and which tends towards a steady-state dependence on the A-B distance, r. Before this steady-state is established, transient terms appear in the diffusion rate constant; these have been recognised theoretically since Smoluchowski's work in 1917, but have only been examined experimentally in recent years. The full time-dependent diffusion equation is

$$\frac{\partial b}{\partial t} = \frac{D}{r} \frac{\partial^2 (br)}{\partial r^2} \tag{1}$$

where b is the time and distance dependent concentration of B and D is the coefficient of relative diffusion. The boundary conditions

$$b(r,t) = 0 \qquad r \leqslant R,\ t \geqslant 0 \tag{2a}$$

$$b(r,t) = b_o \qquad r > R,\ t = 0 \tag{2b}$$

$$b(r,t) \rightarrow b_o \qquad r \rightarrow \infty,\ t \geqslant 0 \tag{2c}$$

are set up, corresponding to a homogeneous distribution of B around A at time zero, and a fully absorbing boundary at $r = R$, the encounter distance. From eqn. (1) the flux, $J(R,t)$, across the absorbing boundary may be determined, and the rate constant, $k(t)$, equated to $J(R,t)/b_o$. This gives:

$$k(t) = 4\pi RD(1 + R/\surd(\pi Dt)) \tag{3a}$$

$$k(t\rightarrow\infty) = 4\pi RD \tag{3b}$$

At short times the rate constant is higher than that found in the steady-state (Fig.1), because the local concentration of B around A is higher than the steady-state value, given by:

$$b(r) = b_o(1 - R/r) \tag{4}$$

Reaction rapidly consumes this excess concentration, and the $b(r)$ profile and the rate constant relax to their steady-state values. For typical parameters ($R \sim 0.5$ nm, $D \sim 3 \times 10^{-9}\ m^2\ s^{-1}$) the term $R/\surd(\pi Dt)$ falls to 0.1 at $t \sim 250$ ps. Thus at normal viscosities the transient terms may only be studied on a picosecond time scale.

RADIATION BOUNDARY CONDITION

The boundary conditions of the Smoluchowski equation have been criticised since they do not permit any probability of reaction on encounter of less than unity [2-6]. If we allow for a partially reflecting boundary, $b(R,t)$ is no longer given by eqn (2a) but by:

$$b(R,t) = \frac{4\pi R^2 D}{k'} \left\{ \frac{\partial b(r,t)}{\partial t} \right\}_{r=R} \tag{2d}$$

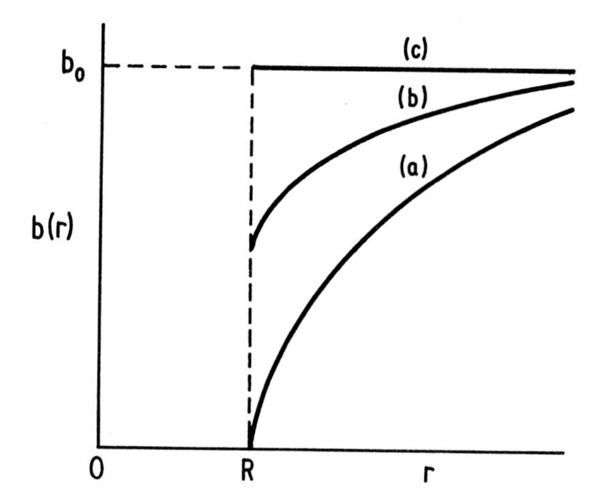

Fig. 1. Concentration profiles for chemical reactions. b(r) vs r for (a) diffusion-controlled reaction, Smoluchowski condition (b) radiation boundary condition (c) activation controlled reaction ≡ random distribution of B around A. R is the encounter distance.

whilst eqn (3) becomes:

$$k(t) = \frac{4\pi RD}{1+4\pi RD/k'} \left\{ 1 + \frac{k'\exp(x^2)\operatorname{erfc}(x)}{4\pi RD} \right\} \tag{4a}$$

where $x = (Dt)^{\frac{1}{2}} (1 + k'/4\pi RD)/R$

For $x \gg 1$ this may be simplified to give:

$$k(t) = \frac{4\pi RD}{1+4\pi RD/k'} \left\{ 1 + \frac{R}{(\pi Dt)^{\frac{1}{2}}(1+4\pi RD/k')} \right\} \tag{4b}$$

In the steady state

$$k(t \to \infty) = 4\pi RD/(1 + 4\pi RD/k') \tag{4c}$$

In these equations, k' is the rate constant which would apply if the equilibrium distribution of encounter pairs were maintained. It is equal to k_cK, where k_c is the first order rate constant for reaction in the encounter pair (i.e. between A and B contained in the same solvent cage), whilst K is the equilibrium constant for formation of the encounter pair. Fig. 1 shows the steady-state concentration profile for a reaction with a partially reflecting boundary.

EXPERIMENTAL INVESTIGATIONS

By studying the transient time domain, R and D may be determined independently, which is not possible for measurements under steady-state conditions (eqn 3b). Provided $b_o >> a(0)$, the zero-time concentration of A, the equation

$$t^{-\frac{1}{2}} \ln\{a(0)/a(t)\} = 4\pi RDb_o\{t^{\frac{1}{2}} + 2R/\sqrt{(\pi D)}\} \qquad (5)$$

applies if the Smoluchowski boundary condition is valid. An equivalent equation may be derived from eqn (4b). D and R may be extracted from plots of $t^{-\frac{1}{2}} \ln a(t)$ vs $t^{\frac{1}{2}}$. This has been achieved in viscous media for reactions of solvated electrons by Buxton et al.[7] and for energy transfer from triplet molecules by Marshall et al. [8]. Nemzek and Ware [9] have used the single photon correlation technique on a nanosecond time scale to study fluorescence quenching in solvents in the viscosity range 50-500 cp. They found R ~ 0.9 - 1.0 nm, whilst their values of D in propane-1,2-diol were somewhat larger than the direct measurements of Mitchell and Tyrrell [10]. They also successfully resolved the further complication of complex formation between the fluorophor and the quencher by comparing continuous excitation and time resolved quenching of fluorescence.

Chuang and Eisenthal [11] have extended the investigation of transient effects into the picosecond range by studying the charge transfer interaction between anthracene $^1B^+_{2u}$ (A^*) and N,N diethylaniline. They monitored the concentration of the charge transfer product, A^-, using a delayed 694 nm probe pulse and found that the kinetics were adequately described by eqn (4a). The second term in the equation, neglected in eqn (4b), falls to 10% of the total bracketed function only after 30 ps, and to 1% after 300 ps. They assumed a relative diffusion coefficient, and derived $R = (0.8 \pm 0.05)$nm and $k' = (11 \pm 1) \times 10^{10}\ M^{-1}\ s^{-1}$. Assuming $K \sim 1\ M^{-1}$[12], k_c, the rate constant for electron transfer in the encounter pair, is $\sim 10^{11}\ s^{-1}$.

Chuang and Eisenthal [11] also applied an alternative theoretical treatment of transient effects derived by Noyes [6,13]. This is not based on Fick's laws, but on a random jump model of diffusion. k(t) is given by:

$$k(t) = k'(1 - \int_0^t h(t')dt')$$

where h(t') is the probability that a pair of reactants that encounter without reaction at time t = 0 will react with each other between t' and t'+ dt'. Provided the molecular displacements are random,

$$k(t) = k'[1 - \beta \mathrm{erfc}(y)], \tag{6a}$$

where

$$y = (\alpha/\beta)\sqrt{(\pi/t)} \tag{6b}$$

where α and β are related to the size and frequency of molecular displacements. For equivalent values of D, k' and R, Chuang and Eisenthal showed that the two theories predict different behaviour, but that the experimental data were too scattered to distinguish between them.

COULOMB INTERACTION BETWEEN REACTANTS

The diffusion equations have also been solved when A and B move in a Coulomb or related field. The steady-state treatment leads to an equation like eqn (3b), but with R replaced by R_{eff} [14]:

$$R_{eff} = [\int_R^\infty \exp(u/kT)(dr/r^2)]^{-1} ,$$

where u is the interaction potential. The divergences from eqn (3a) in the time dependent equations arise essentially from differences between the zero time Boltzmann distribution of reactant ions and the random distribution assumed in eqn (26) [15].

These considerations, and others, have been central to discussions of the picosecond behaviour of excess electrons in aqueous systems. Aldrich et al.[16] found that in the 20 - 350 ps time scale, and with solute concentrations of ~ 1 M, the rate constants for electron reactions differ significantly from those found at longer times for lower solute concentrations. They based their analysis on time independent kinetics; Schwarz [17] reanalysed their data to take into account the short time behaviour predicted by eqn (3a) and obtained satisfactory agreement with experiment for e^-_{aq} + acetone, although he made several simplifying assumptions. The reaction e^-_{aq} + H^+ is slightly slower than diffusion-controlled, and the time-dependent effects are reduced (eqn (4a,b)). Schwarz ascribed the short time results to an inability of the system to establish a Boltzmann distribution of the protons around the electrons on a picosecond time scale. The local concentration of H^+ is thus depressed below the value found at longer times and the short time rate constants are predicted to be smaller than the long time values, in agreement with experiment. Czapski and Peled [18] have extended these arguments to include direct formation of encounter pairs on energy deposition.

LONG RANGE REACTION

Although we have included the effects of a force field, we have not permitted reaction at distances greater than the encounter distance, R. This effect is important in singlet-singlet energy transfer, and eqn (1) becomes:

$$\frac{\partial b}{\partial t} = \frac{D}{r}\frac{\partial^2 (br)}{\partial r^2} - \frac{\alpha b}{r^6} \tag{7}$$

where the term in r^{-6} represents a reaction sink term, which varies with r. At normal viscosities the unquenched singlet lifetime is short, and high concentrations of acceptor must be added to compete. In many systems, long-range energy transfer is so important that static energy transfer takes place, and the diffusion term is omitted from eqn (7). This leads to Forster kinetics [19,20]. The full equation (7) has been solved analytically in the steady-state,[21], and numerically in the time dependent domain [22]. There are comparatively few experimental tests of these equations, since the viscosities required for a significant diffusional contribution are so small. Some experimental work has been done on singlet-triplet energy transfer, where the interactions are less strong [23].

A similar situation arises in triplet-triplet energy transfer, except that now the process is effected by an exchange interaction, which decreases exponentially with distance [24]. The sink term in eqn (7) is replaced by one which depends on r as $\exp(-\beta r)$. The same equation may also be used to describe tunnelling effects in reactions of solvated electrons [25]. The electron is contained in a cavity formed by reorientation of solvent dipoles, and may react with a scavenger molecule by tunnelling. The tunnelling term pushes the fall in the concentration profile (Fig.1) to larger r and this effect is amplified as the diffusion coefficient is reduced. As the relative mobility of the reactants falls, tunnelling from distances increasingly greater than R becomes more able to compete with diffusion, and the effective encounter distance increases. The steady state rate constant is given by

$$k = 4\pi R_{eff} D,$$

where $R_{eff} = \{1.15 + \ln(\alpha/\beta^2 D)\}/\beta$

for $\alpha/\beta^2 D > 2$.

Fig. 2 shows a plot of R_{eff}/R vs D for typical tunnelling parameters for the solvated electron. Even in water at 298 K the effective encounter distance is about twice that expected on the basis of crystallographic radii, in agreement with experiment [26].

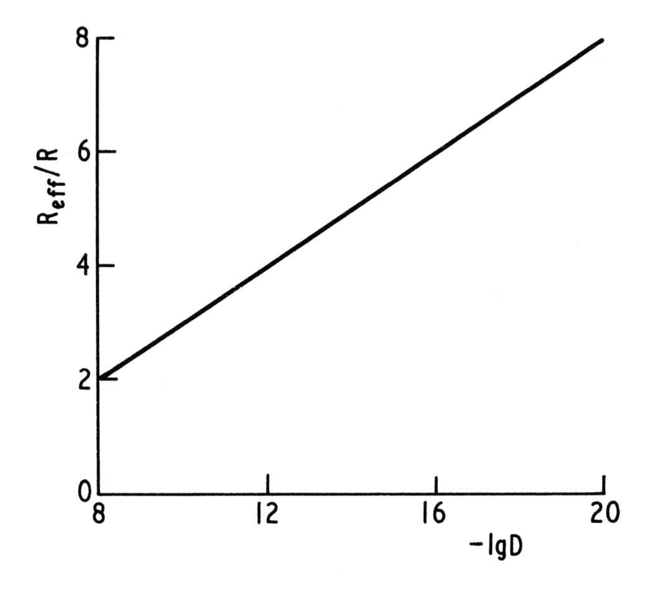

Fig. 2. Plot of R_{eff} vs lgD for electron reactions. The sink term (eqn 7) is $\alpha\exp(-\beta r)$, where $\alpha = 10^{16}\ s^{-1}$, and $\beta = 10\ nm^{-1}$.

The time dependent equations have not been solved analytically, but the increased encounter distance leads to an extension of the transient time domain. Buxton et al. [7] found that, for $CrO_4^{2-} + e^-_{aq}$ in LiCl and NaOH glasses at T ~ 200 K, the encounter distance was increased by a further factor of two over that found at room temperature, in broad agreement with the predictions of Pilling and Rice [25] (Fig. 2).

Tunnelling may be involved in reactions of solvated electrons on picosecond time scales, adding further to the complexities outlined above. The relaxation of the electron trap, observed by Rentzepis and Jones [27], compounds these difficulties, since tunnelling from unrelaxed traps is faster than from the deeper, relaxed traps. This effect could partly explain the very rapid scavenging inferred by Wolff et al.[28] at times shorter than their resolution time.

EFFECTS OF SOLVENT STRUCTURE

The discussion so far has largely presented the solvent as an isotropic medium, and the motion of the reactants as random, except in the specific case of a Coulomb interaction. Fehder and coworkers [29-32] recently published a revealing series of computer simulations of molecular motion in liquid systems. They studied a two dimensional array of Lennard-Jones discs and examined the structure and dynamics of such systems on a picosecond time scale. They found large vacancies in the liquid structure, which persisted for times in excess of 3ps, far longer than the characteristic relaxation time (~ 2.5×10^{-13} s) [29]. Over 2-3 ps, only a very small fraction of the molecules attained an appreciable displacement, and the vast majority showed microdiffusive behaviour around a fixed position. The migrating particles were associated with a group; the motion was cooperative and showed chain- or cluster-like characteristics, with molecules dragged along by the attractive interparticle forces. Thus a large contribution to net singlet diffusion arises from particles that move considerable distances together, and which are unlikely to encounter few, if any, new neighbours during this motion. These cooperative effects slow the approach of reactant molecules when they are within three to four molecular diameters, and the Smoluchowski type of treatment, based on a distance independent relative diffusion coefficient, may overestimate the rate of approach. This effect was recognised and demonstrated experimentally by Noyes [33] for iodine atom recombination.

The hydrodynamic effect of an attractive force between particles has been further discussed by Deutch and Felderhof [34].

There are dangers in extrapolating from a two dimensional simulation to a real three dimensional system. Indeed Emers and Fehder [31] found anomalous behaviour when they applied diffusion theory to a two dimensional model. However, the observations are sufficiently compelling to promote scepticism of any treatment of the relative diffusion of particles separated by small distances which is based on an isotropic continuum or a random jump model.

Experimental observation of short time effects in diffusion-controlled reactions test the behaviour of diffusing particles in exactly this anomalous region. In the steady state the rate constant depends on the relative mobility of the reactants averaged over large distances, although the cooperative deceleration at small r can be invoked, as was done by Noyes [33], in selected instances. The transient time domain is dominated by reaction resulting from the diffusional approach of particles initially separated by only small distances, where the r dependence of D is most evident.

The cage effect [35] provides an even more pertinent experimental test of correlated mobility. Photodisssociation in solution shows a much smaller quantum yield than the same process in the gas phase. The solvent molecules hold the fragments together, prevent their diffusional separation, and instead promote recombination. Noyes [36] showed that the overall recombination may be separated into three types of process:
(i) recombination before any separation can occur, (ii) separation of the photofragments, but recombination of the correlated pair in secondary or subsequent re-encounters, (iii) separation without successful re-encounter, and cross combination with particles from other primary photodissociation events. Processes (i) and (ii) are termed geminate recombination. Noyes [36] showed that scavenging of atoms from the photodissociation of iodine has a complex dependence on scavenger concentration, with the efficiency of the scavenger falling as iodine atoms originally destined to recombine in processes (iii) - (i)are scavenged. Meadows and Noyes [37] found, furthermore, that the quantum yield of photodissociation increases with photon energy. The primary particles are given greater kinetic energy at shorter wavelengths and are more able to escape primary recombination. The wavelength dependence was shown to correlate poorly with a model based

on a hydrodynamic continuum and random molecular motion, and the deviations were in agreement with an r-dependence in the relative diffusion coefficient.

Chuang et al. [38] studied the time dependence of geminate recombination of I_2 using single picosecond pulses at 530 nm both to dissociate the iodine and to probe its subsequent concentration. The recombination occurred over ~ 70 ps in hexadecane and ~ 140 ps in carbon tetrachloride, and any cross combination was unimportant on these time scales. The time dependence differed somewhat from that predicted by the Noyes random walk model [6,13]. The best fits led to values of α and β (eqn 6b) of 0.05 - 0.1 nm and (1-5) x $10^{12} s^{-1}$ for CCl_4 and of 0.01 nm and (2-20) x 10^{12} s^{-1} for hexadecane. If the fit were forced at long times, then the theoretical short time recovery of I_2 was faster than that found experimentally, again illustrating the retardation found in real systems.

The Noyes random walk model [6,13] has also received recent attention in the diffusion model of chemically induced dynamic nuclear polarization (CIDNP)[39]. The observation of CIDNP essentially monitors secondary and subsequent re-encounters of radicals formed by photolysis, but, like studies of scavenging of photodissociation products, the effects are integrated over time. Picosecond measurements permit a time resolved examination of these processes, and clearly their results will find wide application.

REFERENCES

1. M.v. Smoluchowski, Z. Phys. Chem., 92 (1972) 129.
2. F.C. Collins and G.E. Kimball, J. Colloid Sci., 4 (1949) 425.
3. F.C. Collins and G.E. Kimball, Ind. Eng. Chem., 41 (1949) 2551.
4. T.R. Waite, Phys. Rev., 107 (1957) 463, 471.
5. A.T. Barucha-Reid, Arch. Biochem. Biophys., 43 (1952) 416.
6. R.M. Noyes, Prog. Reaction Kinetics, 1 (1961) 129.
7. G.V. Buxton, F.C.R. Cattell and F.S. Dainton, J. Chem. Soc., Faraday Trans. I, 71 (1975) 115.
8. E.J. Marshall, N.A. Philipson and M.J. Pilling, submitted to J. Chem. Soc. Faraday Trans. II.
9. T.L. Nemzek and W.R. Ware, J. Chem. Phys., 62 (1975) 477.
10. M. Mitchell and H.J.V. Tyrrell, J. Chem. Soc. Faraday Trans. II, 68 (1972) 385.

11. T.J. Chuang and K.B. Eisenthal, J. Chem. Phys., 62 (1975) 2213.
12. A.M. North, The Collision Theory of Chemical Reactions in Liquids, Methuen, London, 1964, p.38.
13. R.M. Noyes, J. Chem. Phys. 22 (1954) 1349; J. Am. Chem. Soc., 78 (1956) 5486.
14. P. Debye, J. Electrochem. Soc., 82 (1942) 265.
15. E.W. Montroll, J. Chem. Phys., 14 (1946) 202.
16. J.E. Aldrich, W.B. Taylor, R.K. Wolff and J.W. Hunt, J. Chem. Phys., 55 (1971) 531.
17. H.A. Schwarz, J. Chem. Phys., 55 (1971) 3648.
18. G. Czapski and E. Peled, J. Phys. Chem., 77 (1973) 893.
19. Th. Forster, Disc. Faraday Soc., 27 (1959) 7.
20. M.D. Galanin, Sov. Phys. JETP, 1 (1955) 317.
21. M. Yokota and O. Tanimoto, J. Phys. Soc. Jap. 22 (1967) 779.
22. Y. Elkana, J. Feitelson and E. Katchalski, J. Chem. Phys., 48 (1968) 2399.
23. A.F. Vaudo and D.M. Hercules, J. Am. Chem. Soc., 93 (1971) 2599.
24. M.J. Pilling and S.A. Rice, to be published.
25. M.J. Pilling and S.A. Rice, J. Chem. Soc. Faraday Trans. II, to be published.
26. E.J. Hart and M. Anbar, The Hydrated Electron, Wiley-Interscience, New York, 1970, p.186.
27. P.M. Rentzepis and R.P. Jones, J. Chem. Phys., 59 (1973) 766.
28. R.K. Wolff, J.E. Aldrich, T.L. Penner and J.W. Hunt, J. Phys. Chem., 79 (1975) 210.
29. P.L. Fehder, J. Chem. Phys., 50 (1969) 2617.
30. P.L. Fehder, J. Chem. Phys., 52 (1970) 791.
31. C.A. Emeis and P.L. Fehder, J. Am. Chem. Soc., 92 (1970) 2246.
32. P.L. Fehder, C.A. Emeis and R.P. Futrelle, J. Chem. Phys., 54 (1971) 4921.
33. R.M. Noyes, J. Am. Chem. Soc., 86 (1964) 4529.
34. J.M. Deutch and B.U. Felderhof, J. Chem. Phys., 59 (1973) 1669.
35. J. Franck and E. Rabinowitch, Trans. Faraday Soc., 30 (1934) 120.
36. R.M. Noyes, J. Am. Chem. Soc., 77 (1955) 2042.
37. L.F. Meadows and R.M. Noyes, J. Am. Chem. Soc., 82 (1960) 1872
38. T.J. Chuang, G.W. Hoffman and K.B. Eisenthal, Chem. Phys. Letters, 25 (1974) 201.
39. F.J. Adrian, J. Chem. Phys., 53 (1970) 3374.

DISCUSSION

M.J. PILLING

B.R. Sundheim - As the time scale under consideration becomes shorter, it is necessary to take into account the failure of the diffusion equation to take into account the local structure of the solution. In addition to dielectric relaxation and relaxation of the ionic atmosphere one must also deal with the difference between bulk diffusion and diffusion on an atomic scale, that is, to allow for the fact that the pair distribution function for short distances shows a pronounced structure. One may see this in an extreme form in the interaction of two like charged ions whose approach is describable in terms of macroscopic diffusion equations until the last few angströms where the "collision" rate may be reduced by a factor as large as ten below that deduced from the diffusion equation. The Debye correction or the use of the pair distribution function in a Smoluchowski equation still lead to errors on the picosecond time scale. We have carried out molecular dynamics calculations (computer simulation) on model systems for phosphorescence quenching in ionic liquids and found that the diffusion equation, even when modified in conventional ways, gives an inadequate description for times less than about 10^{-12} seconds but becomes perfectly suitable for longer times.

M.J. Pilling - Your comments are well taken, and emphasis that in analyzing data from picosecond diffusion controlled reactions the traditional methods must be abandoned, and less simple, perhaps numerical approaches used. Perhaps in this way picosecond kinetics can provide an experimental testing ground for your computer simulations.

R. Mohan - I do not know what a psec. chemical reaction means in terms of statistical physics. As far as the random walk phenomena is concerned in this time region, the average displacement is not gaussian and moreover, it is highly peaked. This means one has to use cumulant expansion involving higher order diffusion coefficients, physically this means the velocity correlation is still persistent. This is shown from the beautiful molecular dynamic experiments of B. Alder and Weinwright.

M.J. Pilling - Although the difficulty of the deceleration of the relative approach of reactant molecules certainly arises, for uncharged species the effect corresponds to a reduction in D of about only 50% (see reference 34). Thus within fairly narrow limits we can calculate the lifetime of the encounter pair. For diffusioncontrolled reactions we know that the rate of reaction in the encounter pair is faster than the reciprocal of this lefetime, which at centipoise viscosities puts it well into the picoseconde range. In the case of iodine recombination, for example, the efficiency arises from the effective coupling of energy out of the newly formed I_2^* and into the surrounding molecules, and this presumably takes place at a rate which is comparable with intramolecular energy delocalization in a polyatomic molecule. It should also be noted that many of these reactions take place in highly non-equilibrium, or at least non steady-state situations.

R.M. Hochstrasser - Do you know of experimental evidence suggesting the appropriateness of an exponential form for the distance dependence of the exchange interaction needed for triplet-triplet energy transfer and electron transfer? Do you have theoretical grounds for assuming the exponential form at these moderately large intermolecular separations?

M.J. Pilling - We have made the assumption, originally made by Dexter, that the overlap between wavefunctions falls off exponentially with distance. I know of no theoretical evidence that this idealized picture is valid in practice. In the case of electron reactions, the tunelling rate is determined by the mean potential between the electron and the scavenger, and we have assumed that this is constant. Experimental evidence can be obtained by studying, say, triplet-triplet energy transfer over a range of distances. So far this has only been done in matrices, for static long range quenching, and at low viscosities for dynamic contact quenching. Our aim is to study the distance effect by looking at the transition between these two regions, but, of course, the picture is complicated by the problems of reaction dynamics.

C.B. Harris - If ones uses a realistic potential, particularly for the repulsive portion, and calculates from molecular dynamics the time necessary for molecules to move several molecular diameters, what is the time, say for liquid Ar, and how does it compare to the classical calculations?

M.J. Pilling - I think prof. Sundheim is more able to answer that question than I am.

B.R. Sundheim - The crucial time interval is that below 10^{-12} seconds. It is in this region that the brownian approximation fails so that the particles can no longer be thought of as moving through a uniform resisting medium. All is not lost, however. For any specified potential energy of interaction, quite serviceable descriptions of the short time short distance behaviour can be obtained from molecular dynamics calculations at their present level of sophistication.

S. Ameen - Cranck and Nicholson (1,2) and Spalding et al. (3) (4) have developed a mathematical technique which employes (for the rate of diffusion of substances in direction x) a parabolic partial-differential equation of the following kind using laminar and turbulent boundary layers of fluid mechanics :

$$\frac{\delta \phi}{\delta t} = \frac{1}{r^k} \frac{\delta}{\delta x} \left(r^k D \frac{\delta \phi}{\delta x} \right) + d$$

On the equation index k assumes zero to unity value for plane and axi-symmetrical geometries respectively. The diffusion coefficient D and the source term d are functions of ϕ and the equation is then non-linear. The solution of parabolic equation requires a knowledge of the initial variation of ϕ with x and of the conditions at the boundaries in x.

(1) J.N. Sherwood, A.V. Chadwick, W.M. Miur, F.L. Swinton. Diffusion processes,vol 2,p.562,Gordon & Breach Science publisher, New York, 1971
(2) J. Cranck & P. Nicholson, Proc.Camb.phil Soc., 43 (1974), 50
(3) D.J. Spalding, Heat and Mass Transfer in Boundary Layers, Morgan Grampian, London, 1967

(4) Ph.D.Dissertation of Syed Ameen, Max-Planck-Institut für biophysikalische Chemie, 34 Göttinge, W. Germany

M.J. Pilling - Thank you for the information. My list of reference was intended to be illustrative rather than comprehensive.

SOME CONSIDERATIONS IN THE INTERPRETATION OF LASER FLASH PHOTOLYSIS MEASUREMENTS

Chmouel R. GOLDSCHMIDT
Department of Physical Chemistry, The Hebrew University
JERUSALEM, Israel

SUMMARY

Artifacts are often obtained in laser flash photolysis measurements. Studies have now been made simulating experimental conditions in which incorrect results are to be expected without considering the inherent chemico-physical properties of the analyzed system. In these studies, the impact of artifacts on kinetics, intensity effects and pulse shape is analyzed, taking into account laser profile, pulse intensity and spectral data.

INTRODUCTION

Every photochemist, working with lasers has been concerned with the problem of confidence in instrumental results and in their correct interpretation. It seemed therefore worthwhile to investigate the analysis of flash photolysis measurements through a non-conventional approach by simulating experimental conditions in which incorrect results are to be expected. On this basis, several examples of plausible artifacts, or misinterpretations are presented here, evidently in a non-exhaustive form. For the sake of clarity, these examples are presented in a schematic and sometimes even exaggerated form. Assuming the linearity of the photoelectric detection system, we will analyze the requirement of a very accurate geometrical correlation between exciting and monitoring light source throughout the sample. A discussion of some basic physical processes such as self-focusing and chirping, which are pertinent to very fast photochemistry, will conclude this communication.

First, we will recall some basic formulae concerning flash photolysis : laser flash photolysis is mainly used as a tool to study the history of equilibrium recovery of a system, excited by a laser. When light of intensity I_o falls onto a medium and the portion $\Delta I = I_o - I$ is absorbed, then we have for the optical density.

$$D = \log \frac{I_o}{I} = \log \frac{I}{1 - \frac{\Delta I}{I_o}} \qquad (1)$$

$D = D(\lambda, t)$ indicates the time behavior of the concentration of the excited and unexcited molecules, as well as the absorption spectrum of the excited species.

$$D = (\varepsilon^{*} - \varepsilon_o) C^{*} . l \qquad (2)$$

where ε^{*} and ε_o are the decadic molar absorption coefficients at the monitoring wavelength for the excited and unexcited molecules, respectively (for simplicity we take $\varepsilon_o = 0$), C^{*} is the time dependent concentration ($mole.lit^{-1}$) of the excited molecules, and l denotes the length of the absorbing medium.

It is evident that a perfect linear relationship between the intensity of the monitoring light and the photodetector output is essential. Defoccusing of the light beam which falls onto the photocathode as well as careful wiring of the photomultiplier socket is necessary in order to avoid a time and intensity dependent response of the detection system. If these precautions are not taken, emission or absorption artifacts are likely to appear at high output currents.

Signal to Noise Ratio

If only the detction system is considered, the most accurate results are expected for the highest signal to noise (S/N) ratio. The noise originates from various sources as will be shown for two general cases, where the S/N may be calculated for a given optical density.

a. The S/N for density measurement which is associated with a constant source of noise [1] as e.g. the normal width of the oscilloscope trace, is given by :

$$(S/N)_{C.N.} = \frac{D}{\Delta D} = \frac{X_o}{\Delta X_{const}} \cdot \frac{D}{1 + 10^{D}} \qquad (3)$$

where X_o is the signal height of the non-absorbed monitoring light as measured on the screen of the oscilloscope and

ΔX_{const} is the normal width of the oscilloscope trace. Highest S/N is obtained at D = 0.55 as shown in figure 1a.

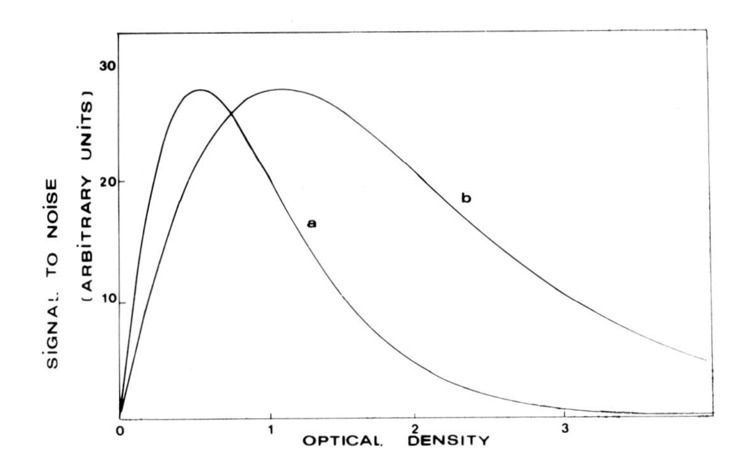

Fig. 1. S/N curves. a) constant noise, b) quantum noise.

b. Similarly, if we assume a Poisson distribution of the photons emitted by a monitoring light source and a binomial distribution probability for the transmission of a photon in each successive optical element and also for the release of photo-electrons from the photocathode, we have for the S/N associated with quantum noise :

$$(S/N)_{Q.N.} = \frac{D}{\Delta D} = 2.3\sqrt{i_o \Delta t}\ \frac{D}{1 + 10^{0.5D}} \qquad (4)$$

where i_o is the cathode photoelectron current (electron/sec) and Δt the resolution time of the detection system. In this case, (Fig. 2b) maximum S/N is obtained for D = 1.1. For composite contributions to the noise, accurate measurements are therefore expected for optical densities in the range $0.5<D<1.2$. At these optical densities however a systematic error has to be taken into account which will now be discussed.

Artifact N°1 : Stray light and laser excitation

Stray light has been recognized as a potential source of errors in spectra and kinetics measurements [2]. We will now show that very often, a stray light equivalent situation may be found in laser flash photolysis work.

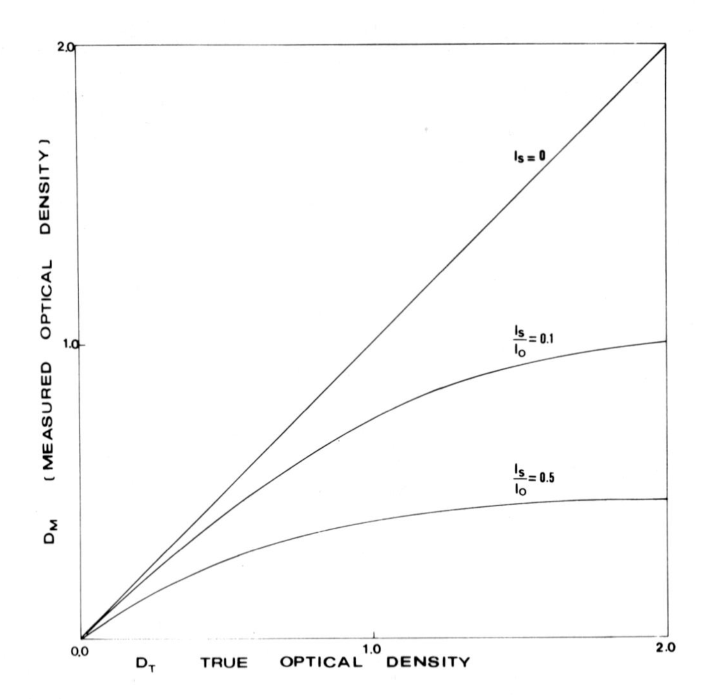

Fig. 2. Measured optical densities in the presence of stray light.

Stray light I_s is the addition of some light of unwanted colour, or some light from an external source, which does not pass through the ensemble of excited molecules under analysis but nevertheless falls onto the light detector. In figure 2, the measured optical density :

$$D_m = \log \frac{I_\circ + I_s}{I + I_s} \qquad (5)$$

is plotted for various $I_s/I_\circ$ ratios against the true optical density $D_T = \log (I_\circ/I)$.

From figure 2, we see immediately that D_M is always smaller than D_T and that this discrepancy is more pronounced for high D_T. Therefore, for every process which is investigated by optical density measurements the results will deviate more or less from the true value in the presence of stray light.

We have evaluated first and second orders kinetics in the presence of stray light. In figure 3, a second order kinetics is plotted in a semi-logarithmic diagram, yielding a quasi straight line when one adds 50% stray light.

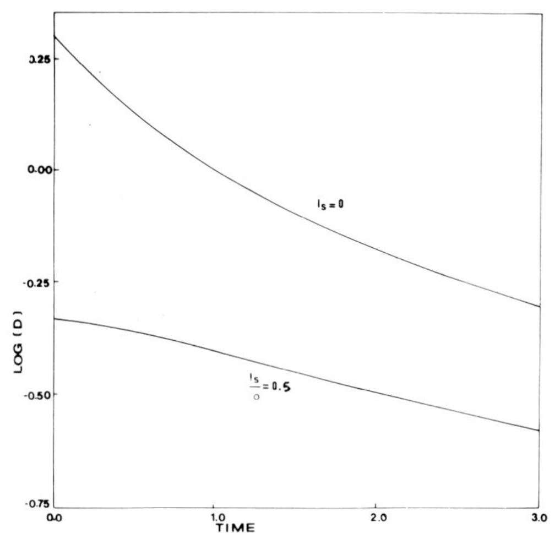

Fig. 3. First order analysis of a second order process in the presence of stray light.

In a pyrene-diethylaniline charge tranfer system, Schomburg et al [3] did not observe the triplet state of pyrene in deaerated polar solutions at very short times (less than 20ns) ; this is at variance with values reported from other laboratories [4]. It is suggested that this discrepancy is due to the charge transfer fluorescence, which in absorption measurements has to be considered as stray light.

The geometry of excitation, i.e., the correlation between the exciting beam and monitoring light for crossed and for collinear arrangements has been extensively analyzed by Bebelaar [5]. Independent of the specific crossed or collinear geometry, we have to be concerned about the correlation between the profile of the exciting beam and the clear aperture to the monitoring light. When a homogeneous rectangular exciting pulse covers only partly the clear aperture a (Fig. 4b), while the monitoring light covers all of it, then this excess light is in essence stray light. More generally, when a pulse of spatial distribution f(x) is used as exciting light, the measured optical density will be given by :

$$D_m = -\log \frac{1}{a} \cdot \int_0^a 10^{-D_T \cdot f(x)} \cdot dx < D_T \qquad (6)$$

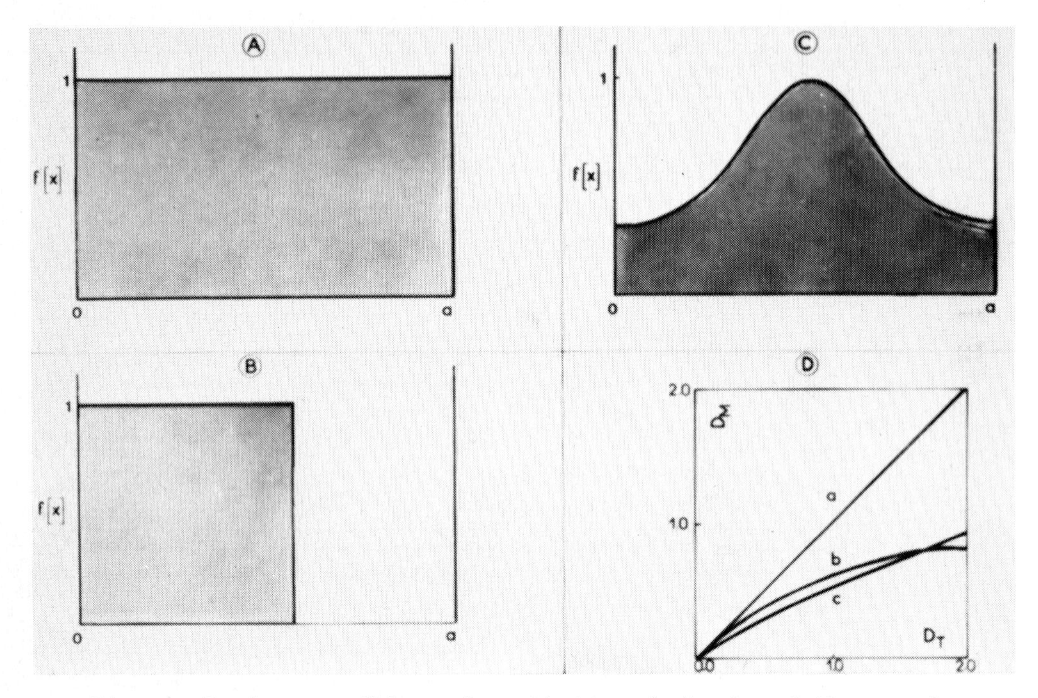

Fig. 4. Various profiles of excitation (a,b,c) and the resulting measured optical densities (d).

Hence even if a perfectly gaussian beam (Fig. 5c) is used as an excitation source, no linear relationship between laser energy and measured optical density is to be expected as seen from figure 5d. Such deviation from linearity will be even more strongly evidenced when a higher - mode or multimode laser beam is used for excitation of the sample. Even when no chemical effect prevails [6] this mode of excitation will lead to saturation - like behavior of the products vs intensity curve. It is clear that reduction of the intensity by neutral filters [6] will not remedy this artifact.

If the integrated output energy of the laser is raised by increasing the pumping voltage of the oscillator, the transversal mode structure of the laser beam will be enhanced. On the contrary, by pumping harder some amplifier stage one may smooth out the final output because of local saturation effect in the amplifier. In both cases, the variation of the spatial distribution f(x) of the light intensity leads in the first case to an increase in the I_s/I_o ratio, while in the second case it leads to a decrease in this ratio. It is thus seen that any variation of intensity through the cross section of the laser beam will be the source of artifacts in the measurement.

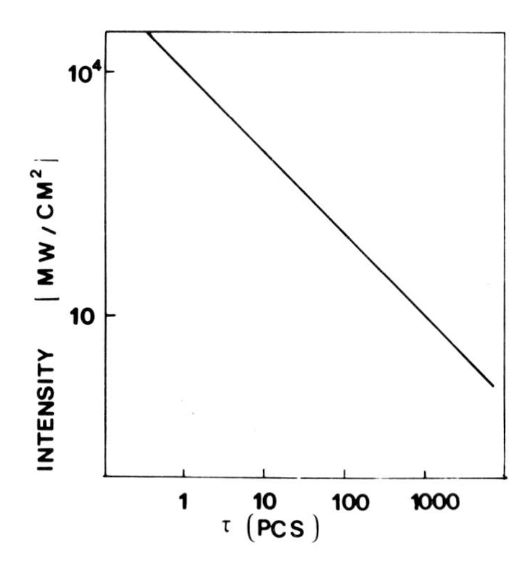

Fig. 5. The intensity of excitation for an absorption measurement : $\varepsilon^{*} = 10^{4}$; λ_{Laser} = 530nm ; D = 0.5.

Even when an entirely homogeneous laser beam is used as an excitation source, such disturbing variations in the intensity distribution may occour because of self-focusing effect. This effect will be significant for high excitation intensities. For a homogeneous and rectangular exciting beam which is totally absorbed in the medium to be analysed the required intensity (power per unit area) will be given by :

$$P = 10^{-2} \frac{D}{\varepsilon^{*}\lambda\tau} \frac{\text{watts}}{\text{cm}^2} \qquad (^7)$$

where τ and λ are the duration and wavelength of the exciting pulse respectively. Figure 5 shows the intensity needed for a standard absorption measurement performed whith a frequency doubled Nd glass laser. The required intensity may be even much higher if the excitation radiation is only partially absorbed by the medium (see below) and if the beam is spatialy structured. At these high intensities, self-focusing indeed occurs, leading to a modification of the spacial energy distribution of the laser beam in the optical components and even more in the medium [7], and hence leading to a modification of the function f(x). The variation in the spatial function f(x) caused either by the laser physics or by self-focusing in the analysed sample is thus a source of variation in the relative amount of equivalent stray light which will generate saturation like effects as well as quadratic or even higher power intensity artifacts.

Artifact N°2 : Deformation of Picosecond Pulses

Light pulses in the picosecond range have a broad longitudinal mode structure. Such polychromatic pulses can be deformed by "chirping" in a dispersive medium. Treacy [8] used a pair of gratings to compress ultra-short pulses to 0.4 picoseconds. Sunak and Gambling [9] measured the time broadening of a picosecond pulse transmitted through a cladded glass fiber as 10 pcs/m. If the same frequency dispersion mechanism is assumed to cause pulse distortion in all dispersive systems, it would seem that the anomalous dispersion associated with the absorption curve of the analysed molecules (which can be calculated by the Kramers-Kronig relations) may be an important source of time distortion of picosecond light pulses. This distortion is expected to be strongly dependent on the precise wavelength region within the absorption curve of the medium at which the picosecond pulse is absorbed.

For a Lorentzian shaped absorption curve, the frequency dependent index of refraction is given by [10] :

$$n(\omega) = n_o + \varepsilon C \cdot \frac{c}{2} \frac{\Delta\omega_o}{\omega} \frac{\omega_o - \omega}{(\omega_o - \omega)^2 + \Delta\omega^2} \qquad (^8)$$

where C is the molar concentration, c the speed of light and $\Delta\omega_o$ the half width of the Lorentzian absorption curve. Time broadening has been calculated for a transform limited pulse of width Δt_o [11]. Using(^{8}) we have then :

$$\frac{\Delta t}{\Delta t_{o}} = \sqrt{1 + 4\left[\frac{D_{max} f(\omega)}{\Delta t_{o}}\right]^{2}} \qquad (9)$$

where $f(\omega) = \frac{\Delta\omega}{\omega}\left(\frac{dn}{d\omega}\Big|_{\omega_{o}} + \frac{\omega}{2}\frac{dn}{d\omega^{2}}\Big|_{\omega_{o}}\right)$

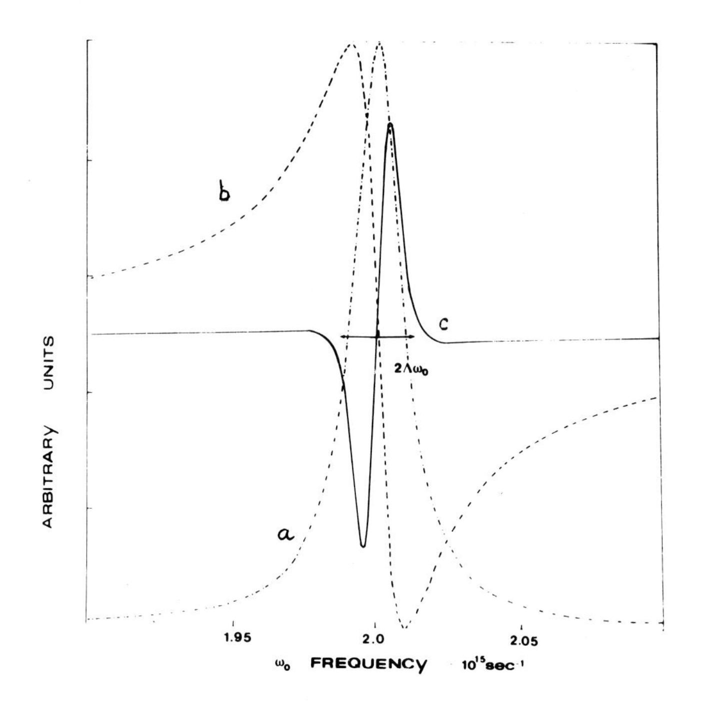

Fig. 6. Frequency dependent absorption (a), refraction (b) and chirping (c) curves.

From figure 6 it is seen that there are two "chirping" peaks, a negative and a positive one, near the center of the absorption curve but no chirping at the center. For small $\Delta\omega_{o}$ and high optical density D_{max} at the maximum of the absorption curve, broadened residual picosecond pulses will emerge from the system after partial absorption by the medium.

Consideration of this effect may settle a reported controversy between two laboratories. MM [12] and AS [13] obtained different rise times of the spontaneous fluorescence of erythrosin using essentially similar methods and equipment. AS used a concentrated solution of erythrosin (2.10^{-4}M, $D_{max} \sim 25$) and measured a 60 pcs rise time whereas MM, whith a low concentration (3.10^{-6}M, D_{max}~0.4) obtained a quasi immediate formation of the fluorescence state. As can be seen from the absorption spectrum of erythrosin [14] the 530nm pulse happens to be absorbed at the wavelength where maximum chirping occurs, therefore, the fluorescence output pulse appears as a time dependent convolution between the exciting pulse - gradually chirped and attenuated - and the true decay curve of the analysed substance. It can be pointed out that for rhodamine 6G, no such effect has been measured [15] nor is to be expected in spite of the rather high concentration (D_{max}~5) used by MM [15]. The 530nm pulse happens to coincide with the peak absorption curve of rhodamine 6G where chirping is negligible. A detailed analysis of the time behavior of an absorption signal when both exciting and monitoring pulses are distorted by ground-state and excited state respectively is now in progress and will be presented elsewhere.

I gratefully acknowledge the helpful discussions with Professors I.B. Berlman, S. Kimel and A. Weinreb.

REFERENCES

[1] Calculations performed by U. Lachish, Jerusalem University.
[2] R.B. Cook and R. Jankow, J. Chem. Educ., 49, (1972), 405.
[3] a. H. Schomburg, H. Steark and A. Weller, Chem. Phys. Lett., 21, (1973), 433.
b. ibid Chem. Phys. Lett., 22, (1973), 1.
[4] a. N. Orbach and M. Ottolenghi, International Exciplex Conference, London, Ontario (May 1974).
b. N. Mataga measured a rise time of 400 pcs of P/DEA in acetonitrile (private communication).
[5] D. Bebelaar, Chem. Physics, 3, (1974), 205.
[6] G. Makkes van der Deijl, J. Dousma, S. Speiser and J. Kommandeur, Chem. Phys. Lett., 20, (1973), 17.
[7] S. Speiser and S. Kimel, Chem. Phys. Lett., 7, (1970), 19.
[8] E.B. Treacy, Phys. Lett., 28A, (1968), 34.
[9] H.R.D. Sunak and W.A. Gambling, Optics Comm., 11, (1974), 277.
[10] O. Kafri, S. Kimel and C.R. Goldschmidt, to be published.
[11] D.J. Bradley and G.H.C. New, Process IEEE, (1974), 313.
[12] G. Mourou and M.M. Malley, Optics Comm., 11, (1974), 282.
[13] R.R. Alfano and S.L. Shapiro, Optics Comm., 6, (1972), 98.
[14] G. Oster, G. Kallman-Oster and G. Karg, J. Phys. Chem., 66, (1962), 2514.
[15] M.M. Malley and G. Mourou, Optics Comm., 10, (1974), 323.

DISCUSSION

C.R. GOLDSCHMIDT

C. Varma - Having Clausius Mossoti's relation for the refractive index in mind, I expect the contribution of solute molecules to dispersion by the sample to be negligible, when their concentration is $\leq 10^{-3}$ mol/l, because their number density is small. I estimate the solute contribution to become effective if at the wavelength of interest, the solute polarizability exceeds that of the solvent molecules by a factor of 10^4.

C.R. Goldschmidt - Our analysis of pulse distortion by chirping effect is based on the so-called anomalous dispersion associated by Kramers-Kronig relations with the absorption curve of the analyzed medium and on the precise position within this band where the laser pulse is absorbed.

M.W. Windsor - You state that one needs 10^8 photons per wave number per cm^2 per psec. in the measuring beam in order to have an adequate S/N ratio to avoid large statistical fluctuations. I have made a rough calculation and find that we have about 10^4 photons per wave number per cm^2 per psec in our system. If you are concerned that this intensity may perturb the excited state population significantly, I would observe that this value is about 10^3 or 10^4 times smaller that our excitation intensity. I do not believe then that our measuring pulse disturbs significantly the system being measured.

C.R. Goldschmidt - The intensity of the monitoring light has to be calculated on the basis of statistical considerations from the total number of photoelectrons per second, released by the photocathode, at the detected wavelength. A similar argument is valid when a photographic plate is used.

S. Ameen - Under experimental conditions, the artifacts produced by non linearity of refractive index or gradients of refractive index or lens-effects do not cause a greater error if the absorption maximum of the transient species under analysis is very sharp. Since $dn/d\lambda$ in the sharp absorption maximum region will be very negligible, the second problem is what I agree with Dr. Hochstrasser, that if the exciting beam and the analysing beams are coaxial, this will also not create problem or error in detection, because the lens-effects are not noticed by the probing beam.

C.R. Goldschmidt - We have analyzed self-focusing in solvents and the resulting inhomogeneous excitation geometry in connection with intense laser beams. On the other hand, the chromatic dispersion $dn/d\lambda$ is connected in our analysis with time distortion of picosecond pulses in absorption bands even at low excitation intensities. The correlation between exciting and analyzing beams has been discussed with prof. Hochstrasser.

R.M. Hochstrasser - Let me say first of all that I agree with the spirit of your scrutiny of flash photolytic methods. Perhaps you could comment on the following three points :

1) In our experiments the transient O.D. was checked to be linear (approximatively) with pulse energy over a range of perhaps an order of magnitude. The normal O.D. at 3530 was between 0.7 and 2.0 per mm.

2)When using a gaussian pulse with probe and pump travelling either in the same or opposite directions does your analysis still hold? Most of our picosecond experiments were done in this fashion.
3) In broad spectra consisting of overlapping bands of vibronic transitions (smearing out oscillator strength) won't the dispersion of the refractive index be quite flat?

C.R. Goldschmidt - 1) At transient optical densities below 0.2 intensity artifacts should be less pronounced. The difference between ground state absorption OD = 0.7 and 2.0 could lead, after proper convolution (chirping and attenuation) to some slight difference in the apparent rise time of the benzophenone triplet. The absorption spectrum of benzophenone is more elaborate than in the case of rhodamine 6G where there is only one sharp absorption peak in the excited region. In fact this difficulty is emphazised by considering the measurement you performed in n-heptane where the absorption spectrum is more structured.
2) In principle any non rectangular exciting beam will lead to non linear apparent OD measurements, whatever the monitoring pulse shape. This can be very easily examplified, using staircaselike pulses.
3)In the case of broad absorption spectra, chirping will not be important. Generally, dispersion and chirping are directly connected to the overall shape of the absorption spectra, whatever the origin of the transitions.

A. Laubereau - If I understand right, you make the point that experimental accuracy states a lower intensity limit for the absorption measurement. Can you please make clear why you arrive at an intensity level (10^{13} W/cm^2 for 1 psec pulse) required and not a power level in order to leave a sufficient number of photons in the measurement.

C.R. Goldschmidt - The exciting beam generates a concentration (molecules per cm^3) of excited molecules which is measured as a transient along a distance ℓ (cm). Hence the number of excited molecules per cm^2 per second (intensity) results as the relevant factor in the measurement. On the other hand, the total number of photons per resolution time unit (power) of the monitoring light should be high enough to overcome relative statistical fluctuations.

M.W. Windsor - I am more concerned with your considerations of the dependence of chirping on optical density of the sample. Could you please elaborate on this with some quantitative examples of what degree of chirping would correlate with what OD?

C.R. Goldschmidt - The density of photons in the monitoring light depends on the specific set up of the photolysis system. When highly focused optics is used, this density might be intense enough to induce disturbing effects.

U.V. NANOSECOND LASER STUDIES OF THE CHEMISTRY OF SOME SYSTEMS OF BIOCHEMICAL INTEREST

GABRIEL STEIN
Department of Physical Chemistry, Hebrew University, Jerusalem, Israel

SUMMARY

Detailed information on dynamic changes in the structure and conformation of enzymes and other biopolymers during their actual functioning are most desirable. Dynamic interpretation derived from detailed static structural studies such as X-ray diffraction often lead to erroneous conclusions. Towards the development of methods which would give dynamic information on biomolecules in vivo or in aqueous solution in vitro results will be described on the use of pulse techniques with lasers in times shorter than the natural dynamic changes of the biomolecule, together with methods for following the changes, in particular fast kinetic spectroscopy in the visible and ultra violet. The formation of hydrated electrons, e^-_{aq}, from excited states of ions and molecules in solution is used to generate one electron equivalent redox reagents. Problems of mono and biphotonic processes and intensity effects in the use of lasers are discussed in connection with the results on e^-_{aq} formation from excited ferrocyanide, tyrosine, tryptophane and other substrates. The results on the redox enzyme, cytochrome c will be discussed, when UV nanosecond laser pulses are used to generate the redox reagents in situ and intramolecular changes in oxidation state and conformation can be followed.

INTRODUCTION

In this paper we report our results using fast laser pulse - kinetic spectroscopy techniques on some systems of biochemical interest. Our aim is to obtain information on dynamic changes in the structure and conformation of macromolecules of biological interest - in particular enzymes - under conditions which are as close as possible to their actual functioning.

Detailed information on the structure of biopolymers, enzymes, nucleic acids and other structural elements has been obtained with great success using static methods such as, for example, X-ray diffraction. For example, detailed structural information was obtained for the enzyme cytochrome-c, both in its oxidized ferri form, and in its reduced ferro form, which are interconvertible by reversible one-electron equivalent redox reaction during the biochemical functioning of this enzyme in the biological oxidation chain. Conclusions about the mechanism of functioning of the enzyme under in vivo conditions were then postulated from such static measurements and - as it turned out - some of these conclusions may have been unreliable [1]. Similarly since the functioning of enzymes in vivo in the biological environment often depends on minute dynamic changes in conformation, bond, length and state of ionization, conclusions from static information alone, however detailed, may lead to erroneous conclusions [2].

In recent years dynamic methods have been increasingly introduced in the study of the mode of functioning of biomolecules under dynamic conditions. These methods include such techniques as n.m.r., e.p.r., Raman scattering and other physical methods [3].

Our research group, parallel with other groups in various laboratories has, over the last few years, developed a technique of pulsed generation of reactive intermediates in short times, of the order of nanosecond and shorter. These reactive intermediates were produced by the method of pulse radiolysis, using mostly particle accelerators and, in particular, electron accelerators. The reactive intermediates, e^-_{aq}, H atoms OH radicals, one-electron equivalent redox reagents, reacted with the biomolecules in times shorter than, for example, the relaxation time of reduction or of configurational changes. Using the fast kinetic spectroscopy in the visible and U.V., it was possible to follow intramolecular dynamic changes as expressed in the spectrum; for example, we observed consecutive intramolecular electron equivalent transfer and configurational changes in the enzyme protein ribonuclease [4] and in the redox enzyme cytochrome-c, as did Land and Swallow [5].

We have recently reviewed this technique and its application to a number of biochemical systems [6].

We have begun to develop a related technique, using U.V. pulsed lasers to obtain the reactive intermediates in times of nanoseconds or shorter and to investigate the conditions under which dynamic information on the functioning of biomolecules can be obtained by this method.

The main step in this technique is the formation in sub-nanosecond time of solvated electrons from excited states of ions and molecules in solution [7], the U.V. laser being used to produce the excited state. Solvated electrons, or other reagents derived from them through well established chemical reactions, are then used to react with the biomolecules, the changes being followed by fast kinetic spectroscopy. There exists also a possibility of direct excitation of the bio-molecule.

Compared to the technique of pulse radiolysis the method offers several advantages : the apparatus is much cheaper and requires less protection and elaborate safety devices; some of the reactive intermediates necessarily produced by ionizing radiation are absent.

On the other hand the use of this laser technique has problems of its own and these we shall discuss briefly in the present paper, together with an outline of our first results. The work was mainly carried out with the cooperation of Mr. Uri Lachish : with him we plan to publish the detailed results. The laser techniques were developed in cooperation with Dr.Ch.R.Goldschmidt.

The Formation of Solvated Electrons from Excited States of Ions and Molecules in Solution

There are several processes by which absorption of quanta causes ions or molecules in solution to undergo a process of electron detachment with the formation of solvated electrons, e^-_{solv}, together with the twin radical from the parent solute. One process is that of direct electron ejection in which the energy of the quantum must be sufficient to photoionize the electron into the solution . There is another monophotonic process which has been investigated in recent years and is a convenient source of hydrated electrons at longer wavelengths in which efficient laser sources are available. In this process a spectroscopically well defined bound excited state of the ion or molecule is formed, which undergoes spontaneous thermal dissociation into a radical and the solvated electron in times shorter than a nanosecond. The process originates in the non-relaxed state of the excited system before thermalization and gains part of its energy from the energy of solvation of the electron in the medium; hence the decrease in the energy requirement compared to the direct photoejection of the electron [7,8].

There are other processes leading to the same result, of which we must emphasize particularly biphotonic processes. In these, two quanta are utilized for the formation of the electron either consecutively - through light absorption by the primary excited (singlet) or the (triplet) state formed from it, or by simultaneous two photon absorption through a virtual intermediate state.

Monophotonic and biphotonic pathways must be carefully separated for a proper understanding of the processes. The utilization of picosecond pulses with their very high instantaneous intensity may introduce unnecessary complications in cases where the elucidation of the photochemical or photobiological process requires time resolutions of a nanosecond or longer only. Indeed there are some cases in the literature already, where the unnecessary employment of picosecond techniques has led to apparently confusing results. Picosecond techniques should be used only where the photo-physics and photo-chemistry of the process studied genuinely demands it and nanosecond techniques are truly inadequate to elucidate mechanistic details.

Once solvated electrons are formed in solution, e.g. e^-_{aq} in water, standard chemical techniques can be used to produce from them selectively other reactive intermediates. The following reactions are some representative ones.

Formation of OH radicals:

$$e^-_{aq} + N_2O \rightarrow {}^{\bullet}OH + N_2 \uparrow + OH^-_{aq}$$

Formation of H atoms:

$$e^-_{aq} + H^+_{aq}\ (\text{or } H_2PO^-_{4aq}) \rightarrow {}^{\bullet}H (+\ HPO_4^{2-})$$

Formation of organic radicals:

$$e^-_{aq} \begin{cases} \xrightarrow{H^+} \dot{H} \\ \xrightarrow{N_2O} {}^{\cdot}OH \end{cases} + RH_2 \rightarrow \begin{cases} H_2 + {}^{\cdot}RH \\ H_2O + {}^{\cdot}RH \end{cases}$$

The utilization of the photochemical technique introduces the following considerations, which are not present in radiation chemistry:

a) The parent substrate may itself serve as an electron scavenger, or may be the source of an oxidation-reduction couple, which may affect the chemical system to be studied.

b) The chemical properties of the twin radical must be taken into account.

In order to optimize the properties of the system we have studied several solutes as sources of hydrated electrons in laser excited photochemistry. We have used two lasers in our studies:

1) A nitrogen laser operating at 337 nm

2) A frequency quadrupled Nd glass laser, operating at 265 nm.

In order to obtain optimum performance the following conditions were sought to be optimized:

a) Maximum possible molar extinction coefficient at the wavelengths used

b) Rate constant of the solvated electron with the parent substance to be as low as possible

c) Parent substance and its associated redox form should be relatively inert in redox reactions in the system

d) Twin radical to be as inert as possible and in particular have relatively low reactivity with the solvated electron.

Of the considerable number of systems we have studied we shall mention the following as sources of solvated electrons in the monophotonic process in which the excited states produce solvated electrons in a spontaneous thermal dissociation:

i) At 337 nm: β-naphthol [8]; α-naphthylamine.

ii) At 265 nm: ferro-cyanide [9] tyrosine; tryptophane.

The work on tyrosine and tryptophane is of intrinsic biochemical interest and has been the subject of a number of recent researches. In the following we shall briefly discuss our results on it.

TECHNICAL DETAILS OF THE LASER SOURCES

1) Nd glass laser, operating in the frequency quadrupled mode at 265 nm.

This laser, manufactured by Laser Associates, is capable of operating at 265 nm the following outputs:

a) 15-20 ns half-width pulse, with up to 10 mJ per pulse;

b) 1 to 3 ns chopped pulse;

c) single picosecond pulse.

The laser has been provided with the required accessories for fast kinetic spectroscopy and is capable of operating in two configurations:

a) Measurements of fluorescence emission from substrates.

b) Configuration for absorption measurements.

Figure 1 shows the scheme of the laser itself;

Figure 2 the fast kinetic spectroscopy set up for absorption measurements.

In the present investigation all experiments were carried out using the 15-20 ns half-time pulse.

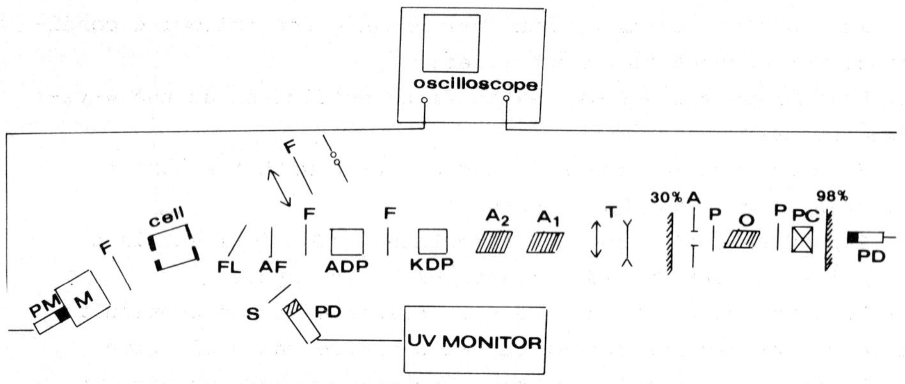

Fig.1. PD - photodiode, PC - Pockels cell, P - Polarizer O - Oscillator rod, A - Aperture, T - Telescope, F - Filter, A - Amplifier, FL - quartz flat, S - fluorescent screen, M - Monochromator, PM - photomultiplier.

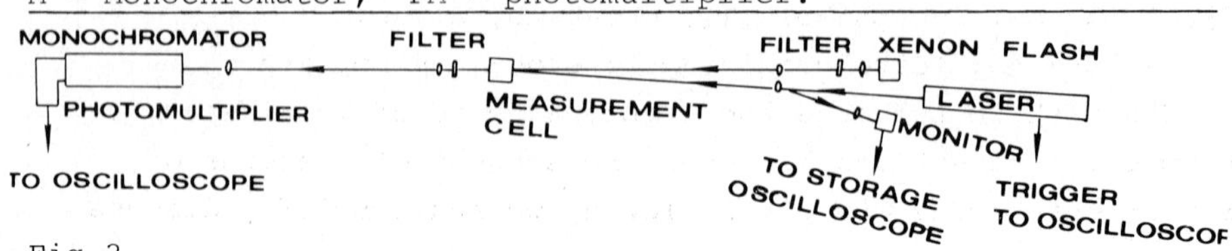

Fig.2

2) N_2 laser operating at 337 nm

This apparatus, capable of higher repetition frequencies, and providing 7-10 ns half-width pulses with 0.2-0.5 mJ per pulse has been described in detail previously [8].

MONO AND BIPHOTONIC PROCESSES IN THE FORMATION OF HYDRATED ELECTRONS FROM FERROCYANIDE, TYROSINE, TRYPTOPHANE.

In our search for an optimal substrate to provide hydrated electrons through the process of activation - thermal dissociation for the study of biochemical systems, we re-investigated tyrosine and tryptophane as possible candidates . These amino acids are themselves among the most important in the photobiology of many systems and their photochemistry has recently been the subject of investigations. Using flash photolysis and laser photolysis techniques conflicting evidence was obtained on the mechanism and formation of hydrated electrons [10].Some of the evidence was interpreted as showing the exclusive role of biphotonic processes, via an intermediate state, while other evidence was clearly consistent with a mechanism involving a monophotonic process, as discussed in the Introduction. From the point of view of photobiology the resolution of this controversy is of some interest. Usually in photobiology under natural conditions light intensities are low. Biphotonic processes may become of overwhelming importance when pulse sources, such as flash or laser, are used, but may be of very small relevance at low continuous light intensities under biological conditions. In our experiments with Mr.Lachish to be published in detail, separately, the results shown in Fig.3 were obtained on ferro-cyanide and tyrosine, respectively, under conditions of varying light intensity in the laser pulse at 265 nm, approximately 20 ns half-width of pulse. The parameter α is an arbitrary normalized measure of the light intensity in the pulse, adjusted by the use of an appropriate filter. On the ordinate hydrated electron production is measured at 550 nm.

The quantum yield of hydrated electron formation from ferrocyanide at 265 nm under steady illumination has been determined previously accurately and found to be = 0.5 [9]

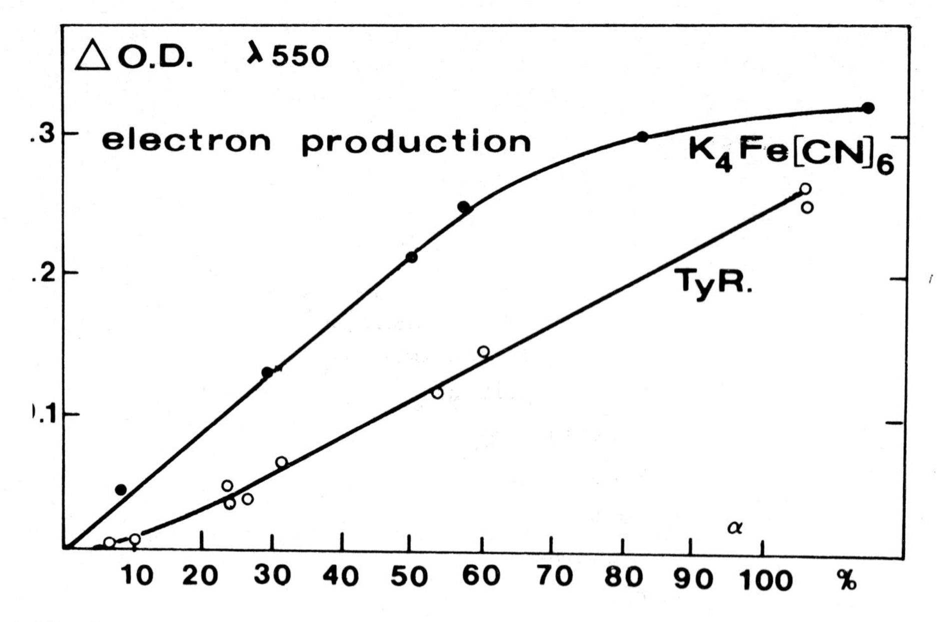

Fig.3.

Fig.3 shows that the formation of hydrated electrons from ferro-cyanide obeys strictly kinetics corresponding to a monophotonic process linear in light intensity. At high light intensities saturation effects are clearly observed under conditions where the light intensity is sufficient to exhaust the absorbing species. Such light saturation effects due to substrate exhaustion may become of central importance in laser pulsed experiments, particularly when high light intensities are used in short times.

By comparison tyrosine shows clearly kinetics which do not correspond to a simple monophotonic process. The quantum yield of hydrated electron formation is of the order of one half of that from ferro-cyanide in the middle portion of the tyrosine curve, i.e. of the order of 0.2 - 0.25. The curve is consistent in much of its course with a biphotonic process in which tyrosine absorbs the first quantum and then a resultant intermediate absorbs a second quantum. From these studies at relatively high light intensities no conclusion can be derived as to the possible contribution of a genuinely monophotonic process, which could be demonstrated only by a precise analysis of the shape of the curve in the region between 0 - 10% α. Under laser flash conditions this cannot be done with sufficient experimental accuracy.

Here independent results obtained from experiments at low light intensity, continuous illumination, utilizing chemical scavenger techniques, provide the required evidence. We have investigated systems, using a disulphide as the chemical scavenger of the hydrated electrons originating from excited tyrosine and tryptophane [11]. The results showed a clearly monophotonic process but with a quantum yield 10 to 100 times smaller now observed for the biphotonic process, using the laser pulse. The details of such systems and the quantitative analysis of the relevant conclusions will be published separately. The point to be emphasized here is that great care and purposeful experiments have to be devoted to every system of biological or biochemical systems before it is concluded reliably that mechanisms and results derived from high intensity laser pulses are directly relevant to the biological or biochemical situation at lower light intensities.

THE STUDY OF DYNAMIC PROCESSES IN THE REDOX ENZYME CYTOCHROME-C USING SOLVATED ELECTRONS

Using pulse radiolysis techniques the reactions of ferricytochrome-c with hydrated electrons has been studied from the sub-microsecond time scale up to seconds. Hydrated electrons react with ferri-cytochrome-c with a very high rate constant, which depends on the ionic strength and varies from 10^{10} to over 10^{11} M^{-1} s^{-1}. Following upon this strictly bimolecular reaction intermolecular configurational changes could be observed within the enzyme molecule, as it took up its equilibrium conformation following on the rapid reduction process [5]. After the bimolecular reaction two first order processes were observed at neutral pH, which were found to be strictly intramolecular. The first process showed a rate of about 10^5 s^{-1} and the second about 10^2 s^{-1}. These conformational changes could be characterized with regard to their pH dependence and activation energies.

When similar experiments were carried out in alkaline solutions an even slower process with a time constant of approximately 0.1 sec could also be demonstrated. The work of Land and Swallow, based on measurements in specific absorbance bands, including measurements in the near U.V. in the Soret band (indicative of protein-heme interaction), and measurements at 695 nm (indicative of the state of S-Fe bond to heme) could be correlated with specific dynamic changes within the molecule. Our observations confirmed and extended these results.

Similar results can be obtained readily using the laser technique for the formation of the hydrated electron, so that the cytochrome-c system may serve as a biochemical comparison between the two methods of fast pulse techniques in the observation of dynamic processes within biomolecules.

Since the ferri-ferro-cytochrome system is itself a redox system, complications arose due to mutual redox processes, when ferrocyanide - otherwise very suitable - was used as the source of electrons at 265 nm. Tyrosine and tryptophane gave better results at this wavelength and had the advantage of being important potential sources of electrons in photo and biochemical processes. However, several factors, including the nature of the intermediate radicals, and the reactivity of tyrosine and tryptophane with electrons, proved drawbacks in this respect. We have now developed the use of the α-naphthylamine as a source of the hydrated electrons using the nitrogen laser at 337 nm. α-naphthylamine has several advantages: its high molar extinction coefficient at 337 nm, relatively low rate of reaction with hydrated electrons, and relatively high quantum yield of electron formation. The use of the 337 nm nitrogen laser minimizes direct light absorption by cytochrome -c itself and enables statistical averaging techniques to be employed. The results which we plan to publish in detail separately confirm the applicability of the laser pulse technique in place of pulse radiolysis under specified conditions,yielding results which are comparable with those obtained using the radiolysis technique on the dynamic behaviour of such enzymes under conditions closer to their biochemical function.

ACKNOWLEDGMENT: The work was carried out in cooperation with Mr.U.Lachish and Dr.Ch.R.Goldschmidt. I thank Dr.A.Shafferman for very valuable comments. The work was supported under contract with US ERDA (AEC).

References

1 R.E.Dickerson and R.Timkovich, The Enzymes-Oxidation-Reduction (1975) in press.

2 B.L.Vallee, in Trace Element Metabolism in Animals Eds. W.G.Hockstra, J.W.Suttie, H.E.Ganther and W.Mertz, Butterworths, London (1974) 5; C.A.Spillburg, J.L.Bethune and B.L.Vallee, Proc.Nat.Acad.Sci.71 (1974) 3922; J.T.Johansen and B.L.Vallee, Proc.Nat.Acad.Sci. 70 (1973) 2006.

3 J.L.Marx, Science 188 (1975) 1002; R.K.Gupta and A.G.Redfield, Science, 169 (1970) 1204; T.G.Spiro, Accts.Chem.Res. 7 (1974) 339.

4 N.N.Lichtin, Y.Ogdan and G.Stein, Biochim.Biophys.Acta.263 (1972) 14; 276 (1972) 124.

5 E.J.Land and A.J.Swallow, Arch.Biochem.Biophys.145 (1971) 365; Biochim.Biophys.Acta,368 (1974) 89.
A.Shafferman and G.Stein, Science 183 (1974) 428; N.N.Lichtin, A.Shafferman and G.Stein, Science, 179 (1973) 680; Biochim,Biophys.Acta, 314 (1973) 117; 357 (1974) 386.

6 A.Shafferman and G.Stein, Biochim.Biophys.Acta, Bioenergetics Rev. (1975) submitted for publication.

7 G.Stein, Advances in Chem. ACS. 50 (1965) 230; in "The Chemical and Biological Actions of Radiations" ed.M.Haissinsky. Masson et Cie.Paris, 13 (1969) 119; H.I.Joschek and L.I.Grossweiner, J.Amer.Chem.Soc.88 (1966) 3261; J.Joussot-Dubien and R.Lesclaux, C.R.Acad.Sci.258(1964) 4260; A.Henglein, in press (1975).

8. Ch.R.Goldschmidt and G.Stein, Chem.Phys.Letters, 2 (1970) 299; Y.Feitelson and G.Stein, J.Chem.Phys.57 (1972) 5378; U.K.Kläning, Ch.R.Goldschmidt, M.Ottolenghi and G.Stein, J.Chem.Phys. 59 (1973) 1753.

9 M.Matheson, W.A.Mulac and J.Rabani, J.Phys.Chem. 67 (1963) 2629; M.Shirom and G.Stein, Nature, 204 (1964) 778; J.Chem. Phys.55 (1971) 3372, 3379.

10 Y.Feitelson, E.Hayon and A.Treinin, J.Amer.Chem.Soc. 95 (1973) 1025; M.P.Pileni, D.Lavalette and B.Muel, J.Amer. Chem.Soc. 97 (1975) 2283.

11 A.Shafferman and G.Stein, Photochem.Photobiol. 20 (1974) 399.